Elements of
Fluid Dynamics

ICP FLUID MECHANICS

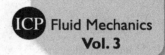

ICP Fluid Mechanics
Vol. 3

Elements of Fluid Dynamics

Guido Buresti

University of Pisa, Italy

Imperial College Press

ICP

Published by

Imperial College Press
57 Shelton Street
Covent Garden
London WC2H 9HE

Distributed by

World Scientific Publishing Co. Pte. Ltd.
5 Toh Tuck Link, Singapore 596224
USA office: 27 Warren Street, Suite 401-402, Hackensack, NJ 07601
UK office: 57 Shelton Street, Covent Garden, London WC2H 9HE

British Library Cataloguing-in-Publication Data
A catalogue record for this book is available from the British Library.

The image on the front cover shows the vorticity field around an impulsively starting airfoil. Direct numerical simulation by Simone Camarri.

ICP Fluid Mechanics — Vol. 3
ELEMENTS OF FLUID DYNAMICS

Copyright © 2012 by Imperial College Press

ISBN-13 978-1-84816-888-6
ISBN-10 1-84816-888-8
ISBN-13 978-1-84816-889-3 (pbk)
ISBN-10 1-84816-889-6 (pbk)

Typeset by Stallion Press
Email: enquiries@stallionpress.com

Printed in Singapore.

*To my family
and to my students*

PREFACE

This book aims at providing, in a simultaneously rigorous and accessible way, an introduction to fluid dynamics for undergraduate and graduate students in different fields of engineering. Nevertheless, it might also be of interest for students in physics and applied mathematics.

The basic point of view adopted in the book is that both the generation mechanisms and the main features of the loads acting on bodies moving in a fluid can be satisfactorily understood only after the equations of fluid motion and all their mathematical and physical implications have been thoroughly assimilated. In other words, the chosen didactic approach is the one in which the material is presented from the general to the particular, rather than starting from simple cases or models and subsequently adding complexity. Obviously, this implies that the reader should already be familiar with some notions of vector operators and algebra and with the writing — but not necessarily with the solution — of differential equations. However, these mathematical tools are usually available to students of courses in fluid dynamics and, in any case, great attention is always devoted throughout the book to carefully describing and discussing the physical meaning of the various mathematical terms appearing in the equations. The advantages of this approach are significant. In particular, the assumptions required to devise simplified models are specified explicitly and consequentially, and this permits a clear definition of the relative merits and limits of applicability of the derived models. In other words, there are no logical jumps or unexplained assumptions, and the mathematical formulations are always connected with physical features of the flow.

The book starts with a brief general introduction on fluid dynamics, and with the description of the main physical properties of fluids and of the mathematical tools that are necessary to characterize their motion. Subsequently, the complete equations of the dynamics of a compressible viscous fluid are derived, their physical and mathematical aspects are discussed in depth, and the necessity of simplified treatments is highlighted. In particular, a detailed analysis is made of the assumptions and range of applicability of the incompressible flow model, which is then adopted for most of the remainder of the book. However, unnecessary assumptions are carefully avoided; for instance, it is shown that, when dealing with problems involving incompressible flows, it is practically never necessary to consider the fluid as non-viscous. On the other hand, great attention is given to the generation and dynamics of vorticity, and to its fundamental role in the production of different types of flows. Furthermore, the importance of the vorticity field as regards the characteristics, magnitude and predictability of the fluid dynamic loads acting on moving bodies is emphasized. The vorticity-based approach is also used to facilitate a deeper comprehension of the evolution of different flows and to permit a straightforward assessment of the validity and limits of applicability of the simplified physical models that may be used for their analysis. Moreover, it is shown that the fluid dynamic forces acting on moving bodies may be directly related to the dynamics of vorticity, thus providing an immediate qualitative justification for the dissimilar loads acting on different types of bodies, as well as formulas for their quantitative prediction.

The book is divided into two parts, which differ in target and method of utilization. The first part contains the basic notions of fluid dynamics that I regard as essential for any student new to the subject, whatever his/her future interests may be. This part of the book is organized in a strictly sequential way; in other words, it is assumed that each chapter is carefully read and studied before the next one is tackled. The objective is to lead the reader to a full understanding of the origin and main characteristics of the fluid dynamic forces acting on various kinds of bodies. Conversely, the second part of the book is devoted to certain selected topics that may be of more specific interest to different categories of students; therefore, a sequential reading is not strictly necessary, even though it is, to a certain extent, advisable. Some theoretical aspects of incompressible flows are first analysed, then classical applications of fluid dynamics — such as the aerodynamics of airfoils, wings and bluff bodies — are described in some detail. Finally, the one-dimensional treatment of compressible flows is considered, together with its application to the study of the motion in ducts. The subjects covered in this second part have been chosen with mainly the needs of engineering courses in mind but, obviously, the choice has also been influenced by my personal interests and teaching experience.

As is practically inevitable in an introductory textbook in fluid dynamics, many important subjects are not covered and others, such as turbulence, are only very briefly outlined. Nonetheless, I hope that this book may serve as a basic reference for those who, being interested in different specific areas of application of fluid dynamics, will subsequently need to also use more specialized texts.

Most of the material contained in the book has been tested in lectures for many years, and the approach adopted has never posed any particular difficulties for the students, in spite of being, in some respects, not completely traditional. I am deeply indebted to my former students, colleagues and friends, whose continuous encouragement has been a fundamental incentive in my decision to write this book. I would also like to express my gratitude to the staff of Imperial College Press, whose kindness and professionalism were essential for the successful completion of this task. Finally, very special thanks are due to Antonella, without whose continuous support and infinite patience this book would have never come to light.

Guido Buresti

CONTENTS

Part I

FUNDAMENTALS

Chapter 1

INTRODUCTION

1.1. Fluid Dynamics and its Tools

The main objective of fluid dynamics is the description of the motion of fluids and the prediction of its effects on surrounding or immersed bodies. It may be seen as a branch of applied physics or of engineering, but it may also be of interest in numerous other fields of science, from medicine to volcanology.

As a matter of fact, the possible applications of fluid dynamics span an extremely wide range of human activities. Aeroplanes fly thanks to the upward lift forces caused by their motion in the atmosphere, but the physical mechanisms producing fluid dynamic forces on an aircraft are quite different from those allowing the flight of a bird or — and even more so — of an insect. The fuel consumption of all road vehicles is directly related to the aerodynamic drag acting on them, which is the force component opposing the motion. Wind forces are fundamental for the design of a tall skyscraper or of a long-span bridge, and their accurate prediction is essential not only to avoid catastrophic failures but also to ensure the correct operation of the structure. A full understanding of the motion of blood in the human body would be invaluable in preventing or treating circulatory problems. Weather forecasting is another challenging application of fluid dynamics.

The above examples illustrate only a few of the possible areas in which the study of the motion of fluids is of relevance. However, apart from the importance of its applications, fluid dynamics is a fascinating subject in itself, and has attracted the interest of brilliant personalities for centuries. Roman waterworks — many of which are still operative — are remarkable examples of the use of significant knowledge of water management, while the works of Leonardo da Vinci on flight or on the motion of water are a source of everlasting admiration.

Curiosity, attentive observation, careful description and critical analysis were certainly the first means used in the study of fluid dynamics, and they still remain fundamental investigation tools for any researcher in the field. Thus, flow visualization methods, measurement techniques and data processing are the basis of experimental fluid dynamics, and have continuously been evolving during past centuries. In fact, experiments and the generalization of their results have always been crucial for the enhancement of physical understanding and for the development of knowledge-based procedures for the prediction of quantities related to fluid motion.

Nonetheless, an essential step forward in the analysis of the motion of fluids — which comprise both gases and liquids — was obtained through progress in their description by means of appropriate mathematical models. In most practical circumstances, this description is based on the assumption that fluids may be considered as continuous substances. Therefore, the equations of motion of a fluid are written by expressing in mathematical terms the fundamental principles of continuum mechanics and by specializing them through *constitutive equations* that are considered to describe, in an adequate manner, the behaviour of the particular fluid being analysed.

Unfortunately, in many (or perhaps most) conditions of practical interest, writing down the 'correct' equations of fluid dynamics is a necessary but insufficient step for describing the motion of a fluid or for predicting the fluid dynamic loads acting on bodies in motion. This derives from the fact that, even when seemingly 'simple' fluids (such as air or water) are considered, the relevant equations of motion are extremely complex, and solutions may be devised in explicit mathematical terms ('closed form' solutions) only in relatively few cases, which are seldom those we are interested in. Even numerical solutions, which can now be obtained through the use of increasingly powerful computers, cannot yet deal — and probably will not do so for years to come — with the 'exact' description of most situations that are of interest in engineering practice or in other fields of science and technology.

It is thus evident that fluid dynamics is a complex subject, which requires both a considerable amount of physical insight and significant mathematical skills. Theory, experiments and numerical simulations are, in practice, the complementary tools that can be used by researchers and engineers for dealing with problems involving the motion of fluids.

A possible logical sequence of analysis would be to start by carefully studying the complete equations expressing mathematically the physical laws describing the motion of fluids, in order to derive all the information that can be obtained by using available theoretical tools. Subsequently, if, as often happens, one recognizes that this is not sufficient for the solution of a certain flow problem, one has to devise simplifications that may permit description, with various degrees of approximation, of the flow features and effects that can be observed experimentally and that are essential for the considered application.

In a sense, this is perhaps the core of both the difficulty and the fascination of fluid dynamics. As a matter of fact, due to the aforesaid complexity of the full equations of motion, the development of simplified physico-mathematical models has always played a central role in fluid dynamics research, in an attempt to find sufficiently accurate solutions to particular problems of practical interest. However, no simplification may be justified by the sole reason that the problem would otherwise be mathematically intractable. Therefore, the key point in this type of effort is to understand which features of the flow are essential and which ones, conversely, may either be disregarded or treated in an approximate way.

In many cases, this is a difficult task indeed, and its accomplishment must rely on all the tools that have already been cited above. In particular, the results of experimental observation may first be used to obtain an idea of the main features of different types of flows, and important indications on the possibility of their description in a simplified form. On the other hand, the analysis of the terms appearing in the equations of motion and of their physical meaning may provide an assessment of their relative importance as a function of the main flow parameters, and thus give clues to the definition of the conditions under which they may be neglected or described in a simpler form. Nowadays, significant help in this type of assessment is also provided by a careful analysis of the results of numerical simulations.

However, due to its complexity, the above-described conceptual process is not the only one that has been used in the development of all the simplified models or prediction procedures presently used in fluid dynamics. In fact, another possible approach may be — and often has been — to make *a priori* assumptions, which lead to simplifications of the equations of motion that may provide solutions in mathematical (or now numerical) terms. These assumptions are usually based on reasonable simplified descriptions both of the physical characteristics and of the behaviour of the fluid. Subsequently, one checks whether, in which conditions, and for which quantities, the obtained solutions describe with 'sufficient' accuracy the experimental results. In other words, the correctness of the assumptions and the utility of the derived models is verified *a posteriori* through comparison with experimental data.

This alternative type of approach is perfectly acceptable, and is actually the basis of many significant successes in most fields of applied physics, including fluid dynamics. Nonetheless, whenever (or as soon as) possible, the physical and mathematical reasons for the success or failure of the assumptions used should be accurately scrutinized. As a matter of fact, this critical analysis is essential for an improved physical understanding and for the development of new and more accurate models, and thus, in short, for the progress of research.

The historical development of fluid dynamics in past centuries offers a good example of the difficulties faced by scientists and of the different (and sometimes tortuous) ways that were followed to cope with such difficulties. In particular, although viscosity, i.e. resistance to flow, had long been realized from experience to be an important physical characteristic of fluids, and Newton (1687) had expressed in words the basic idea of connecting the viscous forces with the velocity of deformation of a fluid element, it was only during the first half of the 19th century that the full equations of motion of a viscous fluid were derived. However, due to the overwhelming difficulty of their solution, for a long period the writing of these equations made practically no contribution to the development of fluid dynamics.

A first and fundamental simplification of the problem was obtained by neglecting the compressibility of the fluid. As we shall see, this assumption is satisfactory not only for liquids but also for gases, provided certain conditions on the motion are satisfied, the most important of which is that the fluid velocity be everywhere sufficiently lower than the speed of sound. The important point is that, within its range of applicability, the *incompressible flow* model does not hinder the possibility of explaining the physical mechanisms producing the flow fields and the fluid dynamic loads connected with the motion of bodies. However, without any further simplification or assumption on the type of motion, the solution of the incompressible flow equations still remains a formidable mathematical task.

On the other hand, well before the full equations of fluid dynamics were derived, another and more powerful approximation had been introduced, which consists in neglecting not only the compressibility but also the viscosity of the fluid. In effect, the motion of such an 'ideal' incompressible and non-viscous fluid could be described by much simpler equations, derived around 1750 by the eminent Swiss mathematician Leonhard Euler. However, around the same time, the famous French scientist Jean-Baptiste le Rond d'Alembert demonstrated mathematically that no resistance force acts on a body moving at constant velocity through an ideal non-viscous incompressible fluid, a result that is now widely known as *d'Alembert's paradox*. In spite of this disappointing conclusion, the ideal non-viscous fluid theory continued to fascinate a significant number of eminent mathematicians for more than 150 years, perhaps also thanks to the attractiveness of its rigorous mathematical approach. Actually, it became so important that it developed into a new autonomous discipline, known as 'hydrodynamics', and elegant mathematical methods were developed for the prediction of the flow of an ideal non-viscous fluid around objects of any shape.

Obviously, the fascination of these theoretical results could not be shared by hydraulic engineers, who struggled in predicting the forces experienced by bodies in motion or the resistance to the flow of water in channels and pipes. Thus, hydraulics developed as an autonomous empirical discipline, in which non-dimensional correlation laws were sought in order to generalize experimental data. This situation is well represented by a famous remark that Lighthill attributed to the Nobel-laureate chemist Sir Cyril Hinshelwood: 'fluid dynamicists were divided into hydraulic engineers who observed things that could not be explained, and mathematicians who explained things that could not be observed' (Lighthill, 1956).

This impasse was partially overcome by Prandtl (1904), who first suggested that, provided a body satisfies certain restrictions regarding both its shape and its motion, the effects of viscosity are 'felt' only within a narrow region near its surface, the *boundary layer*, while outside it the fluid behaves *as if* it were non-viscous. In fact, using physical intuition and an order of magnitude analysis, Prandtl was able to simplify the equations describing the viscous motion in the near-wall region and, more importantly, to show that pressure is practically constant across the boundary layer. This result permits linking the pressure on the body surface to that acting on the flow outside the boundary layer, where the equations to be satisfied coincide with those of a non-viscous fluid. Consequently, with his model, Prandtl recovered the usefulness of the mathematical techniques for solving the outer 'non-viscous' flow and simultaneously justified the experimental fact that (at least) a drag force acts on any body moving in a fluid.

Although perhaps not completely rigorous from the point of view of a pure mathematician, Prandtl's idea was certainly an outstanding one. Nevertheless, it took more than 20 years for boundary layer theory to spread into research and even longer to obtain the deserved recognition in textbooks. Possibly, this was also partly due to the fact that the theory first appeared only in a short paper presented at a mathematical congress in Heidelberg. However, a more appropriate statement would probably be that 'boundary layer shares with most other brilliant concepts the fate of having been at first disregarded and later accepted as obvious' (Liepmann and Roshko, 1957).

Besides providing a sort of reconciliation between experimental evidence and theory, boundary layers may be seen as flow features whose characteristics and dynamics contribute deeply to the physical understanding of fluid motion. In fact, even in his original paper, Prandtl emphasized the fundamental phenomenon of boundary layer *separation*, in order to explain the large wakes characterizing the flow around certain bodies. Prandtl described not only examples of separation and of its large effects on the overall flow, but even how it could be delayed by means of suction at the surface.

The observation of the conditions that lead to boundary layer separation offers a new way of analysing and interpreting the flow past different bodies. In effect, it suggests immediately a fundamental distinction between bodies and flow conditions for which separation occurs and those for which it does not. Indeed, there are important differences between the two cases. For instance, bodies producing large separated wakes (*bluff bodies*) are characterized by much larger drag values than those typical of bodies for which, in certain conditions of motion, the boundary layer remains attached over their whole surface, thus producing only very thin wakes (*aerodynamic bodies*). One might observe that the attached boundary layer condition leads to flows that are more similar to those of an ideal non-viscous fluid and, therefore, it is the condition that would be most desirable in many applications.

Even more importantly, when the boundary layer around a body remains thin and attached, the flow field and the related loads can be analysed and predicted using a much simplified iterative procedure, which implies consecutive solutions of the outer non-viscous flow and of the boundary layer equations. Although a detailed discussion would not be appropriate in an introductory chapter, we may anticipate that when the motion is such that the dynamical effects of the fluid compressibility may be neglected, it is not even necessary to assume that outside the boundary layer the fluid be non-viscous, but only that its motion be *irrotational*. This is a purely kinematical condition, implying that single small elements of fluid do not instantaneously rotate or, as we say, that their associated *vorticity* is zero, and it may be shown that the whole incompressible ideal-fluid theory coincides with the theory of irrotational flows. Furthermore, the thin boundary layers near a solid surface are indeed the flow regions where vorticity is effectively present, while it can be neglected in the outer flow.

Even restricting the analysis to those flow conditions that allow the above simplified approach, before computers were available researchers had to use all their physical insight and mathematical skills to devise ingenious procedures to obtain, in an acceptable time and with a sufficient approximation, useful predictions for particular fluid flow problems. Nowadays, numerical results may be obtained for the same problems with very limited computational effort, even for complex geometries. In fact, the very meaning of the word 'simple', when referred to 'models' or 'prediction procedures', has obviously changed greatly from the first half of the 20th century to the present computer era. In practice, the word 'simple' no longer means 'permits calculations to be carried out by hand', but rather 'requires modest computational resources'. Limited time for solution is thus the actual common feature, which corresponds to growing levels of mathematical and numerical complexity as technical progress gives rise to increasingly powerful computational resources.

Nevertheless, it should be emphasized that, although the present availability of high-performance computers permits the numerical solution of complex equations, simplified models are still extremely useful, for both engineers and researchers. For instance, they may rapidly provide first-order assessments of the fluid dynamic forces and of their variation as a function of the dominant parameters involved in the problem, which are also clearly identified. This is obviously extremely important in engineering practice. Furthermore, a deep analysis of the performance of these models, which is often beyond expectations, helps in the identification of the essential features influencing the flow field, and may give important clues as regards the development of further models for different applications. In reality, simple models normally derive from observation, deduction and the consequent generation of new physical ideas. However, these ideas may arise not only from a careful analysis of the results of experiments and numerical simulations, but also from the recognition of the reasons why certain previously existing models do or do not work.

Aeronautical engineering is perhaps the most representative field as regards the development and successful use of simple prediction procedures. For instance, the forces acting on the wings of an aircraft, in normal flight conditions, may be predicted with sufficient accuracy through the above-mentioned simplified procedure, i.e. without having to solve the full equations of motion. In reality, if compressibility effects are neglected, the lift force on a wing section can be evaluated at first order through the same equations that apply for a non-viscous flow, provided we ignore the presence of the boundary layers and do not investigate the physical process leading to their generation. As for drag, it may be obtained, together with a second-order approximation of lift, by subsequently coupling irrotational flow and boundary layer calculations in an iterative procedure. Moreover, when the dynamical effects of air compressibility become dominant, as happens at very high flight velocities (the so-called *supersonic* conditions), the non-viscous flow model also provides an evaluation of a fundamental part of drag.

In effect, perhaps not all aeronautical engineers fully appreciate how fortunate they are, considering that the bodies they have to design, which must produce high lift with a small drag penalty, are exactly those for which the above-described simplified prediction procedures can be applied. In a sense, if the flow around a body cannot be predicted through those models, the body is not the right one for aeronautical applications. Certainly, the same good luck is not shared by those involved in the prediction of the forces acting on bluff bodies, such as mechanical engineers trying to reduce the drag of road vehicles, or civil engineers facing the hard task of predicting the loads induced by wind on the structures they design and, especially, their often disastrous dynamical effects. In other words, fluid dynamics does not deal just with the particular situations in which thin and attached boundary layers exist. Conversely, a deep understanding of the dynamics of both attached and separated boundary layers, of the conditions for the passage from one situation to the other, and of the effects of separation on the flow features and on the related loads, is essential in many (or perhaps most) applications of fluid dynamics. In practice, the prediction and control of boundary layer separation is an important topic for both engineers and researchers. Although our knowledge on separated flows has increased at a high rate in recent years, the present availability of simple models is scarce in this area, and much research work is still left to be done. Therefore, to improve our predictive ability, a more extensive use of systematic experiments and numerical simulations must be made.

There are many other examples of flow conditions for which our understanding is definitely far from what would be desirable. The most remarkable one is probably turbulence, which is an apparently chaotic type of motion that occurs in many common situations, and which has great effects on the flow features and on the consequent fluid dynamic loads. Decades of research have been devoted to this subject, and probably many more will be necessary to fully unravel the secrets of its mechanisms. In particular, the phenomena leading to transition to turbulence are both extremely complex and vitally important from the point of view of practical applications. In spite of our still overwhelming ignorance as regards turbulence, an entire book would hardly be sufficient to describe, even concisely, all the efforts that have been made over the years to develop theoretical and empirical methods to explain, predict and control turbulent flows.

A common feature of all areas of fluid dynamics in which our present knowledge is still limited is that the synergic use of all the available investigation tools is essential for a significant progress of knowledge. Thus, advances in experimental and numerical techniques, as well as improvements in all the procedures for data analysis that may help in extracting all the valuable information contained in the results of dedicated investigations, may be very advantageous. In particular, great opportunities are now offered by numerical simulation, provided it is correctly used, and it is also seen as an effective aid to carrying out physical and theoretical analyses. Nonetheless, computational fluid dynamics has not reduced the importance of experimental activities, and probably never will. What has occurred is a change in the role and in the type of problems that are now analysed experimentally. In practice, experimental tests are used more and more to verify the adequacy of the most promising solutions derived numerically, or to analyse configurations and flow conditions that cannot be adequately simulated by numerical means. Furthermore, experiments are essential to validate numerical methods, even if, due to the continuous progress in computational power, detailed numerical simulations are now also increasingly used to appraise the performance of experimental techniques. Finally, theoretical analysis, perhaps currently somewhat underestimated, should be viewed as a further precious instrument to deepen our understanding of the motion of fluids. Cooperation between fluid dynamicists and mathematicians may still greatly contribute to the development of original approaches for the solution of many challenging problems.

From all the above, it should be evident that fluid dynamics, rather than being a well-consolidated scientific discipline, is in fact a lively and ever-developing research field. This is probably the reason why it may attract the attention not only of those interested in applications, but also of young researchers, who can be confident that ample room is available for fundamental investigations and for original contributions in many different areas of an important branch of applied physics. Anyhow, in order to attain significant progress in technical and scientific knowledge, profound mastering of the basic notions of fluid dynamics is always essential. Providing this necessary introductory material is the main purpose of this book, whose contents are mostly classical, although the adopted approach may not be considered as totally traditional. Moreover, some of the topics dealt with derive from fairly recent research findings and might also be of interest to a more advanced readership.

1.2. Organization of the Book, Notation and Useful Relations

The first part of the book is entirely dedicated to fundamental elements of fluid dynamics, and all its contents should be viewed as a necessary pre-requisite for deeper studies aimed at either application-oriented or research activities.

In Chapter 2, fluids are defined and their behaviour and main physical properties are described and briefly discussed, together with the basic assumptions for the treatment of fluid dynamics as a branch of the mechanics of continuous media. The methods and mathematical tools for the description of the kinematics of fluids are introduced in Chapter 3, where the deformation of a fluid element in motion is also described. All the material in these two chapters is essential for the subsequent derivation of the equations of motion of fluids, which is given in detail in Chapter 4. This chapter is perhaps the core of Part I, and its final section provides a discussion on the mathematical aspects of the derived equations and on the utility — or, more often, the necessity — of devising simplified models to study the motion of fluids and/or to predict the forces acting on bodies moving inside them. The incompressible flow model is then the object of most of the following chapters of the book, and the conditions for its applicability are presented and discussed thoroughly in Chapter 5, together with the relevant equations. At the end of this chapter the consequences of neglecting viscosity are also critically analysed. Chapter 6 concerns the origin and dynamics of vorticity in incompressible flows, and is a fundamental preliminary for the description of boundary layer theory, which is carried out in detail in Chapter 7. In fact, a thorough knowledge of the dynamics of boundary layers is very helpful for understanding the behaviour of the flow around bodies immersed in a stream, and for identifying the conditions for the applicability of simplified prediction methods. The last section of Chapter 7 also provides a brief outline of the main characteristics of turbulent flows and of the effects of transition to turbulence on the development and general behaviour of boundary layers. In Chapter 8, which is the last chapter of the first part of the book, all the subject matter introduced in the previous chapters is used to describe briefly the fundamental aspects of the fluid dynamic loads acting on bodies in relative motion in a fluid. Particular emphasis is placed on the strict connection between the qualitative and quantitative characteristics of the loads experienced by various types of bodies and the essential features of the related flow fields. Considering its general and practical interest, a detailed description of the physical mechanisms causing the generation of lift on a wing section is also given.

The second part of the book is devoted to some theoretical issues concerning incompressible flows and to particular topics connected with classical applications of fluid dynamics. In particular, Chapter 9 deals with exact solutions of the incompressible-flow equations, some of which are described in more detail, in view of their practical relevance. Chapter 10 illustrates that the energy balance may be useful even when incompressible flows are considered, particularly for the physical interpretation of the drag experienced by different bodies. It is also shown that the dissipation occurring in a flow may often be evaluated in a simple manner. On the other hand, some additional mathematical and physical issues concerning the role of vorticity in incompressible flows are the subject of Chapter 11, where further details on the connection between the fluid dynamic loads and vorticity dynamics are also provided.

The remaining chapters of Part II are more directly related to certain applications. In particular, Chapters 12 and 13 should be of more specific interest to aeronautical engineering students, and describe the main features of the flow, loads and prediction methods, in incompressible flow, for airfoils and finite wings, respectively. Conversely, Chapter 14 provides a brief analysis of the aerodynamics of bluff bodies, which might be useful to those who are expected to deal with the design of road vehicles or civil structures exposed to the wind. Finally, the last chapter of the book is the only one in which the effects of compressibility are considered in some detail. The treatment is restricted to the case of the flow inside ducts using the approximate model of one-dimensional motion, but introduces some concepts and practical prediction methods that are of significant interest in different fields of engineering. In most of Chapter 15 viscosity is neglected, so that the fundamental role of compressibility is illustrated. However, the added effects of friction are considered in the last section, in order to highlight their importance in the development of adiabatic flows in long ducts.

It should be pointed out that in this book the basic notions of continuum mechanics and thermodynamics, as well as the relevant mathematical tools, are assumed to be already familiar to the reader. Nonetheless, care is taken in providing a physical interpretation of the various terms appearing in the numerous equations that are inevitably contained in a textbook on fluid dynamics.

As regards notation, scalars are denoted by lowercase or uppercase characters in italic style (e.g. a, A), vectors by lowercase or uppercase bold-italic characters (e.g. $\boldsymbol{n}, \boldsymbol{V}$), and second-order tensors by lowercase or uppercase bold-roman characters (e.g. $\mathbf{T}, \boldsymbol{\tau}$). The components of a vector $\boldsymbol{a}$ are indicated as a_i (with $i = 1, 2, 3$), and the components of a second-order tensor $\mathbf{T}$ as T_{ij} or T_{ik} (with $i, j, k = 1, 2, 3$).

It must be emphasized that the Einstein summation rule is *not* adopted in this book, and thus summation symbols are always used explicitly. For instance, T_{ii} only indicates a generic term of the diagonal of the matrix representation of a second-order tensor $\mathbf{T}$, and the scalar product between two vectors $\boldsymbol{a} \equiv (a_i)$ and $\boldsymbol{b} \equiv (b_i)$ is written as

$$\boldsymbol{a} \cdot \boldsymbol{b} = \sum_i a_i b_i. \tag{1.1}$$

Using a Cartesian reference system with axes x_i, the divergence of a vector $\boldsymbol{a}$ is then the scalar quantity

$$\operatorname{div} \boldsymbol{a} = \sum_i \frac{\partial a_i}{\partial x_i}, \tag{1.2}$$

and the gradient of a scalar ϕ is the vector

$$\operatorname{grad} \phi = \sum_i \frac{\partial \phi}{\partial x_i} \boldsymbol{e}_i, \tag{1.3}$$

where we have indicated with $\boldsymbol{e}_i$ the unit vectors of the coordinate axes.

A further fundamental quantity is the curl of a vector $\boldsymbol{a}$, which is a vector that in Cartesian coordinates (x_1, x_2, x_3) has the following expression:

$$\operatorname{curl} \boldsymbol{a} = \left(\frac{\partial a_3}{\partial x_2} - \frac{\partial a_2}{\partial x_3} \right) \boldsymbol{e}_1 + \left(\frac{\partial a_1}{\partial x_3} - \frac{\partial a_3}{\partial x_1} \right) \boldsymbol{e}_2 + \left(\frac{\partial a_2}{\partial x_1} - \frac{\partial a_1}{\partial x_2} \right) \boldsymbol{e}_3. \quad (1.4)$$

We recall now the definition of some operators involving vectors and second-order tensors that are used extensively throughout the book and, in particular, in the derivation of the equations of fluid dynamics.

The left and right inner products between a vector $\boldsymbol{a} \equiv (a_i)$ and a tensor $\mathbf{T} \equiv (T_{ij})$ are vectors whose components are respectively defined by the following expressions:

$$(\boldsymbol{a} \cdot \mathbf{T})_i = \sum_j a_j T_{ji}, \quad (1.5)$$

$$(\mathbf{T} \cdot \boldsymbol{a})_i = \sum_j T_{ij} a_j. \quad (1.6)$$

Furthermore, the divergence of a tensor $\mathbf{T} \equiv (T_{ij})$ is the vector with components

$$(\operatorname{div} \mathbf{T})_i = \sum_j \frac{\partial T_{ji}}{\partial x_j}, \quad (1.7)$$

whereas the gradient of a vector $\boldsymbol{a} \equiv (a_i)$ is the tensor whose components are

$$(\operatorname{grad} \boldsymbol{a})_{ij} = \frac{\partial a_j}{\partial x_i}, \quad (1.8)$$

Moreover, we define the contraction (also known as double dot product) of two tensors $\mathbf{T} \equiv (T_{ij})$ and $\mathbf{D} \equiv (D_{ij})$ as the scalar quantity given by the following expression:

$$\mathbf{T} : \mathbf{D} = \sum_{ij} T_{ij} D_{ij}, \quad (1.9)$$

and the dyadic tensor produced by two vectors $\boldsymbol{a} \equiv (a_i)$ and $\boldsymbol{b} \equiv (b_i)$ as the tensor whose components are

$$(\boldsymbol{a}\boldsymbol{b})_{ij} = a_i b_j. \quad (1.10)$$

By using the above definitions, the following relation may be easily obtained:

$$(\text{div}\,\boldsymbol{ab})_i = \sum_j \frac{\partial(a_j b_i)}{\partial x_j} = b_i \sum_j \frac{\partial a_j}{\partial x_j} + \sum_j a_j \frac{\partial b_i}{\partial x_j}$$

$$= (\boldsymbol{b}\,\text{div}\,\boldsymbol{a})_i + (\boldsymbol{a}\cdot\text{grad}\,\boldsymbol{b})_i,$$

so that we may write

$$\text{div}(\boldsymbol{ab}) = \boldsymbol{b}\,\text{div}\,\boldsymbol{a} + \boldsymbol{a}\cdot\text{grad}\,\boldsymbol{b}, \tag{1.11}$$

which is a generalization of the more common expression that can be easily obtained when a scalar ϕ is present in place of vector $\boldsymbol{b}$.

For the sake of completeness, and to highlight the connection with the notation used in other textbooks, it may be useful to introduce the *del* vector operator, denoted by the symbol *nabla*, ∇, which in Cartesian coordinates is defined as

$$\nabla = \sum_i \frac{\partial}{\partial x_i}\boldsymbol{e}_i. \tag{1.12}$$

We may then express the gradient, divergence and curl of a vector also in the forms

$$\text{grad}\,\boldsymbol{a} = \nabla\boldsymbol{a},$$

$$\text{div}\,\boldsymbol{a} = \nabla\cdot\boldsymbol{a},$$

$$\text{curl}\,\boldsymbol{a} = \nabla\times\boldsymbol{a}.$$

Finally, the Laplace operator (or *Laplacian*), which may be applied to both scalar and vector quantities, is defined, in Cartesian coordinates, as follows:

$$\nabla^2 = \nabla\cdot\nabla = \sum_i \frac{\partial^2}{\partial x_i^2}. \tag{1.13}$$

The expressions for the most common operators in cylindrical and spherical coordinates, which may be of interest in certain circumstances, are reported in Appendix B.

Finally, we recall two theorems that will be extensively used in the following chapters. The first is Gauss' theorem, also known as the divergence theorem, which states that, given a volume $\boldsymbol{v}$, bounded by a surface S and

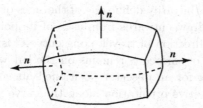

Fig. 1.1. Possible volume for application of Gauss' theorem.

where a differentiable vector field a is defined, then the following relation holds:

$$\int_v \operatorname{div} a \, dv = \int_S a \cdot n dS, \qquad (1.14)$$

where n is the outwards-directed unit vector normal to the bounding surface; note that S may also be piecewise smooth, i.e. the presence of sharp edges is permitted (see Fig. 1.1).

Therefore, Gauss' theorem may be expressed in words by saying that the outflow of a vector field a through a closed surface S is equal to the volume integral of the divergence of the vector field over the region enclosed by S. It thus provides a relation between surface and volume integrals, which is particularly useful when the flow of a certain quantity through a surface appears in the balance equations of the motion of a finite volume of fluid bounded by that surface.

The other theorem that will be of interest is Stokes' theorem (also known as the Kelvin–Stokes theorem) which, conversely, relates surface integrals with line integrals. To formulate the theorem, we first consider a closed curve C in a vector field a, and an arbitrary (generally curved) surface S bounded by curve C (see Fig. 1.2).

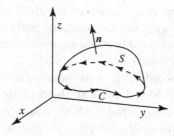

Fig. 1.2. Surface and boundary for application of Stokes' theorem.

Furthermore, we arbitrarily define one of the sides of the surface as the positive one, and we draw outwards from each of its points a unit normal vector n. A positive direction of travel along curve C is then chosen such that the positive side of surface S remains on the left while following the curve. In other words, for an observer positioned above the positive side of the surface, the positive orientation along the curve would be counter-clockwise.

Stokes' theorem is then expressed by the following relation:

$$\int_S \operatorname{curl} \boldsymbol{a} \cdot \boldsymbol{n}\, dS = \oint_C \boldsymbol{a} \cdot d\boldsymbol{l}, \tag{1.15}$$

where $d\boldsymbol{l}$ is an infinitesimal vector element along curve C.

The quantity on the right-hand side of (1.15) is denoted the *circulation* of vector $\boldsymbol{a}$ along the closed curve C. Therefore, Stokes' theorem states that the circulation of a vector $\boldsymbol{a}$ around a closed curve C is equal to the outflow of curl $\boldsymbol{a}$ through an arbitrary surface S bounded by curve C.

However, some care must be paid when applying Stokes' theorem to domains that are not simply connected, i.e. in which 'holes' are present. In effect, the contour to which Stokes' theorem (1.15) may be applied must be 'reducible', i.e. it must be possible to shrink it progressively to zero dimension, always remaining inside a domain in which vector $\boldsymbol{a}$ is defined. In other terms, the curve must always be the boundary of a domain in all points of which $\boldsymbol{a}$ is defined.

Now, in many problems of fluid dynamics we may approximate the flow field as being two-dimensional, which means that a space direction exists along which neither the flow quantities nor the geometry vary, and the relevant velocity component is assumed to be zero. This may be the case of a body that is very long in one dimension, along which its shape does not change (such as a cylinder or long tube), and which is immersed in a cross-flow. In that situation, if we want to apply Stokes' theorem to the vector $\boldsymbol{V}$ representing the velocity of the fluid outside the body, we realize immediately that any curve rounding the body will not be reducible, because if it is shrunk progressively it will eventually exit the space domain where the flow velocity is defined.

We may overcome this difficulty by introducing a 'cut' of the domain, as in Fig. 1.3, linking the considered curve to the contour of the body, and by defining a new curve also comprising the body contour and the cut counted twice. This new curve is then reducible, and we may apply Stokes' theorem considering it to be the boundary of a domain where velocity is completely

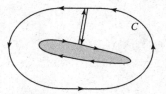

Fig. 1.3. Application of Stokes' theorem to a non-simply connected domain.

defined. Note that the inner curve, coinciding with the body contour, is followed in the clockwise direction for the evaluation of the line integral on the right-hand side of Eq. (1.15). Furthermore, the global contribution of the cut to this line integral is zero, as it is followed twice with opposite orientations.

Chapter 2

PROPERTIES OF FLUIDS

2.1. Definition of Fluids

From the macroscopic point of view, the elementary distinction between *solids, liquids* and *gases* is that solids have a definite volume and shape, liquids have a definite volume but not a definite shape and gases have neither a definite volume nor a definite shape.

Obviously, these definitions are far from satisfactory if we want to proceed to the development of a sound theoretical description of the behaviour of the different substances. However, even if we remain in a macroscopic approach, by describing the various states of matter as *continuous media*, i.e. without (or before) going into an analysis of the differences in their molecular structure, we may differentiate one from the other by observing their response when they are subjected to the action of forces.

In particular, let us consider a generic point lying on a surface bounding a small volume of substance. A force acting at that point may then be decomposed in two components, one in a direction perpendicular to the surface (the *normal force*), and another one tangential to the surface (the *tangential force*). The effect of tangential forces on the behaviour of the considered volume of substance is the fundamental feature that allows us to distinguish between solids and *fluids* (which is a category that comprises both liquids and gases).

In detail, when subjected to finite tangential forces over its surface, a volume of solid will react by *deforming*, but will eventually reach a new condition of static equilibrium in which its shape (or *configuration*) has changed, albeit by a small amount if the applied forces are sufficiently

21

small. In this new equilibrium static condition, tangential forces still act over the surface of the volume of solid — and actually over any surface dividing two parts of the substance.

On the contrary, a fluid may be defined as *a substance that lacks a definite shape and deforms continuously under the action of tangential forces*. In other words, fluids react to tangential forces only with motion, i.e. in dynamic conditions. Conversely, in static equilibrium, a volume of fluid will withstand only forces normal to its surface, which are usually called *pressure* forces.

This definition provides a clear distinction between solids and fluids. On the other hand, the most important difference between liquids and gases lies in their *compressibility*, which is a measure of the change of volume caused by a variation of normal forces. It is indeed a common experience that gases can be compressed much more easily than liquids, and this implies that if a motion causes significant variations in pressure, it will produce much larger variations of volume for a gas than for a liquid. Nonetheless, we shall see that this difference is less fundamental because it does not prevent the motion of liquids and gases being described through the same fundamental equations.

As already mentioned, the above analysis is based on the assumption that both solids and fluids may be considered as continuous media, and this implies that for the description of their macroscopic behaviour it is not necessary to examine in detail the motion of the molecules of which they are composed. The plausibility and limits of this assumption will be further discussed in Section 2.2. However, to achieve a deeper insight as regards the physical origin of the macroscopic properties of solids, liquids and gases, it is convenient to consider briefly the main differences between their molecular structures.

In solids the molecules are positioned orderly at almost constant distances, and their mobility is limited, while in fluids the mobility of the molecules is much higher. In the case of liquids, the molecules are still organized in mutually interacting groups, whose level of 'order' is a decreasing function of the distance between the molecules. On the other hand, in gases, the molecules move almost freely and with very limited mutual interaction.

In more detail, we may consider the trend of the forces that are exchanged reciprocally by two molecules of a generic substance (*intermolecular forces*) as a function of their mutual distance. This trend

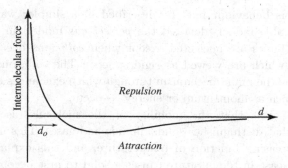

Fig. 2.1. Sketch of the variation of intermolecular forces with the distance d between the centres of two molecules.

is shown schematically in Fig. 2.1 for the case of un-ionized molecules not forming a chemical bound.

As is clear, there is an equilibrium distance d_o (which is of the order of $3 - 4 \times 10^{-4} \, \mu m$) at which the intermolecular force is zero and changes sign. For lower distances the force is increasingly repulsive, whereas for larger distances it is attractive but rapidly falls off with increasing molecular separation.

In solids, the molecules are regularly spaced at mutual distances of the order of d_o, and they undergo extremely limited oscillations around this equilibrium position (with amplitudes that are much less than d_o). The resulting geometrical pattern is very regular, and is known as a *lattice*.

In liquids, the mutual distance between molecules is still approximately d_o, but the oscillations of the single molecules are much larger, namely of the order of d_o; moreover, the regularity of the geometrical pattern is maintained only for space-limited groups of molecules, i.e. for mutual distances below a certain value. In other words, given a specific molecule A, in its immediate neighbourhood the molecules will be positioned in a rather regular pattern (i.e. nearly at multiples of d_o from A), but for increasing distances they progressively lose any geometrical relation with the position of molecule A. For this reason the geometrical structure describing the arrangement of the molecules in a liquid is denoted a *quasi-lattice*.

Finally, in gases, the molecules move much more freely, with mutual distances that are much greater than d_o. The interactions between molecules are then very weak, save for when they happen to come very close to each other. This behaviour may be described in a simpler way through a widely applicable model (denoted the *perfect gas* model) in which the intermolecular forces are neglected except when *collisions* occur between the molecules, which are viewed as rigid spheres. The collisions are then considered to be the main mechanism through which exchanges of physical quantities — such as momentum or energy — occur.

It is thus clear that the mobility of the molecules is the main characteristic that distinguishes solids, liquids and gases. This mobility is, in fact, an increasing function of temperature, and this explains why a progressive increase in temperature causes a solid to first become a liquid and then a gas, provided the ambient pressure is kept constant or within certain limits.

It should be pointed out that the liquid phase is perhaps the least understood of the three states of matter or, in other words, the most difficult to represent theoretically and/or through simple models (see, e.g., Trevena (1975), for an in-depth discussion). This derives from the fact that, in a sense, liquids share some of the properties of both solids and gases. In particular, we have seen that the molecules in a liquid are crowded together in a way that is not very dissimilar to that of a solid, even if they show a greater mobility. Thus we may say that both solids and liquids possess significant *cohesion*, deriving from the action of the intermolecular forces. This is why liquids will not fill up the entire volume available in a containing vessel, as gases do thanks to the free motion of their molecules. On the other hand, at variance with solids, liquids do not withstand tangential forces without moving, and so *flow* occurs. As we have already mentioned, this property is also characteristic of gases and it is so important for the description of motion that we may group together liquids and gases in the same class of substances, namely fluids, in spite of their significantly different molecular structures.

Finally, we must point out that in some cases the distinction between solids and fluids may not be so clear as we have just described. As a matter of fact, there are substances that behave in an intermediate manner, and are known as *dual* substances. Two of them are particularly important: the so-called *creeping fluids*, and the *Bingham fluids*.

Creeping fluids are characterized by the fact that if a force acts on them for a short period they behave like a solid, with the possibility of reaching a static condition with a small deformation. If, however, the force is maintained for a sufficiently long time, they will start to flow, like a liquid. A typical example of a creeping fluid is asphalt.

A Bingham fluid, on the other hand, is able to support a tangential surface force in static conditions, like a solid, as long as the force per unit surface (which is called *stress*) remains below a certain threshold value. Above this value, it behaves like a fluid, i.e. it starts flowing. A very common example of a Bingham fluid is toothpaste.

In this book we shall not consider these intermediate cases, and will only be concerned with *simple fluids* as defined above. Actually, this is not a great restriction, as our treatment will permit description of the behaviour of a large variety of important substances, from practically all gases to apparently quite different liquids, such as water, oil or honey.

2.2. The Continuum Hypothesis

Any substance is composed of myriad molecules in constant motion. However, in most engineering problems one is interested not in the description of the motion of the single molecules but rather in the macroscopic behaviour of the substance, and it is thus useful to think of it as a continuous material. Throughout this book we assume the validity of this *continuum hypothesis* in the treatment of the motion of fluids. Although this obviously implies that we are using a simplified model of reality, in most cases it allows for good predictions, at least whenever the smallest volume of fluid of interest contains enough molecules that statistical averages 'make sense'. We now discuss in more detail the meaning, implications and limits of this statement.

In the continuum description, we are concerned with the values of physical quantities associated with 'material points' of fluid, i.e. with extremely small quantities of fluid that we may regard as point-like. However, this is a vague concept that deserves a much deeper analysis, and we must reconcile it with the fact that we know all matter to be composed of molecules.

A central concept is that of the *elementary fluid particle* or, more simply, *fluid particle*. To define it, let us imagine that we want to measure the density at a point P in the fluid. We may then take a volume δv of fluid around P and measure the mass δm contained in it, which is the sum of the masses of all the molecules that are present within the volume at the precise instant in which the measurement is being taken. The average density of fluid in δv will then be

$$\bar{\rho} = \frac{\delta m}{\delta v}.$$

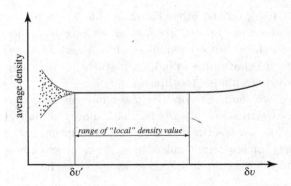

Fig. 2.2.   Variation of the measured average density with the volume δv.

We can now analyse the value that is obtained through this procedure when the volume δv is changed, assuming that an instrument is always available to adequately carry out the measurements. A possible sketch of the qualitative variation of the result is shown in Fig. 2.2.

As may be seen, for large values of δv the average density will generally not be constant when δv is varied (and may possibly increase or decrease), because we are considering such large volumes of fluid that they may contain macroscopic variations of density. On the other hand, as the value of δv is progressively decreased, an asymptotically constant density measure will be obtained. However, if we shrink the volume below a certain limit value $\delta v'$, we find that the measured values for density will start being scattered in an increasingly large interval. In other words, even considering the same value of δv, measurements taken at different times will generally give different results.

The explanation of this behaviour is straightforward. We must remember that the surface bounding the volume of fluid we are considering is permeable to molecules, which are then free to move in and out of δv. This implies that the number of molecules present in the volume is not exactly fixed and may change in time, albeit by a small amount. As a matter of fact, the variation is given by the difference between the number of molecules entering and exiting the volume in the considered time interval. Now, if δv is large enough, the number of molecules present inside it is always so high that its variation in time cannot give rise to significant differences in subsequent measurements. For instance, if the volume is such that, say, 10^7 molecules are present within it, a difference from one instant to another of a few tens or hundreds of molecules will not be measurable. However, if we go on shrinking δv, a limit size will be reached below which the number of molecules inside the volume will become so small that its percentage variations in time will be significant, and thus different values will be measured at different times, giving rise to the scatter shown in Fig. 2.2. In this case we say that the measurement is not *statistically stationary*.

From the above it is clear that $\delta v'$ gives a correct order of magnitude for the definition of an elementary portion of fluid in the context of a continuum description. We may indeed say that *a fluid particle is a volume of fluid that is sufficiently small to be considered infinitesimal with respect to the spatial variations of any macroscopic quantity, but large enough to contain a number of molecules that is sufficiently high to allow the average value of each quantity to be statistically stationary.*

We may now wonder if this definition is a strongly limiting one, or if it allows us to use the continuum hypothesis in many interesting circumstances. To answer this question we may take, for instance, a volume of air of $10^{-12}\,\mathrm{cm}^3$ (e.g. a cube with a side of $1\,\mu m$) in standard conditions, i.e. at sea level pressure and $15°C$. It is reasonable to assume that the size of this volume satisfies the first requirement of the definition, i.e. to be small enough to allow spatial variations of macroscopic quantities within its characteristic dimension to be neglected. As regards the second requirement, the number of molecules contained in the considered volume is easily seen to be approximately 2.7×10^7; therefore, we may safely conclude that such a volume would indeed be adequate for representing a fluid particle.

Having discussed the physical meaning of a fluid particle in the context of the continuum hypothesis, we may define the value of a physical quantity at a point P in the fluid as the average value of that quantity evaluated taking into account all the molecules contained in a fluid particle whose centre of gravity coincides with P. For instance, the density at point P is given by

$$\rho(P) = \lim_{\delta v \to \delta v'} \frac{\delta m}{\delta v}. \tag{2.1}$$

Similarly, the velocity at a point in a homogeneous fluid will be the average of the velocities of all the molecules contained in a volume of the order of $\delta v'$ around the considered point.

It should now be pointed out that the macroscopic behaviour of a fluid, which is of interest in a continuum treatment, is actually dependent on the interactions between neighbouring fluid particles, i.e. on the mechanisms through which physical quantities are exchanged between the particles. From the above it should now be clear that these exchanges (or, as they are called, *transport properties*) are completely determined by the interactions between the molecules that are present in the boundary regions of the various particles. We shall give a somewhat more detailed description of these mechanisms in the following sections.

From the above considerations, the limit of applicability of the continuum hypothesis may also be easily derived. In fact, this limit may be reached when rarefied gases are considered, as may happen, for instance, when studying the motion of a body in the upper atmosphere. An important quantity for this assessment is the *mean free path* of the molecules, which is a measure of the distance travelled by a gas molecule before a collision with another one occurs. The parameter that indicates the level of rarefaction is the *Knudsen number*, which is the ratio between the mean free path of the molecules and a significant reference dimension of the problem, for instance, the characteristic size of a moving body. For values of the Knudsen number that are larger than, say, 10^{-3}, the continuum hypothesis becomes less and less valid, and one must resort to the so-called *kinetic theory of gases* for developing acceptable prediction procedures.

Therefore, the continuum description of the motion of fluids would not be adequate, for instance, to predict the forces acting on a space vehicle during its re-entry in the upper atmosphere. Another example of a situation in which it should be used with caution is the design of MEMS (Micro-Electro-Mechanical Systems), whose typical dimensions may even be of the order of a few molecular mean free paths.

On the other hand, it might be interesting to note that at an altitude of 20 km in the atmosphere the mean free path of the molecules of air is approximately 9.14×10^{-7} m. Therefore, for predicting the fluid dynamic forces acting not only on a road vehicle but also on an aeroplane, the continuum hypothesis may be expected to give very good results. As already pointed out, its validity will always be assumed in our treatment of fluid dynamics, which will thus be considered as belonging to the field of *continuum mechanics*.

2.3. Local Thermodynamic Equilibrium

In our treatment we shall introduce and use quantities normally associated with a thermodynamic meaning, such as temperature or internal energy. This is actually less obvious than it might appear.

Thermodynamics is concerned with equilibrium conditions of uniform matter and the various thermodynamic quantities are normally defined with reference to such conditions. While for fluids at rest this poses no problems, in principle the same is not true for a fluid in motion, in which each particle is generally characterized by conditions of non-equilibrium.

Nonetheless, it may be safely assumed that each fluid particle will be in a condition of *local thermodynamic equilibrium* with the neighbouring ones provided the space and time variations of the various quantities are not too large (in practice larger than those that may be found in almost all conditions of interest). This implies that the characteristic time necessary for the molecules of each particle to reach an equilibrium state with the molecules of the adjacent particles is much smaller than the time in which a macroscopic variation of a quantity may occur due to the fact that the velocity field transports the particle through a space variation (i.e. a *gradient*) of that quantity, or to the presence of time fluctuations at a given frequency.

In practice, save for particular flow conditions, the assumption of local thermodynamic equilibrium is applicable as long as the same is true for the continuum hypothesis. For instance, a situation in which the assumption might not be adequate is the study of the motion inside *shock waves*, which, as will be described in Chapter 15, are abrupt variations of the flow quantities that may occur when the flow is *supersonic*, and compressibility thus plays a fundamental role (see Section 2.5). However, the thickness of a shock wave is usually of the order of a few mean free paths of the molecules, so that even a continuum treatment might become questionable for an analysis of the conditions *inside* the shock.

Assuming the validity of local thermodynamic equilibrium, we may then define the usual thermodynamic variables, which are, in general, functions of both space and time and are associated with the particular fluid particles that occupy the different points in space at a certain time. Then, the thermodynamic state of a system may be completely characterized by using any two of the thermodynamic state variables, while the others may be derived from appropriate thermodynamic relations (*equations of state*). The validity of the first and second laws of thermodynamics is also assumed.

However, it may now be appropriate to examine in more detail the application of the first law of thermodynamics to the analysis of the motion of fluids. This fundamental relation describes the variation of the conditions between two different states of a given uniform mass of fluid in equilibrium conditions, and may be written as follows:

$$\Delta e = Q + W. \tag{2.2}$$

This equation, in which Q is the amount of added heat per unit mass and W the work done per unit mass, is actually a definition of the difference between the values of the internal energy per unit mass, e, in the final and initial equilibrium states of the considered fluid. Obviously, this relation holds for any amount of fluid as long as it is homogeneous. Thus, it may also be used for a particle of fluid, as defined in the previous section, provided it is at rest or moving with a uniform constant velocity (which corresponds to a condition of rest if seen in a reference system moving with the same velocity). However, for a fluid in motion with a velocity that is not space-uniform and constant in time, the work acting on the particle contributes not only to the variation of its internal energy, but also of its kinetic energy. Therefore, when in Chapter 4 the energy balance is used to derive one of the fundamental equations describing the motion of a fluid, Eq. (2.2) will be written in a form that includes the variation of the *total energy* of the fluid, i.e. of the sum of its internal and kinetic energies.

Returning now to the definition of the thermodynamic variables, if we take, say, the density, ρ, and the internal energy per unit mass, e, as basic independent variables of state, it is possible to define other variables by using the relevant equilibrium equations of state written as a function of ρ and e. For instance, the entropy per unit mass, S, is defined as corresponding to the value obtained from the equilibrium relation between ρ, e and S when suitable values of ρ and e are introduced. An analogous procedure may be used for the definition of the remaining fundamental quantities in fluid dynamics, and in particular the temperature, T, or the pressure, p. However, we shall see in Chapter 4 that in fluid dynamics the term *pressure* may also be used with a different meaning, so that, when necessary, we shall refer to the value determined through the above procedure as the *thermodynamic pressure* or *equilibrium pressure*.

2.4. Compressibility and Thermal Expansion

In the analysis of the behaviour of fluids, it is of fundamental importance to estimate the variations of volume caused by changes in pressure or in temperature. These variations may be predicted from the knowledge of certain *thermodynamic derivatives*, i.e. of physical parameters obtained by considering the effect of small variations of the various quantities.

In the discussion that follows, it may be useful to also introduce the *specific volume*, v, which is the volume per unit mass of fluid, i.e. the inverse of the density. Let us now consider the case in which a variation of pressure occurs; we may then define the *coefficient of compressibility*, α, which is an index of the relative variation of specific volume per unit variation in pressure, as follows:

$$\alpha = -\frac{1}{v}\left(\frac{\partial v}{\partial p}\right) = \frac{1}{\rho}\left(\frac{\partial \rho}{\partial p}\right). \tag{2.3}$$

In the above relation, the different signs express the fact that an increase in pressure corresponds to a decrease in specific volume and, therefore, to an increase in density.

Actually, a more accurate definition requires that the conditions in which the variation of pressure takes place be specified. Therefore, if we imagine keeping the temperature constant, we have the *isothermal coefficient of compressibility*:

$$\alpha_T = -\frac{1}{v}\left(\frac{\partial v}{\partial p}\right)_T = \frac{1}{\rho}\left(\frac{\partial \rho}{\partial p}\right)_T. \tag{2.4}$$

On the other hand, if the pressure is varied while keeping the entropy constant (i.e. with an adiabatic reversible process), then we have the *isentropic coefficient of compressibility*:

$$\alpha_S = -\frac{1}{v}\left(\frac{\partial v}{\partial p}\right)_S = \frac{1}{\rho}\left(\frac{\partial \rho}{\partial p}\right)_S. \tag{2.5}$$

Sometimes, it is useful to introduce the reciprocal of the coefficient of compressibility, which is known as the *bulk modulus of elasticity*, E. Again, we have two values, according to whether the process is carried out isothermally or isentropically. Thus, the *isothermal modulus of elasticity* is defined as

$$E_T = -v\left(\frac{\partial p}{\partial v}\right)_T = \rho\left(\frac{\partial p}{\partial \rho}\right)_T, \tag{2.6}$$

and the isentropic modulus of elasticity as

$$E_S = -v\left(\frac{\partial p}{\partial v}\right)_S = \rho\left(\frac{\partial p}{\partial \rho}\right)_S. \tag{2.7}$$

From the relations of thermodynamics, it may be shown that the ratio between the isentropic and isothermal moduli of elasticity is equal to the ratio of the specific heats at constant pressure and constant volume:

$$\frac{E_S}{E_T} = \left(\frac{\partial p}{\partial \rho}\right)_S \bigg/ \left(\frac{\partial p}{\partial \rho}\right)_T = \frac{C_p}{C_v} = \gamma. \tag{2.8}$$

Let us now consider the case in which a variation in temperature occurs. The parameter through which it is possible to estimate the variation in specific volume (or in density) due to a variation in temperature is the *coefficient of thermal expansion*, β, which is defined by imagining that the temperature is varied while keeping the pressure constant:

$$\beta = \frac{1}{v}\left(\frac{\partial v}{\partial T}\right)_p = -\frac{1}{\rho}\left(\frac{\partial \rho}{\partial T}\right)_p. \tag{2.9}$$

Again, the signs in (2.9) are connected with the fact that an increase in temperature produces an increase in the specific volume, and thus a decrease in density.

Therefore, if, for instance, we want to estimate how sensitive the density of a fluid is to variations in pressure and temperature, we may use the following expression:

$$\frac{d\rho}{\rho} = \frac{1}{\rho}\left(\frac{\partial \rho}{\partial p}\right)_T dp + \frac{1}{\rho}\left(\frac{\partial \rho}{\partial T}\right)_p dT. \tag{2.10}$$

With the introduction of definitions (2.4) and (2.9), we then have

$$\frac{d\rho}{\rho} = \alpha_T dp - \beta dT. \tag{2.11}$$

From this relation it becomes immediately clear that changes in density are much more likely to occur in gases than in liquids. Considering, for instance, air and water, which are probably the most common fluids to be encountered in practical problems, we have (at atmospheric pressure and 15°C)

for air: $\alpha_T \simeq 1 \times 10^{-2}/\text{kPa}$, $\beta \simeq 3.5 \times 10^{-3}/\text{K}$;

for water: $\alpha_T \simeq 5 \times 10^{-7}/\text{kPa}$, $\beta \simeq 1.5 \times 10^{-4}/\text{K}$.

From the above values one can immediately see that, for given variations in pressure and temperature, density variations are much greater in air than in water. As a matter of fact, even if the sensitivity to changes in temperature is only one order of magnitude higher in air than in water, the difference in the effect of variations in pressure is definitely impressive. As an example, an increase in pressure of one atmosphere (which corresponds to slightly more than $100\,\text{kPa}$) would strongly increase the value of the density of air, while it would change the density of water by only a few thousandths per cent.

It is then clear that in most circumstances a liquid may be considered as an *incompressible* fluid, while it would seem quite exotic to make the same assumption for gases. However, we shall see that this is not true, and if certain conditions apply (which will be described and discussed in detail in Chapter 5), even the motion of a gas may be studied through a simplified treatment in which its density is assumed to be constant.

In general, however, compressibility must be taken into account when describing the behaviour of a gas, and we must write down an *equation of state*, linking the density with the thermodynamic pressure and with the temperature. In many practical cases (including the study of the motion of air), the *perfect gas* model, already mentioned in Section 2.1, may be used and this leads to the following equation of state:

$$p = \rho RT. \tag{2.12}$$

In this equation R is the *gas constant*, which depends on the molecular mass, m, of the gas through the relation

$$R = \frac{R_0}{m}, \tag{2.13}$$

where $R_0 = 8.31447\,\text{J}\,\text{mol}^{-1}\,\text{K}^{-1}$ is the *universal gas constant*.

In particular, for dry air we have $m = 28.966\,\text{kg/kmol}$, so that from (2.13) its characteristic constant is $R = 287\,\text{m}^2\,\text{s}^{-2}\,\text{K}^{-1}$.

It might be interesting to note that for a perfect gas we may easily check relation (2.8) by using (2.12) and the expression of the link between pressure and density for an isentropic process, namely

$$p\rho^{-\gamma} = \text{constant}. \tag{2.14}$$

In effect, it then follows that

$$E_S = \gamma p; \quad E_T = p. \tag{2.15}$$

2.5. Speed of Sound and Mach Number

It is now appropriate to introduce a quantity that has great importance in fluid dynamics, namely the velocity of propagation of infinitesimal perturbations, which is denoted the *speed of sound*. We must first point out that if, for instance, a pressure perturbation is introduced locally in a fluid, its effect will propagate in the medium through spherical *waves*. In other words, if we imagine that the pressure associated with a particle increases by a very small amount with respect to its unperturbed value, this will create disturbances (i.e. in general, not only a variation in pressure, but also in density, velocity and all other fluid dynamic quantities) that are progressively felt in particles that are placed further and further away, with a time lag that is given by the ratio between the distance and the velocity through which the perturbation is transmitted from one particle to adjacent ones.

If we assume that the perturbation is infinitesimal or, more precisely, that it is so small that the square of the ratio between its intensity and the unperturbed value may be neglected, it can be seen that the value of its velocity of propagation, a, is linked to the modulus of elasticity of the fluid; in effect, the speed of sound is given by the following relation:

$$a^2 = \left(\frac{\partial p}{\partial \rho}\right)_S.   \qquad (2.16)$$

In this expression, the variation is assumed to take place in such a way that entropy remains constant, and the reason for this assumption will become clearer in Chapter 4, where it will be shown that the variations in entropy are proportional to the square of the variations in velocity; as the latter have been assumed to be very small relative to their unperturbed values, the same will be true for the entropy variations.

We may now wonder why the velocity of propagation of very small perturbations, a, is called the speed of sound. This becomes clear when we recall that in standard conditions at sea level the atmospheric pressure is of the order of 10^5 Pa, and we estimate the order of magnitude of the pressure variations connected with a sound of a given intensity. The measurement of the intensity of sound is usually carried out in acoustics using the *decibel* (dB), which is a unit directly connected with the characteristics of the human ear; it is defined as

$$\mathrm{dB} = 20 \log_{10}\left(\frac{p_{rms}}{p_{ref}}\right),   \qquad (2.17)$$

where p_{rms} is the root mean square value of the pressure fluctuations caused by the sound, and p_{ref} is the relevant value for a reference sound intensity, which is chosen to be the human auditory threshold level of a sound at 2 kHz.

Now, p_{ref} is 2×10^{-5} Pa, so that normal talking at a distance of 1 m, producing a sound level of 60 dB, corresponds to 2×10^{-2} Pa, and a sound of 120 dB (which is the hearing damage level for short-term exposure) corresponds to 20 Pa. It is then clear that 'normal' sound levels are associated with pressure perturbations that are definitely small with respect to the unperturbed atmospheric pressure, so that the assumptions leading to relation (2.16) may safely be considered to apply.

For a perfect gas, by using relations (2.14) and (2.12), we have

$$a^2 = \gamma \frac{p}{\rho} \Rightarrow a = \sqrt{\gamma R T}. \tag{2.18}$$

Therefore, in this case the speed of sound is a function only of the physical characteristics of the gas and of the absolute temperature. For air at 15°C we then have (assuming $\gamma \simeq 1.4$)

$$a \simeq 340.5 \text{ m/s.}$$

An important parameter in fluid dynamics is the *Mach number*, M, which is defined as the ratio between the modulus of the velocity, V, and the speed of sound, a:

$$M = \frac{V}{a}. \tag{2.19}$$

The importance of the Mach number derives from the extremely different features that characterize flows at velocities that are lower or higher than a. To clarify this point, one may observe the different conditions that occur when an infinitesimal source of perturbation moves in a fluid with different velocities (see Fig. 2.3).

If the source is at rest, the perturbation waves progressively propagate in all directions, symmetrically with respect to the source position, as is shown in Fig. 2.3(a). In this figure, the circles represent the positions occupied at the considered time (denoted by 0) by the waves that were produced by the source at previous time intervals (denoted by the numbers −1, −2 and −3). On the other hand, let us assume that the source is moving with a velocity that is lower than a, or, as we say, with a *subsonic* velocity. This condition is represented in Fig. 2.3(b), where the motion is leftwards, and the positions occupied by the source at different times are also shown. As can be seen, in this case the circles representing the positions reached by the waves generated at the corresponding times are not symmetrical with respect to the source position at time 0. As a matter of fact, the distance

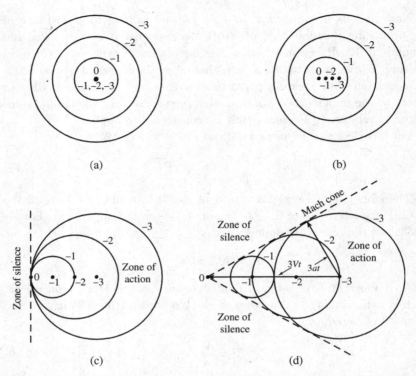

Fig. 2.3.   Waves produced by a source moving at different Mach numbers. (a) $M = 0$, (b) $0 < M < 1$, (c) $M = 1$, (d) $M > 1$.

between subsequent waves is now reduced in front of the moving source, while it is increased behind it. This is the origin of the well-known *Doppler effect*, i.e. the increase in the frequency of a sound in front of a moving source and the decrease behind it (a common experience when an observer listens to the whistle of a train passing by). The important issue, from the fluid dynamic point of view, is that the effect of the moving source is still felt both in front and behind it.

   With the increase in the velocity of the source, the waves will become more and more crowded in front of it; therefore, a limit condition will be reached when the source is moving with a velocity that is equal to the speed of sound ($M = 1$, as in Fig. 2.3(c)). In that case, a completely new situation arises, in which ahead of the source no perturbation is felt at all. In other words, the whole space is now divided in two parts: a *zone of silence*, ahead of the moving source, and a *zone of action* (or, more correctly, a *domain of influence*) behind it.

Finally, when $M > 1$, i.e. the source is moving with a *supersonic* velocity, its perturbation waves remain confined within a downstream cone, whose semi-apex angle θ is given by the relation

$$\sin \theta = \frac{a}{V} = \frac{1}{M}. \tag{2.20}$$

This angle is called the *Mach angle*, and the *Mach cone* represents the domain of influence of the moving source. Obviously, the situation is somewhat different if, instead of an infinitesimal perturbation source, we consider a finite-size moving body. In particular, its domain of influence will no longer coincide with the Mach cone, but will be slightly larger. Nonetheless, the important feature will remain that the perturbation produced by the body is not 'felt' in front of it, but remains confined within a region behind it. In other words, no insect or bird would be able to avoid a vehicle reaching it at a supersonic speed!

From the above it should be clear that compressibility plays a crucial role in the definition of the features characterizing the motion of a fluid. In fact, due to its connection with the elasticity of the fluid, the speed of sound is much higher in liquids than in gases (for instance, it is approximately 1500 m/s for water at 25°C). Therefore, in practical problems with plausible values of the fluid velocity, Mach numbers will usually be much lower in liquids than in gases.

Alternatively, the Mach number can be seen as the main parameter indicating how large the effects of compressibility are. As a matter of fact, if we consider the change in density *linked only to the variations in pressure due to the motion of a fluid* (for instance, caused by a body moving within it), we shall see that its relative value is strictly connected with the Mach number. More precisely, if we indicate with $\Delta \rho$ and Δp the orders of magnitude of the variations in density and pressure, we may obtain a first-order estimate of their connection from

$$\Delta \rho = \frac{\Delta p}{a^2}. \tag{2.21}$$

It will be shown in the following chapters that the order of magnitude of the variations in pressure due to the motion is

$$\Delta p \sim \frac{1}{2} \rho V^2, \tag{2.22}$$

and thus the order of magnitude of the consequent relative variation in density may be estimated to be

$$\frac{\Delta\rho}{\rho} \sim \frac{1}{2}\frac{V^2}{a^2} = \frac{1}{2}M^2. \tag{2.23}$$

In spite of the approximations involved in the derivation, we may anticipate that the validity of this result is quite general, and that the variations in density of a fluid due to its motion may be neglected if the Mach number is everywhere sufficiently small (say, $M < 0.3$), *irrespective of the fluid being a liquid or a gas.*

As we have already mentioned in Section 2.4, the full conditions for a fluid to be considered as incompressible will be discussed in detail in Chapter 5. However, it might be interesting to point out here that assuming the density of a fluid to remain constant in space and time corresponds to assuming that the speed of sound is infinite. In other words, it implies that $M = 0$ everywhere for any finite value of V, and that the perturbations are instantaneously 'felt' in the whole flow field. Obviously, there is no real fluid for which this happens, so the use of this approximation may only be justified by the fact that, within its limits of applicability, it permits reasonable descriptions of the essential physical features of fluid motion and, in more practical terms, it provides predictions whose agreement with experiments is considered satisfactory for given purposes. Actually, this is the basic justification for using any type of simplified physico-mathematical model.

2.6. Viscosity

Generally speaking, viscosity is the resistance of a fluid to flow. More precisely, one can say that viscosity is the physical property by which a fluid resists deformation by shear stresses. All fluids are viscous,[a] but it is a common experience that significant differences in viscosity may occur between one fluid and another. Thus, glycerine, olive oil, water and air are characterized by very different and definitely decreasing degrees of viscosity. However, we shall see that these differences are not essential in our treatment, and that all these fluids have common properties — strictly

---

[a]We do not consider here the so-called *superfluids*, such as liquid helium at temperatures of the order of 2 K, which have a very peculiar molecular structure and behave as if their viscosity were negligible.

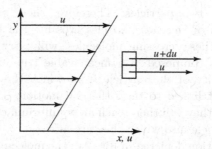

Fig. 2.4. Fluid particles in a unidirectional flow with varying velocity.

connected with their being viscous — that allow their motion to be studied by using the same theoretical models.

Viscosity is one of the so-called *transport properties* of fluids, and its effects can be observed and measured within the continuum (i.e. macroscopic) description. However, in order to understand its physical origin one has to analyse the molecular mechanisms through which *momentum* (i.e. the product of mass and velocity) is transported from one fluid particle to an adjacent one.

To do so, we refer to the most simple flow condition for which the effects of viscosity may be highlighted, leaving for the following chapters the explanation of the role of viscosity in a generic motion. We thus consider a *perfect gas* moving in a two-dimensional unidirectional flow, in which only one component of velocity, say u in the x direction, is different from zero and varies only in the orthogonal y direction (see Fig. 2.4). We analyse two adjacent fluid particles (shown magnified in the figure), which have a side in common and whose centres of gravity are at a distance dy apart. Let the velocities of the particles be u and $u + du$, respectively; recalling what we have already pointed out in Section 2.2, u and $u + du$ are the average values of the velocities of an extremely high number of molecules instantaneously contained in the two particles. The velocity of each molecule is then given by this average value plus a random *thermal velocity*, whose magnitude is an increasing function of temperature. The thermal velocity of the molecules contained in a particle has thus a zero mean value, and has equiprobable components in all directions.

Furthermore, we recall that the surfaces bounding the particles are obviously non-physical, i.e. they may be crossed by the molecules in their random motion, and thus an interaction takes place between the molecules instantaneously present in the neighbourhood of the dividing surface between the two particles. Therefore, there will be a certain number of molecules of the two particles crossing the bounding surface in both directions. These 'crossing molecules' will carry with them their momentum, which is composed of a mean value (the one corresponding to the average velocity of the particle from which they originate), plus a random component linked to their thermal motion. Through a certain number of collisions, they will then reach an equilibrium condition with the molecules of the particle into which they are moving.

Now, in the situation depicted in Fig. 2.4, the molecules flying from the upper particle to the lower one will have an average momentum component in the x direction that is higher than that of the molecules of the lower particle, which will then tend to be increased. Obviously, the opposite happens for the molecules flying from the lower to the upper particle, and the momentum of the latter will tend to be decreased. We may then say that the global effect is that the lower particle tends to be dragged by the upper one, which, conversely, tends to be slowed down. Therefore, if we now look at this mechanism from the macroscopic point of view, we may describe the momentum exchange taking place at the boundary between the particles by saying that a dragging force is acting on the upper surface of the lower particle and an opposite slowing force is acting on the lower surface of the upper particle. Actually, we shall refer, rather than to forces, to *stresses*, i.e. to forces per unit surface, which can be defined at any point in a flow, provided we specify the surface over which they act, for instance by means of a unit vector normal to it.

In this example, the generated stresses will be tangent to the bounding surface, i.e. they will be in the x direction. Furthermore, it is quite evident from the above description that these stresses will increase with increasing difference in velocity between the two particles. In fact, it becomes immediately clear that if the velocities of the two particles were equal, there would be no average transport of momentum between them. If we now suppose that the link between the tangential stress, say τ, and the rate of variation of u in the y direction is as simple as possible, we may assume a linear relationship (which corresponds to neglecting the effects of second and higher degree terms in a Taylor expansion), and write

$$\tau = \mu \frac{du}{dy}. \tag{2.24}$$

In this relation τ has the same dimensions of pressure, i.e. a force (corresponding to mass times acceleration) per unit surface, so that its units are $kg/(m\,s^2)$. The coefficient of proportionality, μ, between tangential stress and velocity gradient appearing in Eq. (2.24), whose dimensions are $kg/(m\,s)$, is called the *coefficient of viscosity* (also known as *dynamic viscosity*, or simply *viscosity*), and is a function of both the type of fluid and its thermodynamic state. In reality, one may see that μ varies, in practice, only with the absolute temperature, T, and is not dependent on density (or pressure).

From the above-described molecular mechanism of momentum exchange, it should be evident that we may expect the coefficient of viscosity of a gas to increase with temperature. As a matter of fact, the net flux of momentum through a unit surface element is proportional to the number of molecules crossing the unit area in unit time, which is directly linked to the random thermal velocity. In turn, the latter is an increasing function of temperature and, for a gas, it varies approximately as the square root of T.

On the other hand, the independence of viscosity from density is somewhat less evident and its detailed explanation requires the use of concepts from the kinetic theory of gases (see, e.g., Vincenti and Kruger, 1965). However, one can observe briefly that the momentum transport is produced by the molecules contained in a layer, adjacent to the considered surface, having a thickness of the order of the mean free path of the molecules, say l. In fact, the molecules will come into equilibrium with the neighbouring ones after a limited number of collisions. The variation of momentum in the process will then be proportional to the density of the gas times the difference Δu in average velocity corresponding to the length l, whose first-order evaluation is then $\Delta u = l\,\partial u/\partial y$. But, as can be shown, and may probably be easily grasped, the mean free path l is inversely proportional to density. Consequently, for a fixed value of temperature, the global variation of momentum is not a function of density.

If we now turn to the analysis of the behaviour of liquids, the origin of viscosity cannot be explained through the mechanism that justifies the viscosity of gases. Actually, the momentum transport at the molecular level occurs in liquids in a much more complex manner, not yet completely understood in detail. What can be said is that, due to the fact that in liquids the molecules are much more closely spaced than in gases, intermolecular forces play a fundamental role in the process. In other words, besides the momentum exchange between molecules due to their random thermal motion, a more important effect is connected with the fact that a velocity gradient tends to tear apart the neighbouring molecules, and this is resisted by the intermolecular forces, which produce an action that is not dissimilar to that of spring forces.

In any case, in spite of the significant differences in the molecular mechanisms producing viscosity in gases and liquids, there is ample experimental evidence demonstrating that the relation expressed in (2.24) is formally valid also for liquids. However, although the viscosity coefficient is, again, dependent only on the absolute temperature, in the case of liquids this dependence is the opposite of that of gases. In other words, the viscosity coefficient in liquids is a decreasing function of the temperature (and often a rapidly decreasing one). This fact can easily be verified by heating up some olive oil in a frying pan, and can be explained by considering that with increasing temperature the mean distance between molecules increases, so that the cohesive effects of the intermolecular forces decrease.

It should now be pointed out that if we consider a generic fluid motion instead of the particularly simple case of Fig. 2.4, the linear expression (2.24) may be generalized, leading to a formally more complex relationship that links the components of the *viscous stress tensor* with those of the *rate of strain tensor* and characterizes the behaviour of the so-called *Newtonian fluids*. This will be described in detail in Chapter 4, where we shall show that for such fluids the viscous stress components may be expressed as a linear function of the spatial variations of all the velocity components.

In order not to generate incorrect ideas from the particularly simple flow condition analysed in this section, we may here anticipate and emphasize that the physical mechanisms that are at the basis of viscosity may produce not only tangential stresses, but also stresses that are normal to the surface of a moving fluid particle. These stresses are then additive to the thermodynamic pressure; however, at variance with the pressure stresses, the viscous normal stresses are not *isotropic*, which means that, in general, they are different for different orientations of the considered surface.

2.7. Diffusion, Heat Conduction and Prandtl Number

The transport of momentum at the basis of viscosity is only one of the molecular transport mechanisms that may take place in a fluid. When any quantity happens to be non-uniform in a fluid flow, a tendency to equalization, i.e. to reaching an equilibrium condition, occurs. This characteristic behaviour is directly linked to the existence of interaction mechanisms at the molecular level between contiguous portions of fluid and, in particular, between neighbouring fluid particles.

Therefore, we may have, for instance; transport of mass whenever the fluid is not homogeneous or when we consider a mixture of different fluids. In these cases migration of molecules of the various types will occur when a gradient of concentration is present. In other words, if we have a mixture of two fluids, and we imagine the molecules of each constituent to be marked with different colours, a flux of marked molecules will occur if their proportion is different on the two sides of an element of an imaginary surface drawn in the fluid. In this case we may say that there is *mass transport*, or *diffusion* of matter. This phenomenon may be easily seen in action by letting a drop of milk fall in a cup of coffee.

As this book concerns essentially the motion of homogeneous fluids, we shall be more interested in another transport phenomenon at the molecular level, namely *heat conduction*. Heat is connected with the kinetic energy of molecules, and temperature is a measure of the energy level. Thus, if the temperature in a fluid is not uniform and, for instance, it varies linearly in the y direction (as was the case for velocity in Fig. 2.4), the molecules contained in the fluid particles occupying different y coordinates have different levels of kinetic energy. Then, transport of heat will occur at the molecular level through the exchange of kinetic energy between the molecules of neighbouring particles. From the macroscopic point of view, we may say that heat conduction is taking place, and that this is due to the fact that the temperature is not constant in the y direction.

Again, we may express this dependence in a simple way by writing the following relation:

$$q = -k\frac{dT}{dy}. \tag{2.25}$$

In this formula, q is the component in the y direction of the *heat flux*, i.e. of the amount of heat that crosses the unit surface in unit time, whose units are $J/(m^2 s)$. The coefficient of proportionality between heat flux and temperature variation, k, is the *coefficient of thermal conductivity* (or, simply, *thermal conductivity*) and its dimensions are then $J/(m s K)$. The minus sign appearing in relation (2.25) derives from the fact that heat flows from high to low temperature regions.

In fact, for a generic temperature field, the heat flux is a vector and expression (2.25) is a particular case of the *Fourier law*, which links the heat flux to the gradient of temperature and which will be further discussed in Chapter 4.

As might be expected, in gases, the molecular mechanism producing heat conduction is analogous to that of momentum transport, i.e. the migration of molecules and the exchange of kinetic energy between them through collisions. Therefore, in a gas, the variation of the thermal conductivity with temperature is similar to that of the coefficient of viscosity, i.e. it is a similarly increasing function of T.

On the other hand, the mechanism of heat transport in liquids is different from the one causing the exchange of momentum. In particular, heat transfer occurs primarily due to the direct exchange of kinetic energy between adjacent molecules, and the global effect is a moderate increase of conductivity with temperature. This is at variance with the significant decrease in viscosity with increasing temperature that is caused in liquids by the increase of the molecular spacing and by the consequent reduction of the intermolecular forces.

As we have already mentioned, these momentum and heat transport processes are examples of *diffusive* phenomena. Now, it may be seen that the time evolution of a certain diffusing quantity, say C, may adequately be described through the following *diffusion equation*:

$$\frac{\partial C}{\partial t} = \kappa_c \nabla^2 C, \tag{2.26}$$

where C may be either a scalar or vector quantity, t is time, κ_c is the so-called *diffusion coefficient* or *diffusivity* of quantity C, and ∇^2 is the Laplace operator defined in (1.13).

In later chapters we shall discuss the main features of the diffusion process described by Eq. (2.26). However, it may be noted here that the diffusing quantity C appears in both terms of the equation; hence, the dimensions of *any* diffusion coefficient are those given by the differential operators, i.e. m^2/s.

The equations of motion of a fluid will also show that the diffusivity of momentum is given by the so-called *kinematic viscosity*, which is defined as

$$\nu = \frac{\mu}{\rho}. \tag{2.27}$$

Actually, it will soon become clear that in the dynamics of fluids the kinematic viscosity is usually much more important than the viscosity coefficient, and the knowledge of the value of ν is thus a primary issue for the characterization of the conditions of motion.

The values of viscosity, density and kinematic viscosity for various fluids are reported in Table 2.1 for a temperature of 15°C and a pressure of one

Table 2.1. Values of μ, ρ and ν for various fluids at 15°C and 1 bar.

	Air	Water	Mercury	Olive oil	Glycerol
$\mu\,(\mathrm{kg/(m\,s)})$	1.78×10^{-5}	1.14×10^{-3}	1.58×10^{-3}	0.99×10^{-1}	2.33
$\rho\,(\mathrm{kg/m^3})$	1.225	999.2	13.61×10^3	918	1260
$\nu\,(\mathrm{m^2/s})$	1.45×10^{-5}	1.14×10^{-6}	1.16×10^{-7}	1.08×10^{-4}	1.85×10^{-3}

bar. It is interesting to note that fluids that are more viscous than others may actually have a lower kinematic viscosity. For instance, water has a viscosity coefficient that is $\simeq 64$ times that of air, but its kinematic viscosity is less than 8% of that of air, due to its much higher density.

As for the diffusion of heat, it may be described through an equation like (2.26) in which the temperature T appears instead of C, and the relevant diffusion coefficient, or *thermal diffusivity*, is given by

$$\kappa_H = \frac{k}{\rho C_p}, \qquad (2.28)$$

where C_p is the specific heat at constant pressure.

The ratio between the diffusivities of momentum and heat is a very important parameter in fluid dynamics; it is known as the *Prandtl number*, Pr, and is given by the relation

$$Pr = \frac{\nu}{\kappa_H} = \frac{C_p \mu}{k}. \qquad (2.29)$$

The specific heats of many gases depend very weakly on temperature, so that a simplified model may be adopted in which they are assumed to be constant and the gas is then said to be *calorically perfect*. Conversely, we have seen that both μ and k vary significantly with temperature; however, in a gas, the same molecular mechanism is responsible for the transport of momentum and heat, namely the collisions between neighbouring molecules. Therefore, it is reasonable to infer that the variation of the Prandtl number for a gas is negligible in a wide range of temperatures, and this is indeed the result that is experimentally found. For instance, the Prandtl number for air varies between 0.72 and 0.70 by varying the temperature from 0°C to 100°C. The constancy of the Prandtl number for gases is actually an advantageous feature when the thermal effects of viscosity are considered in compressible flows.

On the other hand, the situation is obviously very different for liquids, in which, as we have already pointed out, the molecular mechanisms causing transfer of momentum and heat are different, and so is the dependence on temperature of μ and k. As a consequence, the Prandtl number for a liquid is a strongly decreasing function of temperature, and this is mainly due to the rapid decrease of viscosity with increasing temperature.

Tables providing the variation with temperature of the most significant properties of air and water are reported in Appendix A.1. It may be interesting to note that, by varying the temperature from 10°C to 20°C, the viscosity of air increases by less than 3%, whereas the viscosity of water decreases by more than 23%. When considering motions in which variations of temperature may be expected, this greater variability of viscosity for liquids than for gases must always be remembered. For instance, assuming a constant viscosity coefficient may be more questionable if the fluid is a liquid than if it is a gas.

2.8. Fluid Behaviour at Solid Surfaces

From the point of view of the physical development of a fluid flow, as well as from that of the solution of the equations of motion, it is fundamental to describe in an accurate way the behaviour of a fluid moving over a solid surface. Obviously, this description should be coherent with the continuum hypothesis we have adopted.

Let us consider a fluid particle to which, at a certain instant of time, we associate the same coordinates as those of a point on a solid surface. From what has been said in Section 2.2, we know that this means we are considering a volume of fluid, adjacent to that point of the surface, which is sufficiently small to allow its characteristic dimension to be considered negligible in the continuum description, but contains an extremely high number of molecules. The velocity V of that particle is then the instantaneous average value of the velocities of the molecules contained in its small volume; in effect, we know that all these molecules also have an additional random thermal velocity with zero mean value.

If we now denote as n the unit vector normal to the surface at a certain point, we may separately consider the scalar component of V in the n direction, say $V_n = V \cdot n$, and the velocity vector component tangential to the surface, say V_t.

Now, the condition that the surface is solid implies that the fluid cannot penetrate inside it. Therefore, if V_w is the velocity of the surface, we may express this *non-penetration condition* by saying that the difference between the normal components of the particle and surface velocities must be zero, so that we have

$$(\boldsymbol{V} - \boldsymbol{V}_w) \cdot \boldsymbol{n} = 0. \tag{2.30}$$

In principle, the condition to be assigned to the tangential velocity component is much less obvious, and it was the object of significant controversies within the fluid dynamic community during the 19th century (see, e.g., Goldstein, 1957). However, it is now fully recognized that, at least as long as the continuum hypothesis is valid, the *no-slip condition* applies, which means that no discontinuity exists at the boundary between a fluid and a solid surface as regards the velocity component tangential to the surface. To express this condition in mathematical terms, we may recall that, given a generic vector $\boldsymbol{u}$, its vector component lying on a plane tangent to a surface at a point where the normal unit vector is $\boldsymbol{n}$ may be written as follows:

$$\boldsymbol{u}_t = \boldsymbol{u} - (\boldsymbol{u} \cdot \boldsymbol{n})\boldsymbol{n} = \boldsymbol{n} \times (\boldsymbol{u} \times \boldsymbol{n}). \tag{2.31}$$

In effect, it easy to check that the vector $\boldsymbol{u} \times \boldsymbol{n}$ is equal to vector $\boldsymbol{u}_t$ rotated clockwise by 90° around $\boldsymbol{n}$. Therefore, the no-slip condition corresponds to the following relation:

$$(\boldsymbol{V} - \boldsymbol{V}_w) \times \boldsymbol{n} = 0. \tag{2.32}$$

The physical soundness of this condition has been completely confirmed by experimental evidence. For instance, we may have the impression that water slides over an inclined solid plate, but if we look carefully at the conditions within a small layer of fluid in the close neighbourhood of the surface, we can observe that in reality the velocity varies from zero at the surface to a value that is a function of the inclination of the plate. We shall see that the behaviour of this layer is crucial for the establishment of the flow features, and the dependence of its thickness and dynamics on the conditions of motion will be described and discussed at length in the following chapters.

In general, one might say that the validity of the no-slip condition is justified by the fact that if we use it as a boundary condition for the equations of motion of a fluid we obtain results that are in good agreement with carefully conducted experiments. However, it should be pointed out that it is also consistent with our picture of the behaviour of a viscous fluid and, in particular, with the origin of viscosity. In fact, as pointed out by Stokes, the existence of slip would imply that the friction between a fluid and a solid wall is of a different nature from, and infinitely lower than, the friction between layers of fluid.

Therefore, the no-slip condition corresponds to assuming that the molecules contained in a fluid particle tend to reach mechanical equilibrium with those of the adjacent solid surface. In other words, this implies that the characteristic times necessary for the molecules of fluid to come into mechanical equilibrium (through collisions or other types of interaction) with those of the solid surface are much smaller than the characteristic times necessary for a particle to change its velocity in the macroscopic (continuum) description of the motion.

Deeper analyses have shown that the slip velocity of a fluid at a solid surface is proportional to the mean free path of the molecules. Therefore, we may say that as long as we are allowed to use the continuum hypothesis (see Section 2.2), we may also apply the no-slip boundary condition for the description of the motion of a fluid. For instance, as pointed out in the detailed discussion by Gad-el-Hak (1999), in the design of MEMS there are situations in which this condition should be relaxed and more appropriate models should be used.

It might be useful to emphasize, once again, that the no-slip condition is not a constraint on the identity and velocity of single molecules, but on the velocity of a fluid particle. Therefore, it does not imply that if, for instance, an aeroplane flies from London to New York, the same molecules of air that were adjacent to the wing surface in London will still be there in New York. It simply means that, during the whole journey, the velocity of a fluid particle adjacent to the wing will be the same as that of the aeroplane. However, as already pointed out, this velocity is the average of the velocities of the molecules that are present, at any instant of time, within the extremely small volume of fluid we regard as a particle. These molecules may — and actually will — change from one instant to another, because the surface bounding the volume of fluid corresponding to a particle is permeable to their random motion; the latter is equiprobable in all directions and is superposed on the average value that represents the particle velocity.

The other quantity whose behaviour in relation to a solid boundary is of interest is temperature. If we accept the fact that, essentially, the no-slip condition originates from the same mechanisms that produce viscosity, it is clear that it is plausible to expect that a fluid particle adjacent to a solid wall will also have the same temperature as the wall. This stems from the fact that, as we have seen, heat conduction, like viscosity, is produced by molecular interaction mechanisms leading to a tendency to equilibrium. Thus, if the no-slip condition is a manifestation of the mechanical equilibrium achieved between a fluid and a solid surface, the condition of no temperature jump is an expression of the tendency to thermal equilibrium.

Therefore, if T is the temperature of a fluid particle lying over a solid surface where the temperature is T_w, we may say that the following boundary condition applies:

$$T - T_w = 0. \tag{2.33}$$

As was the case for the no-slip condition, this relationship is in very good agreement with experimental evidence as long as the continuum hypothesis retains its validity.

Chapter 3

CHARACTERIZATION OF THE MOTION OF FLUIDS

3.1. Lagrangian and Eulerian Descriptions of Motion

The motion of a fluid may be described in two different ways, which are called the *Lagrangian* and *Eulerian* descriptions. In the Lagrangian approach, the motion of the single particles is followed in time, together with the variation of the quantities that are associated with them. This implies that all the particles must be recognizable, and this may be achieved by suitably 'tagging' them. For instance, we may use the positions they occupied at a certain reference time t^*, which are supposed to be known, together with the values of the various associated quantities at that time. In other words, at any generic time t the problem is to find the coordinates x, y, z and the velocity V — as well as all the remaining quantities of interest — of the particle that at time t^* occupied, say, the position identified by coordinates x^*, y^*, z^*, and whose velocity was V^*.

In this description, as well as in any type of analysis of the motion of fluids, a central concept is that of a *material volume of fluid*. This is defined as a volume of fluid that is composed always of the *same* particles. Thus, a material volume will generally deform during the motion and, possibly, change the measure of its volume, but the global amount of mass contained in it will always remain the same.

In order to figure out the dynamics of such a volume, perhaps one might observe a flying flock of starlings or a swimming shoal of sardines, which, even if composed of discrete rather than continuous elements, when seen as a whole may indeed give an approximate idea of the variations in shape and volume of a set of fluid particles.

In principle, this type of description would seem the most natural one for studying the motion of a fluid because, as we shall see, the equations of motion are derived by expressing, in mathematical terms, physical laws (or *balances*) that refer to material volumes of fluid or to their smallest elements, i.e. the single particles. However, the use of the Lagrangian description leads to complex equations (see, e.g., Lamb, 1932); moreover, it is impossible to avoid the explicit dependence of all quantities on time and to derive directly the spatial gradients of velocity. For this reason, it is seldom used except for particular applications, such as the analysis of the dispersion of pollutants in the atmosphere, and the Eulerian description is generally preferred. In this approach, which is also denoted a 'field' description, a reference frame is first chosen and the various flow variables are then assumed to be functions of the position vector $\boldsymbol{x} \equiv (x, y, z)$ and of time, t. Thus, from a purely mathematical point of view, we may write, for instance,

$$\boldsymbol{V} = \boldsymbol{V}(x, y, z, t); \quad \rho = \rho(x, y, z, t); \quad T = T(x, y, z, t). \qquad (3.1)$$

However, it might be useful to point out that these quantities are actually associated with the moving fluid particles. The connection between the two different approaches, the Lagrangian and the Eulerian, is that in the latter the values of the quantities described in (3.1) are those of the particular fluid particle that at time t happens to pass through point $\boldsymbol{x}$. Therefore, if we describe, for instance, the velocity variation in time at point $\boldsymbol{x}$, we are actually referring to the velocities of the *different* particles that occupy that point in different times.

There are great advantages in using the Eulerian description. Firstly, as already pointed out, the mathematical analysis turns out to be simpler. Furthermore, it may sometimes be possible to choose a suitable reference frame such that all the quantities of interest become dependent on the spatial coordinates only, and not on time. If this happens, the motion is said to be *steady*.

As an example, this may be the case for a body moving through a fluid with a constant translational velocity. In effect, if the flow field has certain characteristics — which imply that the shape of the body and its motion satisfy particular conditions — the flow variables will always be the same at fixed points in a reference frame moving with the body. For instance, if a point in the flow around the wing of an aeroplane is observed from a cabin window, the velocity at that point may be considered to remain invariant in time as long as the flight remains in cruise conditions. On the other hand, if the same motion is seen from the still air through which the aircraft is flying or from the ground, it is clearly time dependent (or, as we say, *unsteady*). In effect, the air particles are continuously displaced as the aeroplane moves along its trajectory, and thus the velocity will be seen to vary in time at points that are fixed relative to the new reference frame. This means that, as already pointed out, the same motion may be steady or unsteady depending on the reference system that is used to describe it.

It should also be evident that there is a direct connection between the values derived from an Eulerian analysis and those that may be obtained from experiments. In fact, measurements are practically always taken at fixed points in space — as, for instance, in a wind tunnel — and then the results may be directly expressed in the form (3.1). On the other hand, Lagrangian measurements are very difficult because, in principle, one should have an instrument that flies with a fluid particle. An approximate measurement of this type is given by meteorological balloons, which are dragged by the motion of the atmosphere and which may be considered to be sufficiently small with respect to the dimensions within which variations of the various quantities occur.

Nonetheless, the Eulerian description also has a significant drawback. We shall see that it is often necessary to express, in mathematical terms, the variation of a physical quantity associated with a fluid particle — or with a material volume — that is moving in space. In a Lagrangian description this would pose no problem, as it would simply be given by the time derivative of the considered quantity for each 'tagged' particle. Conversely, the expression of such a variation is not immediate in the Eulerian approach, thus implying the introduction of a particular differential operator and of a theorem that refer to the time variation of a quantity associated with a fluid particle and with a material volume respectively. All these aspects will be considered in greater detail in the following sections.

3.2. The Material Derivative

Let A be a (scalar or vector) quantity associated with the fluid particle that at time t occupies the position described by the coordinates (x, y, z), and let us further assume that the Eulerian description $A(x, y, z, t)$ of the quantity is available. We now want to find the expression, in terms of an Eulerian approach, for the variation of A experienced by that particle in an elementary time interval dt, i.e. the variation that would be 'measured' if we could move with the particle.

In order to solve the problem, we start by expressing the total differential of function $A(x, y, z, t)$:

$$dA = \frac{\partial A}{\partial t} dt + \frac{\partial A}{\partial x} dx + \frac{\partial A}{\partial y} dy + \frac{\partial A}{\partial z} dz. \tag{3.2}$$

Now, the variations of the coordinates that we must consider in (3.2), i.e. dx, dy and dz, are not generic ones but those associated with the displacement, in the small time interval dt, of the particle that at time t is at point (x, y, z). Therefore, if that particle is moving with the velocity $\boldsymbol{V}(x, y, z, t) \equiv (u, v, w)$ corresponding to the same point, we may write

$$dx = \frac{\partial x}{\partial t} dt = u dt; \quad dy = \frac{\partial y}{\partial t} dt = v dt; \quad dz = \frac{\partial z}{\partial t} dt = w dt. \tag{3.3}$$

We thus obtain the following relation:

$$dA = \frac{\partial A}{\partial t} dt + \frac{\partial A}{\partial x} u dt + \frac{\partial A}{\partial y} v dt + \frac{\partial A}{\partial z} w dt. \tag{3.4}$$

The instantaneous time variation of quantity A for a particle, following its motion, may then be expressed in terms of Eulerian description as follows:

$$\frac{DA}{Dt} \equiv \frac{\partial A}{\partial t} + \frac{\partial A}{\partial x} u + \frac{\partial A}{\partial y} v + \frac{\partial A}{\partial z} w = \frac{\partial A}{\partial t} + \boldsymbol{V} \cdot \operatorname{grad} A, \tag{3.5}$$

where we have introduced the *material derivative* differential operator (also denoted the *substantial derivative*)

$$\frac{D}{Dt} = \frac{\partial}{\partial t} + \boldsymbol{V} \cdot \operatorname{grad}. \tag{3.6}$$

We shall soon appreciate that this operator is a characteristic and fundamental tool in the description of the motion of fluids, and we may say that it represents the Eulerian mathematical expression of a Lagrangian instantaneous time variation.

As already pointed out, the material derivative can be applied to both scalar and vector quantities. Obviously, if a vector field is concerned, it should be recalled that the gradient of a vector $\boldsymbol{a} \equiv (a_i)$ is a second-order tensor whose components are given by (1.8), which is here rewritten,

$$(\mathrm{grad}\,\boldsymbol{a})_{ij} = \frac{\partial a_j}{\partial x_i},$$

and that the left inner product between a vector $\boldsymbol{a}$ and a second-order tensor $\mathbf{T} \equiv [T_{ij}]$ is a vector with the following components:

$$(\boldsymbol{a} \cdot \mathbf{T})_i = \sum_j a_j T_{ji}.$$

In practice, we may say that when (3.6) is applied to a vector $\boldsymbol{a}$, it produces a new vector whose components can be obtained by applying the material derivative operator to each component of vector $\boldsymbol{a}$.

It may be readily verified that all the fundamental derivation rules are also valid for the material derivative. For instance, we have

$$\frac{D(AB)}{Dt} = A\frac{DB}{Dt} + B\frac{DA}{Dt}. \tag{3.7}$$

It could now be useful to thoroughly discuss the physical meaning of the terms appearing in the material derivative. The first term on the right-hand side of (3.6) is called the *unsteady term*, and represents the variation linked to a possible *unsteadiness* of the flow field, i.e. to the existence of an explicit time dependence of the various quantities for each fixed point in space. As already pointed out, in an Eulerian description a motion may be unsteady with respect to a certain reference frame and steady with respect to another one; in the latter case the unsteady term in the material derivative is obviously zero.

The second term $(\boldsymbol{V} \cdot \mathrm{grad})$ is denoted the *convective term*, and is connected with the fact that the particle is being carried with velocity $\boldsymbol{V}$ through a spatial variation of the considered flow variable, which is given, in magnitude and direction, by the gradient operator. Its physical interpretation should become clearer by recalling the definition and significance of a scalar product. In effect, a particle will experience a time variation of a certain quantity due to the existence of a gradient of that quantity only if it is moving with a velocity that has a non-zero component in the direction of the gradient. Furthermore, the magnitude of the variation will obviously be proportional to both the value of that velocity component and to the intensity of the gradient.

A simple example may help in understanding the meaning and utility of the material derivative. Let us imagine we want to describe the time variation of the temperature in a train carriage, assuming that it always coincides with the ambient temperature of the location occupied by the train at that instant and that no measuring instrument inside the carriage is present. Let the train connect two towns A and B, with B located 800 km south of A, and assume that an extensive network of instruments is available outside the train along the whole track it follows, so that the temperature field is known at each instant of time. In other words, we have a complete space and time description of the temperature field. Now, during the day the temperature will almost certainly vary in time at each location, with a minimum just before dawn, a rapid increase after sunrise, a plateau in the early afternoon, and then a decrease at different rates before and after sunset. Furthermore, at each time during the day we may imagine the temperature to increase in the north–south direction, for instance by 1°C/100 km; in other words, a positive gradient of temperature exists between the two towns.

Therefore, provided we also know the velocity of the train at each instant of time, in magnitude and direction, we are able to evaluate the rate of variation of temperature that would be measured by a passenger in the train at that instant. In fact, we have all the elements appearing in the expression of the material derivative. By the way, it will be immediately realized that the unsteady term in (3.6) will generally be non-zero along the whole journey. On the other hand, the convective term will be maximum only when the track of the train is exactly in the north–south direction, and directly proportional to the speed of the train in those periods. Conversely, the convective variation of temperature will be zero when the winding of the track makes the train travel, at whatever speed, eastwards or westwards.

3.3. Pathlines, Streamlines and Streaklines

Several curves, which are directly related to the velocity field, are of interest in the description of the motion of fluids, and their definitions are now considered.

A *pathline* (or simply a *path*) is the trajectory followed by a particle in motion. In other words, it is the locus of the positions occupied by the particle at subsequent instants of time. If an Eulerian velocity description $V(x, y, z, t) \equiv (u, v, w)$ is available, a pathline may then be obtained for each particle P by integrating in time, starting from a reference time t_0, the following equations:

$$dx_P = u(\boldsymbol{x}_P, t)dt, \quad dy_P = v(\boldsymbol{x}_P, t)dt, \quad dz_P = w(\boldsymbol{x}_P, t)dt, \qquad (3.8)$$

where $\boldsymbol{x}_P \equiv (x_P, y_P, z_P)$ is the vector defining the position of the particle at time t.

On the other hand, the *streamlines* are the curves that, at a given instant of time, are tangent to the velocity vectors. In other words, at each point a streamline is tangent to the velocity vector at that point. We may also say that the streamlines are the *field lines* of the velocity vector field.

An element $d\boldsymbol{s} \equiv (dx, dy, dz)$ of the streamline passing through point $\boldsymbol{x} \equiv (x, y, z)$ is then linked to the velocity vector $\boldsymbol{V} \equiv (u, v, w)$ at that point by the following equations:

$$w(\boldsymbol{x}, t)dy - v(\boldsymbol{x}, t)dz = 0,$$
$$u(\boldsymbol{x}, t)dz - w(\boldsymbol{x}, t)dx = 0, \tag{3.9}$$
$$v(\boldsymbol{x}, t)dx - u(\boldsymbol{x}, t)dy = 0,$$

which can also be recast in a more concise form as

$$\frac{dx}{u(\boldsymbol{x}, t)} = \frac{dy}{v(\boldsymbol{x}, t)} = \frac{dz}{w(\boldsymbol{x}, t)}. \tag{3.10}$$

It can be easily verified that these differential equations express the condition that the cross product between the vectors $d\boldsymbol{s}$ and $\boldsymbol{V}$ is zero. However, only two of the equations in the set (3.9) are actually independent. In fact, it may be shown that a streamline could be described as the curve of intersection of two surfaces and is thus determined from the knowledge of two functions of the spatial coordinates (for a deeper discussion, see Karamcheti, 1966).

Streamlines are defined at given instants, hence they generally vary in time. Only if the motion is steady do they always remain the same, and in this particular case they coincide with the paths of the particles. From the discussion in Section 3.1, it should then be clear that streamlines are dependent on the reference system that is used. For instance, in Fig. 3.1 the streamlines around a body moving at constant velocity through a fluid are drawn in two reference frames, one fixed to the fluid and the other to the body. In the first one the motion is unsteady and the streamline pattern seen in Fig. 3.1(a) corresponds to a particular instant of time and moves leftwards with the body. Conversely, in Fig. 3.1(b) the motion is steady and the streamlines do not vary in time.

Finally, a *streakline* is the locus, at a given instant t, of the positions of the particles that, starting from a certain reference time t_0, have passed through a given point in space, say $\boldsymbol{x}_0$. To obtain a streakline mathematically, one should then find the positions occupied at time t by

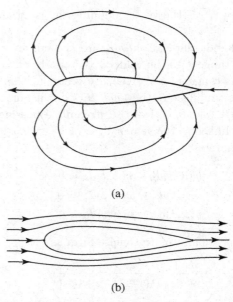

(a)

(b)

Fig. 3.1.   Streamlines around a moving body. (a) Reference frame fixed to the fluid; (b) reference frame fixed to the body.

the particles that were at x_0 at times t', with $t_0 < t' < t$. This can be done by integrating Eqs. (3.8) with the appropriate initial conditions.

The real importance of streaklines is that they may be readily visualized by continuously introducing smoke or dye from fixed points into a flow. Furthermore, in general, streaklines vary in time and are different from streamlines and pathlines; however, if the flow is steady, then the streaklines do not vary and they coincide with both the pathlines and the streamlines. This is a great advantage, because in this case the visualization of the streaklines allows the streamlines — which are usually much more important — to be characterized. However, the significant difference between streaklines and streamlines must always be recalled when dealing with unsteady flows, like those that may occur, for instance, in the wake of several types of bodies. As a matter of fact, in those cases it is not uncommon to misinterpret the visualizations of streaklines, and to take them to be connected with the direction of the instantaneous velocity field, which is often far from true.

In many cases, it is interesting to consider *streamtubes*, which may be constructed by drawing the streamlines passing through closed curves, as

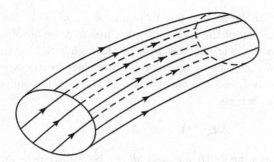

Fig. 3.2. An example of streamtube.

shown in Fig. 3.2. The importance of streamtubes is that, by definition, the instantaneous flow of fluid through their lateral surfaces is zero.

3.4. Deformation of a Fluid Element

As we shall see in the next chapter, the forces acting on the surface of an elementary volume of fluid are connected directly with its deformation in a small interval of time. Therefore, it is expedient to derive the mathematical expression of this deformation for the case of a velocity field that varies generically in space.

To this end, let us consider two fluid particles that at time t occupy the points $\boldsymbol{x} \equiv (x, y, z)$ and $\boldsymbol{x} + \delta\boldsymbol{x} \equiv (x + \delta x, y + \delta y, z + \delta z)$. These points may be taken to define an elementary volume of fluid with sides $(\delta x, \delta y, \delta z)$, as shown in Fig. 3.3.

Let the velocities of the particles positioned at the two points be $\boldsymbol{V}(\boldsymbol{x}) \equiv (u, v, w)$ and $\boldsymbol{V}(\boldsymbol{x} + \delta\boldsymbol{x}) = \boldsymbol{V} + \delta\boldsymbol{V} \equiv (u + \delta u, v + \delta v, w + \delta w)$. Due to their different velocities, during a small interval of time dt the two particles will move in different ways, with the first particle displacing to

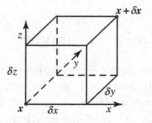

Fig. 3.3. Elementary volume of fluid.

point $x + V dt$ and the second one to point $(x + \delta x) + (V + \delta V)dt$. Thus, the new configuration of the elementary volume may be obtained from that corresponding to the original time instant through a rigid translation $V\,dt$ plus a *deformation,* which is completely described by the vector defining the new distance between the two particles. The latter is indeed no longer δx, but $\delta x + \Delta x$, where

$$\Delta x = \delta V dt \equiv (\Delta x, \Delta y, \Delta z). \tag{3.11}$$

If we now assume both δx and dt to be sufficiently small, we can evaluate Δx through a first-order expansion, which gives

$$\Delta x = \delta u dt = \left(\frac{\partial u}{\partial x}\delta x + \frac{\partial u}{\partial y}\delta y + \frac{\partial u}{\partial z}\delta z\right) dt,$$

$$\Delta y = \delta v dt = \left(\frac{\partial v}{\partial x}\delta x + \frac{\partial v}{\partial y}\delta y + \frac{\partial v}{\partial z}\delta z\right) dt, \tag{3.12}$$

$$\Delta z = \delta w dt = \left(\frac{\partial w}{\partial x}\delta x + \frac{\partial w}{\partial y}\delta y + \frac{\partial w}{\partial z}\delta z\right) dt.$$

Expressions (3.12) show that the deformation of the elementary volume is defined, at first order, by the nine components of the tensor grad V, which can be written explicitly as follows:

$$\mathrm{grad}V \equiv \begin{bmatrix} \dfrac{\partial u}{\partial x} & \dfrac{\partial v}{\partial x} & \dfrac{\partial w}{\partial x} \\[2mm] \dfrac{\partial u}{\partial y} & \dfrac{\partial v}{\partial y} & \dfrac{\partial w}{\partial y} \\[2mm] \dfrac{\partial u}{\partial z} & \dfrac{\partial v}{\partial z} & \dfrac{\partial w}{\partial z} \end{bmatrix}. \tag{3.13}$$

More precisely, we may express the vector describing the deformation in a concise form by using the inner product definition (1.5), so that we may write

$$\Delta x = (\delta x \cdot \mathrm{grad}V)dt. \tag{3.14}$$

In order to get a clearer physical insight into the deformation, it is now expedient to express the tensor grad$V$ as the sum of various tensors that can be associated more readily with elementary deformations.

To this end, we start by decomposing grad$V$ in its symmetric and anti-symmetric parts, respectively denoted by $\mathbf{E}$ and $\mathbf{\Omega}$. This is always possible and the result is the following:

$$\mathbf{E} \equiv [E_{ij}] = \begin{bmatrix} \dfrac{\partial u}{\partial x} & \dfrac{1}{2}\left(\dfrac{\partial u}{\partial y} + \dfrac{\partial v}{\partial x}\right) & \dfrac{1}{2}\left(\dfrac{\partial u}{\partial z} + \dfrac{\partial w}{\partial x}\right) \\ \dfrac{1}{2}\left(\dfrac{\partial v}{\partial x} + \dfrac{\partial u}{\partial y}\right) & \dfrac{\partial v}{\partial y} & \dfrac{1}{2}\left(\dfrac{\partial v}{\partial z} + \dfrac{\partial w}{\partial y}\right) \\ \dfrac{1}{2}\left(\dfrac{\partial w}{\partial x} + \dfrac{\partial u}{\partial z}\right) & \dfrac{1}{2}\left(\dfrac{\partial w}{\partial y} + \dfrac{\partial v}{\partial z}\right) & \dfrac{\partial w}{\partial z} \end{bmatrix}, \quad (3.15)$$

$$\mathbf{\Omega} \equiv [\Omega_{ij}] = \begin{bmatrix} 0 & \dfrac{1}{2}\left(\dfrac{\partial v}{\partial x} - \dfrac{\partial u}{\partial y}\right) & \dfrac{1}{2}\left(\dfrac{\partial w}{\partial x} - \dfrac{\partial u}{\partial z}\right) \\ \dfrac{1}{2}\left(\dfrac{\partial u}{\partial y} - \dfrac{\partial v}{\partial x}\right) & 0 & \dfrac{1}{2}\left(\dfrac{\partial w}{\partial y} - \dfrac{\partial v}{\partial z}\right) \\ \dfrac{1}{2}\left(\dfrac{\partial u}{\partial z} - \dfrac{\partial w}{\partial x}\right) & \dfrac{1}{2}\left(\dfrac{\partial v}{\partial z} - \dfrac{\partial w}{\partial y}\right) & 0 \end{bmatrix}. \quad (3.16)$$

An interpretation of the role of the anti-symmetric tensor $\mathbf{\Omega}$ may be obtained by introducing the *vorticity* vector $\boldsymbol{\omega} \equiv (\omega_x, \omega_y, \omega_z)$, defined as

$$\boldsymbol{\omega} = \operatorname{curl} \boldsymbol{V}, \quad (3.17)$$

so that

$$\omega_x = \frac{\partial w}{\partial y} - \frac{\partial v}{\partial z}; \quad \omega_y = \frac{\partial u}{\partial z} - \frac{\partial w}{\partial x}; \quad \omega_z = \frac{\partial v}{\partial x} - \frac{\partial u}{\partial y}.$$

We may then recast $\mathbf{\Omega}$ in terms of vorticity components:

$$\mathbf{\Omega} \equiv \frac{1}{2}\begin{bmatrix} 0 & \omega_z & -\omega_y \\ -\omega_z & 0 & \omega_x \\ \omega_y & -\omega_x & 0 \end{bmatrix}, \quad (3.18)$$

so that we can immediately check that the portion of (3.14) associated with $\mathbf{\Omega}$ may be expressed as

$$(\delta \boldsymbol{x} \cdot \mathbf{\Omega})dt = \left(\frac{\boldsymbol{\omega}}{2} \times \delta \boldsymbol{x}\right) dt. \quad (3.19)$$

As can be easily realized, this relation describes the effect of a *rigid rotation* of point $\boldsymbol{x} + \delta\boldsymbol{x}$ around point $\boldsymbol{x}$ with angular velocity $\boldsymbol{\omega}/2$. Thus, the vorticity vector, whose primary importance in the characterization of fluid motion will become evident in the following chapters, may be given a clear physical interpretation. In effect, we may say that the vorticity at a certain point in a flow is equal to twice the instantaneous angular velocity of a very small volume of fluid around that point.

Now, even if we may still say that (3.19) represents a deformation, in the sense that it produces a new configuration of the elementary volume after the elapsed time dt, it is actually a peculiar one, because it is a rigid motion producing no variation in shape or volume.

Conversely, the changes in shape and volume of the considered elementary volume of fluid (and thus what we might call the 'real' deformation) are connected with $\mathbf{E}$, i.e. with the symmetric part of grad $\boldsymbol{V}$, which is called the *rate of strain tensor*. This is indeed the reason why, as we shall see, $\mathbf{E}$ is the fundamental tensor that appears in the constitutive equations expressing the link between the viscous stresses and the velocity field.

Nonetheless, even tensor $\mathbf{E}$ may be usefully expressed as the sum of two components, each of them connected with particular deformations. To this end we put

$$\mathbf{E} = \mathbf{A} + \mathbf{D}, \tag{3.20}$$

where

$$\mathbf{A} \equiv [A_{ij}] = \begin{bmatrix} \dfrac{1}{3}\mathrm{div}\boldsymbol{V} & 0 & 0 \\[2mm] 0 & \dfrac{1}{3}\mathrm{div}\boldsymbol{V} & 0 \\[2mm] 0 & 0 & \dfrac{1}{3}\mathrm{div}\boldsymbol{V} \end{bmatrix}, \tag{3.21}$$

$$\mathbf{D} \equiv [D_{ij}] = \begin{bmatrix} \left(\dfrac{\partial u}{\partial x} - \dfrac{1}{3}\mathrm{div}\boldsymbol{V}\right) & \dfrac{1}{2}\left(\dfrac{\partial u}{\partial y} + \dfrac{\partial v}{\partial x}\right) & \dfrac{1}{2}\left(\dfrac{\partial u}{\partial z} + \dfrac{\partial w}{\partial x}\right) \\[3mm] \dfrac{1}{2}\left(\dfrac{\partial v}{\partial x} + \dfrac{\partial u}{\partial y}\right) & \left(\dfrac{\partial v}{\partial y} - \dfrac{1}{3}\mathrm{div}\boldsymbol{V}\right) & \dfrac{1}{2}\left(\dfrac{\partial v}{\partial z} + \dfrac{\partial w}{\partial y}\right) \\[3mm] \dfrac{1}{2}\left(\dfrac{\partial w}{\partial x} + \dfrac{\partial u}{\partial z}\right) & \dfrac{1}{2}\left(\dfrac{\partial w}{\partial y} + \dfrac{\partial v}{\partial z}\right) & \left(\dfrac{\partial w}{\partial z} - \dfrac{1}{3}\mathrm{div}\boldsymbol{V}\right) \end{bmatrix}. \tag{3.22}$$

Tensors $\mathbf{A}$ and $\mathbf{D}$ are respectively denoted the *isotropic part* and the *deviatoric part* of tensor $\mathbf{E}$. It must be recalled that a tensor is said to be isotropic if its components with respect to some coordinate system do not vary if the coordinate system is changed through a generic rotation. The fundamental isotropic tensor of order two is the *identity tensor* $\mathbf{I} \equiv [\delta_{ij}]$, where δ_{ij} is the Kronecker delta, which is a function defined by the following relations:

$$\delta_{ij} = \begin{cases} 1 & \text{if } i = j \\ 0 & \text{if } i \neq j \end{cases}. \tag{3.23}$$

Therefore, tensor **A** may be recast in a concise form as follows:

$$\mathbf{A} = \left(\frac{1}{3}\text{div}\mathbf{V}\right)\mathbf{I}. \tag{3.24}$$

It should be noted that div $\mathbf{V}$ is the *trace* of tensor **E**, i.e. its first invariant, given by the sum of its diagonal components. It is then immediately realized that the trace of tensor **D** is zero, and this is indeed the characteristic feature of a deviatoric tensor.

It can now be observed that the portion of (3.14) associated with **A**, i.e. $(\delta \mathbf{x} \cdot \mathbf{A})dt$, represents an equal relative variation in length of each component of vector $\delta \mathbf{x}$ in its own direction, with a value $1/3(\text{div } \mathbf{V})$. Therefore, this deformation corresponds to a change in volume of the element without variation of its shape, and may then be denoted an *isotropic dilatation*.

It must here be pointed out that div $\mathbf{V}$ has a clear physical meaning. As a matter of fact, it is easy to check that, considering the deformation corresponding *to the whole tensor* **E**, the variation of volume of the element of fluid that has taken place after the (very small) time interval dt is equal (at first order) to

$$\Delta v = (\text{div}\mathbf{V})\delta x\delta y\delta zdt. \tag{3.25}$$

This relation shows that div $\mathbf{V}$ represents the variation in time of the volume of a fluid element, per unit volume.

It is useful to express this finding in a more general and concise form. In effect, by recalling the physical meaning of the material derivative operator introduced in Section 3.2, and observing that Δv represents the variation, in the small time interval dt, of the volume of a generic element of fluid $dv = \delta x\delta y\delta z$ *that moves with the flow*, we may recast relation (3.25) as follows

$$\text{div}\mathbf{V} = \frac{1}{dv}\frac{D(dv)}{Dt}. \tag{3.26}$$

Now, the variation in volume given in (3.25) is readily seen to be the same as that given by the sole action of **A**, i.e. of the isotropic part of tensor **E**. Therefore, decomposition (3.20) corresponds to dividing the rate of strain tensor into one part, **A**, producing the whole variation in volume but no change in shape, and another one, $\mathbf{D}$, producing the whole variation in shape, but no change in volume.

Let us now look in more detail at the deformation associated with tensor $\mathbf{D}$, i.e. $(\delta\boldsymbol{x}\cdot\mathbf{D})dt$, and analyse, in particular, whether it can be further divided into elementary components.

To do this, we may first observe that the components $[D_{ij}]$ in (3.22) provide the matrix representation of tensor $\mathbf{D}$ in a generic reference system (in this case denoted as x, y, z). Now, when carrying out the inner product $(\delta\boldsymbol{x}\cdot\mathbf{D})$, using expression (1.5), we may distinguish between the effects produced by the diagonal terms $[D_{ij}\delta_{ij}]$ and by the off-diagonal terms $[D_{ij} - D_{ij}\delta_{ij}]$.

The action of the diagonal terms is a variation of the length of the line elements having the direction of the coordinate axes; in other words, these line elements are stretched or shortened by the relative amount given by the corresponding component D_{ii}. However, from what has been said above, this deformation does not produce (at first order) a variation in volume, and thus the signs of all the D_{ii} components cannot be equal. In other words, if, for instance, we have a lengthening in two directions, a shortening in the third one must occur in such a way that the volume remains unchanged. This deformation is called a *pure strain* and the motion producing it a *pure straining motion*.

Finally, the deformation produced by the off-diagonal terms of $[D_{ij}]$ may be seen to correspond to a variation of the angles between the line elements lying in the i- and j-directions. This variation is caused by equal and opposite rotations of any two of these line elements around the axis that is orthogonal to the plane defined by them, so that the bisector between the elements does not change. This deformation is termed a *pure distortion*, in which the angles between line elements change but, at first order, their lengths remain unaltered, as does the volume of the fluid element defined by them.

The single elementary terms into which we have decomposed the global deformation of a fluid element are exemplified in Fig. 3.4, in which, for the sake of simplicity, a two-dimensional motion is considered, i.e. a motion in which, say, both the velocity component w and all the derivatives with respect to z are zero.

For completeness, it should be noted that, thanks to the fact that tensor $\mathbf{D}$ is symmetric, it can be diagonalized. Therefore, one may find three orthogonal directions — the so-called *principal axes* — forming a particular reference system such that the representation of the tensor in this system is

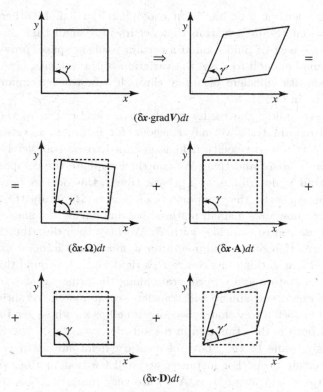

$(\delta x \cdot \mathrm{grad} V)dt$

$(\delta x \cdot \Omega)dt$ $(\delta x \cdot A)dt$

$(\delta x \cdot D)dt$

Fig. 3.4. Decomposition of the deformation of an elementary volume of fluid (two-dimensional case).

a diagonal matrix. This may be seen to be true for any symmetric matrix, and the directions of the principal axes coincide with those of the so-called *eigenvectors* of the matrix.

Without going into details, it can be seen that the diagonal elements of the matrix representation of tensor $\mathbf{D}$ in the principal axes reference system are the *eigenvalues* of $[D_{ij}]$. Therefore, line elements parallel to the principal axes, when subjected to the deformation associated with $\mathbf{D}$, may change in length but not in direction; in other words, they may be stretched or shortened, but do not rotate. We may then say that the whole deformation associated with tensor $\mathbf{D}$ becomes a pure strain when it is described using a reference system whose axes coincide with the principal axes of $\mathbf{D}$.

It must be emphasized that the above description of the deformation of a fluid element is a local one, and is not restricted to the particular shape of the element that has been chosen in Fig. 3.3. In other words, it may represent the deformation that occurs in a small time interval to any volume element of fluid around a certain point in space, provided the element is small enough to allow the variation of the velocities inside it to be expressed with sufficient accuracy through a first-order expansion, as has been done in (3.12).

In order to allow the reader to obtain a clearer idea of the three-dimensional deformation, we may consider, for instance, a certain point P inside a flow whose velocity field is assumed to be completely known as a function of space and time. We can then evaluate the components of the tensor grad V describing, at a generic time t, the local motion around point P; consequently, the components of tensors Ω, A and D are also known. Let us now take a small material volume having the shape, say, of a sphere whose centre coincides with P. We may then describe the most general deformation of this volume, after a small time interval dt, as the sum of a rigid translation of its centre P, a rigid rotation around this point, an isotropic dilatation and a pure strain along the principal axes of tensor D. In other words, our sphere will translate, rotate, swell (or shrink) and be distorted in such a way that it becomes an ellipsoid whose semi-axes are well defined both in direction and in magnitude.

Obviously, some of the above elementary deformations may not be present in certain flows. For instance, we shall see that motions in which div V is zero — and, thus, so is A — are very important, because they describe the behaviour of fluids that may be assumed to be incompressible. Moreover, the presence or not of vorticity ω in the flow — and, thus, of non-zero components of tensor Ω — will be shown to be a fundamental feature, which influences both the flow development and the level of difficulty of its prediction.

3.5. The Transport Theorem

In Section 3.1, we defined a material volume of fluid as a volume that is always composed of the same particles, so that it may be seen as the natural reference volume in a Lagrangian description. As already mentioned, the equations of fluid dynamics are derived from balances that require the variation in time of quantities associated with material volumes of fluid to be expressed mathematically. However, we shall be using an Eulerian description, and it is thus necessary that the relation between the expressions of this variation in the two approaches be available.

Let us denote by $v(t)$ the configuration, at a certain time t, of a material volume of fluid, which may be identified by the positions occupied at that time by the particles composing the volume. If we consider one of these particles, we may use its coordinates with respect to a chosen reference system, say $x_i \equiv (x_1, x_2, x_3)$, to identify it. These coordinates will be a function of time and of the coordinates, say $X_i \equiv (X_1, X_2, X_3)$, that the particle occupied at a previous reference time t^*, at which the configuration of the material volume was v^*. We may then express this dependence by writing

$$x_i = f_i(X_j, t). \tag{3.27}$$

We assume that the transformation leading from v^* to $v(t)$ is sufficiently regular and that each particle maintains its identity. This implies that function f_i must not only be continuously differentiable but also bijective, i.e. with the property that a one-to-one correspondence exists between the elements — which in this case correspond to the fluid particles — of the original and derived sets. Consequently, relation (3.27) is invertible, so that we can write

$$X_j = F_j(x_i, t). \tag{3.28}$$

Furthermore, it may be shown that each element dv of the configuration of the material volume at time t is linked to the corresponding element dv^* in the reference configuration through the following relation:

$$dv = J dv^*, \tag{3.29}$$

where J is the Jacobian determinant of the transformation (3.27), which may be explicitly written as follows

$$J = \begin{vmatrix} \dfrac{\partial x_1}{\partial X_1} & \dfrac{\partial x_1}{\partial X_2} & \dfrac{\partial x_1}{\partial X_3} \\[2mm] \dfrac{\partial x_2}{\partial X_1} & \dfrac{\partial x_2}{\partial X_2} & \dfrac{\partial x_2}{\partial X_3} \\[2mm] \dfrac{\partial x_3}{\partial X_1} & \dfrac{\partial x_3}{\partial X_2} & \dfrac{\partial x_3}{\partial X_3} \end{vmatrix}. \tag{3.30}$$

It is possible to show that the regularity and bijectivity of f_i imply that $0 < J < +\infty$. Let us now assume that we want to express the variation in time of a scalar or vector quantity associated with a material volume whose

Eulerian description is, say, $G(x_i, t)$. The expression to be evaluated is then

$$\frac{D}{Dt} \int_{\boldsymbol{v}(t)} G(x_i, t) d\boldsymbol{v}, \tag{3.31}$$

where the material derivative operator has been used in order to emphasize that the instantaneous time variation is to be evaluated following the material volume during its motion.

The fundamental difficulty in the evaluation of integral (3.31) is the time dependence of the integration domain. However, the Eulerian and Lagrangian descriptions are connected by relations (3.27) and (3.29), and one may thus write

$$\frac{D}{Dt} \int_{\boldsymbol{v}(t)} G(x_i, t) d\boldsymbol{v} = \frac{D}{Dt} \int_{\boldsymbol{v}^*} G(X_i, t) J d\boldsymbol{v}^*. \tag{3.32}$$

In the expression on the right-hand side of (3.32) the integration domain is now fixed in time, so that it is possible to take the derivative operator inside the integral sign, and write (omitting for conciseness the functional dependences and recalling that the usual derivation rules are also valid for the material derivative)

$$\frac{D}{Dt} \int_{\boldsymbol{v}(t)} G d\boldsymbol{v} = \int_{\boldsymbol{v}^*} \frac{D(GJ)}{Dt} d\boldsymbol{v}^* = \int_{\boldsymbol{v}^*} \left(J \frac{DG}{Dt} + G \frac{DJ}{Dt} \right) d\boldsymbol{v}^*. \tag{3.33}$$

Furthermore, it may be seen that the following relation holds:

$$\frac{DJ}{Dt} = J \operatorname{div} \boldsymbol{V}. \tag{3.34}$$

This relation can be demonstrated in several ways, but it may also be readily obtained by using one of the results of the previous section. By applying relation (3.29) twice, and recalling that $d\boldsymbol{v}^*$ does not depend on time, we may write

$$\frac{D(d\boldsymbol{v})}{Dt} = \frac{D(J d\boldsymbol{v}^*)}{Dt} = \frac{DJ}{Dt} d\boldsymbol{v}^* = \frac{DJ}{Dt} \frac{d\boldsymbol{v}}{J}. \tag{3.35}$$

Rearranging terms, we then obtain

$$\frac{DJ}{Dt} = \frac{J}{d\boldsymbol{v}} \frac{D(d\boldsymbol{v})}{Dt}, \tag{3.36}$$

so that (3.34) immediately follows from relation (3.26).

By introducing (3.34) into (3.33), and using again (3.29), we finally obtain the so-called *Reynolds transport theorem*:

$$\frac{D}{Dt} \int_{v(t)} G dv = \int_{v(t)} \left(\frac{DG}{Dt} + G \operatorname{div} \boldsymbol{V} \right) dv. \tag{3.37}$$

As already mentioned, and as will become clear in the following chapter, this theorem is absolutely fundamental for the derivation of the equations of motion. It allows the variation in time of a quantity integrated over a (time-varying) material volume to be expressed by using a relation in which only quantities evaluated at a generic time t are present, *integrated over the volume corresponding to the configuration taken by the material volume at time t.* In other words, by writing the right-hand side of relation (3.37) and introducing in the integrand the desired quantity G, we are expressing the instantaneous time variation of the integral of this quantity over a volume that is changing in time, without apparently using an explicit form for the time variation of the volume. Actually, this variation is expressed by the second term in the integrand, in which div $\boldsymbol{V}$ appears.

One of the main consequences of the transport theorem is that it allows *fixed control volumes* to be used when writing the physical balances from which the equations of fluid dynamics are derived. In effect, these equations must be valid for *any* material volume and thus, when using (3.37), it is possible to consider any fixed control volume in a given reference frame and imagine that it corresponds to the instantaneous image, at time t, of a *suitable* material volume, i.e. of the one that at time t coincides with the chosen control volume. To express this fact, when using the expression on the right-hand side of (3.37) in the following chapters, we shall drop the indication of the dependence of the integration volume on time.

Several other forms of the transport theorem may be derived by using algebraic and integral identities. First of all, the explicit expression of the material derivative may be introduced, so that we have

$$\frac{D}{Dt} \int_{v(t)} G dv = \int_{v(t)} \left(\frac{\partial G}{\partial t} + \boldsymbol{V} \cdot \operatorname{grad} G + G \operatorname{div} \boldsymbol{V} \right) dv. \tag{3.38}$$

Furthermore, from (1.11) we find that the following relation holds, with G either a scalar or vector function,

$$\operatorname{div}(\boldsymbol{V} G) = G \operatorname{div} \boldsymbol{V} + \boldsymbol{V} \cdot \operatorname{grad} G, \tag{3.39}$$

and we thus obtain

$$\frac{D}{Dt} \int_{v(t)} G dv = \int_{v(t)} \left[\frac{\partial G}{\partial t} + \operatorname{div}(\boldsymbol{V} G) \right] dv. \qquad (3.40)$$

Finally, by using the Gauss divergence theorem, the following form may be derived:

$$\frac{D}{Dt} \int_{v(t)} G dv = \int_{v(t)} \frac{\partial G}{\partial t} dv + \int_{S(t)} \boldsymbol{n} \cdot (\boldsymbol{V} G) dS, \qquad (3.41)$$

where we have indicated with $S(t)$ the surface bounding volume $v(t)$, and with $\boldsymbol{n}$ the unit vector normal to this surface at a generic point, directed outwards.

Any of the above forms may be used, choosing the most suitable one according to the problem considered or to the method used for its analysis. For instance, form (3.41) is perhaps the most useful one when a balance is directly written and used in integral form but, as we shall see, it is not adequate to derive the differential form of the equations of motion.

3.6. Examples of Flows with and without Vorticity

Vorticity is defined by relation (3.17) as the curl of the velocity vector, and is thus a purely kinematic quantity. Nonetheless, it will be shown in later chapters that many dynamical aspects of fluid motion are strongly influenced by the presence or absence of vorticity in a flow. For instance, it will be seen that the type, magnitude and predictability of the forces acting on bodies immersed in a fluid stream are strictly connected with the space extent of the regions where vorticity is present. In practice, the evolution of the whole flow is dependent on the amount and organization of the vorticity present in the field, and this is why ample space will be devoted in this book to an in-depth mathematical and physical analysis of the origin and dynamics of vorticity. In this section we anticipate some simple examples of flows in which vorticity is or is not present, in order to better highlight its relation to the deformation of fluid elements. For the sake of simplicity, we consider only two-dimensional flows, in which the only non-zero component of vorticity is thus in the direction normal to the plane of motion.

We have already pointed out that the vorticity vector at a point may be interpreted in physical terms as twice the instantaneous angular velocity of an elementary volume of fluid around that point. Flows with vorticity are consequently denoted as *rotational*, and those without vorticity as *irrotational.* However, confusion should not arise between the rotation of an elementary volume of fluid in a very small time interval, which is a local property of the flow, and the existence of a macroscopic rotation of finite volumes of fluid in finite times. Thus, vorticity may be present in flows that are globally rotating, and also in purely translating flows. Conversely, flows may rotate as a whole and be irrotational, i.e. devoid of vorticity.

Let us start with the simplest rotational flow, namely a global rigid rotation with angular velocity Ω around a point O, which we take as the origin of our reference system (see Fig. 3.5). In this case it is more appropriate to use a cylindrical coordinate system (r, θ, z), so that the velocity components may be written as

$$v_r = 0; \quad v_\theta = \Omega r; \quad v_z = 0. \tag{3.42}$$

All fluid particles rotate with the same angular velocity Ω, and thus their vorticity will have the same value at every point, equal to 2Ω. This can also be derived from the expression of all the vorticity components in cylindrical coordinates, which we write here for the sake of completeness:

$$\omega_r = \frac{1}{r}\frac{\partial v_z}{\partial \theta} - \frac{\partial v_\theta}{\partial z}, \tag{3.43a}$$

$$\omega_\theta = \frac{\partial v_r}{\partial z} - \frac{\partial v_z}{\partial r}, \tag{3.43b}$$

$$\omega_z = \frac{1}{r}\frac{\partial(rv_\theta)}{\partial r} - \frac{1}{r}\frac{\partial v_r}{\partial \theta}. \tag{3.43c}$$

By introducing the velocity components given by (3.42) in these relations, we indeed find that $\omega_z = 2\Omega$ is the only non-zero component of the vorticity vector, as expected.

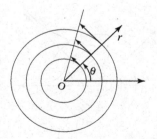

Fig. 3.5. Rigidly rotating flow.

Now let Γ denote the circulation of the velocity vector $\boldsymbol{V}$ around the circle C representing a streamline at a generic radius r. By introducing the azimuthal velocity component from (3.42), we have

$$\Gamma = \oint_C \boldsymbol{V} \cdot d\boldsymbol{l} = \Omega r 2\pi r = 2\Omega\pi r^2, \tag{3.44}$$

where $d\boldsymbol{l}$ is the infinitesimal element vector along C. From relation (3.44) we immediately recognize that the value of the circulation could have been evaluated by applying Stokes' theorem (1.15). In effect, it is equal to the flux of the constant vorticity vector (which is perpendicular to the plane of motion) through the surface of the disk enclosed by C. Therefore, in this flow the circulation Γ increases as the square of the radius r.

Let us now consider another flow, which coincides with the previous one up to a certain radius $r = r_0$, while for larger values of r the only non-zero velocity component is still the azimuthal one, but is now given by the expression

$$v_\theta = \frac{\Omega r_0^2}{r}. \tag{3.45}$$

It may be seen from relations (3.43) that the velocity field defined by (3.45) is irrotational, because the vorticity component in the z direction is now zero, as are the remaining ones. Actually, any two-dimensional flow in which the only velocity component is of the form $v_\theta = A/r$, where A is a constant, is irrotational; in this case the value $A = \Omega r_0^2$ was chosen for the constant, so that the velocity is continuous at $r = r_0$. The complete flow field is depicted in Fig. 3.6, and corresponds to what is known as a *Rankine vortex*, which is characterized by a value of the vorticity that is finite and constant for $r \leq r_0$, and zero for $r > r_0$. The inner constant-vorticity region is denoted as the *core* of the vortex.

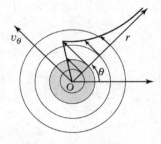

Fig. 3.6.   Rankine vortex.

We shall see in later chapters that the velocity field corresponding to a Rankine vortex cannot exist in an unlimited fluid domain due to the discontinuity in vorticity that is present at $r = r_0$. However, it can be considered as a useful model flow, through which one can represent, with satisfactory approximation, a real flow in which vorticity is mainly concentrated in a limited circular region of space, while the outer field is effectively irrotational. Such flows are found in many common situations, for instance downstream of various types of bodies, and their description, even if in an approximate way, is thus a very important issue in fluid dynamics. These concentrated-vorticity flows are usually referred to as *vortices*, although it must be mentioned that the rigorous definition of a vortex is a source of everlasting academic discussions, which need not be considered in an introductory book.

The velocity field outside the vortex core, given by relation (3.45), is the prototype of an irrotational flow with circular streamlines and is often exactly or approximately encountered in many situations. This flow is thus an example of a case in which a global rotation of the fluid is present, but there is no instantaneous rotation of small elements of fluid. This may better be understood with the help of Fig. 3.7, where a highly magnified depiction of the variation of the position and shape of a rectangular element of fluid after a small time interval dt is shown for the two above-described flows with circular streamlines, namely a purely rotational flow and an irrotational one.

As can be seen, in the first case the fluid element does not change its shape (i.e. no real deformation occurs), but it rotates, as may be

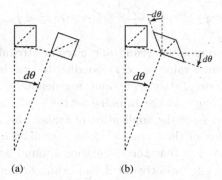

(a)   (b)

Fig. 3.7. Deformation of a fluid element in a flow with circular streamlines. (a) Rigid rotation flow; (b) irrotational flow.

deduced from the variation of the direction of its diagonal. Conversely, in the irrotational flow the two sides of the fluid element rotate by the same amount in opposite directions, and thus the direction of the diagonal remains unchanged and we may say that no rotation occurs. On the other hand, a significant distortion of the element is present. Obviously, as already pointed out, although finite elements and displacements are shown in Fig. 3.7, one should recall that we are actually referring to a first-order description of the deformation of a very small element of fluid in a small interval of time. Therefore, the objective of the figure is just to give an idea of the significant difference existing between the two types of flows.

A practical example might further elucidate the difference between the two motions. Imagine that a Rankine vortex is an approximation of a whirlpool in a water stream, around which a small floating body moves, for instance, a cork with the same shape as the above fluid element and with a cross drawn on it. The body is rigid, so that its shape cannot change during the motion. In other words, it cannot distort but it is free to rotate around an axis. If the body is within the core region where vorticity is present, then during its circular motion it will spin and the cross will appear to rotate around its axis. On the other hand, if it is placed in the outer irrotational region, it will not rotate around its axis, and during its motion in a circular trajectory the cross will keep its direction unchanged (apart perhaps from small oscillations due the finiteness of the motion).

Let us now calculate the circulation Γ of the velocity vector along circular streamlines of a Rankine vortex. If we remain inside the core, i.e. if $r \leq r_0$, then the circulation is still given by relation (3.44). Conversely, if $r > r_0$ the circulation turns out to be

$$\Gamma = \oint_C \boldsymbol{V} \cdot d\boldsymbol{l} = \frac{\Omega r_0^2}{r} 2\pi r = 2\Omega \pi r_0^2 = \Gamma_0. \tag{3.46}$$

Therefore, the circulation around any circular streamline outside the vortex core has the same value Γ_0 corresponding to the circulation around the external boundary of the core, and we denote this value of the circulation as the *strength* of the Rankine vortex. This result might have been obtained directly from the application of Stokes' theorem, considering that Γ_0 is also the flux of the whole vorticity contained in the core. More generally, we may observe that the circulation around any generic-shape closed curve (and thus irrespective of it coinciding or not with a circular streamline) enclosing the rotational core will always be equal to Γ_0. On the other hand, the circulation around any closed curve that does not

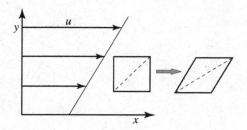

Fig. 3.8. Unidirectional shear flow.

enclose the core and which, therefore, surrounds a completely irrotational flow region, is always zero.

Finally, in order to show that vorticity, and thus rotation of elementary volumes of fluid, may be also present when the streamlines are perfectly rectilinear, we consider the simple unidirectional shear flow shown in Fig. 3.8. As it can be appreciated from the figure, in this case the deformation of a square element of fluid comprises both a rotation and a distortion. In fact, the deformation is completely determined by the only non-zero term, namely $\partial u/\partial y$, appearing in both the symmetric and anti-symmetric parts of the tensor grad $\boldsymbol{V}$, which, as we have seen in Section 3.4, describes the whole deformation of an elementary volume of fluid. In particular, in this flow, the vorticity component in the z direction is constant and equal to $-\partial u/\partial y$, and the circulation around any closed curve is equal to this value multiplied by the area enclosed by the curve.

Chapter 4

THE EQUATIONS OF MOTION OF FLUIDS

4.1. General Remarks

The equations that describe the motion of fluids may be derived by expressing in mathematical terms physical 'laws' or 'principles' that are assumed to be valid and capable of providing a 'good' representation of the dynamics of fluids. In other words, the validity of the laws we are assuming rests on the fact that they permit us to derive a set of equations whose solutions describe, with sufficient accuracy, the real behaviour of a fluid in motion. Obviously, this means that the validity of the description must be checked, at least for a sufficient number of cases, through experimental measurements having an accuracy that is consistent with our requirements for the particular problems we are considering. Essentially, we may say that our task is to construct a physico-mathematical model that provides a satisfactory description of the behaviour of fluids within certain limits of applicability.

As already pointed out in Chapter 2, we choose to assume the validity of the continuum hypothesis, which implies that we are giving up the possibility of describing the motion of the single molecules of a fluid or of evaluating the forces acting on a body in motion inside a rarefied gas. Furthermore, we also assume that the velocities of interest will always be much smaller than the velocity of light, so that relativistic effects may be safely neglected. This already sets some limits on the applicability of the physico-mathematical model we are going to use, in which fluid dynamics is considered as a particular chapter of classical continuum mechanics, or, more precisely, of the classical mechanics of deformable continua. Nonetheless, this model is still a wide-ranging one, and allows extremely diverse situations to be considered.

In the following sections we thus start by deriving the equations of the dynamics of any deformable continuum. However, we shall see that this leads to a set of equations containing a number of unknown functions that is greater than the number of equations. Therefore, it will be necessary to introduce further relations, the so-called constitutive equations, which specify the physical behaviour of the continua we are dealing with, namely fluids. We shall confine our treatment to particular types of fluids, the so-called Newtonian fluids, thus restricting further the range of application of the derived equations. However, this restriction will not be a very serious one, as the derived set of equations are still capable of describing accurately the motion of most common fluids encountered in practical applications.

The physical laws whose validity is assumed, and which must be formulated in mathematical terms, express balances of certain quantities for a generic material volume of fluid in motion. Therefore, the *integral form* of the equations, in which integrals appear extended to material volumes or to their bounding surfaces, will naturally arise. In practice, the above-mentioned balances express that the variation in unit time of the integral of a certain (scalar or vector) quantity over a material volume of fluid is equal either to zero or to one or more integral terms. Hence, the Reynolds transport theorem, introduced in Section 3.5, plays a fundamental role in the writing of the equations, and we shall apply it to derive different but equivalent forms of the integral equations, to be used according to the particular case or to the chosen solution procedure.

The integral forms of the equations of motion may be directly used to solve particular problems, by choosing the most suitable control volume — which we have seen in Chapter 3 to represent the instantaneous image of a material volume — according to the considered case. For instance, if we are analysing the motion in a duct, it is natural to choose a control volume that is bounded by the (impermeable) inner surface of the duct and by two surfaces orthogonal to its axis, which will then be the only surfaces through which the fluid may flow.

Nonetheless, in many (or perhaps most) cases, it is useful to express the equations of motion in *differential form*. This may be obtained either by referring to a very small elementary material volume of fluid — as we have done in Chapter 3 to describe its deformation — and by applying the balance laws directly to it, or by starting from the integral form of the equations written for a generic volume of fluid. This second procedure has some advantages and is the one that is used in this book; therefore, its fundamental steps will now be described.

The first essential point is to express all the balance equations in the following integral form:

$$\int_v f(x, y, z, t) dv = 0, \tag{4.1}$$

where $f(x, y, z, t)$ is a function comprising differential operators applied to scalar or vector quantities, and v is the instantaneous configuration of a generic material volume. Obviously, to attain form (4.1), all the integrals involved in the expression of the balances must be extended over the considered volume. However, as we already mentioned, integrals that are extended over the surface bounding the volume may also appear, and must then be expressed as fluxes of certain quantities, so that we can apply Gauss' theorem and obtain the corresponding volume integrals.

We now observe that, once an integral balance is expressed as in (4.1), it must be valid for *any* volume, because it represents a physical law having general validity. This implies that we may use the so-called *localization lemma*, which states that the necessary and sufficient condition for a relation such as (4.1) to be satisfied for any arbitrary volume of integration v is that

$$f(x, y, z, t) = 0. \tag{4.2}$$

The differential form of the equations is thus immediately derived. However, we must point out that the application of this lemma requires that $f(x, y, z, t)$ be a continuous function. Therefore, it cannot be used if regions where the integrand function is discontinuous are present inside the volume. Now, as briefly mentioned in Section 2.3, if the flow is supersonic (i.e. the Mach number is greater than 1) abrupt changes of the flow quantities may occur within regions whose dimension in the direction of the flow is of the order of a few molecular mean free paths. These are the so-called shock waves, which, in the continuous treatment, may normally be considered as surfaces of discontinuity. Therefore, if such discontinuities are present inside the integration volume, the localization lemma cannot be used, the balances must be kept in integral form, and a proper mathematical treatment must be used to derive the variations of the various quantities through the discontinuity surfaces. This will be done, for instance, in Chapter 15 to obtain the jumps of the various flow variables through a normal shock wave.

In conclusion, the differential equations cannot be used to describe the flow variations through shock waves, but they are valid in front and behind them. In fact, one of the advantages of the derivation of the differential form of the equations of motion from their integral form is precisely the possibility of pointing out the reduced range of validity of the differential equations; the other advantage is the simplicity of the mathematical treatment.

We can now derive the equations of motion of a fluid by expressing, in mathematical terms, the following fundamental balances of the mechanics of continua:

- the balance (or conservation) of mass;
- the balance of momentum;
- the balance of total energy.

Further balances will also be written for other flow quantities, such as kinetic energy, internal energy, entropy, enthalpy and total enthalpy. The equations expressing the balances of these quantities are useful in many circumstances, but we shall see they may all be derived from the above-mentioned fundamental balances, so that they do not provide additional equations of motion, but only alternative ones.

4.2. Balance of Mass

The principle of mass conservation implies that *the mass of a material volume of fluid does not change in time*. This is equivalent to saying that no creation or destruction of mass may occur inside the material volume.

Bearing in mind that density represents the mass contained in unit volume of fluid, the mathematical expression of the above physical law may then be written as follows:

$$\frac{D}{Dt} \int_{v(t)} \rho \, dv = 0, \qquad (4.3)$$

where the functional dependence of the density ρ on position and time has been omitted for the sake of conciseness.

The following equivalent integral forms of the balance of mass can then be derived by using the Reynolds transport theorem:

$$\int_v \left(\frac{D\rho}{Dt} + \rho \operatorname{div} \boldsymbol{V} \right) dv = 0, \tag{4.4a}$$

$$\int_v \left(\frac{\partial \rho}{\partial t} + \operatorname{div}(\rho \boldsymbol{V}) \right) dv = 0, \tag{4.4b}$$

$$\int_v \frac{\partial \rho}{\partial t} dv + \int_S \boldsymbol{n} \cdot (\rho \boldsymbol{V}) dS = 0. \tag{4.4c}$$

As anticipated in Section 3.5, in the above equations we have dropped the indication of the dependence of the integration domain on time, to express the fact that, although the integrals represent the time variation of the mass contained in a material volume (which changes in time), they may be extended to *any* fixed control volume (or to its bounding surface), which may be considered as the instantaneous configuration of a suitable material volume of fluid.

Incidentally, relation (4.4c) might have been directly written even by starting our analysis from a generic fixed control volume. In effect, it expresses the fact that the variation in time of the mass contained in the considered volume, given by the first term in (4.4c), is equal to the opposite of the flux of mass through the bounding surface. Due to our choice of the normal to the surface as being directed outwards, the flux is indeed positive if the mass that leaves the volume is more than the mass that enters it. Then Eqs. (4.4a) and (4.4b) could have been obtained by successively using Gauss' theorem and the definition of material derivative.

The differential forms of the equation of mass may now be derived starting from (4.4a) or (4.4b) and applying the localization lemma, so that we obtain

$$\frac{D\rho}{Dt} + \rho \operatorname{div} \boldsymbol{V} = 0, \tag{4.5a}$$

$$\frac{\partial \rho}{\partial t} + \operatorname{div}(\rho \boldsymbol{V}) = 0. \tag{4.5b}$$

By expanding the operators in the above relations, the following more explicit form is obtained:

$$\frac{\partial \rho}{\partial t} + \boldsymbol{V} \cdot \operatorname{grad} \rho + \rho \operatorname{div} \boldsymbol{V} = 0. \tag{4.5c}$$

As can be easily verified, Eq. (4.5a) is equivalent to Eq. (3.26) and again provides the physical interpretation of div V as the time variation of the volume of a fluid element, per unit volume.

The differential form of the mass balance equation is commonly called the *equation of continuity*, and its validity allows the transport theorem to be expressed in a simplified form. In effect, consider the case in which relation (3.37) is applied to a function $G = \rho F$. For instance, F might be a certain quantity per unit mass, so that G would be the same quantity per unit volume. We then have

$$
\frac{D}{Dt} \int_{v(t)} \rho F \, dv = \int_v \left(\frac{D(\rho F)}{Dt} + (\rho F) \operatorname{div} V \right) dv
$$
$$
= \int_v \left(\rho \frac{DF}{Dt} + F \left(\frac{D\rho}{Dt} + \rho \operatorname{div} V \right) \right) dv,
$$

(4.6)

and thus, from (4.5a), we get

$$
\frac{D}{Dt} \int_{v(t)} \rho F \, dv = \int_v \rho \frac{DF}{Dt} \, dv.
$$

(4.7)

We shall use this form of the transport theorem extensively in the following sections, as the quantities involved in the remaining balances may be expressed in terms of their values per unit mass and, consequently, their values per unit volume may be obtained through multiplication by density. However, when the form (4.7) is used to write an integral balance equation, one should always remember that the validity of the continuity equation in differential form has been assumed. Therefore, all the conditions of applicability of the localization lemma, described in the previous section, must be valid even if we are writing an integral form of a balance equation.

4.3. Balance of Momentum

The balance of momentum may be considered as the fundamental principle of the dynamics of fluids and can be expressed as follows: *the variation in unit time of the (linear) momentum of a material volume of fluid is equal to the resultant of the forces acting on the volume.*

In mathematical terms, we may then write

$$
\frac{D}{Dt} \int_{v(t)} \rho V \, dv = F,
$$

(4.8)

where F is the resultant of all the forces acting on the volume of fluid.

Equation (4.8) is a vector equation, so that three scalar equations may be derived from it, in the form

$$\frac{D}{Dt} \int_{v(t)} \rho u_i dv = F_i \quad (i = 1, 2, 3), \tag{4.9}$$

where u_i and F_i are respectively the components of the velocity vector and of the resultant force vector. Unless otherwise specified, in the following we shall use a Cartesian coordinate system whose axes are denoted as x_i, with unit vectors e_i.

The forces acting on the volume of fluid may be decomposed into *body* forces (also called *volume* forces) and *surface* (or *contact*) forces, whose resultants we respectively denote by $\boldsymbol{F_V}$ and $\boldsymbol{F_S}$.

Body forces are those that are characterized as having a long range of influence and acting on all the material particles of fluid, i.e. they are proportional to the mass contained in any elementary volume of fluid. They are normally connected with the existence of force fields, the most common one being the gravitational field; other examples of this class of forces are the electromagnetic forces and the apparent forces, such as centrifugal force. The body forces per unit mass will be indicated in the following by the symbol $\boldsymbol{f}$, and their components by f_i, so that the corresponding values per unit volume are $\rho \boldsymbol{f}$ and ρf_i. Therefore, the resultant vector of the body forces over the considered volume of fluid and its components may be obtained directly by integration, as follows:

$$\boldsymbol{F_V} = \int_v \rho \, \boldsymbol{f} dv, \tag{4.10}$$

$$(\boldsymbol{F_V})_i = \int_v \rho f_i dv. \tag{4.11}$$

In relation (4.11) we have dropped the indication of the variation of the indices that was present in (4.9), and the same will be done in the following when reference is made to vector components or to the relevant equations. It should be pointed out that the body forces are assumed to be known functions in our analysis, and their dependence on space and time is implicitly considered in the above relations.

On the other hand, surface forces are those that are produced by short-range actions, such as molecular interactions; consequently, they significantly decrease within distances of the order of the molecular separation. They may thus be considered to act on any elementary area of a surface bounding a volume of fluid and to be produced by the surroundings.

Therefore, instead of being proportional to the volume, like the body forces, they are actually proportional to the extent of the area and are usually described through their values per unit surface, which are called *stresses*. Thus, a stress may be considered as the limit of the ratio between a surface force and the area over which it acts when the latter is made infinitesimally small.

The stress acting on a point of a surface is a vector that depends not only on the vector defining the position of the point in space, x, and on time, t (as all the remaining quantities we have encountered so far), but also on the orientation of the element of surface rounding the considered point. This orientation may be specified by the direction of the unit vector normal to the surface, say n directed outwards from the volume of fluid bounded by the surface. Obviously, the stress acting on an element of this surface does not, in general, have the same direction as n, and will be denoted in the following by the symbol t_n, as a short form of $t(n, x, t)$.

With the above notation, the resultant of the surface forces acting on the surface S bounding the volume of fluid, and its component in the ith direction, may be written as follows:

$$F_S = \int_S t_n dS, \qquad (4.12)$$

$$(F_S)_i = \int_S (t_n)_i dS. \qquad (4.13)$$

As just mentioned, in order to characterize the stress vector at a generic point in the fluid, it should be considered that different stresses will act over the infinitely many elementary surfaces passing through that point. Therefore, each component of the stress vector t_n acting on an elementary surface passing through a certain point is, in general, a function of the three components of the vector n defining the orientation of the surface. Consequently, at least nine scalars are necessary and they are also sufficient if the dependence of the stress components on the normal vector is a linear one. The demonstration that this is indeed the case was first given by Cauchy, who showed that the stress state can be completely characterized by the components of a second-order tensor, which is called the *stress tensor*.

In particular, Cauchy derived the following relation (known as *Cauchy's theorem*), which is absolutely fundamental for our treatment and may be considered as one of the most important contributions to continuum mechanics:

$$t_n = n_1 t_1 + n_2 t_2 + n_3 t_3, \qquad (4.14)$$

where n_i are the components of n with respect to a chosen reference system, and t_i are the stress vectors acting on the three elementary surfaces passing through the considered point and having as normal unit vectors the unit vectors of the coordinate axes, e_i.

Relation (4.14) may be demonstrated by applying the momentum balance to a tetrahedral volume element having its vertex at the point where we want to evaluate t_n, its lateral faces parallel to the coordinate planes, and its base parallel to the plane that is tangent to the considered surface element with unit normal n (see Fig. 4.1).

It is easy to see that the variation of momentum and the body forces within the element are proportional to the volume of the tetrahedron, while the surface forces are proportional to the areas of its faces. Therefore, if we let the linear dimensions of the tetrahedron tend to zero without change of its shape, and use the mean theorem for each of the faces, we realize that the volume of the tetrahedron becomes of lower order than its bounding surface. Consequently, the momentum balance can be satisfied only if the resultant of the surface forces vanishes, which may be shown to correspond to relation (4.14).

Therefore, thanks to Cauchy's theorem, the generic ith component of the stress vector can be expressed as follows:

$$(t_n)_i = t_n \cdot e_i = n_1 (t_1 \cdot e_i) + n_2 (t_2 \cdot e_i) + n_3 (t_3 \cdot e_i)$$

$$= \sum_k n_k T_{ki} = (n \cdot \mathbf{T})_i. \qquad (4.15)$$

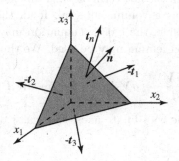

Fig. 4.1.   Elementary tetrahedral volume of fluid.

In (4.15) we have used (1.5) and introduced the stress tensor $\mathbf{T}$, whose generic component T_{ki} is immediately seen to represent the component in the direction of $\mathbf{e}_i$ of the stress vector acting on an element of surface having unit normal $\mathbf{e}_k$. It is now evident that we can express the stress vector acting on an element of surface having normal $\mathbf{n}$ as

$$t_n = \mathbf{n} \cdot \mathbf{T}. \tag{4.16}$$

Summarizing, we can evaluate the stress vector acting on *any* element of surface passing through a point in the fluid provided we know the components — with respect to a chosen reference system — of the normal unit vector of the surface element and of the stress vectors that act on three orthogonal planes passing through the same point and having normal unit vectors coinciding with the unit vectors of the coordinate axes. In brief, we need to have the components of $\mathbf{n}$ and of the stress tensor $\mathbf{T}$ at the considered point in the fluid.

We are now able to write the vector and scalar integral equations representing the balance of momentum by introducing the appropriate terms in Eqs. (4.8) and (4.9). To this end, we first use the form (3.37) of the Reynolds transport theorem to express the time variation of the momentum contained in the considered volume of fluid:

$$\int_v \left(\frac{D(\rho \mathbf{V})}{Dt} + \rho \mathbf{V} \operatorname{div} \mathbf{V} \right) dv = \int_v \rho \, \mathbf{f} dv + \int_S \mathbf{n} \cdot \mathbf{T} dS, \tag{4.17}$$

$$\int_v \left(\frac{D(\rho u_i)}{Dt} + \rho u_i \operatorname{div} \mathbf{V} \right) dv = \int_v \rho f_i dv + \int_S (\mathbf{n} \cdot \mathbf{T})_i dS. \tag{4.18}$$

In order to derive the associated differential equations, we must first use the divergence theorem in the last term on the right-hand side of Eq. (4.17), so that it is expressed as a volume integral. If all the integrals are then moved to the left-hand side, the balance equation may be put in the form (4.1) and the localization lemma may be used. We thus obtain

$$\frac{D(\rho \mathbf{V})}{Dt} + \rho \mathbf{V} \operatorname{div} \mathbf{V} = \rho \, \mathbf{f} + \operatorname{div} \mathbf{T}. \tag{4.19}$$

By projection on the coordinate axes, the scalar differential equations are then derived:

$$\frac{D(\rho u_i)}{Dt} + \rho u_i \operatorname{div} \mathbf{V} = \rho f_i + (\operatorname{div} \mathbf{T})_i. \tag{4.20}$$

Recall here that the divergence of a second-order tensor is a vector, whose components are defined by relation (1.7), namely,

$$(\operatorname{div} \mathbf{T})_i = \sum_j \frac{\partial T_{ji}}{\partial x_j}.$$

In Eq. (4.19), which is also known as *Cauchy's first law of motion*, the terms on the left-hand side represent the variation in unit time of the momentum of an elementary volume of fluid, and the terms on the right-hand side are the resultants, *per unit volume*, of the body and surface forces.

It may be useful to show that, to first order, div $\mathbf{T}$ is indeed the resultant, per unit volume, of the surface forces acting on a fluid element. To this end, let us consider the elementary volume shown in Fig. 4.2, chosen for simplicity such that the unit vectors normal to its surfaces are parallel to the coordinate axes, now denoted by x, y, z, and let us evaluate one of the components of the resultant of the surface forces, say the x component.

In Fig. 4.2 the surfaces of the volume element are numbered, and we denote by $[(t_n)_x]_i$ the component in the x direction of the stress acting on the generic ith face. The x component of the resultant surface force acting on the small elementary volume is then

$$(\boldsymbol{F_S})_x = ([(\boldsymbol{t_n})_x]_1 + [(\boldsymbol{t_n})_x]_2)\delta y \delta z + ([(\boldsymbol{t_n})_x]_3 + [(\boldsymbol{t_n})_x]_4)\delta x \delta z$$
$$+ ([(\boldsymbol{t_n})_x]_5 + [(\boldsymbol{t_n})_x]_6)\delta x \delta y.$$

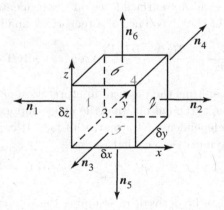

Fig. 4.2. Elementary volume of fluid.

Now, by applying relation (4.15), we have

$$[(\boldsymbol{t}_n)_x]_i = [\boldsymbol{t}_n \cdot \boldsymbol{e}_x]_i = \sum_k [n_k T_{kx}]_i = [n_x T_{xx} + n_y T_{yx} + n_z T_{zx}]_i.$$

Furthermore, the unit vectors normal to the various faces are defined by the following components:

$$\boldsymbol{n}_1 \equiv (-1, 0, 0); \quad \boldsymbol{n}_2 \equiv (1, 0, 0); \quad \boldsymbol{n}_3 \equiv (0, -1, 0); \quad \boldsymbol{n}_4 \equiv (0, 1, 0);$$
$$\boldsymbol{n}_5 \equiv (0, 0, -1); \quad \boldsymbol{n}_6 \equiv (0, 0, 1).$$

Therefore, the stress components in the x direction acting on the various faces are the following:

$$[(\boldsymbol{t}_n)_x]_1 = -[T_{xx}]_1; \quad [(\boldsymbol{t}_n)_x]_2 = [T_{xx}]_2; \quad [(\boldsymbol{t}_n)_x]_3 = -[T_{yx}]_3;$$
$$[(\boldsymbol{t}_n)_x]_4 = [T_{yx}]_4; \quad [(\boldsymbol{t}_n)_x]_5 = -[T_{zx}]_5; \quad [(\boldsymbol{t}_n)_x]_6 = [T_{zx}]_6.$$

Considering the small dimensions of the elementary volume, we may use a first-order expansion for the variation of the components of the stress tensor between the various faces, so that we get

$$[T_{xx}]_2 = [T_{xx}]_1 + \frac{\partial T_{xx}}{\partial x} \delta x; \quad [T_{yx}]_4 = [T_{yx}]_3 + \frac{\partial T_{yx}}{\partial y} \delta y;$$

$$[T_{zx}]_6 = [T_{zx}]_5 + \frac{\partial T_{zx}}{\partial z} \delta z.$$

If we now introduce all the above relations into the expression of the x component of the resultant surface force on the elementary volume, and recall the definition of the divergence of a tensor, we finally obtain

$$(\boldsymbol{F_s})_x = \left(\frac{\partial T_{xx}}{\partial x} + \frac{\partial T_{yx}}{\partial y} + \frac{\partial T_{zx}}{\partial z} \right) \delta x \delta y \delta z = (\operatorname{div} \mathbf{T})_x \delta x \delta y \delta z,$$

which immediately confirms the physical interpretation of div $\mathbf{T}$.

It is now important to point out that the nine components of the stress tensor are not all independent. As a matter of fact, it can be shown that the stress tensor is symmetric, i.e. that we have

$$T_{ij} = T_{ji}. \tag{4.21}$$

Relation (4.21) can be derived by assuming the validity of the balance of angular momentum, which may be expressed by saying that the variation in unit time of the angular momentum of a material volume of fluid is equal to the resultant moment of the forces acting on the volume.

In mathematical terms, this postulate may thus be expressed in the following form:

$$\frac{D}{Dt} \int_{v(t)} \boldsymbol{x} \times \rho \boldsymbol{V} \, dv = \int_{v} \boldsymbol{x} \times \rho \boldsymbol{f} \, dv + \int_{S} \boldsymbol{x} \times \boldsymbol{t}_n \, dS, \qquad (4.22)$$

where $\boldsymbol{x}$ is a position vector.

If we introduce Cauchy's theorem and the stress tensor, we may easily see that Eq. (4.22) can be recast in a form which implies that if the balance of linear momentum (4.19) is satisfied, then relation (4.21) must be necessarily true. Conversely, we could have assumed the stress tensor to be symmetric, and the balance of angular momentum (4.22) would have been derived consequently from the linear momentum balance. The result that, assuming the validity of the balance of linear momentum, the balance of angular momentum may be satisfied only if the stress tensor is symmetric is known as *Cauchy's second law of motion.*

The symmetry of the stress tensor provides the obvious advantage that only six scalar components, instead of nine, are sufficient for the complete characterization of the stress at a point. Furthermore, it allows the last term in Eq. (4.20) to be expressed in a slightly different form.

In effect, by recalling the above-described physical meaning of the components of the stress tensor, we may first express the stress vector $\boldsymbol{t}_i$, which acts on a surface element with unit normal vector $\boldsymbol{e}_i$, as a function of its components:

$$\boldsymbol{t}_i = T_{i1} \boldsymbol{e}_1 + T_{i2} \boldsymbol{e}_2 + T_{i3} \boldsymbol{e}_3. \qquad (4.23)$$

By using relations (1.5), (1.7) and (4.15), the symmetry of the stress tensor allows the following expressions to be derived:

$$(\boldsymbol{t}_n)_i = \boldsymbol{t}_n \cdot \boldsymbol{e}_i = (\boldsymbol{n} \cdot \mathbf{T})_i = \sum_j n_j T_{ji} = \sum_j T_{ij} n_j = \boldsymbol{t}_i \cdot \boldsymbol{n}, \qquad (4.24)$$

$$(\text{div } \mathbf{T})_i = \sum_j \frac{\partial T_{ji}}{\partial x_j} = \sum_j \frac{\partial T_{ij}}{\partial x_j} = \text{div } \boldsymbol{t}_i. \qquad (4.25)$$

If we now introduce (4.25) and (4.7) (i.e. we use the differential form of the mass balance equation), the vector and scalar forms of the momentum balance differential equation may be recast as follows:

$$\rho \frac{DV}{Dt} = \rho \left(\frac{\partial V}{\partial t} + V \cdot \mathrm{grad} V \right) = \rho f + \mathrm{div}\, \mathbf{T}, \qquad (4.26)$$

$$\rho \frac{Du_i}{Dt} = \rho \left(\frac{\partial u_i}{\partial t} + V \cdot \mathrm{grad}\, u_i \right) = \rho f_i + \mathrm{div}\, t_i. \qquad (4.27)$$

The momentum balance equations we have introduced so far describe the dynamics of any homogeneous deformable continuum. However, when the motion of fluids is considered, it is customary to decompose the stress tensor into its *isotropic* and *deviatoric* parts, i.e. into a tensor whose components do not change under any rotation of the coordinate system and a tensor whose trace is zero. Actually, this decomposition is not essential at this point in our discussion, and could be considered to originate more from historical reasons than from a real necessity. Nonetheless, we introduce it here because, as will be shown in the following, a specific physical meaning may be associated with the two parts according to which fluid is considered.

To better understand the role of the decomposition, we may first refer to the particular case of a fluid at rest, and recall that, by definition, in that condition a fluid can only withstand normal stresses (the diagonal terms of the stress tensor matrix representation) and not tangential stresses (the off-diagonal terms). Furthermore, in static conditions the normal stresses acting on any plane passing through a point are all equal, and their value is given by the opposite of the pressure in the fluid p, which in this case coincides with the thermodynamic pressure.

Therefore, for a fluid at rest, the following relations hold:

$$\mathbf{T} = -p\mathbf{I}; \quad T_{ij} = -p\delta_{ij}; \quad t_n = -p\mathbf{n}. \qquad (4.28)$$

Consequently, in static conditions we may write

$$p = -\frac{1}{3} \sum_i T_{ii}. \qquad (4.29)$$

In dynamic conditions, the situation is different, not only because non-zero tangential stresses are usually present, but also because the normal stresses acting on mutually orthogonal elementary surfaces passing through the considered point are, in general, no longer equal.

However, we may now introduce a *mechanical definition of pressure*. In other words, we define the (mechanical) pressure, denoted by the symbol p, as the quantity given by relation (4.29), i.e. the opposite of the *mean value* of the normal stresses acting on any three orthogonal elementary surfaces passing through the considered point, or, more correctly, as minus one-third of the trace of the stress tensor. As a matter of fact, the trace is one of the invariants of a second-order tensor, and thus it does not change on changing the reference system; this ensures that our definition identifies a single scalar quantity relating to the point we are considering in the flow field. Note, also, that this 'pressure' does not necessarily coincide with the thermodynamic pressure; however, we shall see in the following that this may be the case if certain further conditions are assumed to be valid.

We may now decompose the stress tensor as follows:

$$\mathbf{T} = -p\mathbf{I} + \mathbf{\tau}, \tag{4.30}$$

or, in terms of components,

$$T_{ij} = -p\delta_{ij} + \tau_{ij}. \tag{4.31}$$

The tensors $-p\mathbf{I}$ and $\mathbf{\tau}$ are thus the isotropic and the deviatoric parts of the stress tensor respectively, and their components are *defined* by relations (4.29) and (4.31). It might be useful to point out that, considering the stress tensor $\mathbf{T}$ is symmetric, the deviatoric stress tensor is symmetric too, and its normal components, which are given by the expression

$$\tau_{ii} = T_{ii} + p = T_{ii} - \frac{1}{3}\sum_j T_{jj}, \tag{4.32}$$

are, in general, non-zero; on the other hand, their sum, i.e. the trace of tensor $\mathbf{\tau}$, is obviously zero.

With this decomposition, and operating in analogy with (4.24), we may then express the stress vector and its components as follows:

$$\mathbf{t}_n = -\mathbf{n} \cdot p\mathbf{I} + \mathbf{n} \cdot \mathbf{\tau} = -p\mathbf{n} + \mathbf{\tau}_n, \tag{4.33}$$

$$(\mathbf{t}_n)_i = \mathbf{t}_n \cdot \mathbf{e}_i = -p(\mathbf{n} \cdot \mathbf{e}_i) + \mathbf{\tau}_n \cdot \mathbf{e}_i = -pn_i + \mathbf{\tau}_i \cdot \mathbf{n}, \tag{4.34}$$

so that their integrals over the surface bounding a volume of fluid become

$$\int_S \mathbf{t}_n dS = -\int_S \mathbf{n} \cdot p\mathbf{I}\, dS + \int_S \mathbf{n} \cdot \mathbf{\tau} dS$$

$$= -\int_v \operatorname{grad} p\, dv + \int_v \operatorname{div} \mathbf{\tau} dv, \tag{4.35}$$

$$\int_S (t_n)_i dS = -\int_S p\boldsymbol{n} \cdot \boldsymbol{e}_i dS + \int_S \boldsymbol{\tau}_n \cdot \boldsymbol{e}_i dS$$

$$= -\int_v \frac{\partial p}{\partial x_i} dv + \int_v \operatorname{div} \boldsymbol{\tau}_i dv, \qquad (4.36)$$

where the following identities have been used:

$$\operatorname{div}(p\mathbf{I}) = \operatorname{grad} p; \quad \operatorname{div}(p\boldsymbol{e}_i) = \frac{\partial p}{\partial x_i}.$$

Therefore, taking the decomposition of the stress tensor into account, the integral and differential forms of the momentum balance equation may now be recast as follows:

$$\int_v \rho \left(\frac{\partial \boldsymbol{V}}{\partial t} + \boldsymbol{V} \cdot \operatorname{grad} \boldsymbol{V} \right) dv = \int_v \rho \boldsymbol{f} dv - \int_S p\boldsymbol{n}\, dS + \int_S \boldsymbol{\tau}_n dS, \quad (4.37)$$

$$\int_v \rho \left(\frac{\partial u_i}{\partial t} + \boldsymbol{V} \cdot \operatorname{grad} u_i \right) dv = \int_v \rho f_i dv - \int_S p n_i dS + \int_S \boldsymbol{\tau}_i \cdot \boldsymbol{n}\, dS, \quad (4.38)$$

$$\rho \left(\frac{\partial \boldsymbol{V}}{\partial t} + \boldsymbol{V} \cdot \operatorname{grad} \boldsymbol{V} \right) = \rho \boldsymbol{f} - \operatorname{grad} p + \operatorname{div} \boldsymbol{\tau}, \qquad (4.39)$$

$$\rho \left(\frac{\partial u_i}{\partial t} + \boldsymbol{V} \cdot \operatorname{grad} u_i \right) = \rho f_i - \frac{\partial p}{\partial x_i} + \operatorname{div} \boldsymbol{\tau}_i, \qquad (4.40)$$

where

$$\boldsymbol{\tau}_i = \tau_{i1} \boldsymbol{e}_1 + \tau_{i2} \boldsymbol{e}_2 + \tau_{i3} \boldsymbol{e}_3. \qquad (4.41)$$

4.4. Balance of Total Energy

As anticipated in Section 2.3, the first law of thermodynamics is valid for a moving fluid provided we take the variation of kinetic energy into account. Therefore, the energy balance refers to the variation of the total energy, i.e. of the sum of the internal and kinetic energies of a material volume of fluid, and can be formulated as follows: *the variation in unit time of the total energy of a material volume of fluid is equal to the sum of the work done in unit time by the forces acting on the volume and of the heat introduced in unit time through its bounding surface.*

This is expressed in mathematical terms by the equation

$$\frac{D}{Dt} \int_{v(t)} \rho(e + V^2/2) dv = W_f + W_p + W_\tau + Q, \qquad (4.42)$$

where e is the internal energy per unit mass, W_f is the work done in unit time by the body forces, W_p and W_τ are the works done in unit time by the surface forces connected respectively with the isotropic and the deviatoric parts of the stress tensor, and Q is the heat introduced in unit time inside the considered volume of fluid.

The above-defined quantities may now be explicitly written as follows:

$$W_f = \int_v \rho \boldsymbol{f} \cdot \boldsymbol{V} dv, \qquad (4.43)$$

$$W_p = -\int_S p\boldsymbol{n} \cdot \boldsymbol{V} dS, \qquad (4.44)$$

$$W_\tau = \int_S \boldsymbol{\tau}_n \cdot \boldsymbol{V} dS, \qquad (4.45)$$

$$Q = -\int_S \boldsymbol{q} \cdot \boldsymbol{n} \, dS, \qquad (4.46)$$

where $\boldsymbol{q}$ is the *heat flux vector*, defined as the amount of heat that crosses the unit surface in unit time.

The work W_τ can be expressed in the form of a flux through the surface by considering that, thanks to the symmetry of the deviatoric part of the stress tensor, we have

$$\boldsymbol{\tau}_n \cdot \boldsymbol{V} = \sum_i (\boldsymbol{\tau}_n \cdot \boldsymbol{e}_i) u_i = \sum_i (\boldsymbol{\tau}_i \cdot \boldsymbol{n}) u_i = \sum_i u_i \boldsymbol{\tau}_i \cdot \boldsymbol{n}. \qquad (4.47)$$

Any form of the Reynolds transport theorem may now be used to express the left-hand side of the energy balance (4.42), and in the following we shall use relation (4.7), recalling once again that this implies the applicability of the differential form of the mass balance equation. We thus obtain the following integral form of the total energy balance:

$$\int_v \rho \frac{D}{Dt}(e + V^2/2) dv = \int_v \rho \boldsymbol{f} \cdot \boldsymbol{V} dv - \int_S p\boldsymbol{V} \cdot \boldsymbol{n} \, dS$$
$$+ \sum_i \int_S u_i \boldsymbol{\tau}_i \cdot \boldsymbol{n} \, dS - \int_S \boldsymbol{q} \cdot \boldsymbol{n} \, dS. \qquad (4.48)$$

This is clearly a scalar equation. If we now use the divergence theorem and the localization lemma, we obtain the following differential form of the energy balance, in which the material derivative is written explicitly:

$$\rho\left[\frac{\partial}{\partial t}(e + V^2/2) + \boldsymbol{V} \cdot \mathrm{grad}(e + V^2/2)\right]$$
$$= \rho\boldsymbol{f} \cdot \boldsymbol{V} - \mathrm{div}(p\boldsymbol{V}) + \sum_i \mathrm{div}(u_i\boldsymbol{\tau}_i) - \mathrm{div}\,\boldsymbol{q}. \tag{4.49}$$

Before proceeding in our exposition, it may be expedient to introduce the derivation of a different integral form of the energy balance equation, which may be very useful for the solution of particular problems, for instance the analysis of the motion in ducts.

To this purpose, we first rewrite the integral energy balance using form (3.41) of the Reynolds transport theorem, so that the flux of the total energy through the bounding surface appears:

$$\int_v \frac{\partial \rho(e + V^2/2)}{\partial t}dv + \int_S \rho(e + V^2/2)\boldsymbol{V} \cdot \boldsymbol{n}\,dS$$
$$= \int_v \rho\boldsymbol{f} \cdot \boldsymbol{V}dv - \int_S p\boldsymbol{V} \cdot \boldsymbol{n}\,dS + \int_S \boldsymbol{\tau}_n \cdot \boldsymbol{V}dS - \int_S \boldsymbol{q} \cdot \boldsymbol{n}\,dS. \tag{4.50}$$

Let us now assume that the body forces are *conservative*, i.e. that they may be derived from a scalar potential, say $-\Psi(\boldsymbol{x})$, which is a function of the position vector only, so that we may write

$$\boldsymbol{f} = -\mathrm{grad}\,\Psi. \tag{4.51}$$

The work of the body forces in unit time may then be expressed as

$$\rho\boldsymbol{f} \cdot \boldsymbol{V} = -\rho\boldsymbol{V} \cdot \mathrm{grad}\,\Psi = -\rho\mathrm{div}(\Psi\boldsymbol{V}) + \rho\Psi\mathrm{div}\,\boldsymbol{V}. \tag{4.52}$$

The last term in (4.52) may be manipulated by using the mass balance differential equation in the form (4.5c), so that we get

$$\rho\Psi\mathrm{div}\,\boldsymbol{V} = -\Psi\left(\frac{\partial\rho}{\partial t} + \boldsymbol{V} \cdot \mathrm{grad}\rho\right) = -\frac{\partial(\rho\Psi)}{\partial t} - \Psi\boldsymbol{V} \cdot \mathrm{grad}\rho, \tag{4.53}$$

where the fact that Ψ is not dependent on time has also been considered.

By substitution of (4.53) into (4.52) we then obtain

$$\rho\boldsymbol{f} \cdot \boldsymbol{V} = -\rho\mathrm{div}(\Psi\boldsymbol{V}) - \frac{\partial(\rho\Psi)}{\partial t} - \Psi\boldsymbol{V} \cdot \mathrm{grad}\rho$$
$$= -\frac{\partial(\rho\Psi)}{\partial t} - \mathrm{div}(\rho\Psi\boldsymbol{V}). \tag{4.54}$$

By using this relation and the divergence theorem, the first integral on the right-hand side of Eq. (4.50) may then be recast in the form

$$\int_v \rho \boldsymbol{f} \cdot \boldsymbol{V} \, dv = -\int_v \frac{\partial(\rho \Psi)}{\partial t} dv - \int_v \text{div}(\rho \Psi \boldsymbol{V}) dv$$
$$= -\int_v \frac{\partial(\rho \Psi)}{\partial t} dv - \int_S \rho \Psi \boldsymbol{V} \cdot \boldsymbol{n} \, dS. \tag{4.55}$$

By introducing this expression into Eq. (4.50) and reorganizing the various terms, we finally obtain

$$\int_v \frac{\partial \rho(e + V^2/2 + \Psi)}{\partial t} dv + \int_S \rho(e + p/\rho + V^2/2 + \Psi)\boldsymbol{V} \cdot \boldsymbol{n} \, dS$$
$$= \int_S \boldsymbol{\tau}_n \cdot \boldsymbol{V} dS - \int_S \boldsymbol{q} \cdot \boldsymbol{n} \, dS = W_\tau + Q. \tag{4.56}$$

This form of the energy balance equation is particularly useful when the steady adiabatic motion in a duct is analysed and a control volume is chosen such that it is bounded by the inner surface of the duct and by two consecutive duct cross-sections. In that case, the first and last terms are zero, while the work W_τ may normally be neglected, because the corresponding integral is zero on the duct surface (due to the boundary condition $\boldsymbol{V} = 0$) and on the duct cross-sections it is generally negligible compared to the work done by the pressure forces (which appears in the second left-hand side term).

4.5. Constitutive Equations

4.5.1. *The need for constitutive equations*

By expressing the balances of mass, momentum and energy in mathematical terms we have been able to derive, either in integral or in differential form, five scalar equations: one from the mass balance, three from the momentum balance and one from the energy balance. We have also seen that these equations may be written in slightly different but substantially equivalent ways, according to the form of the Reynolds transport theorem used for their derivation. As already pointed out in Section 4.1, the laws that have been expressed mathematically to derive this set of equations are assumed to be adequate to describe the dynamics of *any* deformable continuum. In other words, no relations that refer to the particular properties of fluids, discussed in Chapter 2, have been introduced so far. It is then appropriate to wonder whether these further relations are indeed necessary.

To this end, we must first identify the unknown functions that are present in the equations. For instance, by examining Eqs. (4.5), (4.26) and (4.49) we obtain the following list of unknown functions: the density ρ, the three components of the velocity vector V, the six independent components of the symmetric stress tensor T, the internal energy per unit mass e, and the three components of the heat flux vector q. Note that we have not introduced the isotropic and deviatoric stress tensors separately because they are derived directly from T by using the mechanical definition of pressure (4.29) and relation (4.30).

In summary, there are 14 unknown scalar functions, and thus the five available equations are obviously insufficient to solve the problem. Therefore, as already discussed in Section 4.1, it is necessary to introduce further relations specifying the characteristic behaviour of the continua we are dealing with, namely fluids. In other words, we need to write appropriate *constitutive equations*, so that we finally end up with a mathematical problem described by the same number of equations and unknowns.

It should be emphasized that, by doing so, we are actually restricting the validity of the derived set of equations to continuous media whose behaviour is satisfactorily described, with a level of approximation that is acceptable for our purposes, by the introduced constitutive equations. Actually, we shall see that even the constitutive equations we are going to introduce may not be considered completely satisfactory for *all* fluids (as defined in Section 2.1), but only for a certain class of fluids. Fortunately, this class is a wide-ranging one, so that most fluids of common interest may be considered to fall within it.

4.5.2. *Thermodynamic relations and Fourier law*

As already discussed briefly in Section 2.3, we assume that a local thermodynamic equilibrium applies for a fluid in motion or, in other words, that a fluid particle moving in a flow in which neither velocity nor temperature are uniform may be considered to pass through a succession of states in which the departure from equilibrium is negligibly small. This assumption is acceptable under quite general conditions and implies that the usual quantities introduced in thermodynamics may be safely used also in fluid dynamics. We have already taken advantage of this fact in the previous sections, where we have dealt with concepts such as density and internal energy, which are known to be thermodynamic state functions.

However, one might observe that the above is not completely true. The definition of density as the ratio between the mass and the instantaneous volume of a fluid particle does not require the introduction of the concept of equilibrium. Furthermore, in writing the energy balance equation, we have used the first law of thermodynamics, although we modified it with the addition of the kinetic energy, which is a mechanical quantity. As anticipated in Section 2.3, this law is effectively a definition of the difference between the internal energy per unit mass of a fluid particle in two different equilibrium states, linking it to the work done on the particle and the heat added to it. In principle, the latter are 'observable' quantities, whose definitions are independent of the existence or not of equilibrium. Therefore, as suggested by Batchelor (1967), we may define the internal energy per unit mass e of a fluid particle at any instant by means of the first law, if we specify that the equilibrium state to which e refers instantaneously is achieved by suddenly isolating the particle from the surrounding fluid and allowing it to come to equilibrium without work being done on it and without introduction of heat. Nevertheless, local equilibrium will certainly be assumed in an explicit form in the following, because thermodynamic equations of state will be used not only to define new equilibrium quantities, but also to introduce constitutive equations to complement the equations of motion derived in the previous sections.

The first equation of this type is the one linking density, internal energy and temperature, which may formally be written as

$$e = e(\rho, T). \tag{4.57}$$

Note that, if this relation is seen as a constitutive equation for the internal energy, it does not actually decrease the number of unknown functions in the equations of motion because it introduces another quantity, namely the absolute temperature T, which was not present before. In practice, Eq. (4.57) simply implies a change of one unknown function with another one. Nonetheless, temperature is indeed usually introduced, both because of its direct measurability, and because other quantities will be linked to it through further constitutive equations.

In particular, the heat flux vector is assumed to be connected with the variation in temperature through the *Fourier law*:

$$q = -k\text{grad}T, \qquad (4.58)$$

where k is the *coefficient of thermal conductivity*, already introduced in Section 2.7, and is, in general, a function of the thermodynamic state, but, in most practical cases, only of temperature and of the type of fluid.

This relation adequately describes heat transport through *conduction* within the fluid and between the fluid and a solid wall. Perhaps it is appropriate to recall here that, should other types of heat transport or sources of heat be present in the flow (for instance, radiation or exothermal reactions), they should be explicitly inserted in the energy balance equation through further terms.

Other thermodynamic functions may be introduced by means of equations of state similar to (4.57), and in particular the *entropy* per unit mass S and the *enthalpy* per unit mass h. In the following section the balance equations for these quantities will also be derived. However, we may anticipate that these equations cannot be added to the set of equations of motion obtained in the previous sections, but only used as an alternative for the total energy balance, from which we shall actually derive them.

In order to proceed in the derivation of appropriate constitutive equations, an essential quantity that must be introduced now is the *thermodynamic pressure* or *equilibrium pressure*, which we denote here by the symbol p_e to distinguish it from the mechanical pressure defined by relation (4.29). This pressure is the one that corresponds, through a thermodynamic equilibrium equation of state, to the local values of density and internal energy (or temperature) that may be instantaneously associated with a fluid particle. In other words, it would be the value of the pressure if the particle were at rest, in equilibrium conditions, in a fluid with those values of density and temperature.

The equation of state defining the equilibrium pressure may then be formally written, for instance, as

$$p_e = p_e(\rho, T). \qquad (4.59)$$

The difference between the mechanical pressure p and the equilibrium pressure p_e will be discussed at length in the following sections, in connection with the derivation of the constitutive equations for the stress tensor. In particular, we shall provide a detailed description of the assumptions required for these two pressures to coincide, pointing out their physical significance and their limits of applicability.

4.5.3. *Stress tensor of Newtonian viscous fluids*

The derivation of the relations linking the stress tensor with the velocity field is one of the core issues in the mathematical treatment of fluid dynamics. These relations must express, with sufficient accuracy, the physical properties and the behaviour of fluids, whose main features have already been briefly reviewed in Chapter 2.

Nonetheless, it is practically impossible to devise relations that are suitable for all types of fluids, so that different relations are used to describe the behaviour of different classes of fluids. In the following we shall introduce constitutive equations that are applicable to a particularly wide-ranging class, so that the motion of most commonly encountered fluids may be analysed. However, the derivation will also give clues to understanding the possible modifications that must be introduced to deal with fluids that show different behaviour.

The starting point of the analysis is another decomposition of the stress tensor $\mathbf{T}$, alternative to the one defined by relation (4.30), namely

$$\mathbf{T} = -p_e \mathbf{I} + \boldsymbol{\tau}_v. \qquad (4.60)$$

In this new decomposition, p_e is the equilibrium pressure defined as a function of density and temperature through the thermodynamic relation (4.59). Thus, the tensor $-p_e \mathbf{I}$, which obviously corresponds to equal normal stresses whatever the reference system, coincides with the stress tensor that would act on the particle if it belonged to a fluid at rest with the same values of density and temperature that characterize the particle when it is actually moving with a certain velocity. Therefore, it depends on motion only through the values of density and temperature. On the other hand, tensor $\boldsymbol{\tau}_v$ is completely dependent on motion, and derives from the momentum transport mechanisms at the molecular level that we have seen in Section 2.6 to be the basis of viscosity. For this reason, it will be called, from now on, the *viscous stress tensor*.

It must be emphasized that, while tensor $-p_e \mathbf{I}$ is clearly isotropic by construction, there is no reason to assume at this stage that, with the above decomposition, $\boldsymbol{\tau}_v$ is a deviatoric tensor, i.e. a tensor whose trace is zero. Therefore, $\boldsymbol{\tau}_v$ generally does not coincide with the tensor $\boldsymbol{\tau}$ of decomposition (4.30), and the recognition of the conditions for this coincidence to occur is actually one of the important issues that will be discussed in the following.

The derivation of the constitutive equations for the viscous stress tensor that we shall follow rests upon certain assumptions on the dependence of the complete stress tensor $\mathbf{T}$ on the local velocity of deformation of an elementary volume of fluid, which has been analysed in Section 3.4. These assumptions can be traced back to the work of Sir George Stokes (1845), and may be formulated as follows

1. $\mathbf{T}$ is a continuous function of the rate of strain tensor $\mathbf{E}$, and is independent of all the remaining kinematical quantities.
2. When $\mathbf{E} = 0$, $\mathbf{T}$ reduces to $-p_e\mathbf{I}$.
3. $\mathbf{T}$ does not depend explicitly on the position $\boldsymbol{x}$ (spatial homogeneity of the fluid).
4. There are no preferential directions in space (isotropy of the fluid).

We now discuss briefly the significance of these assumptions. The first one implies that a change of shape or volume of a fluid element must occur in unit time for a stress to be produced over its surface. Consequently, a rigid rotation, linked to the anti-symmetrical part, $\boldsymbol{\Omega}$, of the tensor grad $\boldsymbol{V}$ defined by (3.16), does not contribute to the stress tensor, because it only produces a change of orientation of an elementary volume of fluid. Note, however, that the independence of $\mathbf{T}$ from $\boldsymbol{\Omega}$ may also be seen as a consequence of the fact that the stress tensor would otherwise depend on the chosen positive direction of the rotations.

Assumption 2 is actually the basis of decomposition (4.60), and directly connects the viscous stress tensor, $\boldsymbol{\tau}_v$ with the motion of the fluid and, in particular, with the velocity of the deformation it produces on an elementary volume of fluid.

Assumption 3 expresses the fact that the fluid is homogeneous, i.e. that there are no differences in its characteristics according to the spatial position. On the other hand, assumption 4 refers to the fact that there are no preferential directions in the fluid, i.e. the viscous stress tensor does not depend on the chosen reference system.

A fluid satisfying all the above assumptions is usually called a Stokesian fluid. However, in this book we almost exclusively consider the so-called *Newtonian fluids*, which also satisfy a further condition, namely that the dependence of $\boldsymbol{\tau}_v$ on $\mathbf{E}$ be *linear*.

It may be demonstrated (see, e.g., Serrin, 1959, or Batchelor, 1967) that the most general form of the viscous stress tensor of a fluid satisfying all the above-described conditions is the following:

$$\tau_v = (\lambda \operatorname{div} \boldsymbol{V})\mathbf{I} + 2\mu \mathbf{E}. \tag{4.61}$$

In this relation (in which the multiplying factor 2 is introduced only for convenience) λ and μ are two coefficients that depend, in principle, on the thermodynamic state of the fluid. The coefficient μ is the *dynamic viscosity coefficient* (or, simply, *viscosity coefficient*) already introduced in Section 2.6, while the coefficient λ is usually called the *second viscosity coefficient*.

In order to achieve a deeper comprehension of the link between the viscous stress tensor and the various components of deformation connected with the rate of strain tensor, it may now be useful to express tensor $\mathbf{E}$ as the sum of its isotropic and deviatoric parts, $\mathbf{A}$ and $\mathbf{D}$, so that, with the help of (3.20) and (3.24), we get

$$\tau_v = (\lambda \operatorname{div} \boldsymbol{V})\mathbf{I} + 2\mu(\mathbf{A} + \mathbf{D}) = \left(\lambda + \frac{2}{3}\mu\right)(\operatorname{div} \boldsymbol{V})\mathbf{I} + 2\mu \mathbf{D}. \tag{4.62}$$

Let us now introduce a new coefficient, κ, the *bulk viscosity coefficient* (or, sometimes, *volume viscosity coefficient*), which is defined as

$$\kappa = \left(\lambda + \frac{2}{3}\mu\right). \tag{4.63}$$

With this definition, we may rewrite (4.62) as

$$\tau_v = (\kappa \operatorname{div} \boldsymbol{V})\mathbf{I} + 2\mu \mathbf{D}. \tag{4.64}$$

As can be appreciated, the viscous stress tensor has, in general, an isotropic part and a deviatoric part, which are respectively expressed by the two terms appearing in (4.64).

From (4.60) and (4.64), the complete stress tensor becomes

$$\mathbf{T} = (-p_e + \kappa \operatorname{div} \boldsymbol{V})\mathbf{I} + 2\mu \mathbf{D}. \tag{4.65}$$

Therefore, we can now compare the two different decompositions of the stress tensor we have introduced with relations (4.30) and (4.60). In effect, the isotropic part of tensor $\mathbf{T}$ is now

$$-p\mathbf{I} = (-p_e + \kappa \operatorname{div} \boldsymbol{V})\mathbf{I}, \tag{4.66}$$

while the deviatoric part of $\mathbf{T}$ turns out to be

$$\tau = 2\mu\mathbf{D} = 2\mu(\mathbf{E} - \mathbf{A}). \tag{4.67}$$

We can thus identify the term characterizing the isotropic part of the viscous stress tensor, $\kappa\mathrm{div}\,\mathbf{V}$, as the difference between the thermodynamic pressure, p_e, and the mechanical pressure, p, defined by (4.29). Furthermore, as may be readily deduced from the description of the deformation reported in Section 3.4, this is the only viscous term linked to the isotropic dilatation of an elementary volume of fluid.

By introducing in (4.65) the components of tensor $\mathbf{D}$ given in (3.22), the components of the complete stress tensor of a Newtonian fluid may be written explicitly as

$$T_{ik} = (-p_e + \kappa\mathrm{div}\,\mathbf{V})\delta_{ik} + \mu\left(\frac{\partial u_i}{\partial x_k} + \frac{\partial u_k}{\partial x_i}\right) - \frac{2}{3}\mu\mathrm{div}\,\mathbf{V}\delta_{ik}. \tag{4.68}$$

Most normally encountered fluids, such as almost all gases (including air) and liquids like water, oil or gasoline, satisfy the above assumptions defining a Newtonian fluid. However, there are certain particular types of fluid that do not fall within this category. For instance, polymers may be non-isotropic due to the existence of particular preferential directions of their chain-like molecules. Certain emulsions and mixtures may fail to be homogeneous, and there are fluids for which the relation between τ_v and $\mathbf{E}$ is non-linear, with either an increasing or a decreasing rate of variation of the stresses with increasing rate of strain. All these different types of fluid are termed *non-Newtonian fluids*, and although some of them may be of significant practical importance, they will not be further considered in this book. Nonetheless, we point out once again that, should it be required to study their dynamical behaviour, all the general equations described in the previous sections from 4.2 to 4.4 still apply, even if different constitutive equations for the stress tensor should be used.

4.5.4. *Stokes' hypothesis*

As already observed, the viscous stress term $\kappa\,\mathrm{div}\,\mathbf{V}$ may be physically interpreted as the difference between the thermodynamic pressure and the mechanical pressure, defined as the opposite of the average of the normal stresses acting on any three orthogonal planes passing through a point in the fluid. This difference is usually considered to be due to the time lag with which the thermodynamic equilibrium condition is reached in a motion that implies an isotropic dilatation of a fluid element.

However, following a suggestion by Stokes (1845), it is customary to assume that κ is negligible or, alternatively, that the two coefficients of viscosity are linked by the relation

$$\lambda = -\frac{2}{3}\mu. \tag{4.69}$$

This is known as *Stokes' hypothesis*, and its use has become common practice in the analysis of the motion of compressible fluids. In effect, it renders the mathematical treatment considerably easier; in particular, as may be seen from relations (4.64)–(4.67), it implies that the isotropic part of the stress tensor is completely determined by the thermodynamic pressure and that its deviatoric part coincides with the viscous stress tensor. In other words, to assume the validity of relation (4.69) is equivalent to stating that isotropic dilatations of an elementary volume of fluid do not produce viscous stresses.

The use of this hypothesis has been the object of long-lasting discussions (see, for instance, the papers presented at the workshop chaired by Rosenhead, 1954, or the discussion by Gad-el-Hak, 1995). In effect, it may be seen from the kinetic theory of gases that, for a monatomic gas, $\kappa = 0$; however, for the more interesting case of polyatomic gases, the available data for κ, though not numerous due to the complexity of its experimental evaluation, show that κ is certainly not zero and actually often far from negligible. For instance, κ is of the same order as μ for nitrogen and oxygen (the main components of air), but other gases, such as carbon dioxide, are characterized by much larger values of κ, of the order of $10^3 \mu$. Furthermore, the bulk viscosity is also significantly higher than the dynamic viscosity for many liquids (see, e.g., Rosenhead, 1954, and Dukhin and Goetz, 2009); as an example, the value of κ/μ for water is approximately 2.5.

Therefore, such evaluations of the order of magnitude of the bulk viscosity seem to be in contradiction with the fact that, excluding very particular conditions, in most applications good results may indeed be obtained using Stokes' hypothesis. We are thus left with the rather puzzling situation in which one of the viscosity coefficients appearing in relation (4.64) may be neglected, but not the other one, which may be of the same order of magnitude or even much smaller.

A reasonable explanation of this circumstance can be given by analysing, in more detail, the contribution of $\kappa \operatorname{div} \boldsymbol{V}$ to the stress tensor, and the admissibility of the operation that is done when a term is neglected in a relation or in an equation. To this end, it is useful to rewrite the expressions of the generic tangential and normal components of the complete stress tensor separately, and to recall the physical interpretation of the elementary components of the rate-of-strain tensor, described in Section 3.4.

Tangential stresses $(i \neq k)$

$$T_{ik} = \mu \left(\frac{\partial u_i}{\partial x_k} + \frac{\partial u_k}{\partial x_i} \right) = 2\mu D_{ik}.$$

Normal stresses $(i = k)$

$$T_{ii} = (-p_e + \kappa \text{div}\,\mathbf{V}) + 2\mu \left(\frac{\partial u_i}{\partial x_i} - \frac{1}{3} \text{div}\,\mathbf{V} \right) = (-p_e + 3\kappa A_{ii}) + 2\mu D_{ii}.$$

The essential point to be observed now is that the terms A_{ii} are all equal, while this cannot happen for the terms D_{ii}, because tensor $\mathbf{D}$ is the deviatoric part of the rate of strain tensor $\mathbf{E}$, and thus its trace is zero.

Thus, the effect of the term $\kappa \text{div}\,\mathbf{V}$ is perfectly additive to that of the thermodynamic pressure, at variance with what happens for the terms of the normal stresses in which the viscosity coefficient, μ, appears. In other words, we may say that $\kappa \text{div}\,\mathbf{V}$ is associated with the same deformation, i.e. isotropic dilatation of a fluid element, that is connected with the thermodynamic pressure; however, the latter is generally higher than $\kappa \text{div}\,\mathbf{V}$ (in absolute value) by several orders of magnitude.

Conversely, the coefficient μ, however small it may be, is associated with stresses causing deformations that cannot be justified without taking viscosity into account, namely pure strain (i.e. non-isotropic normal deformation without change of volume) and pure distortion.

Therefore, it seems more correct to say that Stokes' hypothesis does not imply that $\kappa = 0$, but rather that the absolute value of $\kappa \text{div}\,\mathbf{V}$ is negligible with respect to the thermodynamic pressure. Actually, the rationale for this different statement of Stokes' hypothesis stands upon the obvious fact that a term appearing in a relation or equation cannot be neglected just because it is small, but *only if it is small compared to another one that has an effect of the same type*, i.e. which is *qualitatively similar*.

If this different point of view is adopted, there are good reasons for Stokes' hypothesis to be a largely acceptable approximation. As a matter of fact, only in very particular conditions will the term $\kappa \text{div}\,\mathbf{V}$ be comparable to the thermodynamic pressure. This may happen, for instance, when the fluid is characterized by large values of κ (e.g. carbon dioxide) *and* the motion is such that extremely large values of div V occur, as for instance in hypersonic flows (i.e. flows at Mach numbers larger than, say, 5) or in flows through shock waves (see Emanuel, 1992, and Emanuel and Argrow, 1994).

Nonetheless, there is at least one other situation in which, although the above particular conditions do not occur, Stokes' hypothesis leads to results that are in contradiction with experimental evidence. In effect, as we shall see in the next section when discussing the internal energy balance equation, it is necessary to reintroduce the viscous stresses associated with an isotropic dilatation to justify the damping of acoustic waves at very high frequencies (*ultrasounds*). In fact, measuring the attenuation of these waves is one of the few experimental means of estimating the bulk viscosity.

Having discussed the physical interpretation and the limits of applicability of Stokes' hypothesis, we shall adopt it throughout this book. Therefore, unless otherwise specified, it will be assumed that the mechanical pressure defined in (4.29) coincides with the thermodynamic pressure given by (4.59), and the symbol p will be used to indicate it. Consistently, the viscous stress tensor is assumed to coincide with the deviatoric stress tensor, and will be indicated with the symbol τ. The fundamental equations of the momentum and energy balances, derived in Sections 4.3 and 4.4, will then remain formally unchanged, with the new meaning of the relevant symbols.

With these assumptions and definitions, the full stress tensor and its components become

$$\mathbf{T} = -p\mathbf{I} + \tau = -p\mathbf{I} + 2\mu\mathbf{D}, \tag{4.70}$$

$$T_{ik} = -p\delta_{ik} + \tau_{ik} = -p\delta_{ik} - \frac{2}{3}\mu \text{div } \boldsymbol{V}\delta_{ik} + \mu\left(\frac{\partial u_i}{\partial x_k} + \frac{\partial u_k}{\partial x_i}\right). \tag{4.71}$$

4.6. Supplementary Equations

4.6.1. *Introductory remarks*

The equations of motion and the constitutive equations derived in the previous sections, which express in mathematical terms the physical laws valid for all continua and describe the particular behaviour of the fluids we shall be dealing with, are sufficient to formulate a problem in which the number of equations is equal to the number of unknowns.

Nonetheless, further equations, expressing the balances of other quantities of interest for a fluid in motion, may be advantageously used in the solution of particular problems. These equations, which may all be derived from the fundamental ones, refer to kinetic energy, internal energy, entropy, enthalpy and total enthalpy, and one scalar equation is obtained for each of these quantities. With the exception of the kinetic energy equation, which is directly derived from the momentum balance equation, any of these supplementary equations may be used in place of the total energy balance equation.

In the following, these further equations will be obtained in their differential form through straightforward manipulations of the same form of the fundamental balance equations. The corresponding integral forms may be easily derived using analogous procedures or by integrating the relevant differential equations over a generic volume of fluid.

4.6.2. *Kinetic energy balance*

The differential equation expressing the balance of kinetic energy per unit mass, $V^2/2$, may be obtained in a straightforward manner by multiplying scalarly the momentum equation by V.

Therefore, starting from Eq. (4.39) and using the material derivative operator, we get

$$\rho V \cdot \frac{DV}{Dt} = \rho f \cdot V - V \cdot \operatorname{grad} p + V \cdot \operatorname{div} \tau, \qquad (4.72)$$

which may be rewritten as

$$\rho \frac{DV^2/2}{Dt} = \rho f \cdot V - V \cdot \operatorname{grad} p + \sum_i u_i \operatorname{div} \tau_i. \qquad (4.73)$$

We thus find that the variation of kinetic energy is connected with the whole work done, in unit time, by the body forces acting on an elementary volume of fluid, and with only a part of the work done by the surface forces acting over its bounding surface, namely the work of the resultants (per unit volume) of the pressure and viscous stresses.

As already pointed out, the above form of the kinetic energy balance, being directly derived from the momentum balance equation, does not introduce any further physical content with respect to that equation.

Therefore, although sometimes useful, it cannot be used as an alternative to the total energy balance. Conversely, for that purpose we may use the balance equations derived in the following sections, which are all obtained — either directly or indirectly — from the total energy balance, and thus share its physical significance. Nonetheless, it will be seen that Eq. (4.73) is also useful for deriving those equations.

4.6.3. *Internal energy balance*

Once the kinetic energy equation is available, it may be subtracted from the total energy balance equation to obtain the differential equation for the balance of internal energy.

Therefore, if we use Eqs. (4.49) and (4.73), and write in explicit form the total work of the surface stresses, we get

$$
\rho \frac{D(e + V^2/2)}{Dt} - \rho \frac{D(V^2/2)}{Dt} = \rho \boldsymbol{f} \cdot \boldsymbol{V} - p\mathrm{div}\,\boldsymbol{V} - \boldsymbol{V} \cdot \mathrm{grad}\,p
$$
$$
+ \sum_i u_i \mathrm{div}\,\boldsymbol{\tau}_i + \sum_i \boldsymbol{\tau}_i \cdot \mathrm{grad}\,u_i - \mathrm{div}\,\boldsymbol{q}
$$
$$
- \rho \boldsymbol{f} \cdot \boldsymbol{V} + \boldsymbol{V} \cdot \mathrm{grad}\,p - \sum_i u_i \mathrm{div}\,\boldsymbol{\tau}_i,
$$

from which we obtain the following differential equation expressing the internal energy balance:

$$
\rho \frac{De}{Dt} = -p\mathrm{div}\,\boldsymbol{V} + \sum_i \boldsymbol{\tau}_i \cdot \mathrm{grad} u_i - \mathrm{div}\,\boldsymbol{q}. \tag{4.74}
$$

This equation is usually recast in a more concise form by introducing the *dissipation function*, Φ, defined as

$$
\Phi = \sum_i \boldsymbol{\tau}_i \cdot \mathrm{grad}\,u_i = \boldsymbol{\tau} : \mathrm{grad}\,\boldsymbol{V}, \tag{4.75}
$$

so that we finally have

$$
\rho \frac{De}{Dt} = -p\,\mathrm{div}\,\boldsymbol{V} + \Phi - \mathrm{div}\,\boldsymbol{q}. \tag{4.76}
$$

It may be useful to express the dissipation function in explicit form, by using the components of the viscous stress tensor derived from (4.71) and a reference system whose axes are (x, y, z) and in which the corresponding components of the velocity vector are u, v and w:

$$\Phi = -\frac{2}{3}\mu(\text{div } \boldsymbol{V})^2 + 2\mu \left[\left(\frac{\partial u}{\partial x}\right)^2 + \left(\frac{\partial v}{\partial y}\right)^2 + \left(\frac{\partial w}{\partial z}\right)^2 \right]$$
$$+ \mu \left[\left(\frac{\partial u}{\partial y} + \frac{\partial v}{\partial x}\right)^2 + \left(\frac{\partial v}{\partial z} + \frac{\partial w}{\partial y}\right)^2 + \left(\frac{\partial w}{\partial x} + \frac{\partial u}{\partial z}\right)^2 \right]. \tag{4.77}$$

The comparison of the kinetic energy and internal energy equations provides a clear characterization of the role of the various parts of the work done in unit time by the surface forces. In effect, as we have already pointed out, the variation of the kinetic energy is due to the work done in unit time by the vectors grad p and div $\boldsymbol{\tau}$, which represent the *resultants* per unit volume of the surface pressure and viscous stresses acting on the elementary volume of fluid that is moving with velocity $\boldsymbol{V}$. Conversely, the variation of the internal energy is caused by the work done by the surface forces due to the deformation of the volume element. This can be understood by realizing that the single forces acting on each element of surface bounding an elementary volume of fluid are displaced by the deformation of this volume that occurs in unit time.

In more detail, the dissipation function Φ represents the work done in unit time on an elementary volume of fluid by the viscous stresses acting over its surface due to the non-isotropic part of the deformation, namely the straining and the distortion of the volume element. Furthermore, it corresponds to the *irreversible* (or *dissipative*) part of the work done in unit time by the surface stresses. This is due to the fact that, as will become clearer from the analysis of the entropy balance equation, Φ is non-negative, which thus implies that any motion with space variations of velocity is connected with an increase in internal energy.

On the other hand, the term $-p \, \text{div } \boldsymbol{V}$, also appearing in (4.76), represents the work done in unit time by the pressure forces due to the isotropic part of the deformation, i.e. the isotropic dilatation of the volume element. Now, having assumed the validity of Stokes' hypothesis (which implies that p coincides with the thermodynamic pressure), this term corresponds to the *reversible* (or *non-dissipative*) part of the work done in unit time by the surface stresses.

To better clarify this statement, let us assume that the only deformation that occurs is precisely an isotropic dilatation. In this case, the sign of the variation of internal energy that is associated with the term $-p \, \text{div } \boldsymbol{V}$ is determined by div $\boldsymbol{V}$, so that if, for instance, we have a cyclic variation of volume — which corresponds to a variation of div V between opposite values — the internal energy will remain unaltered in an entire cycle of volume fluctuation.

As a corollary of the above interpretations, we can now explain why, as anticipated, Stokes' hypothesis does not allow the damping of high-frequency acoustic waves to be justified. An acoustic wave may be represented by — or, in any case, be associated with — an oscillatory isotropic change of volume between opposite values which, from the above analysis, would not imply any increase in internal energy or, in other words, any dissipation. However, if we do not use Stokes' hypothesis, then it is easily seen that a further term $\kappa(\text{div } V)^2$ would appear in Eq. (4.76). Considering that κ cannot be negative (as may be shown from the entropy equation without application of Stokes' hypothesis), it is evident that this additional term is always associated with an increase in internal energy, and thus with dissipation. Obviously, if $\kappa \text{div } V$ is small (as is usually the case in normal conditions), only acoustic waves at very high frequency will produce a sufficient number of cycles for this term to sum up, in an acceptable time interval, to a measurable value of increased internal energy.

4.6.4. *Entropy balance*

The differential equation expressing the balance of entropy may be obtained from the fundamental thermodynamic equation linking entropy with internal energy, pressure and density:

$$T dS = de + p dv, \tag{4.78}$$

where $v = 1/\rho$ is the specific volume.

In a flowing fluid, the variations appearing in (4.78) are those experienced by a fluid particle during its motion, so that they may be expressed through the material derivative operator. Using obvious manipulations and the continuity equation (4.5a), we thus obtain

$$T\frac{DS}{Dt} = \frac{De}{Dt} + p\frac{D(1/\rho)}{Dt} = \frac{De}{Dt} - \frac{p}{\rho^2}\frac{D\rho}{Dt} = \frac{De}{Dt} + \frac{p}{\rho}\text{div } V; \tag{4.79}$$

so that, from (4.76), we finally get the following form of the entropy balance equation:

$$\rho\frac{DS}{Dt} = \frac{1}{T}(\Phi - \text{div } q). \tag{4.80}$$

This equation shows that the variation of the entropy of a moving particle is essentially connected with the viscosity and the conductivity of the fluid. Furthermore, considering that the second law of thermodynamics requires that the entropy of an elementary volume of fluid cannot decrease in an adiabatic flow, Eq. (4.80) also implies that Φ is non-negative. As it can be easily derived from a closer analysis of expression (4.77), this is equivalent to saying that the viscosity coefficient μ must be non-negative, which, although physically reasonable, had not been explicitly assumed or demonstrated in the previous derivations.

It may be useful to express the link between the variations of internal energy and of entropy by means of the following relation, which is derived from Eqs. (4.76) and (4.80),

$$\rho\frac{De}{Dt} = \rho T\frac{DS}{Dt} - p\operatorname{div}\boldsymbol{V}. \tag{4.81}$$

4.6.5. *Enthalpy balance*

The enthalpy per unit mass (or *specific enthalpy*) is a thermodynamic state function defined as

$$h = e + p/\rho. \tag{4.82}$$

Therefore, the variation of enthalpy for a fluid particle may be expressed as a function of the variations of internal energy, pressure and density, so that, with the help of Eq. (4.5a) for the mass balance, we get

$$\begin{aligned}\rho\frac{Dh}{Dt} &= \rho\frac{De}{Dt} + \rho\frac{D(p/\rho)}{Dt} = \rho\frac{De}{Dt} + \frac{Dp}{Dt} - \frac{p}{\rho}\frac{D\rho}{Dt}\\ &= \frac{De}{Dt} + \frac{Dp}{Dt} + p\operatorname{div}\boldsymbol{V}.\end{aligned} \tag{4.83}$$

By introducing Eq. (4.76), we obtain

$$\rho\frac{Dh}{Dt} = \Phi - \operatorname{div}\boldsymbol{q} + \frac{Dp}{Dt}. \tag{4.84}$$

As we did for the internal energy, it may be expedient to express, by using Eqs. (4.80) and (4.84), the direct connection between the variations of enthalpy and entropy:

$$\rho\frac{Dh}{Dt} = \rho T\frac{DS}{Dt} + \frac{Dp}{Dt}. \tag{4.85}$$

4.6.6. *Total enthalpy balance*

The total enthalpy (per unit mass) is defined as

$$h_0 = h + V^2/2. \tag{4.86}$$

The balance equation for total enthalpy may then be derived either from (4.73) and (4.84), or directly from the total energy balance, similar to what has been done for the enthalpy balance:

$$\rho \frac{Dh_0}{Dt} = \rho \frac{D(e + V^2/2)}{Dt} + \frac{Dp}{Dt} - \frac{p}{\rho} \frac{D\rho}{Dt}. \tag{4.87}$$

Using Eq. (4.49) and the continuity equation, we then obtain, after simple manipulations,

$$\rho \frac{Dh_0}{Dt} = \rho \boldsymbol{f} \cdot \boldsymbol{V} + \sum_i \operatorname{div}(u_i \boldsymbol{\tau}_i) - \operatorname{div} \boldsymbol{q} + \frac{\partial p}{\partial t}. \tag{4.88}$$

As already pointed out in Section 4.6.1, although the supplementary balance equations were derived in the present section in differential form, it is useful to introduce here the integral form of the total enthalpy balance that derives from the particular total energy balance for conservative body forces expressed by Eq. (4.56). In effect, if in that equation the definitions of enthalpy and total enthalpy are used, and the term

$$\int_v \frac{\partial \rho(p/\rho)}{\partial t} dv$$

is added and subtracted on the left-hand side, we easily obtain

$$\int_v \frac{\partial \rho(h_0 + \varPsi)}{\partial t} dv + \int_S \rho(h_0 + \varPsi) \boldsymbol{V} \cdot \boldsymbol{n} \, dS$$
$$= \int_S \boldsymbol{\tau}_n \cdot \boldsymbol{V} dS - \int_S \boldsymbol{q} \cdot \boldsymbol{n} \, dS + \int_v \frac{\partial p}{\partial t} dv = W_\tau + Q + \int_v \frac{\partial p}{\partial t} dv. \tag{4.89}$$

As already highlighted in the discussion regarding Eq. (4.56), this integral form of the total enthalpy equation is especially advantageous when studying the motion of compressible fluids inside ducts. In effect, it will be used in Chapter 15, albeit in a simplified form, when the one-dimensional approximation will be applied to the analysis of duct flows.

4.7. Discussion on the Equations of Motion

4.7.1. *The Navier–Stokes equations*

The fundamental balance laws valid for all continuous media and the constitutive equations introduced in the previous sections allowed us to derive a set of equations that, in principle, close the problem of the mathematical description of the motion of Newtonian fluids. Unfortunately, it will be seen that this does not allow predictions to be made for generic flow problems, due to the intrinsic difficulties connected with the solution of the obtained system of equations.

In order to briefly discuss some of the mathematical aspects of the problem at hand, it may be useful to rewrite and renumber the complete set of available equations. To this end, we choose to express the energy balance through the internal energy equation, and we use the explicit form of the material derivative.

Fundamental equations

$$\frac{\partial \rho}{\partial t} + \boldsymbol{V} \cdot \text{grad}\rho + \rho \,\text{div}\,\boldsymbol{V} = 0, \tag{4.90}$$

$$\rho \left(\frac{\partial \boldsymbol{V}}{\partial t} + \boldsymbol{V} \cdot \text{grad}\boldsymbol{V} \right) = \rho \boldsymbol{f} - \text{grad}\,p + \text{div}\,\boldsymbol{\tau}, \tag{4.91}$$

$$\rho \left(\frac{\partial e}{\partial t} + \boldsymbol{V} \cdot \text{grad}\,e \right) = -p\,\text{div}\,\boldsymbol{V} + \Phi - \text{div}\,\boldsymbol{q}. \tag{4.92}$$

Constitutive equations

$$p = p(\rho, T), \tag{4.93}$$

$$e = e(\rho, T), \tag{4.94}$$

$$q = -k\text{grad}T, \tag{4.95}$$

$$\tau_{ik} = -\frac{2}{3}\mu\,\text{div}\,\boldsymbol{V}\delta_{ik} + \mu \left(\frac{\partial u_i}{\partial x_k} + \frac{\partial u_k}{\partial x_i} \right). \tag{4.96}$$

In (4.92) we also introduced the dissipation function, defined as

$$\Phi = \boldsymbol{\tau} : \text{grad}\,\boldsymbol{V}. \tag{4.97}$$

Equation (4.93) and the components of the viscous stress tensor $\boldsymbol{\tau}$ given in (4.96) characterize a Newtonian fluid with Stokes' hypothesis. Furthermore, recall that the coefficients k and μ in (4.95) and (4.96) are, in general, functions of the thermodynamic state but, in almost all cases, they may be considered to be functions of the type of fluid and of the absolute temperature T only. For the sake of completeness, it may be useful to explicitly write the most general form of the viscous force appearing in (4.91). If the validity of Stokes' hypothesis is not assumed, i.e. if Eq. (4.61) is used for the viscous stress tensor, the relevant expression is (see also Petrila and Trif, 2005)

$$\operatorname{div} \boldsymbol{\tau} = \mu \nabla^2 \boldsymbol{V} + (\lambda + \mu) \operatorname{grad} (\operatorname{div} \boldsymbol{V}) + (\operatorname{div} \boldsymbol{V}) \operatorname{grad} \lambda$$

$$+ 2(\operatorname{grad} \mu) \cdot (\operatorname{grad} \boldsymbol{V}) + (\operatorname{grad} \mu) \times \boldsymbol{\omega}, \qquad (4.98)$$

whereas with Stokes' hypothesis it becomes

$$\operatorname{div} \boldsymbol{\tau} = \mu \nabla^2 \boldsymbol{V} + \frac{1}{3}\mu \operatorname{grad} (\operatorname{div} \boldsymbol{V}) - \frac{2}{3}(\operatorname{div} \boldsymbol{V}) \operatorname{grad} \mu$$

$$+ 2(\operatorname{grad} \mu) \cdot (\operatorname{grad} \boldsymbol{V}) + (\operatorname{grad} \mu) \times \boldsymbol{\omega}. \qquad (4.99)$$

The above relations, where $\boldsymbol{\omega}$ is the vorticity vector defined by (3.17), are particularly expressive. In effect, they clearly highlight the simplifications that are obtained when the viscosity coefficients may be assumed to be constant or when the time variation of the volume of a fluid element, given by the term $\operatorname{div} \boldsymbol{V}$, may be neglected.

It may be interesting to recall here that Newton himself did not actually derive the mathematical expression of the stress–strain relations that we now consider as characterizing a Newtonian fluid. Nonetheless, in his *Principia Mathematica* he expressed in words, as a hypothesis, the fundamental physical idea on which those relations are based: '*The resistance which arises from the lack of lubricity in the parts of a fluid, other things being equal, is proportional to the velocity with which the parts of the fluid are separated from each other*' (Newton, 1687). It was only many years later that the equations of fluid motion were derived, using various physical assumptions, by Navier (1827), Poisson (1831), Saint-Venant (1843) and, finally, Stokes (1845), who also gave an account of the work by himself and by previous researchers in a report published one year later (Stokes, 1846).

Let us now analyse the equations of motion in more detail. After introducing the constitutive equations (4.93)–(4.96), and considering that the momentum balance (4.91) is a vector equation, Eqs. (4.90)–(4.92) form a system of five scalar partial differential equations, which are usually called the *Navier–Stokes equations* (although, from a historical point of view,

this denomination should more correctly refer to the momentum equation
only). The unknowns in this set of equations are the density, ρ, the three
components of the velocity vector, V, and the temperature, T. All these
quantities are, in general, functions of the space position vector, x, and of
time, t.

Suitable initial and boundary conditions must be provided for the
equations to properly define a mathematical problem. As for the initial
conditions, it may be shown (see, e.g., Zeytounian, 2001) that at the initial
time $t = 0$ the values of all the unknown functions must be given in the
whole fluid domain; in other words, we must assign the functions

$$\rho(x, 0) = \rho^*(x); \quad V(x, 0) = V^*(x); \quad T(x, 0) = T^*(x). \quad (4.100)$$

We now consider the boundary conditions to be assigned at a solid,
non-permeable wall. This might be the surface of a body moving in an
unlimited fluid domain, or the contour of a vessel containing moving fluid.
From the mathematical analysis of the Navier–Stokes equations it may be
shown that in this case the necessary boundary conditions concern the three
components of the velocity vector and the temperature.

In agreement with the discussion on the behaviour of fluids at solid
walls reported in Section 2.8, we assume that the fluid is in mechanical and
thermal equilibrium with the surface, where the velocity must then satisfy
both the non-penetration and the no-slip conditions. Therefore, if V_w is
the velocity of the solid wall, the velocity V of the fluid at points lying
over its surface is given by the relation

$$V - V_w = 0. \quad (4.101)$$

Analogously, the wall condition for temperature is

$$T - T_w = 0, \quad (4.102)$$

where T_w is the temperature of the wall.

Furthermore, when a body starts moving in an infinite domain of still
fluid, the motion of the fluid and the variations in time of the different
quantities in the flow field are completely due to the perturbations caused
by the moving body, which die away at infinite distance from its surface.
In this case, the unperturbed values of the unknown functions must also
be given; in particular, in a reference system fixed to the fluid, the velocity
must go to zero at infinity and the density and the temperature must tend
to their unperturbed values.

Other sets of boundary conditions must be provided when an interface exists between two different fluids, such as a liquid and a gas (water and air being the most usual situation) or between two immiscible liquids. In these cases the velocity must be continuous at the interface and proper relations must apply for the stress tensor, taking the *surface tension* of the liquids into account. Note that the problem may be further complicated when the position and shape of the interface surface is not known in advance, and is itself a solution of the problem. In this book we are not concerned with cases involving the presence of fluid interfaces, and the interested reader may refer, for instance, to the deeper discussions on the relevant boundary conditions reported by Batchelor (1967) and Zeytounian (2001).

Once the problem of the required initial and boundary conditions has been analysed, we may say that a system of equations that accurately describes the dynamics of many (or perhaps most) fluids of practical interest is available. However, this does not mean that we are able to find solutions of these equations in explicit mathematical terms (*closed-form solutions*) for generic problems, such as the prediction of the forces acting on bodies of any shape moving through a fluid. In fact, it may be readily seen that the problem of solving the Navier–Stokes equations for generic initial and boundary conditions is a formidable one indeed.

The main difficulties arise from the mathematical structure itself of Eqs. (4.90)–(4.92), which form a system of *coupled* and *non-linear* partial differential equations. The fact that the equations are coupled implies that all the unknowns derive from the simultaneous solution of the whole set of equations. However, the feature entailing the greatest mathematical complexity is certainly the non-linearity of the equations.

It may be useful to briefly recall that a differential equation is non-linear if products of n-order derivatives of the unknown functions (including the $n = 0$ case, i.e. the functions themselves) are present. If this is the case, the fundamental property of linear differential equations, namely that a linear combination of solutions is still a solution, does not apply. The main consequence is the impossibility of using many effective mathematical and numerical procedures that exploit this property.

A rapid analysis soon shows that several non-linear terms are indeed present in Eqs. (4.90)–(4.92). In particular, all the convective parts of the material derivatives, which appear on the left-hand side of all equations, are non-linear, and the same is clearly true for the dissipation function in the energy balance. Furthermore, density multiplies the material derivatives in the momentum and energy balances, and is actually the main quantity responsible for a strict coupling of all the equations. The constitutive equations are also a source of non-linear terms, because the coefficients of viscosity and thermal conductivity are (at least) functions of the temperature. In effect, as may be readily seen from relation (4.98), the fact that the viscosity coefficients are not constant is the only source of non-linearity in the viscous term appearing in the momentum equation.

The mathematical problem is actually so complex that even the existence and uniqueness of the solution of the Navier–Stokes equations for generic initial and boundary conditions has not yet been fully demonstrated, in spite of the considerable efforts made in recent years, particularly for the two-dimensional case (see, e.g., Zeytounian, 2001). However, it must be stressed that this does not mean that there are no solutions for these equations. A large number of solutions do exist, particularly for the incompressible flow case (see, e.g., the reviews by Berker, 1963, and Wang, 1989 and 1991, and the book by Drazin and Riley, 2006); however, only a limited number of them satisfy boundary conditions that may be of interest for applications, and we shall describe some of these solutions in Chapter 9.

All the above discussion naturally introduces the fundamental issue of the simplification either of the equations themselves or of the analysed problems involving the dynamics of fluids. This point deserves deeper analysis, which is the object of the following section.

4.7.2. *Simplified models*

By deriving the Navier–Stokes equations we have developed a physico-mathematical model that we have good reasons to regard as well founded and sufficiently wide ranging to permit the description of the motion of many fluids. On the other hand, the difficulties inherent in obtaining general solutions from this model lead to enquiring whether and when simplifications may be introduced in order to render it more tractable mathematically, and thus to develop solution procedures providing predictions with the level of accuracy that is required for problems of interest.

Whenever dealing with the simplification of a problem in fluid dynamics, the first point that must be emphasized is that the result of this procedure will practically always be the development of further models with narrower ranges of validity. The identification of the new limits of applicability of the simplified models is, then, absolutely essential because their use in cases in which they are not fully valid might have unpredictable — and sometimes unfortunate — consequences on the physical significance and utility of the obtained results.

The first class of simplifications that may be attempted concerns the constitutive equations describing the behaviour of the fluid. This implies neglecting or describing in a simplified form one or more of the physical properties that we have shown in Chapter 2 to characterize the behaviour of fluids. The most important simplifying assumption of this type is neglecting compressibility. In fact, as we have seen in Section 2.4, many fluids are characterized by very small coefficients of compressibility, so that this assumption seems to be a reasonable one, at least for liquids. As anticipated in Section 2.5 and as will be described in more detail in the next chapter, even gases may be assumed to be incompressible provided certain conditions are satisfied, the most important of which is that the Mach number be sufficiently small in the whole flow field. On the other hand, there are circumstances in which compressibility cannot be neglected even for liquids, as for instance in the analysis of sound propagation.

In order to emphasize the fact that, strictly speaking, no fluid is exactly incompressible, we shall employ the term 'incompressible flows' to indicate the motions of a fluid that satisfy the conditions for using the incompressibility assumption. Note that this assumption applies to a number of situations of great interest, such as the prediction of the aerodynamic loads acting on almost all road vehicles, or on civil structures subjected to winds. But it may also be used in the design of aeroplanes, and this is true not only for general aviation aircraft travelling at relatively low velocities, but more generally. As a matter of fact, the flight of any aeroplane will be at low Mach numbers at least during take-off and landing. Furthermore, even for higher velocities, the physical mechanisms at the origin of the fluid dynamic loads remain qualitatively similar as long as the flow is subsonic everywhere (although compressibility may start playing a role), and these loads may sometimes be predicted through suitable corrections of those derived from an incompressible flow analysis.

Unfortunately, even when the conditions for its applicability are well satisfied, the assumption of incompressible flow in itself does not lead to an easily solvable mathematical problem. We shall see that the equations governing the motion are still non-linear and that their solution in general cases is extremely complex.

Therefore, a further step towards the development of an even simpler model might be the introduction of stronger assumptions regarding the constitutive equations; in particular, the viscosity of the fluid might be completely neglected. The assumption of the fluid being non-viscous is indeed a powerful one as regards mathematical simplification. This can be realized immediately from a rapid scrutiny of the equations of motion reported in the previous section; in effect, with such an assumption the highest-order term in the momentum equation disappears. Furthermore, if one neglects viscosity there is no reason to take heat conduction into account, due to the similarity of the relevant physical mechanisms acting at the molecular level, which were briefly described in Chapter 2. Consequently, by using the entropy balance (4.80), the energy equation becomes quite simple, reducing to the statement that the entropy of a fluid particle remains constant during its motion.

In spite of its favourable mathematical consequences, the assumption of a non-viscous fluid might seem very drastic from the point of view of physical description. Nonetheless, it may be justified by considering the ratio between the orders of magnitude of the various terms appearing in the momentum balance Equation (4.91). In effect, if the considered problem is such that a reference velocity U and a reference length L may be identified (for instance, the velocity and the length of a body moving in the fluid), we show in the next chapter that, with the further assumption of incompressible flow, the ratio between the orders of magnitude of the inertial term (i.e. the rate of variation of momentum) and the viscous term in Eq. (4.91) is given by the *Reynolds number*, Re, defined as

$$Re = \frac{\rho U L}{\mu} = \frac{U L}{\nu}. \tag{4.103}$$

Now, it is easy to check that the Reynolds numbers involved in quite common situations are very high, at least as long as we consider bodies of ordinary dimensions moving at even moderate velocities in low-viscosity fluids, such as air and water. For instance, if we have $L = 1.5\,\mathrm{m}$ and $U = 10\,\mathrm{m/s}$, then the Reynolds number would be more than 1×10^6 in air and one order of magnitude larger in water. Therefore, it is clear that in many situations the viscous term would indeed seem to be negligible.

However, as was anticipated in Chapter 1 and will be further discussed in Chapter 5, the use of the simultaneous assumptions of non-viscous fluid and incompressible flow leads to the unphysical result that no forces would act on a body moving with uniform velocity in a fluid, which is obviously a very discouraging conclusion, in evident contrast with everyday experience. This result derives from the different boundary conditions that must be satisfied by the flows of viscous and non-viscous fluids. While the no-slip boundary condition applies at a solid wall for a viscous fluid, *however high the value of the Reynolds number*, a non-viscous fluid is allowed to slide along a solid surface. In effect, for such an ideal fluid the physical mechanism at the molecular level leading to mechanical equilibrium between fluid and solid surface is absent, and thus only the non-penetration condition (2.30) applies. From the mathematical point of view, the necessity of only one scalar boundary condition regarding velocity for a non-viscous fluid, instead of the three that are necessary for a viscous fluid, derives directly from the lowering of the order of the differential equations that is a consequence of the absence of the viscous term. This difference in the boundary conditions at solid walls may actually be seen as the origin of all the main discrepancies between the incompressible flow treatment of viscous and non-viscous fluids. Therefore, taking viscosity into account is essential to explain the physical mechanisms that are responsible for the generation of the fluid dynamic loads acting on bodies moving in incompressible flows.

Nonetheless, it can be shown that neglecting viscosity is not really necessary for the analysis of incompressible flows, and that considerable simplifications may be obtained by introducing further assumptions involving not the physical characteristics of the fluid but the geometry of the problem and the kinematics of the flow. In effect, it will become clear in the next chapters that a fundamental role is played by the behaviour of vorticity, which is a purely kinematical quantity. In particular, it will be shown that if an incompressible flow of a viscous fluid is such that vorticity is either zero or negligible, its velocity field is governed by the same linear differential equation that applies for a non-viscous fluid, and this apparently opens the way to the solution of the problem.

However, it will also be seen that no incompressible viscous flow can be devoid of vorticity if solid surfaces are present. Therefore, the generation and dynamics of vorticity will have to be analysed in detail, in order to identify the cases in which vorticity is confined in thin layers of fluid close to the solid surfaces. When this happens, the problem is amenable to a simplified mathematical treatment, and an iterative procedure may be developed for the prediction of the flow field and of the fluid dynamic loads on moving bodies. It will be seen, for instance, that the forces acting on airfoils (i.e. the cross-sections of an aeroplane wing) may be predicted in such a way. Note that in this case the simplification of the problem derives from the identification of appropriate geometrical and dynamical restrictions to the flow, rather than from disregarding certain terms in the equations of motion.

The picture is definitely different for flows at high Mach numbers, and particularly for supersonic flows. When the involved velocities are high enough for supersonic flow conditions to be reached (which in practice may happen almost exclusively in the motion of gases), then compressibility — rather than viscosity — becomes the crucial physical feature causing the generation of the aerodynamic forces, and neglecting it would lead to absolutely unacceptable results. This fact has already been anticipated in Section 2.5, in which the fundamental differences between the perturbations induced by a body moving at subsonic or supersonic velocities were briefly described.

Consequently, in supersonic conditions the solution of the equations derived by assuming the fluid to be compressible but non-viscous may indeed be useful, both for first-order engineering evaluations and for the comprehension of many fundamental physical features of the flow. This is the case, for instance, for the prediction of the forces on sharp-edged thin airfoils in supersonic flow, or for the description of the flow in short expansion nozzles. The main reason is that, as already pointed out, at high Mach numbers compressibility is the fundamental physical property playing a role in flow development. In particular, the fluid dynamic forces on a moving body may be explained by the action of the pressure field associated with the wave system produced by the motion of the body, which can be described with satisfactory approximation through a non-viscous compressible flow model. Thus, the production of the forces on an airfoil may be physically justified, and good approximations of their values may be obtained, even neglecting the presence of the viscosity-induced vorticity layers, at least as long as they remain thin and close to the body surface. In fact, in these cases the effects of viscosity only produce second-order corrections to the values of the forces.

In most of the remaining chapters of this book, we shall consider and analyse in detail the dynamics of incompressible flows, and try to highlight the role of viscosity and the conditions that must be introduced in order to derive prediction procedures with different levels of approximation. Only in the last chapter will compressibility be taken into account, to analyse the motion of gases inside ducts. However, we shall do so by introducing an additional and strong simplification, namely that the flow may be described with sufficient accuracy by means of a one-dimensional treatment. This is a powerful approximation, whose significance, utility and limits of applicability will be discussed in detail, together with the conditions in which the effects of viscosity may also be neglected.

Chapter 5

THE INCOMPRESSIBLE FLOW MODEL

5.1. The Incompressibility Hypothesis

We shall now discuss in some detail the conditions that must be fulfilled for a flow to be considered as incompressible. We stress again that the point is not whether a fluid in itself is or is not compressible, as all fluids are compressible, albeit with different degrees of compressibility. What is of interest here is to understand whether and when a fluid in motion may behave *as if* it were incompressible. In other words, we want to investigate under what conditions the main features and effects of the motion of a fluid may be estimated with sufficient accuracy by neglecting one of its physical characteristics, namely compressibility.

More precisely, we must point out that here we follow the usual practice of denoting as *incompressible* a flow in which we may neglect the variations of density occurring in a fluid *due to its motion*. The latter is an important specification, because density differences in the fluid might also be the consequence, for instance, of temperature variations that are independent of motion. An example of such a situation would be the case in which there are large temperature differences between various boundaries of the flow, or between a boundary and the ambient fluid. In those circumstances the temperature would be a function of position even if the fluid were not in motion, and so would the density and the other physical properties of the fluid, such as viscosity.

There are situations in which the motion itself is produced by density differences present in the field due to an imposed gradient of temperature. This is the case, for instance, in *natural convection* (also known as *free convection*), in which temperature differences produce variations in density,

121

causing the fluid to rise due to buoyancy forces. As examples, we may cite the rising of air above sun-heated land or sea, or of water inside a pot placed over a cooker, in which the variations in the fluid density certainly cannot be neglected.

We do not consider conditions of this type, nor any situation in which significant temperature gradients exist in the fluid irrespective of its motion. In practice, if a fluid with unperturbed temperature T_∞ is moving past a solid boundary whose surface temperature is T_w, we suppose that the following condition is satisfied:

$$\left| \frac{T_w - T_\infty}{T_\infty} \right| \ll 1. \tag{5.1}$$

A flow may then be assumed to be incompressible if the density variation, $\Delta\rho$, experienced by a fluid element during a typical time interval characterizing its motion is not large with respect to the unperturbed value of the density, i.e. if we have

$$\left| \frac{\Delta\rho}{\rho} \right| \ll 1. \tag{5.2}$$

Let us now take L to be a characteristic length scale of the motion, which means that the various quantities characterizing the flow vary only slightly over distances that are small compared to L. Furthermore, we assume that U is a reference velocity of the flow, for instance, the free-stream velocity in which an object is immersed. Considering that points where the velocity is zero may be present in the field, U also corresponds to the order of magnitude of the velocity variations, with respect to both position and time. Consequently, a characteristic time in which significant variations of the flow quantities may occur is L/U. Therefore, we may estimate the order of magnitude of the relative density variation as

$$\frac{\Delta\rho}{\rho} \sim \frac{1}{\rho} \frac{D\rho}{Dt} \frac{L}{U}, \tag{5.3}$$

so that condition (5.2) becomes

$$\left| \frac{1}{\rho} \frac{D\rho}{Dt} \right| \ll \frac{U}{L}. \tag{5.4}$$

As we are considering homogeneous fluids, we may now choose the density ρ and the entropy per unit mass S as the two independent thermodynamic state variables, so that the equation of state for pressure may be written as $p = p(\rho, S)$, and we have

$$\frac{Dp}{Dt} = \left(\frac{\partial p}{\partial \rho}\right)_S \frac{D\rho}{Dt} + \left(\frac{\partial p}{\partial S}\right)_\rho \frac{DS}{Dt}.$$
(5.5)

Consequently, by recalling expression (2.16) for the speed of sound, condition (5.4) is equivalent to

$$\left| \frac{1}{\rho a^2} \frac{Dp}{Dt} - \frac{1}{\rho a^2} \left(\frac{\partial p}{\partial S}\right)_\rho \frac{DS}{Dt} \right| \ll \frac{U}{L}.$$
(5.6)

As, in general, we cannot expect the terms on the left-hand side to mutually cancel out, this condition implies that each of them must be sufficiently small compared to U/L. Let us now consider the first (and most important) of the two conditions arising from (5.6),

$$\left| \frac{1}{\rho a^2} \frac{Dp}{Dt} \right| \ll \frac{U}{L}.$$
(5.7)

To estimate the order of magnitude of the term Dp/Dt, we use the kinetic energy equation (4.73) and neglect the effects of the viscous term, which normally is not expected to influence the order of magnitude of the pressure variations significantly. Therefore, also recalling the definition (3.6) of the material derivative, we get

$$\left| \frac{1}{\rho a^2} \frac{\partial p}{\partial t} - \frac{1}{2a^2} \frac{DV^2}{Dt} + \frac{\boldsymbol{f} \cdot \boldsymbol{V}}{a^2} \right| \ll \frac{U}{L}.$$
(5.8)

Again, we cannot rely on cancellation of terms, so that (5.8) gives rise to three different conditions, one for each term on the left-hand side. Let us first analyse the second one:

$$\left| \frac{1}{a^2} \frac{DV^2}{Dt} \right| \ll \frac{U}{L}.$$
(5.9)

Now, the order of magnitude of the variations of V^2 given by the material derivative operator is U/L times U^2, i.e. U^3/L, so that condition (5.9) becomes

$$\frac{U^2}{a^2} = M^2 \ll 1,$$
(5.10)

where M is the Mach number characterizing the flow, whose connection with the magnitude of the relative variations of density has already been mentioned in Section 2.5.

We consider now the condition arising from the first term in (5.8), which is directly connected with the unsteadiness of the flow,

$$\left| \frac{1}{\rho a^2} \frac{\partial p}{\partial t} \right| \ll \frac{U}{L}. \tag{5.11}$$

To estimate the magnitude of the time fluctuations of pressure, let us consider a very simple case, namely a unidirectional horizontal flow with a velocity that is constant in space for a distance of order L, but oscillates in time between opposite values of order U, with a typical dominant frequency f. For the sake of simplicity, we can describe the oscillation by a sinusoid with amplitude U and frequency f. As an example, we imagine the flow in a tube at whose extremities a pressure difference with a sinusoidal time variation is applied. Furthermore, we neglect the effects of the body forces — as can be done if they are only forces due to gravity and the motion is horizontal — and of the viscous forces, which cannot invalidate our order-of-magnitude analysis.

By applying the momentum balance equation (4.91) to this simple case, and considering that the convective term is zero due to the space-constancy of the velocity, we determine that the order of magnitude of the pressure difference driving the motion of a volume of fluid of typical length L and oscillating with a time period $\Delta t = 1/f$ is $\Delta p = \rho U f L$. Therefore, the order of magnitude of the unsteady pressure term becomes

$$\frac{\Delta p}{\Delta t} \sim \rho U f^2 L, \tag{5.12}$$

and thus condition (5.11) may be expressed as

$$\frac{f^2 L^2}{a^2} \ll 1. \tag{5.13}$$

It should be noted that if the order of magnitude of the frequency f is U/L, then condition (5.13) reduces to (5.10), i.e. to the requirement that the Mach number be low. However, if the oscillations occur at higher frequencies, then (5.13) is an independent and more demanding condition than (5.10). In particular, if a sound wave with frequency f is considered, then the time variations will be characterized by a wavelength $L = a/f$, so that condition (5.13) can never be satisfied. This explains why it is essential to take compressibility into account in acoustics, even when the fluid is a liquid and the Mach numbers are low.

Let us now assume that the body force arises from the gravitational acceleration g. The order of magnitude of the last term in (5.8) will then be gU/a^2, so that the condition that it be small compared to U/L is

$$\frac{gL}{a^2} \ll 1, \tag{5.14}$$

which implies that the typical length scale of the motion should be small compared to the characteristic length a^2/g. In view of the high values of the speed of sound in liquids, this condition may be restrictive only for gases. If we consider, for instance, air under normal conditions, a^2/g is almost 12 km. Therefore, although vertical temperature variations in the atmosphere may somewhat reduce this value, it is evident that condition (5.14) will be satisfied for any motion occurring in layers not exceeding a few hundred meters in depth. Conversely, it is also clear that, generally, this is not the case in meteorological applications.

We return now to (5.6), and in particular to the condition arising from the second term:

$$\left| \frac{1}{\rho a^2} \left(\frac{\partial p}{\partial S} \right)_\rho \frac{DS}{Dt} \right| \ll \frac{U}{L}. \tag{5.15}$$

Taking the entropy balance (4.80) into account and carrying out an order of magnitude analysis of the various terms (see, e.g., Batchelor, 1967), we can see that this condition is largely satisfied in almost all cases of practical interest, provided the Mach number is low and the externally imposed temperature gradients are not significant.

Therefore, the flow may be confidently considered as incompressible if conditions (5.10), (5.13) and (5.14) are all satisfied. From the mathematical point of view, an incompressible flow may then be defined as a flow satisfying the following relation:

$$\frac{1}{\rho} \frac{D\rho}{Dt} = 0. \tag{5.16}$$

It might be appropriate to observe that (5.16) does not actually require the fluid density to be constant everywhere in the flow field, but only that changes in the density of individual fluid particles be negligible. Therefore, in some cases the flow may still be considered as incompressible even when the density of adjacent particles is different, provided that any single particle keeps its density unchanged during the motion. For instance, this may be the case in the ocean, where density differences due to variations in salinity and temperature with depth may be present. These situations belong to the class of *stratified flows* which, albeit interesting for several applications, will not be considered in the present book; thus, the incompressible flow assumption will correspond, in practice, to the constancy of the density of all fluid particles.

To conclude this section, it may be instructive to derive an estimate of the dependence on Mach number of the variations of temperature, pressure and density that are exclusively due to compressibility in the steady motion of a perfect gas. To this end, we utilize the total enthalpy balance, neglecting the effects of body forces, viscous forces and heat conduction. With such assumptions, Eq. (4.88) reduces to

$$\boldsymbol{V} \cdot \text{grad}(h_0) = \boldsymbol{V} \cdot \text{grad}(h + V^2/2) = 0, \tag{5.17}$$

which implies that the total enthalpy of a fluid particle is constant along its motion, and can also be put in the simpler form

$$h_0 = h + V^2/2 = \text{constant.} \tag{5.18}$$

For a perfect gas, i.e. a gas whose equation of state is $p = \rho RT$, it can be shown that the specific internal energy and enthalpy are functions of temperature only, through the differential relations

$$de = C_v dT, \tag{5.19}$$

$$dh = C_p dT, \tag{5.20}$$

where C_v and C_p are the specific heats at constant volume and pressure respectively. The latter quantities may be often assumed to be constant, in which case the gas is denoted as *calorically perfect*, and the internal energy and enthalpy may be simply expressed as

$$e = C_v T, \tag{5.21}$$

$$h = C_p T. \tag{5.22}$$

Assuming the validity of (5.22), we can recast Eq. (5.18) as follows

$$C_p T_0 = C_p T + V^2/2, \tag{5.23}$$

which, in practice, defines the *total temperature*, T_0, as the temperature that would be obtained if the velocity were brought to zero in a flow satisfying the above assumptions.

Now, for a perfect gas the following relation applies:

$$R = C_p - C_v, \qquad (5.24)$$

so that, with the definition $\gamma = C_p/C_v$, we immediately obtain

$$C_p = \frac{\gamma R}{\gamma - 1}. \qquad (5.25)$$

By substitution of (5.25) into (5.23), and recalling expression (2.18) giving the speed of sound of a perfect gas, after simple manipulations we obtain the following fundamental relation for the ratio between the total temperature, T_0, and the temperature, T, as a function of Mach number:

$$\frac{T_0}{T} = 1 + \frac{\gamma - 1}{2} M^2. \qquad (5.26)$$

If we take the quantity $\Delta T = (T_0 - T)$ as a representative value of the differences in temperature that may be encountered in the flow, it is easy to see that if the Mach number is sufficiently low the relative variations in temperature are definitely small. For instance, if the gas is air (for which $\gamma = 1.4$) at a temperature of 300 K moving at $M = 0.3$, $\Delta T/T$ would be 0.018, thus giving a temperature variation only slightly above 5 K. On the other hand, relation (5.26) shows that flows at very high Mach numbers may produce significant temperature variations. In effect, ΔT would be about 240 K when $M = 2.0$ and 540 K when $M = 3.0$. Even if the temperature experienced at the surface of a moving body is slightly less than T_0, these estimates explain why considerable heating problems occur for aeroplanes flying at supersonic velocities.

In order to obtain the estimates of the variations in pressure and density, we may observe that, considering Eq. (4.80), the assumptions leading to (5.18) also imply that the flow is isentropic. Consequently, we may use the thermodynamic relations linking pressure, density and temperature in an isentropic process, so that from (5.26) we get

$$\frac{p_0}{p} = \left(\frac{T_0}{T}\right)^{\frac{\gamma}{\gamma-1}} = \left(1 + \frac{\gamma - 1}{2} M^2\right)^{\frac{\gamma}{\gamma-1}}, \qquad (5.27)$$

$$\frac{\rho_0}{\rho} = \left(\frac{T_0}{T}\right)^{\frac{1}{\gamma-1}} = \left(1 + \frac{\gamma - 1}{2} M^2\right)^{\frac{1}{\gamma-1}}. \qquad (5.28)$$

These relations define the *total pressure*, p_0, and the *total density*, ρ_0, as the corresponding quantities that would be obtained if the velocity were brought to zero with an adiabatic and isentropic process. Conversely, the values of temperature, pressure and density denoted by the symbols T, p and ρ are usually called *static* values, and are those that would be measured by a hypothetical instrument attached to a fluid particle moving at a Mach number equal to M.

A table giving the values of the inverse of relations (5.26), (5.27) and (5.28), as a function of M, is reported in Appendix A.2 (for $\gamma = 1.4$). From an analysis of that table it may be easily verified that, if we again take the difference between the total and static values as representative of the order of magnitude of the variations of the various quantities, $\Delta\rho/\rho$ would be about 2% at $M = 0.2$ and 4.4% at $M = 0.3$. This explains why even the flow of a gas may be considered as incompressible if the motion is characterized by low values of the Mach number.

Relations (5.27) and (5.28) will be extensively used in Chapter 15, where the effects of compressibility will be described using a simplified one-dimensional treatment. As these relations are of the form $(1 + x)^k$, it may be useful to expand them in a binomial series for low values of M, so that the dependence of $\Delta p = p_0 - p$ and $\Delta\rho = \rho_0 - \rho$ on Mach number is more directly highlighted. After simple calculations, we find

$$p_0 - p = \frac{1}{2}\rho V^2 \left[1 + \frac{M^2}{4} + \frac{(2-\gamma)}{24} M^4 + O(M^6) \right], \qquad (5.29)$$

$$\rho_0 - \rho = \rho\frac{M^2}{2} \left[1 + \frac{(2-\gamma)}{4} M^2 + \frac{(2-\gamma)(3-2\gamma)}{24} M^4 + O(M^6) \right].$$
$$(5.30)$$

These expressions may be used for a quick evaluation of the effects of compressibility when viscosity and heat conduction are neglected.

5.2. Incompressible Flow Analysis

5.2.1. *Equations of incompressible flow*

Having discussed the conditions necessary to regard a flow as incompressible, we may now proceed to the derivation of the relevant equations of motion. The first simplification — and perhaps the most notable one — deriving from the hypothesis of incompressible flow concerns the mass balance equation. By introducing (5.16) into (4.90), and considering that density cannot be zero, we get the much simpler equation

$$\operatorname{div} \boldsymbol{V} = 0. \tag{5.31}$$

In other words, the velocity field must be solenoidal, and this is a powerful constraint on the motion. Note that, at variance with Eq. (4.90), Eq. (5.31) is a *linear* partial differential equation; however, it still contains three unknown functions, namely the three components of the velocity vector. This can be better appreciated by expressing Eq. (5.31) explicitly in a Cartesian (x_1, x_2, x_3) reference system:

$$\frac{\partial u_1}{\partial x_1} + \frac{\partial u_2}{\partial x_2} + \frac{\partial u_3}{\partial x_3} = 0. \tag{5.32}$$

As for the momentum balance, from a strictly formal point of view it is still described by Eq. (4.91). However, the density ρ is now a constant, and the viscous force term, $\operatorname{div} \boldsymbol{\tau}$, may be expressed in a simpler form by exploiting the properties of an incompressible flow.

To this end, let us first observe that the components of the viscous stress tensor, which in the general case are given by expression (4.96), thanks to (5.31) become

$$\tau_{ik} = \mu \left(\frac{\partial u_i}{\partial x_k} + \frac{\partial u_k}{\partial x_i} \right). \tag{5.33}$$

We are assuming that no external heat sources are present, so we may safely assert that the temperature variations, which are only those due to motion, are very limited in the whole flow field. As the variations deriving from compressibility, expressed by (5.26), are neglected, the temperature may change, in practice, only due to variations in internal energy caused by dissipation. However, as already pointed out in the previous section when discussing the validity of relation (5.15), these variations are definitely small if the remaining incompressibility conditions are satisfied.

The important consequence of the above is that the viscosity and conductivity coefficients, which are not, in general, strong functions of temperature, may be assumed to remain constant during the motion and to be a function of the type of fluid only. This fact greatly simplifies the expression for the viscous term in the momentum balance equation. In effect, the constancy of the viscous coefficients and the solenoidality of the velocity field allow the following simple linear relation to be derived:

$$\operatorname{div} \boldsymbol{\tau} = \mu \nabla^2 \boldsymbol{V}, \tag{5.34}$$

which can be obtained immediately either from the general expression of
the viscous force per unit volume, provided in (4.98), or by applying the
divergence operator to the viscous stress tensor whose components are given
in (5.33).

If we now introduce (5.34) into (4.91) and divide the whole equation
by density, the following momentum balance equation characterizing an
incompressible flow is obtained:

$$\frac{\partial V}{\partial t} + V \cdot \mathrm{grad} V = f - \mathrm{grad}\left(\frac{p}{\rho}\right) + \nu \nabla^2 V, \qquad (5.35)$$

where the fact that the gradient of density is zero has been taken into
account and we have introduced the kinematic viscosity, $\nu = \mu/\rho$.

The energy balance is, again, expressed using the internal energy
equation (4.92), which, after the introduction of (5.31), becomes

$$\rho\left(\frac{\partial e}{\partial t} + V \cdot \mathrm{grad} e\right) = \Phi - \mathrm{div} q. \qquad (5.36)$$

The dissipation function Φ is still defined by (4.97), but its expression,
which is given for a compressible fluid in (4.77), is slightly simplified by the
absence of the first term connected with div V and by the constancy of the
viscosity coefficient.

It may be useful to highlight that, as may be derived from (4.81), for an
incompressible flow the internal energy balance is directly connected with
the entropy balance by the relation

$$\frac{De}{Dt} = T\frac{DS}{Dt}. \qquad (5.37)$$

It should be also noted that the heat flux vector is still given by
the Fourier law (4.95) but, considering that k may now be taken as a
characteristic constant of the fluid and that we have $\mathrm{div}(\mathrm{grad} T) = \nabla^2 T$,
Eq. (5.36) may be recast as

$$\rho\frac{De}{Dt} = \Phi + k\,\nabla^2 T. \qquad (5.38)$$

The final constitutive equation needed is the one for the specific
internal energy. In incompressible flow analysis, variations of density,
and thus of specific volume, are neglected. We may safely assume, then,
that the internal energy is a function of temperature only and that
relation (5.19), which applies for a compressible perfect gas, is still valid in

the incompressible flow case, irrespective of whether the considered fluid is a gas or a liquid.

Equations (5.31), (5.35) and (5.38) constitute the system of five scalar partial differential equations describing the incompressible flow of a fluid. We shall now discuss in more detail the problem of the identification of the unknown quantities in this set of equations. There is no doubt that, as was the case for a generic compressible flow, the three components of the velocity vector, V, and the temperature, T, are still unknown functions in the problem. However, the last unknown can no longer be density, which does not change in the motion and, neglecting the case of stratified fluids, is effectively constant in the whole flow field. As can be appreciated from the previous derivation of the equations, the quantity that replaces density as an unknown function is the pressure p, but its physical meaning deserves further comment.

In an incompressible flow the quantity denoted pressure can no longer be considered as a thermodynamic state function, and thus its variations are not linked to those of density and temperature through a relation like (4.93). However, we can still give it a physical meaning by using the mechanical definition (4.29); in other words, the pressure is interpreted as the opposite of the mean value of the normal stresses acting on three mutually orthogonal elementary surfaces passing through a point in the fluid.

5.2.2. *Physical and mathematical features*

Let us now analyse the consequences, from the physical and the mathematical points of view, of the incompressible flow assumption, i.e. of the fact that the motion of a fluid is described through a mathematical model defined by the above-derived system of equations.

Considering we have used Eq. (5.16) to derive the equations of an incompressible flow, the first point to be made is that, whenever applying this model, we are assuming that the density variations of a particle in motion are effectively zero, and not just sufficiently small, as was required by conditions (5.2) or (5.4). As regards the physical behaviour of the fluid we are describing, this fact has several consequences. One of these is that, as already observed in Chapter 2, the speed of sound is infinite everywhere and, theoretically, the Mach number is always zero, irrespective of the value of velocity. Note that this also means that any perturbation at a point is instantaneously 'felt' in the whole flow field, which is obviously a physical feature that we know not to be representative of the behaviour of any fluid.

The problem now is to understand when we can accept this physical model. For instance, we have already seen that this is not possible if we want to study sound propagation.

More generally, the problem we are concerned with is whether the motion of a fluid satisfying the conditions described in Section 5.1 can be studied using the incompressible flow model irrespective of the fact that, from a rigorous point of view, it may imply some types of behaviour that we recognize as being non-physical.

In practice, as pointed out by Panton (1984), using the incompressible flow equations corresponds to assuming that the solution of the limit of the complete Navier–Stokes equations for M tending to zero is the same as the limit for M tending to zero of the solution of the Navier–Stokes equations. As may be easily imagined, there is no reason, in principle, to expect this to be generally true, and it would be a hard task to demonstrate mathematically under what conditions it would be true. Therefore, all we can do is hope that the solutions we obtain with the incompressibility hypothesis do provide satisfactory approximations to the real solutions of the Navier–Stokes equations, provided the motion of the fluid satisfies the conditions described in Section 5.1.

Fortunately, all the available experimental and numerical evidence suggests that this is indeed the case, and that we may use the solutions of the incompressible flow model to obtain many valuable results as regards both the features of fluid motion and the fluid dynamic loads acting on moving bodies. Nonetheless, it must always be remembered that in certain circumstances these solutions may be characterized by features (for instance, values of velocity or pressure in restricted regions of the flow field) that may be locally inconsistent with the conditions of applicability of the incompressible flow model. However, in almost all cases this fact does not give rise to significant consequences as regards the accuracy of the global flow.

From the mathematical point of view, one of the most interesting features of the equations of motion of an incompressible flow is that, due to the assumed constancy of μ (and thus also of ν), the temperature appears in the energy equation only. Consequently, the mass and momentum equations are decoupled from the energy equation, and form a separate system of four scalar equations that, in principle, can be solved for the three velocity components and for the pressure. Subsequently, and only if necessary, the obtained functions can be substituted into the energy equation to derive the temperature field. Obviously, this is not the case for a compressible flow, in which all equations are strictly coupled by density.

The consequence of the above is, for instance, that if the focus is on the evaluation of the forces acting on a body due to a motion satisfying the incompressible flow conditions, the energy equation is not usually taken into account in the analysis. In effect, considering that the viscous stresses acting over the body surface can be derived from the velocity gradient, all that is needed is knowledge of the pressure and velocity fields. In fact, in most cases, the energy balance equation is not even written down when an incompressible flow is considered, and the problem is thus restricted to the solution of the system given by Equations (5.31) and (5.35). Nevertheless, it will be shown in Chapter 10 that the energy equation may sometimes be useful (especially in an integral form) to obtain a deeper understanding of the flow features and to interpret the physical origin of the drag force component.

We now give a very brief account of the problem of the initial and boundary conditions to be assigned to the incompressible flow equations. In the unsteady case, a solenoidal velocity field must be prescribed at the initial time, as well as the temperature field. In principle, the pressure field should also be given, but a closer analysis of the equations shows immediately that only the *gradient* of pressure appears in the momentum balance equation. Therefore, the motion depends only on pressure differences, and would not be changed by a constant variation of pressure involving the whole flow domain. Consequently, the solution of the equations only provides the difference between the pressure at various points of the flow field and an assigned reference pressure, which might be representative of an initial condition or of the value in an unperturbed region.

As for the boundary conditions, they are the same as those discussed for the compressible flow case in Section 4.7.1; therefore, at solid walls both the velocity and the temperature are assumed to be the same as those of the bounding surfaces. In particular, as regards the velocity vector, the different roles played by the non-penetration condition ($\boldsymbol{V} \cdot \boldsymbol{n} = \boldsymbol{V}_w \cdot \boldsymbol{n}$) and by the no-slip condition ($\boldsymbol{V} \times \boldsymbol{n} = \boldsymbol{V}_w \times \boldsymbol{n}$), already distinguished in Chapter 2, will become evident in the following chapters. When a boundary at infinity is present, both velocity and temperature are assumed to tend to the relevant unperturbed values in a suitable way.

However, the problem of the definition of adequate initial and boundary conditions is often more complex, especially if the flow domains are finite, as is nearly always the case when numerical simulations are used to solve the equations of motion. In those cases, *in-flow* and *out-flow* boundaries are present, and the choice of the number and type of conditions to be imposed therein on the various quantities may be a non-trivial problem. We shall go no further in this discussion, and the interested reader may refer, for instance, to the reviews by Gresho (1991) or Zeytounian (2001).

We may now enquire whether, after the effort in simplification we made in order to arrive at the incompressible flow model, straightforward solutions of the relevant mathematical problem can be obtained. Unfortunately, this is not the case and, similar to what occurs for the equations of motion of a compressible fluid, even the demonstration of the existence and uniqueness of solutions of the incompressible flow equations for generic initial and boundary conditions has not yet been fully attained. Non-linearity is still the main source of difficulties, and the convective term in the momentum balance equation (5.35) is indeed non-linear, even if all the remaining terms, as well as the mass balance equation (5.31), are now linear. Again, closed-form solutions to particular problems do exist (see Chapter 9), and some of them may be directly applied to practical cases or may give clues to obtaining approximate solutions. However, the problem of predicting a generic flow field remains a formidable one.

In spite of this, it is possible to carry out a deeper analysis of the equations and to derive important relations, which may be helpful both to increase the understanding of the flow features and to devise even more simplified models for the solution of certain types of problems. In particular, we shall see that the fluid dynamic loads acting on a moving body may be predicted at various levels of approximation provided both the body geometry and its motion satisfy certain further conditions.

5.3. Manipulation of the Equations of Motion

5.3.1. *Expression of momentum equation in terms of vorticity*

In order to discuss some properties of the solutions of the incompressible flow equations, it is convenient to recast the momentum equation (5.35) in a different form, in which the vorticity vector $\boldsymbol{\omega} = \operatorname{curl} \boldsymbol{V}$ appears. To this end, it is useful to introduce the following vector identities, which, as may be easily checked, are valid for any vector $\boldsymbol{a}$:

$$\boldsymbol{a} \cdot \operatorname{grad} \boldsymbol{a} = \operatorname{curl} \boldsymbol{a} \times \boldsymbol{a} + \operatorname{grad} \frac{a^2}{2}, \tag{5.39}$$

$$\nabla^2 \boldsymbol{a} = \operatorname{grad}(\operatorname{div} \boldsymbol{a}) - \operatorname{curl}(\operatorname{curl} \boldsymbol{a}). \tag{5.40}$$

In (5.39) we have indicated by a the modulus of vector $\boldsymbol{a}$. If we now apply these relations to the velocity vector $\boldsymbol{V}$, recall the definition of vorticity, and use Eq. (5.31), we immediately obtain

$$\boldsymbol{V} \cdot \operatorname{grad} \boldsymbol{V} = \boldsymbol{\omega} \times \boldsymbol{V} + \operatorname{grad} \frac{V^2}{2}, \tag{5.41}$$

$$\nabla^2 \boldsymbol{V} = -\operatorname{curl}\boldsymbol{\omega}. \tag{5.42}$$

By introducing the above relations into Eq. (5.35), and further assuming the body forces to be conservative (so that $\boldsymbol{f} = -\operatorname{grad}\varPsi$), we finally obtain the following form of the momentum balance equation

$$\frac{\partial \boldsymbol{V}}{\partial t} + \boldsymbol{\omega} \times \boldsymbol{V} = -\operatorname{grad}\left(\frac{p}{\rho} + \frac{V^2}{2} + \varPsi\right) - \nu\operatorname{curl}\boldsymbol{\omega}. \tag{5.43}$$

Equation (5.43) can probably be regarded as the most important relation for understanding some of the fundamental features of an incompressible flow. It allows the conditions of validity of widely used relations to be clearly defined and shows that the kinematics of the flow, and in particular the existence or not of a non-zero value of the vorticity vector, may have a profound influence on the forces acting on a fluid particle, and thus on its motion. In effect, if the motion is *irrotational*, i.e. such that vorticity is zero, then the last term in Eq. (5.43), connected with viscosity, vanishes. We can observe that this is true even if vorticity is present, but its curl is zero. This may happen if vorticity is constant but also, more generally, if its value can be derived from a scalar potential σ, i.e. if $\boldsymbol{\omega} = \operatorname{grad}\sigma$, because the curl of the gradient of any scalar function is zero.

The extremely important consequence of the above result is that, if the motion is irrotational (or is such that $\boldsymbol{\omega} = \operatorname{grad}\sigma$), the momentum equation of the incompressible flow of a *viscous* fluid *coincides* with the momentum equation of the incompressible flow of a *non-viscous* fluid. This finding may come as a surprise, as we have seen in Section 4.5.3 that the viscous stress tensor does not depend on the anti-symmetrical part of the deformation tensor $\operatorname{grad} \boldsymbol{V}$, and is thus independent of $\boldsymbol{\omega}$. Furthermore, it may appear as particularly unexpected that a kinematical feature of the flow, namely the behaviour of the vorticity field, can influence the possible presence, in a governing equation, of a force term linked to viscosity, i.e. to a physical characteristic of the fluid.

The explanation of this apparent paradox is that, as already pointed out, the term $\operatorname{div}\boldsymbol{\tau} = \mu\nabla^2 \boldsymbol{V} = -\mu\operatorname{curl}\boldsymbol{\omega}$ does not represent the viscous

force acting on a surface element, but the *resultant*, per unit volume, of
the viscous forces acting over the surface bounding an elementary volume
of fluid. Thus, it is just a fortunate circumstance that, thanks to identity
(5.42), the viscous forces sum up to a zero resultant and the viscous term
drops out from the momentum balance equation whenever $\mathrm{curl}\,\omega = 0$.
Therefore, in all the flow regions where this happens, the fluid *locally*
behaves *as if* it were non-viscous. In other words, a fluid particle follows the
same paths that it would follow if the fluid were non-viscous. Perhaps this
is better appreciated by substituting relation (5.42) directly into Eq. (5.35)
and observing that, if $\mathrm{curl}\,\omega = 0$, the acceleration of a fluid particle, given
by the left-hand side of that equation, only depends on the body force and
on the pressure gradient acting on the particle.

Nonetheless, it should be always kept in mind that even when the
incompressible flow of a viscous fluid is such that $\mathrm{curl}\,\omega = 0$, and the
resultant viscous forces per unit volume are thus zero, the viscous forces
acting upon any elementary surface in the fluid are generally *not* zero. For
instance, a viscous fluid in irrotational motion is still *dissipative* because,
in general, the viscous stresses do work in deforming a fluid element and,
consequently, the dissipation function is not zero.

5.3.2. *Bernoulli's theorem and its applicability*

Let us now define the following *Bernoulli trinomial*:

$$H = \frac{p}{\rho} + \frac{V^2}{2} + \Psi. \tag{5.44}$$

From Eq. (5.43), we see that if the following relation applies:

$$\frac{\partial V}{\partial t} + \omega \times V + \nu\,\mathrm{curl}\,\omega = 0, \tag{5.45}$$

then we have

$$\mathrm{grad}\,H = 0, \tag{5.46}$$

which means that H does not vary in space, so that it may only be either
a constant or, at most, a function of time. This may be expressed as

$$\frac{p}{\rho} + \frac{V^2}{2} + \Psi = f(t). \tag{5.47}$$

Furthermore, if we denote by e_v the unit vector in the direction of the
velocity vector, the condition for the trinomial H to be constant, or at most

a function of time, *along a streamline* (even if its value may change from one streamline to another) is

$$\left(\frac{\partial \boldsymbol{V}}{\partial t} + \nu \mathrm{curl}\boldsymbol{\omega}\right) \cdot \boldsymbol{e}_v = 0. \tag{5.48}$$

In effect, the vector $\boldsymbol{\omega} \times \boldsymbol{V}$ is perpendicular to both $\boldsymbol{\omega}$ and $\boldsymbol{V}$, and thus its projection in the direction of $\boldsymbol{e}_v$ is always zero.

Relation (5.47) is a generalized form of the relation that is obtained when we also assume the motion to be steady, which may be expressed by saying that H is constant, either in the whole space or along a streamline. This is widely known as *Bernoulli's theorem*, after the famous Swiss–Dutch mathematician Daniel Bernoulli, even if it was explicitly expressed in mathematical form only by Leonhard Euler, another great Swiss mathematician (see, e.g., Truesdell, 1953). .

As we shall see, Bernoulli's theorem is absolutely fundamental, not only for obtaining a deeper understanding of the effects of fluid motion, but also for devising prediction methods for practical applications. For instance, let us assume, for a moment, that the steady form of Bernoulli's theorem applies and that the body force is due to gravity, so that if z is a vertical upwards coordinate, $\Psi = gz$. We then have

$$p + \frac{1}{2}\rho V^2 + \rho g z = \text{constant}, \tag{5.49}$$

which implies that variations in pressure, velocity and vertical coordinate are not independent. In particular, if the motion is horizontal, or if variations in z may be neglected, then an increase (decrease) in velocity corresponds to a decrease (increase) in pressure. Equation (5.49) also has the consequence that the pressure field may be easily obtained once the velocity field is known.

Incidentally, as our equations are also valid if $\boldsymbol{V} = 0$, in which case (5.45) is obviously satisfied, (5.49) then shows that a column of fluid of height h causes a pressure difference between the bottom and the top of the column equal to $\rho g h$, which is known as *hydrostatic pressure*.

Let us now examine the conditions of applicability of the various forms of Bernoulli's theorem in more detail. We start by noting that (5.45) and (5.48) express the necessary and sufficient conditions for a generalized Bernoulli theorem to apply, respectively in the whole space or along a streamline. As can be seen, relation (5.48) requires that the sum of two vectors (the first representing the local time variation of velocity and the second the opposite of the viscous force per unit mass acting on a fluid element) be perpendicular to the velocity vector, whereas (5.45) is satisfied when the sum of those two vectors and vector $\boldsymbol{\omega} \times \boldsymbol{V}$ is zero. As can be easily understood, the classification of all the possible flows that satisfy those conditions is an intriguing and non-trivial problem of applied mathematics (see, e.g., Truesdell, 1950, 1954, or Remorini, 1989, and the references therein). However, we shall not pursue this task, and our main interest will remain in trying to understand which particular flows that may have a physical significance and may be encountered in practice do satisfy conditions that are sufficient for Bernoulli's theorem to apply.

We start by considering steady flows, which are certainly interesting for practical applications and in which the first term in (5.45) and (5.48) disappears. Consequently, for such flows our main concern is the behaviour of the viscous term $\nu \operatorname{curl} \boldsymbol{\omega}$. It is clear that when this term is zero, or may be reasonably assumed to be negligible, (5.48) is satisfied, and the Bernoulli trinomial is constant along any streamline. Furthermore, if $\nu \operatorname{curl} \boldsymbol{\omega}$ is zero but vorticity is present, it is seen from (5.43) that H varies only in the direction of the vector $\boldsymbol{\omega} \times \boldsymbol{V}$, which is perpendicular to both velocity and vorticity.

The first idea that comes to mind regarding the term $\nu \operatorname{curl} \boldsymbol{\omega}$ is that it would be identically zero if we could neglect viscosity, and Bernoulli's theorem was indeed first derived by assuming the fluid to be non-viscous. However, we have already seen that this assumption is not strictly necessary and, before considering it in more detail and deriving its consequences, it is instructive to first study the possible alternative, namely that the motion be such that $\operatorname{curl} \boldsymbol{\omega}$ is zero. We have already seen that, in that case, the incompressible flow of a viscous fluid would have the very important feature of behaving as if the fluid were non-viscous, which avoids the necessity of neglecting another physical property of the fluid besides compressibility.

A particular case in which $\operatorname{curl} \boldsymbol{\omega}$ is zero is when $\boldsymbol{\omega}$ is constant. This condition may occasionally occur in certain types of flow and may be used as an approximation in some simplified models of real flows. However, in the following we shall analyse the more interesting situation in which vorticity is zero or its value is so small that it may be reasonably neglected.

5.3.3. *Irrotational flows*

As we have already observed, if the motion is irrotational, i.e. vorticity is zero, the velocity vector may be expressed as the gradient of a scalar function of position and time, say $\varphi(\boldsymbol{x}, t)$, and we may then write

$$\boldsymbol{V}(\boldsymbol{x}, t) = \text{grad}\varphi(\boldsymbol{x}, t). \tag{5.50}$$

The function $\varphi(\boldsymbol{x}, t)$ is called the *velocity potential*, and a flow in which $\boldsymbol{\omega} = 0$, and thus such that the velocity field may be expressed in the form (5.50), is denoted as a *potential flow*.

The first consequence of the irrotationality assumption is that, if we introduce (5.50) into Eq. (5.43), and consider that the time and space derivatives may be interchanged, we obtain

$$\text{grad}\left(\frac{\partial \varphi}{\partial t} + \frac{p}{\rho} + \frac{V^2}{2} + \Psi\right) = 0, \tag{5.51}$$

which leads to the following unsteady form of Bernoulli's theorem

$$\frac{\partial \varphi}{\partial t} + \frac{p}{\rho} + \frac{V^2}{2} + \Psi = f(t). \tag{5.52}$$

Furthermore, if the motion is steady, then the Bernoulli trinomial (5.44) has the same constant value in the whole irrotational field.

However, from the point of view of the mathematical description of incompressible flows, the most important consequence of irrotationality is that, by introducing (5.50) into (5.31), the mass balance equation reduces to a Laplace equation for the velocity potential:

$$\nabla^2 \varphi = 0, \tag{5.53}$$

or, in a Cartesian (x, y, z) reference system,

$$\frac{\partial^2 \varphi}{\partial x^2} + \frac{\partial^2 \varphi}{\partial y^2} + \frac{\partial^2 \varphi}{\partial z^2} = 0. \tag{5.54}$$

This result is a fundamental breakthrough for the solution of the incompressible flow problem because Eq. (5.53) is linear and has one single unknown scalar function; moreover, it is one of the most famous and studied equations of mathematical physics, for whose solution many effective mathematical and numerical tools are available. Therefore, if the whole flow were indeed irrotational, the problems of obtaining the velocity and the pressure fields would be totally decoupled, and their difficulty significantly reduced. In practice, we might devise a procedure in which the whole velocity field is derived by first solving Eq. (5.53) and then applying the gradient operator to the potential φ. Subsequently, the pressure field would be obtained from the application of Bernoulli's theorem.

In order to understand whether this simplified procedure may be applied or not, we must now face the problem of recognizing whether, and under which conditions, an incompressible flow may be irrotational. To this end, let us consider, for instance, a finite body in uniform translational motion with velocity U in an unlimited fluid at rest at infinity. Then, assuming the flow to be everywhere irrotational and using a reference frame fixed to the fluid, the differential problem defining the velocity field would be given by Eq. (5.53), with the boundary conditions grad $\varphi = U$ on the body surface (due to the viscous no-slip condition) and grad $\varphi = 0$ at infinity. Unfortunately, it may be shown that no solution to this mathematical problem exists unless $U = 0$, which corresponds to a condition of rest of body and fluid. This demonstrates that a *completely* irrotational flow of a viscous incompressible fluid, in the presence of solid bodies in uniform translational motion, is impossible.

However, this result, which seems to hinder the possibility of applying the above-described simplified calculation procedure to obtain the velocity and pressure fields, should not be misunderstood. It does not mean that vorticity must necessarily be present everywhere in the flow domain. In other words, flows in which vorticity is non-zero only in limited parts of the field are still possible. As we shall see in the following chapters, the identification of the conditions that must be fulfilled for this to be true is one of the core problems for developing useful calculation procedures for many important applications of fluid dynamics. We shall find that there are situations in which the solution of the Laplace equation, which governs potential flows, may still be useful for the prediction of the fluid dynamic forces acting on certain bodies. To understand when this is the case, we must first analyse in much more detail the problem of the generation and dynamics of vorticity in an incompressible flow, and indeed this is the main subject of Chapter 6. However, before doing so, it is convenient to examine the motivations and consequences of the assumption that the viscosity of the fluid may be completely neglected, which is a more drastic — but widely used — option for simplifying our problem.

5.4. Effects of Neglecting Viscosity

5.4.1. *Non-viscous flow and d'Alembert's paradox*

We have already pointed out in Section 4.7.2 that neglecting viscosity definitely seems a reasonable assumption in many circumstances. This may be better explained now by analysing the expression (5.35) of the momentum balance equation for an incompressible flow, and by assessing the order of magnitude of the ratio between the inertial term and the viscous term.

To this end, let us assume that L and U are a reference length and a reference velocity value of the problem respectively (for instance, a typical dimension and the translational velocity of a body in a fluid). The order of magnitude of the above-mentioned ratio may then be evaluated as follows (with the order of magnitude of the various terms indicated by square brackets):

$$\frac{[\partial \boldsymbol{V}/\partial t + \boldsymbol{V} \cdot \mathrm{grad} \boldsymbol{V}]}{[\nu \nabla^2 \boldsymbol{V}]} = \frac{UUL^{-1}}{\nu UL^{-2}} = \frac{UL}{\nu} = \frac{\rho UL}{\mu} = Re. \qquad (5.55)$$

Thus, the Reynolds number directly gives the value of the ratio between the orders of magnitude of the inertia and viscous forces in the momentum balance equation. Consequently, considering that most flows of practical interest are characterized by very high values of the Reynolds number, as can be easily checked, neglecting viscosity does indeed seem a plausible choice.

Let us then assume the fluid to be non-viscous and the flow to be irrotational. The latter assumption is, in fact, perfectly consistent with the former one, as it can be shown that no vorticity is generated in the incompressible flow of a non-viscous fluid starting from rest. Even if we do not go through the mathematical demonstration of this result, it should not come as a surprise if we recall the physical interpretation of vorticity given in Section 3.4. It was shown that vorticity at a point is equal to twice the instantaneous angular velocity of a very small element of fluid around that point. Now, in a non-viscous fluid there would be no tangential stresses acting over the surface of a fluid element, but only the pressure normal stresses. Therefore, there would be no forces that can produce the torque that is necessary for an element starting from rest to acquire rotation.

If we now use the above assumptions in the problem of a three-dimensional body moving with uniform translational velocity $\boldsymbol{U}$, we again end up with the Laplace equation (5.53) for the velocity potential. However, as already discussed in Section 4.7.2, the velocity boundary conditions at a solid surface are drastically different for a non-viscous fluid with respect to the viscous case, for both mathematical reasons (the lowering of the order of the differential equations) and physical reasons (the absence of the no-slip condition). Therefore, only the non-penetration boundary condition (2.30) applies at a solid surface, over which the velocity is free to slip.

In terms of velocity potential, and in a reference system fixed to the fluid, the boundary conditions for a non-viscous fluid would then be

$$\frac{\partial \varphi}{\partial n} = \boldsymbol{U} \cdot \boldsymbol{n}, \qquad (5.56)$$

where $\boldsymbol{n}$ is the normal unit vector at the considered point over the body surface and n is the scalar coordinate in the direction of $\boldsymbol{n}$. Furthermore, adequate vanishing of velocity at infinite distance from the body must be imposed. Obviously, in a reference system fixed to the moving body, the right-hand side of (5.56) would be zero and the velocity at infinity would tend to the undisturbed value $\boldsymbol{U}$. In both reference systems the equation for the velocity potential would always be (5.53).

Now, it may be shown that the solution of this problem exists, is unique, and corresponds to a non-zero velocity field, which then describes the irrotational incompressible flow of a non-viscous fluid around a solid body and is generally characterized by a non-zero value of the velocity component tangent to the body surface.

The above mathematical problem can be solved without great difficulty for a generic body shape, using either theoretical or numerical means. Therefore, we may use the derived solution for the potential function φ to obtain, by differentiation, the velocity field and then, through Bernoulli's theorem, the pressure field. The latter is actually all we need to evaluate the forces acting on the body, because no tangential forces act over its surface due to the assumption of non-viscosity. Therefore, we can integrate the elementary pressure forces over the body surface, and obtain the resultant force acting on it, as well as its components with respect to any reference system.

Unfortunately, it can be demonstrated that the outcome of all this procedure is that the resultant force acting on the body is zero. This is known as *d'Alembert's paradox*, after the French scientist Jean le Rond d'Alembert who first derived it, although for a slightly simplified case. This disappointing result is in evident contrast with the incontestable everyday experience that a body moving in a fluid is subjected, at least, to a force component acting in the direction of its motion and opposing it (which is commonly denoted *drag*).

There is a physical explanation for this apparently surprising result. In fact, no mechanisms producing dissipation are present in the flow of an ideal non-viscous fluid, and it may be shown that the kinetic energy of the whole flow field does not change with the translation of the body. This corresponds to no work being done by the body on the fluid; therefore, considering that, by definition, the opposite of drag is the only force component that can do work on the fluid, no drag acts on the body. In reality, it will be shown in Chapter 10 that drag may appear even in non-viscous flow, but only when the body is accelerating, and not translating with uniform velocity. In that case, the kinetic energy of the whole flow field does change in time, and this corresponds to the appearance of a drag force on the body that is proportional to its acceleration.

Returning to d'Alembert's paradox, the above-cited formulation is strictly correct only for a three-dimensional body moving in an infinite domain of fluid. Conversely, for the two-dimensional case (say a motion in the $x-y$ plane) the conclusion is slightly different. If we consider a body immersed in a steady flow with velocity U in the x direction, we again find that the force component in that same direction, i.e. the drag force, is zero, but a force in the cross-flow direction may exist, and is given by the following formula:

$$F_y = -\rho U \Gamma. \tag{5.57}$$

The quantity Γ in (5.57) is the circulation of the velocity vector around a generic closed curve C rounding the body and, in agreement with the definition introduced in Section 1.2, is given by the expression

$$\Gamma = \oint_C \boldsymbol{V} \cdot d\boldsymbol{l}, \tag{5.58}$$

where $d\boldsymbol{l}$ is a line element along the curve, which must be followed in the counter-clockwise direction.

The result given by relation (5.57) shows that, in principle, a force in the cross-flow direction might arise even in a non-viscous flow. Apparently, this opens the way to important applications such as the evaluation of the *lift* on a wing section, albeit with no prediction of the drag force. However, it may be shown that, with the assumptions of incompressible flow and non-viscous fluid, the generation of a non-zero circulation around a body moving from rest is impossible. In other words, there is no way of justifying the appearance of circulation if we remain within the limits of a non-viscous flow model. In fact, the existence of a circulation is strictly connected with the presence of vorticity which, as already pointed out, cannot be produced in the incompressible flow of a non-viscous fluid starting from rest. Therefore, we find that an apparently reasonable simplified model gives rise to results that either are in contradiction with experience or, in any case, do not seem to be able to lead to the development of satisfactory procedures for the prediction of the flow features and of the consequent fluid dynamic forces. Nevertheless, this deduction deserves a deeper analysis, to which we devote the following section.

5.4.2. *Further discussion on ideal and real flows*

The result that the incompressible flow of a non-viscous fluid corresponds to zero global forces acting on bodies moving in uniform translational motion is certainly a discouraging one. However, before discarding the non-viscosity assumption as leading to useless mathematical exercises, with no reference to physical reality, it may be convenient to verify whether the details of the relevant results are always far from those that may be experimentally observed.

To this end, let us first consider the uniform translational motion of a body whose dimension in the direction of motion is much greater than its dimension in the cross-flow direction. Furthermore, we assume that the geometry of the body is such that it is symmetrical with respect to the direction of motion, rounded on its fore part, and characterized by a sharp edge at the end of its rear part. We will call a body of this type a *streamlined* body, and the reasons for its particular shape will become clear in the following chapters.

If the body is moving leftwards with velocity U in a still fluid, where the unperturbed pressure is p_∞, we define the *pressure coefficient* as

$$C_p = \frac{p - p_\infty}{\frac{1}{2}\rho U^2}. \tag{5.59}$$

The pressure coefficient is thus a non-dimensional scalar quantity giving the variation between the local pressure at a generic point in the flow and the unperturbed pressure.

A qualitative comparison between the distributions of the pressure coefficient over the surface of the above-described body that derive from a calculation assuming the fluid to be non-viscous and from experimental measurements is shown in Fig. 5.1. Note that the values of the pressure coefficient are reported, as is usual practice, using arrows pointing towards the body surface if they are positive, and outwards if they are negative. In practice, these arrows indicate the differential pressure forces (i.e. those due to motion) acting on the local unit elements of surface. The global pressure force acting on the body can be obtained by integrating these local pressure forces over the body surface.

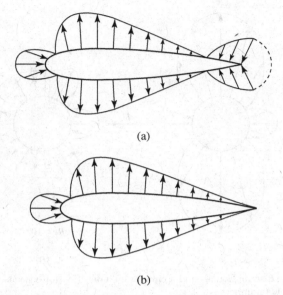

(a)

(b)

Fig. 5.1. Pressure distribution around a streamlined body. (a) Non-viscous flow; (b) real flow.

As would be expected, the two pressure distributions in Fig. 5.1 are different. In the non-viscous flow the pressures must sum up to a zero resultant force, as predicted by d'Alembert's paradox, while this is not true for the real flow situation, for which a non-zero resultant pressure force is clearly present in the direction of motion. It should also be recalled that, in the latter case, a drag force due to the tangential stresses acting over the body surface is also present, while it is absent in the ideal non-viscous flow condition.

Despite these differences, it can also be appreciated that, in practice, the two pressure distributions only differ in their values on the rear part of the body, say, on the last 20% of the body surface. Conversely, the pressures over the fore and mid portions of the body predicted by the non-viscous flow model are in good agreement with the experimental ones.

Let us now consider a similar motion for a different body, namely a circular cylinder. The relevant pressure distributions for the non-viscous and real cases are reported in Figs. 5.2(a) and (b) respectively. As can be seen, the situation is definitely different from the previous one. Apart from the very fore part of the body surface, the pressure distributions for the non-viscous and real cases are now completely different, both quantitatively

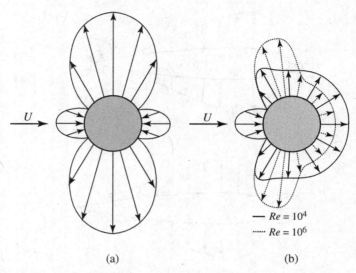

Fig. 5.2.   Pressure distribution around a circular cylinder. (a) Non-viscous flow; (b) real flow at two Reynolds numbers.

and qualitatively. Furthermore, even as regards the experimental data, significantly different pressure distributions are found for the two values of Reynolds number in Fig. 5.2(b).

The conclusion that can be drawn from the above examples is that the geometry of the moving body may have a profound influence on the level of approximation provided by the results of an ideal non-viscous flow model. Even if this model can never predict the global forces acting on a body, in the first case it provides values of the pressures acting on most of the body surface with an accuracy that might be considered as satisfactory for many practical applications. On the other hand, this is certainly not the case for the example in Fig. 5.2.

The main purpose of the following chapters in Part I of the present book will be explaining the physical reasons leading to such dissimilar results for different classes of body geometry and different types of motion. This will naturally lead us to identifying the geometrical features that a body must possess to belong to the first class, i.e. to the one exemplified by Fig. 5.1. We shall also show that, for this type of body, the solution of the mathematical problem that corresponds to the non-viscous flow model may be used as a first step in a procedure leading, eventually, to satisfactory predictions of the actual pressure distributions and global forces found in experiments.

Before proceeding in our treatment, it may be interesting to analyse, in some more detail, an apparently plausible justification that is often given for the use of the non-viscous flow model. We start by neglecting the effect of the body forces, which are not essential for our discussion. We then assume, as previously done, that we can identify a reference length L and a reference velocity U for our problem, and we express in non-dimensional form the incompressible flow momentum equation, (5.35), by defining a non-dimensional velocity vector $\mathbf{V}' = \mathbf{V}/U$, a non-dimensional time $t' = tU/L$ and a non-dimensional pressure $p' = p/(\rho U^2)$. After simple manipulations, Eq. (5.35) can be recast as

$$\frac{\partial \mathbf{V}'}{\partial t'} + \mathbf{V}' \cdot \mathrm{grad}\,\mathbf{V}' = -\mathrm{grad}\,p' + \frac{1}{Re}\nabla^2\mathbf{V}', \qquad (5.60)$$

where we have introduced the Reynolds number, $Re = UL/\nu$, and the gradient and Laplace operators should be interpreted as implying derivatives with respect to space coordinates that have been made non-dimensional using length L.

Equation (5.60) clearly highlights the role of the Reynolds number as a *similarity parameter*. In other words, all flow problems characterized by a similar geometry and by the same value of the Reynolds number are described by an identical non-dimensional differential equation. From its solution, the dimensional quantities corresponding to each flow may then be obtained from the non-dimensional ones by introducing the relevant values of U and L. Furthermore, it is easily realized that the non-dimensional velocities and pressures relating to different (geometrically similar) flows can be expressed as a function of the Reynolds number by means of so-called *similarity laws*, which can be derived not only theoretically, but also experimentally.

In perfect agreement with our interpretation of the Reynolds number as a parameter providing the ratio between inertia and viscous forces, we also see from Eq. (5.60) that the larger the Reynolds number the smaller the importance of the viscous term in the momentum equation. This naturally leads to the conclusion that, apparently, the non-viscous flow equations may be interpreted as the limit of the viscous flow equations for Re tending to infinity, rather than as the equations of motion of a non-viscous fluid.

However, there are mathematical and physical difficulties in this interpretation. First of all, the statement that the viscous term in the momentum equation vanishes for Re tending to infinity tacitly implies that, in the whole flow, the Laplacian of velocity does not simultaneously tend to infinity. As will become clearer in the following chapter, this is not generally true, at least in the neighbourhood of solid surfaces. Furthermore, from the mathematical point of view, the limit process gives rise to a definitely different equation, due to the disappearance of the highest order term in the equation. As already mentioned, this entails that the number of necessary boundary conditions decreases, and that only the non-penetration condition may be imposed for velocity at a solid wall. Consequently, the solution of the limit equation does not have the same mathematical and physical properties as the solution corresponding to finite Reynolds numbers (however high they may be), for which the no-slip boundary condition always applies. In other words, a qualitative discontinuity occurs when the limit is reached.

Therefore, even if we try to justify the use of the non-viscous flow model with the infinite Reynolds number argument, when using this model what we are actually doing is describing the motion of an effectively non-viscous fluid. Following the same rationale we used when discussing the validity of the incompressible flow model, we might then say that using the non-viscous flow model corresponds to assuming that the solution of the limit of the viscous flow equations for Re tending to infinity coincides with the limit of the solution of the viscous flow equations for Re tending to infinity. Again, in general, there is no reason to expect this to be true, and in most cases we may indeed say that it is not, at least as long as the conditions for using the incompressible flow model are also satisfied. Conversely, as we have anticipated in Section 4.7.2, when compressibility plays an essential role, as in supersonic flows, the performance of the non-viscous flow model may be much more satisfactory in a wider range of cases.

Nonetheless, we have seen that, even in incompressible flows, there are situations in which the solution of the non-viscous flow equations provides, at least locally, a first approximation to the experimental evidence. Therefore, an essential point in our analysis will be to understand when and, especially, why this occurs. As already mentioned, the above points should become clearer in the following chapters, thanks to an in-depth examination of the role played by viscosity and vorticity in incompressible flows. In particular, we shall show that, when dealing with certain incompressible flow problems, it is actually not necessary to make the assumption of non-viscosity to advantageously use solutions of the Laplace equation (5.53) with the non-penetration boundary condition (5.56), in spite of the fact that this mathematical problem apparently characterizes the flow of a non-viscous fluid.

Chapter 6

VORTICITY DYNAMICS
IN INCOMPRESSIBLE FLOWS

6.1. Origin of Vorticity in Incompressible Flows

The importance of the presence of vorticity in the flow, as well as the impossibility of a completely irrotational flow of an incompressible viscous fluid in the presence of solid boundaries, have already been pointed out in the previous chapters. Therefore, it is now essential to understand how vorticity is introduced in a homogeneous incompressible flow. We neglect, for the moment, the possible effect of body forces, which will be analysed in the following section.

Let us start by considering an extremely simple two-dimensional case, namely an infinite plate positioned at $y = 0$ that is impulsively set in motion in a direction parallel to its surface, say in the negative (leftwards) x direction with velocity $-U$. In a reference system fixed to the plate the situation will correspond to the whole fluid moving instantaneously in the positive x direction with velocity U, as shown in Fig. 6.1. Obviously, this is an idealized flow condition, which cannot occur in reality, but it serves well the purpose of showing the physical mechanisms that take place when a solid wall is set in relative motion with respect to a fluid.

Let us now consider a very small fluid element, say $\delta x - \delta y$, adjacent to the plate. If there were no viscosity, this element would not be influenced by the motion of the plate at all, as the fluid particles would be allowed to slip over the solid surface. Therefore, in a reference system fixed to the fluid this element would not move, whereas, in a reference system fixed to the plate, it would simply slide over it (see Fig. 6.2(a)), without changing its shape.

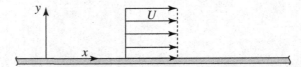

Fig. 6.1.   Impulsive start of an infinite flat plate in a reference system fixed to the plate.

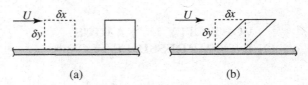

Fig. 6.2.   Impulsive start of an infinite flat plate in a reference system fixed to the plate.
(a) Non-viscous flow; (b) viscous flow.

On the other hand, for any real viscous fluid, the no-slip boundary condition will act immediately at the onset of the motion, so that the particles of the fluid element in contact with the surface of the plate will acquire its velocity, which corresponds to no motion in a reference system fixed to the plate. However, the fluid particles at a distance δy from the surface will move rightwards with velocity U. Therefore, after a very small time dt from the start of the motion, the above-mentioned fluid element will be distorted, as shown in Fig. 6.2(b).

It is now important to recall that the vorticity vector is equal to twice the angular velocity of an elementary volume of fluid. Therefore, looking at Fig. 6.2(a) one can see that, in the ideal case of non-viscous flow, no rotation of any fluid element adjacent to the plate will occur and, consequently, vorticity will remain zero everywhere. Conversely, for a real viscous fluid the deformation of the element corresponds not only to a distortion but also to a rotation, as may be appreciated by observing the variation of the direction of the diagonal of the element in Fig. 6.2(b); therefore, the fluid element acquires vorticity.

Moreover, we may even estimate the amount of vorticity that has been introduced in the flow by applying Stokes' theorem to the deformed fluid element of Fig. 6.2(b), immediately after the start of the motion. Considering that the total contribution from the lateral sides of the contour is zero, the result is that the circulation around the boundary of the element — which must be evaluated by following the contour in the counter-clockwise direction — is equal to $-U\delta x$. Consequently, a flux of vorticity, globally equal to this value of the circulation, is present through the surface of the fluid element. If $\bar{\omega}$ is the mean value of the vorticity crossing the element surface, we may then write

$$\bar{\omega}\delta x \delta y = -U\delta x. \tag{6.1}$$

Therefore, the global amount of vorticity, say γ, normal to the $x-y$ plane, i.e. in the z direction, that is present above a point over the plate surface (per unit length in the z direction) is equal to

$$\gamma = \bar{\omega}\delta y = -U. \tag{6.2}$$

The negative sign is consistent with the fact that the relevant rotation is clockwise, i.e. that the vorticity vector is directed in the negative z direction. It is easy to check that the sign would not change by using a reference system fixed to the fluid. Note also that as δy tends to zero, the value of $\bar{\omega}$ tends to infinity, but the global value γ remains finite and always equal to $-U$.

In conclusion, we may say that the onset of motion causes the appearance of an extremely thin layer of vorticity over the surface of the plate and that the total amount of this vorticity may be determined from knowledge of the value of the velocity outside the layer itself. Furthermore, as is evident from the above description, vorticity appears in the flow due to the viscous no-slip boundary condition, which prevents the presence of a velocity difference between fluid particles and solid walls. Note that, if we analyse this problem in a reference system fixed to the fluid, we observe that the impulsive motion of the plate generates an equal motion of the fluid particles adjacent to its surface, while all the remaining fluid is not affected and does not move.

Obviously, if we assume the plate to be infinitely thin, and consider the fluid to be present on both its sides, the vorticity produced over the lower surface of the plate after the onset of the motion will still be given by relation (6.2), but with the opposite sign. Indeed, the vorticity over the lower surface will be counter-clockwise, and thus positive.

Once produced, the subsequent development of vorticity in the flow is governed by the equation of vorticity dynamics which will be derived and discussed in Section 6.2; we focus here on the mechanism of vorticity generation for some further cases.

In particular, let us analyse a variation of the above problem, i.e. the flow produced by a flat plate of length L and negligible thickness, which again moves impulsively from rest with velocity U in the negative

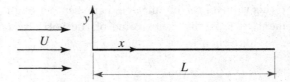

Fig. 6.3.   Impulsive start of a finite flat plate in a reference system fixed to the plate.

Fig. 6.4.   Impulsive starting flow of a finite-thickness body.

(leftwards) x direction. In a reference frame fixed to the plate, the situation at the start of the motion is then equivalent to the plate being suddenly immersed in a flow with constant velocity U directed in the rightwards direction, as shown in Fig. 6.3.

Again, it is reasonable to infer that, at the start of the motion, two thin layers of vorticity are generated over the opposite surfaces of the plate due to the no-slip boundary condition. However, the finiteness of the plate implies that the amount of vorticity of each sign (per unit length in the z direction) that is generated is now finite. In particular, the total vorticity values are $-UL$ and UL, respectively over the upper and lower surfaces. Note that the global amount of vorticity is still zero, as it was before the start of the motion.

We now consider the more realistic case of a thick two-dimensional body, symmetrical with respect to the direction in which it is moving impulsively from rest with velocity $-U$; the starting flow relative to the body is qualitatively shown in Fig. 6.4. The shape of this body is not really a generic one, as it is characterized by a length in the direction of motion that is sufficiently larger than its cross-flow thickness and by a sharp downstream end. The implications of such a choice will become clear in the next chapter.

With respect to the previous cases, the important difference now is that, due to its thickness, the body will also displace the fluid particles, which, obviously, cannot penetrate its solid surface. Consequently, as no vorticity is initially present in the domain, the velocity field that tends to be produced at the start of the motion must satisfy the irrotational flow problem given by Eq. (5.53) with the non-penetration boundary condition (5.56) at the body surface, i.e. the same problem that would apply if the fluid were non-viscous. Two further conditions must be added: that the circulation around the body is zero and that the velocity suitably vanishes at infinity.

Therefore, if φ is the velocity potential in a reference system fixed to the fluid, the velocity is $\boldsymbol{V} = \boldsymbol{U} + \text{grad } \varphi$ in a reference system fixed to the body, and the initial flow field that tends to be produced is the solution of the following mathematical problem (where $\boldsymbol{n}$ is the normal unit vector at a point on the body surface):

$$\begin{cases} \nabla^2 \varphi = 0 \text{ in the flow domain}, \\ \dfrac{\partial \varphi}{\partial n} = -\boldsymbol{U} \cdot \boldsymbol{n} \text{ over the body contour}, \\ \varGamma = 0 \text{ around any circuit rounding the body}, \\ \varphi \to 0 \text{ at infinite distance from the body}. \end{cases} \qquad (6.3)$$

As shown in Fig. 6.4, in a reference system fixed to the body, the corresponding potential flow is tangent to the surface and is characterized by two stagnation points (say A and B), positioned at the front and rear extremities of the body respectively. Now, let x be a clockwise curvilinear coordinate along the body surface with origin at A, and $V(x)$ the velocity component of this flow tangent to the body surface at a generic point. At the start of the motion the no-slip boundary condition will immediately act to bring this velocity to zero over the body surface, similar to what was described for the flat plate. Therefore, two layers of vorticity of opposite sign will be generated over the upper and lower surfaces of the body, which may be evaluated with a similar procedure as that used for the flat plate. In effect, the local value of vorticity over a generic point on the surface (per unit length in the direction normal to the flow) is equal to $-V(x)$ but, considering the flow shown in Fig. 6.4 and the direction of increasing values of x, $V(x)$ is positive over the upper surface of the body while it is negative over the lower surface. We thus observe that the value of the generated vorticity at the start of the motion is no longer given simply by the velocity of the body, but by the solution of the potential flow problem (6.3).

Therefore, the global amounts of negative and positive vorticity initially generated by the no-slip boundary condition may be obtained, in this case, from the integrals of $V(x)$ from point A to point B, evaluated over the upper and the lower surfaces of the body respectively. If we consider that the circulation of the velocity around the body contour is zero, we may also note that the integral over the whole flow field of the generated vorticity is zero, as in the case of the finite plate; as we shall see later on in this same chapter, this is not fortuitous.

The above description of the generation of vorticity at moving solid surfaces allows the different roles of the non-penetration and no-slip conditions (which obviously act simultaneously) to be highlighted. In particular, even if vorticity is generated by the viscous no-slip condition at the wall, it is seen that its value at the initial time may be evaluated from the solution of the potential flow problem, with the boundary condition that the normal relative velocity component be zero at the wall. Incidentally, we may note that, even in the previous example relating to the finite flat plate, the flow field that tends to be produced at the start of the motion is the solution of problem (6.3). However, in that case the non-penetration condition does not introduce any velocity perturbation in the flow, and the potential flow solution is the trivial one with either zero or constant velocity in the fluid, depending on the reference system.

When the body set in impulsive motion is a generic three-dimensional one, the flow field representing the irrotational flow solution satisfying the non-penetration condition will also be three-dimensional. In that case formula (6.2) giving the local intensity of the vorticity contained in a thin layer of fluid adjacent to the body surface may be generalized as follows (see also Chapter 11 for further details):

$$\gamma = n \times V. \tag{6.4}$$

In this formula, γ is the vector representing the local global vorticity over a point on the body surface, where n is the normal unit vector (now directed from the surface into the fluid), and V is the tangential vector component of the velocity deriving from the potential flow solution. Thus, the vector γ is tangent to the body surface and orthogonal to the velocity V.

Finally, we may observe that if a time variation of the body velocity occurs, then further vorticity will be produced at its surface, with a sign determined by the sign of the body acceleration. Thus, the acceleration of a solid boundary may be seen as a source of vorticity in the flow. This also suggests that the assumption of impulsive motion of the body, which has been made up to now for simplicity, is actually not indispensable to explain the origin of vorticity. In fact, in all the above cases, the motion could have started with any acceleration law, which might be considered as a succession of infinitesimal velocity increments. However, during the transient in which the velocity varies, the vorticity generated at previous times would have evolved in the field following dynamical processes that can be described only if the equation ruling the dynamics of vorticity is already available.

6.2. Vorticity Dynamics Equation and its Physical Interpretation

Having briefly explained the origin of vorticity from solid boundaries in an incompressible flow, it is now necessary to derive the equation describing the dynamics of vorticity, so that its subsequent evolution and its connection with the main features of the velocity field can be examined in more detail.

We start our analysis by rewriting the momentum balance equation for a homogeneous incompressible flow, Eq. (5.35), in a slightly different form in which we use the expression for the convective acceleration given by identity (5.41):

$$\frac{\partial \boldsymbol{V}}{\partial t} + \boldsymbol{\omega} \times \boldsymbol{V} = \boldsymbol{f} - \text{grad}\left(\frac{p}{\rho} + \frac{V^2}{2}\right) + \nu \nabla^2 \boldsymbol{V}. \tag{6.5}$$

Next, it is expedient to recall the following vector identity, which is valid for any two vectors $\boldsymbol{a}$ and $\boldsymbol{b}$:

$$\text{curl}(\boldsymbol{a} \times \boldsymbol{b}) = \boldsymbol{a}\,\text{div}\,\boldsymbol{b} - \boldsymbol{b}\,\text{div}\,\boldsymbol{a} + \boldsymbol{b} \cdot \text{grad}\,\boldsymbol{a} - \boldsymbol{a} \cdot \text{grad}\,\boldsymbol{b}. \tag{6.6}$$

We now put $\boldsymbol{a} = \boldsymbol{\omega}$ and $\boldsymbol{b} = \boldsymbol{V}$ in (6.6) and note that both velocity and vorticity are solenoidal vectors, i.e. their divergence is zero. In effect, the continuity equation (5.31) applies to the velocity field and, as may be easily verified, the divergence of any curl (and thus also of vorticity) is zero. Consequently, we get the following expression

$$\text{curl}(\boldsymbol{\omega} \times \boldsymbol{V}) = \boldsymbol{V} \cdot \text{grad}\,\boldsymbol{\omega} - \boldsymbol{\omega} \cdot \text{grad}\,\boldsymbol{V}. \tag{6.7}$$

The fundamental equation describing the dynamics of vorticity can now be obtained by taking the curl of both sides of Eq. (6.5). To this end, we recall that the curl of the gradient of a scalar function is always zero and note that the curl can be interchanged with the time derivative and the Laplacian, because all these operators are commutative. By using relation (6.7), we obtain the following *vorticity dynamics equation*:

$$\frac{\partial \boldsymbol{\omega}}{\partial t} + \boldsymbol{V} \cdot \text{grad}\,\boldsymbol{\omega} = \text{curl}\,\boldsymbol{f} + \boldsymbol{\omega} \cdot \text{grad}\,\boldsymbol{V} + \nu \nabla^2 \boldsymbol{\omega}. \tag{6.8}$$

It is easily seen that the two terms on the left-hand side of Eq. (6.8) represent the material derivative of vorticity, i.e. the variation of vorticity

experienced by a fluid particle during its motion. Therefore, the terms on
the right-hand side express the possible sources of variation of vorticity
for a moving particle, and we now discuss the physical meaning of each of
these terms.

The first one, i.e. the curl of the body force, is non-zero only if f is non-
conservative, i.e. if it cannot be expressed as the gradient of a potential
function. This is the case, for instance, for the Coriolis force, which has
great importance in large-scale geophysical flows, such as the motion of
the atmosphere. Indeed, the Coriolis force connected with the rotation
of the Earth may introduce vorticity in large masses of moving air, thus
producing the cyclonic rotation of hurricanes, which is counter-clockwise
in the northern hemisphere and clockwise in the southern hemisphere.
However, the Coriolis force due to Earth's rotation is very small indeed
and, in normal laboratory conditions, its vorticity production effect may be
detected only through specific and carefully conducted experiments. The
Coriolis force can hardly influence, for instance, the direction of rotation of
flows that may be observed in common everyday situations, such as water
flowing through a draining sink. In that case, the direction of rotation is
essentially determined by previously introduced vorticity, which is, at least,
a function of the shape of the sink and the way it was filled.

On the other hand, many common body forces, such as the gravitational
force or the centrifugal force, are conservative, and can thus be derived from
a potential. Therefore, if only this type of body forces act on the flow, they
do not contribute to variations of the vorticity of a moving fluid particle,
and the vorticity dynamics equation becomes

$$\frac{\partial \boldsymbol{\omega}}{\partial t} + \boldsymbol{V} \cdot \mathrm{grad}\,\boldsymbol{\omega} = \boldsymbol{\omega} \cdot \mathrm{grad}\,\boldsymbol{V} + \nu \nabla^2 \boldsymbol{\omega}. \tag{6.9}$$

In order to explain the physical process through which the term
$\boldsymbol{\omega} \cdot \mathrm{grad}\,\boldsymbol{V}$ can produce vorticity variations, it is useful to assume, for
the moment, that it is the only one acting, i.e. that the body forces are
conservative and that the viscous term is zero or negligible. Furthermore,
we should remember that a *vortex-line* is a line that, at a certain time
instant, is tangent to the vorticity vectors. If we now take all the vortex-
lines passing through a closed circuit inside the flow, we obtain a *vortex-tube*.
By construction, then, there is no flux of the vorticity vector through the
lateral surface of a vortex-tube, while the flux is non-zero through any cross-
section S of the vortex-tube. This flux corresponds to a non-zero circulation
Γ of the velocity vector around the contour C of the cross-section. In effect,

if n is the unit vector normal to the cross-section, Stokes' theorem gives

$$\Gamma = \oint_C \boldsymbol{V} \cdot d\boldsymbol{l} = \int_S \boldsymbol{\omega} \cdot \boldsymbol{n} \, dS. \qquad (6.10)$$

The circulation Γ is referred to as the *strength* of the vortex-tube. With the above assumptions, i.e. no effect of body forces and viscosity, the following *Helmholtz theorems* apply to the vorticity field:

1. the strength of a vortex-tube is constant along the tube;
2. vortex-lines are material lines;
3. the strength of a vortex-tube remains constant as the tube moves with the fluid.

Note that the first Helmholtz theorem derives directly from the solenoidality of the vorticity field and the application of Gauss' theorem. Thus, the assumption of non-viscosity is not necessary for its validity. Conversely, this assumption is needed for the remaining theorems, which concern the dynamics of the fluid. Specifically, the second theorem implies that the particles composing a vortex-line at a certain instant will continue to form the same vortex-line, generally displaced and distorted, at later instants, while the third one is a consequence of the first two.

Let us now consider a material volume coinciding with a vortex-tube with small cross-section and length. The strength Γ of the vortex-tube may be expressed as ωS, where S is a generic cross-section of the tube and ω the average value of the vorticity crossing it. Note that, in general, the velocity vector will be neither parallel nor orthogonal to the vorticity vector. Without loss of generality, we imagine the portion of vortex-tube to be both rectilinear and with constant cross-section, as shown in Fig. 6.5, and decompose the velocity vector into two components, one parallel to $\boldsymbol{\omega}$, say $\boldsymbol{u}$, and the other orthogonal to $\boldsymbol{\omega}$, say $\boldsymbol{v}$.

Now, in order to analyse the effect of the vector $\boldsymbol{\omega} \cdot \mathrm{grad}\boldsymbol{V}$, we first consider its component $\boldsymbol{\omega} \cdot \mathrm{grad}u$, where u is the modulus of $\boldsymbol{u}$. If this

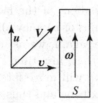

Fig. 6.5.   Example of small vortex-tube and velocity field.

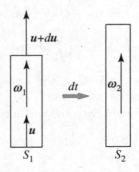

Fig. 6.6.   Effect of a gradient of the velocity component parallel to vorticity.

component is not zero, then the vector $\mathrm{grad}\,u$ has a non-zero component in the direction of $\boldsymbol{\omega}$, i.e. there is a difference between the normal velocities of the two extreme cross-sections of the vortex-tube, as shown in the left part of Fig. 6.6.

Therefore, in a small time interval dt these two cross-sections will displace differently. For instance, if u increases in the $\boldsymbol{\omega}$ direction, then the length of the vortex-tube will increase and its cross-section will simultaneously decrease, as in Fig. 6.6. This reduction of the cross-section is a consequence of the fact that the second Helmholtz theorem implies that a vortex-tube is a material volume, and this implies — due to the assumption of incompressibility — that the measure of this volume must remain constant during the motion. However, in agreement with the third Helmholtz theorem, the strength of the vortex-tube must also remain constant and thus the vorticity will increase correspondingly, so that the following relation is satisfied (see Fig. 6.6):

$$\omega_1 S_1 = \omega_2 S_2. \tag{6.11}$$

Obviously, if the velocity decreases along the vortex-tube, the opposite occurs, i.e. the cross-section increases and the vorticity value decreases.

Let us now see how the gradient of the lateral velocity, i.e. the term $\boldsymbol{\omega} \cdot \mathrm{grad}\,v$ (where v is the modulus of $\boldsymbol{v}$), may also vary the vorticity vector. By referring to Fig. 6.7, where, again, we assume that v increases in the $\boldsymbol{\omega}$ direction, we can indeed see that, in a time interval dt, the two extreme cross-sections of the vortex-tube will move laterally by different amounts, producing a *tilting* of the vortex tube and thus a change in the *direction* of the $\boldsymbol{\omega}$ vector.

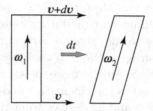

Fig. 6.7. Effect of a gradient of the velocity component orthogonal to vorticity.

This mechanism of variation of the magnitude and direction of the vorticity vector, due to the stretching and bending of vortex-tubes by the gradients of the velocity components that are parallel and orthogonal to ω, is extremely important for explaining the production, in certain cases, of high-intensity and concentrated vorticity. For instance, regions where a vertical component of vorticity is present may be stretched in the atmosphere by the action of thermally induced velocity gradients, thus possibly causing the generation of strong tornados.

More generally, we can say that that in three-dimensional flows the term $\omega \cdot \operatorname{grad} V$, which is completely independent of viscosity, may make a significant contribution to the vorticity dynamics. On the other hand, it is clear that this term is zero when the flow is two-dimensional. In effect, if the flow takes place, for instance, in the $x-y$ plane, no variation in the z direction exists, so that the gradients of all velocity components will lie in the same plane of motion, whereas the vorticity vector will be orthogonal to it.

Therefore, for two-dimensional motion with conservative body forces, the vorticity dynamics equation reduces to the simpler form

$$\frac{\partial \omega}{\partial t} + V \cdot \operatorname{grad} \omega = \nu \nabla^2 \omega. \tag{6.12}$$

Through this equation, it is possible to analyse thoroughly the specific contribution of viscosity to the dynamics of vorticity. To this end, we start by rewriting Eq. (6.12) with the introduction of the material derivative operator

$$\frac{D \omega}{D t} = \nu \nabla^2 \omega. \tag{6.13}$$

It is now easily seen that this equation coincides, in practice, with Eq. (2.26) describing the diffusion of a quantity, the only difference being that the partial derivative with respect to time is replaced by the material derivative. We may thus say that the term $\nu\nabla^2\omega$ represents the variation of vorticity experienced by a fluid particle during its motion due to diffusion by viscous effects. In other words, given a certain particle, vorticity may diffuse to or from neighbouring particles if their vorticity is different from that of the considered particle. Furthermore, it is seen from (6.13) that the diffusivity of vorticity is equal to the kinematic viscosity of the fluid, which, as anticipated in Section 2.7 and as derives from the form of the viscous term in the momentum equation of an incompressible fluid, also coincides with the diffusivity of momentum.

Therefore, the contribution of viscosity to the development of an incompressible flow may be described in various, but equivalent, ways. In particular, we may say that viscosity produces a diffusion of momentum, by causing a variation of the velocity of adjacent particles if a gradient of velocity exists. Alternatively, we may say that the quantity that is diffusing is vorticity, which is directly related to the velocity field through a differential operator. Finally, we may look at the flow evolution in terms of dragging or retarding actions by viscous forces due to the presence of spatial variations of velocity.

As for the variation of the vorticity of a fluid particle in a generic three-dimensional viscous flow, diffusion due to viscosity will occur together with the stretching–bending mechanism and the action of non-conservative body forces, if present. Thus, in general, the strength of a vortex-tube will not remain constant, as required by the Helmholtz theorems, because vorticity may diffuse to or from that tube due to viscosity.

Obviously, the various mechanisms may act together in such a way that the vorticity of a fluid particle remains constant during its motion. For instance, the increase due to stretching might be balanced by a decrease due to diffusion to lower-vorticity adjacent particles. On the other hand, if no stretching of the vortex-tubes occurs (as happens in two-dimensional flows) and the body forces are conservative, the vorticity of a fluid particle can remain constant only if the viscous term in (6.13) is zero, which happens when either the vorticity of the adjacent particles has the same value (locally constant vorticity and no diffusion) or it is different, but diffusion causes the vorticity entering a fluid particle to be equal to that exiting it.

However, if one is interested in the time variation of the vorticity at a fixed point in space, rather than of the vorticity of a fluid particle, attention should be focused on the first term on the left-hand side of Eq. (6.8), i.e. the partial derivative of vorticity with respect to time. It is apparent, then, that another mechanism must be taken into account, namely convection by the velocity field, which is represented by the term $V \cdot \mathrm{grad}\omega$. In other words, vorticity may remain unchanged in time, at a certain fixed point in space, only when the amount of vorticity carried instantaneously to or from that point by convection is balanced by the actions of diffusion, vortex stretching and non-conservative body forces.

Another interesting observation that derives from Eq. (6.9) is that when the body forces are conservative, any vorticity present in the flow may only disappear by annihilation with vorticity of opposite sign. In other words, there is no term in the equation that corresponds to 'dissipation' of vorticity, which may thus be convected, diffused or stretched (in three-dimensional flows), but can actually disappear only if it is brought into contact (basically by diffusion) with vorticity of opposite sign.

6.3. Conservation of Total Vorticity

As we have seen in the previous sections, in an incompressible flow of a homogeneous fluid subjected to conservative body forces, vorticity is produced at solid surfaces and subsequently evolves in the flow as described by Eq. (6.9); some examples of this evolution will be given in Section 6.4. However, we first show that in most practical situations the vorticity field must satisfy an important integral constraint.

In particular, we assume that the flow is due to the motion of one or more finite solid bodies in an infinite domain of fluid otherwise at rest. In other words, the solid bodies are initially at rest, together with the fluid, and are located within finite distances of one another. Subsequently, they move with a prescribed motion, causing motion of the fluid and production of vorticity over their contours, which are the inner boundaries of the fluid domain (the outer one being at infinity). This vorticity will then spread into the fluid by diffusion and will be progressively transported away from the boundaries by both convection and diffusion. At the same time, the no-slip condition provides a mechanism for the continuous generation of vorticity at the surface of the solid bodies. Therefore, vorticity is present in regions surrounding the solid bodies and trailing behind them, while elsewhere it is zero or, more precisely, negligible.

Let us now consider a reference system fixed to the fluid and with its origin located at finite distances from the initial position of all the solid bodies, and let r be the distance of a generic point from the origin. As we have seen, vorticity is effectively confined within finite distances of the origin after the onset of the motion. This feature of the initial vorticity field permits us to show that at any finite time t after the start of the motion vorticity approaches zero exponentially with increasing distance from the origin (see, e.g., Wu, 1981, 2005).

We may now ask if a relation exists providing, in the above conditions, the time variation of the total vorticity present in the flow. This is indeed the case, but to express this relation we must first introduce an extension of the velocity and vorticity fields, so that they also comprise the volume occupied by the solid bodies. This may be done by including the velocity of the points inside the bodies in the velocity field, and by defining vorticity within the bodies as the curl of velocity, which corresponds to twice their angular velocity. Note that the solid bodies are obviously incompressible, and thus the velocity vector inside them is solenoidal everywhere. Moreover, thanks to the no-slip condition, the velocity is continuous at the surface of the bodies, while vorticity may be discontinuous, but remains at least piecewise continuous in the whole extended field.

With the above extension, the vorticity inside the bodies is zero if they are moving with translational velocity, but equal to twice their angular velocity when they are rotating. The whole region jointly occupied by the fluid and the bodies is now denoted by R_∞, and is bounded at infinity by a closed boundary B_∞. The region occupied by the fluid is designated by R_f, and is internally bounded by the surfaces of all the solid bodies, collectively designated by B_s. We may further denote by R_j the region occupied by the jth solid body, and its boundary as B_j.

With all the above definitions and assumptions on the motion, it may be demonstrated that the following relation applies:

$$\frac{d}{dt} \int_{R_\infty} \omega dR = 0. \tag{6.14}$$

This relation, whose proof is given in Chapter 11, is valid for both three-dimensional and two-dimensional flows, and is usually referred to as the *principle of total vorticity conservation*. It implies that if a motion is started from rest, i.e. from a condition in which vorticity is zero everywhere, then the total amount of vorticity in the whole flow field (including the solid bodies) remains zero for all subsequent times. This means that if vorticity of a given sign is introduced into the field, then an equal amount of vorticity of opposite sign will also be introduced.

We have already seen that this is indeed true in the cases of Section 6.1 in which a finite plate or a finite body were set in impulsive translational motion from rest; in both cases, the total vorticity in the fluid region

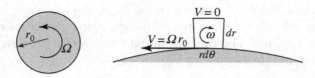

Fig. 6.8. Circular cylinder in impulsive rotational motion.

remained zero after the start of the motion, in agreement with the fact that no rotation of the moving bodies was present. On the other hand, relation (6.14) was obviously not satisfied in the example of a one-sided infinite plate, which was a situation not satisfying the assumptions leading to its derivation. However, it may be of interest to note that the total vorticity was also zero in the case of a two-sided infinite plate. Therefore, at the start of the motion the amounts of vorticity of opposite sign are always equal for finite bodies, but this happens only in particular cases for infinite bodies.

In order to show that relation (6.14) is also valid in the case of a rotating body, let us consider a simple two-dimensional case, i.e. a solid circular cylinder of radius r_0 that is impulsively set in rotational motion with constant counter-clockwise angular velocity Ω (see Fig. 6.8). This implies that the cylinder surface moves impulsively with a counter-clockwise velocity $V(r) = \Omega r_0$, and so also will all the fluid particles in contact with the surface. Consequently, following the line of reasoning of Section 6.1, it is easy to see that an amount of clockwise (and thus negative) vorticity equal to $-\Omega r_0$ will be generated over each point of the cylinder surface, within an extremely thin layer of fluid. Then, the total vorticity present in this layer over the whole circumference of the cylinder will be

$$-\Omega r_0 (2\pi r_0) = -2\Omega \pi r_0^2. \qquad (6.15)$$

Considering that at every point within the solid cylinder the vorticity caused by the impulsive rotation is equal to twice the angular velocity, the total amount of counter-clockwise (and thus positive) vorticity inside the cylinder is equal to 2Ω times the area of the cylinder cross-section, i.e. exactly the opposite of the value given in (6.15).

In summary, if $\mathbf{\Omega}_j$ is the angular velocity of the jth body immersed in the flow, relation (6.14) can also be written in the following form:

$$\frac{d}{dt} \int_{R_f} \boldsymbol{\omega} dR_f = -\frac{d}{dt} \sum_j 2\mathbf{\Omega}_j R_j, \qquad (6.16)$$

which shows that the total vorticity contained in the fluid may vary only if the angular velocity of the contained bodies varies. Furthermore, if the motion starts from rest, and thus the initial total vorticity is zero, then at any subsequent instant in time we have

$$\int_{R_f} \omega dR_f = -\sum_j 2\Omega_j R_j. \tag{6.17}$$

6.4. Flow Evolution in Terms of Vorticity Dynamics

6.4.1. *Infinite plate*

Let us consider again the first case analysed in Section 6.1, namely an infinite plate set in impulsive motion at time $t = 0$ with velocity U in the leftward direction; we now analyse the evolution of the vorticity and velocity fields.

For this particularly simple two-dimensional problem, an exact solution of the equations of motion, with the above-prescribed initial condition, exists. In a reference system fixed to the fluid, and with the positive x axis in the rightwards direction, this solution is given by the following relation:

$$u = -U\left(1 - \mathrm{erf}\left(\frac{y}{2\sqrt{\nu t}}\right)\right), \tag{6.18}$$

where the function

$$\mathrm{erf}(\eta) = \frac{2}{\sqrt{\pi}} \int_0^\eta e^{-\eta^2} d\eta \tag{6.19}$$

is called the 'error function' and varies from 0 for $\eta = 0$ to 1 for $\eta = \infty$.

In a reference system fixed to the body, the fluid is set in motion at time $t = 0$ with velocity U in the rightwards (positive) direction, so that the corresponding solution for the velocity field is obtained by adding U to (6.18); we thus get

$$u = U\mathrm{erf}\left(\frac{y}{2\sqrt{\nu t}}\right). \tag{6.20}$$

Obviously, the vorticity vector is completely defined by its sole non-zero component, i.e. the one in the z direction (say ω), which, in this case, is equal to $-\partial u/\partial y$ and is readily found to be given by

$$\omega = -\frac{U}{\sqrt{\pi \nu t}} e^{-\frac{y^2}{4\nu t}}. \tag{6.21}$$

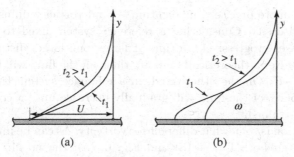

Fig. 6.9.   Velocity (a) and vorticity (b) profiles for impulsively started plate.

As may be easily verified, the vorticity field does not depend on the chosen reference system. This is obvious because, contrary to velocity, vorticity is Galilean invariant, i.e. it has the same value for any reference system in uniform translational motion. Incidentally, this is one of the advantages of the analysis of fluid flows in terms of vorticity.

The above solutions are valid for any time $t > 0$, and describe the variation in time of the velocity and vorticity space distributions. The progressive variation of velocity and vorticity at two subsequent time instants is shown in Fig. 6.9. Note that the two fields do not depend on y and t separately but on the variable $\eta = y/(2\sqrt{\nu t})$, and thus each of them can be plotted using a single curve in terms of this variable.

After having described the mathematical solution for this problem, we now discuss its main features from a more physical point of view. This is a two-dimensional case and, therefore, vorticity may vary at any point of the flow due to convection and diffusion only, as may be deduced from Eq. (6.12). However, convection does not play a role in this case, as may be mathematically seen by observing that the second term in Eq. (6.12) is zero. More physically, we may note that this derives from the fact that convection simply moves the same quantity of vorticity in a direction parallel to the plate. In other words, if we consider the region between two vertical lines at different values of x, the same quantity of vorticity is continuously introduced through one line and removed from the other. This is due to the fact that the plate is considered to be infinite both upstream and downstream.

The consequence is that the only physical mechanism causing the flow evolution is diffusion by viscosity, of both vorticity and velocity. In fact, relation (6.18) is indeed a solution of a pure diffusion equation, with a diffusion coefficient equal to the kinematic viscosity. Therefore, both velocity — or, more precisely, momentum — and vorticity diffuse into the flow due to viscosity. Considering a reference system fixed to the fluid, velocity increases progressively in time at larger and larger distances from the plate and, after a theoretically infinite time, all the fluid will be moving with velocity $-U$. On the other hand, in a reference system fixed to the plate, the velocity of the fluid will gradually decrease until a condition of global rest is attained.

The situation is somewhat different for vorticity. As can be immediately appreciated from both Eq. (6.18) and Fig. 6.9(a), the no-slip boundary condition is always satisfied for any $t > 0$. Therefore, no further production of vorticity is necessary besides the amount generated at the start of the motion, which we have seen in Section 6.1 to be equal to $-U$ per unit length in the x and z directions. Consequently, this vorticity diffuses into the infinite field, and its local value at any point of the flow progressively changes, as can also be seen in Fig. 6.9(b). However, a fundamental point to be stressed is that the global amount of vorticity present along a vertical line above a point on the plate surface always remains the same, i.e. $-U$. This can be readily verified by integrating (6.21) from $y = 0$ to $y = \infty$, and recalling the following fundamental integral:

$$\int_0^\infty e^{-a\eta^2}\,d\eta = \frac{1}{2}\sqrt{\frac{\pi}{a}}. \tag{6.22}$$

Therefore, it is not strictly correct to say that vorticity at a generic point in the flow field becomes zero as time tends to infinity. We should rather say that its value becomes infinitesimal, but in such a way that this infinitesimal value, once integrated along an infinite-length line above the plate, gives a total constant value equal to $-U$. For all practical purposes, there will be no difference between a field in which vorticity is zero and one in which it is infinitesimal. However, there is a conceptual difference, which becomes important when analysing and discussing how a certain vorticity field has been generated and how it may evolve.

For instance, let us imagine that, at the end of the above-described transient phenomenology, the plate is suddenly stopped. This may be considered as a new starting condition in which the whole fluid is moving in the negative x direction with velocity $-U$, while the plate surface and the fluid particles therein are brought to rest. In a reference frame fixed to the fluid, this corresponds to the same initial condition as above, but with the plate suddenly starting in the opposite direction with velocity U. Then, in order to satisfy the no-slip boundary condition, an infinitely thin layer of positive (counter-clockwise) vorticity will form over the surface of the plate. Once again, it will progressively diffuse into the flow, annihilating, in the meantime, the vorticity of opposite sign that was present (albeit with infinitesimal local value) in the field. At the end of this second transient, the whole fluid will come to rest, along with the plate, and the local vorticity will become exactly zero, not an infinitesimal.

A variation of the above problem is when another plate is present, parallel to the one that suddenly moves with velocity $-U$, and placed at a distance $y = d$ away from it. Also in this case, a local vorticity layer of intensity $-U$ will be produced at the start of the motion over each point on the lower plate. We do not enquire, for the moment, as to the availability of an exact solution for this case, but we can infer that the produced vorticity will again start to diffuse in the fluid. However, the available space region where vorticity may diffuse above the plate no longer extends infinitely, now being confined between $y = 0$ and $y = d$. Therefore, this region progressively fills up with the vorticity generated at the start of the motion, and in the final condition, reached after a sufficiently long time, the local vorticity at any coordinate x between the plates will be $\omega = -U/d$. Then, the corresponding final velocity field is the solution of the following simple equation:

$$\omega = -\frac{\partial u}{\partial y} = -\frac{U}{d}, \qquad (6.23)$$

whose general solution is

$$u = \frac{U}{d}y + c. \qquad (6.24)$$

In (6.24) c is a constant that must be chosen so that the boundary conditions are satisfied. Therefore, c will depend on the chosen reference system, and will be equal to $-U$ if the upper plate is taken to be at rest, while it will be 0 in a reference system fixed to the lower plate, with respect to which the upper plate moves rightwards with velocity U.

This type of simple unidirectional motion is referred to as the *Couette problem* and will be thoroughly described in Chapter 9, from the point of view of the relevant steady and unsteady solutions of the equations of motion. However, the above derivation is an instructive example of how an analysis in terms of vorticity dynamics may sometimes be helpful for understanding the features and the evolution of a flow field.

Nonetheless, it may be appropriate to point out that, in the above, examples, the physical process leading to the establishment of the flow field could have been described in different ways. In fact, we particularly concentrated on the diffusion of vorticity by viscosity, but we could also have observed that the diffusing quantity is momentum, or that viscous tangential stresses progressively cause layers of fluid at increasing distances from the plate to move. These different ways of interpreting the evolution of a certain flow may be used not only alternatively, choosing the one that seems more enlightening or that leads to more straightforward predictions, but also, and especially, simultaneously. In effect, our physical interpretations must be consistent with all types of analysis, and this may sometimes be invaluable for reaching a deeper understanding of fluid motions.

To end this section, we now use the simple case of a suddenly moving plate to analyse, more closely, the concept of a layer of vorticity being produced at the initial time over the plate, and also to provide an assessment of its subsequent evolution. A closer look at expression (6.21) shows that, from a strictly mathematical point of view, vorticity is actually present everywhere in the flow, i.e. up to $y = \infty$, for any time $t > 0$. However, it can also be seen that, for very small values of t, vorticity is concentrated in the immediate neighbourhood of the plate surface, and rapidly tends to zero with increasing y. Therefore, it may be expedient to introduce a distance δ from the plate within which a very large percentage of vorticity — say 99% or even 99.9% — is contained instantaneously. For all practical purposes, δ may then be regarded as the thickness of the vorticity-containing region in the fluid. From (6.21) it is clear that if we consider a very small time instant just after the start of the motion, δ will be very small indeed, and we may then safely say that a layer of vorticity has been produced in the flow. Another important result deriving from (6.21), which we shall find to be very useful in later analyses, is that the time evolution of the above-defined thickness is given by the relation

$$\delta = A\sqrt{\nu t}, \tag{6.25}$$

where A is a numerical coefficient depending on the chosen percentage of vorticity.

It is interesting to use relation (6.25) to evaluate the order of magnitude of the time that is necessary for δ to reach a given value. If we take δ to be the distance containing 99% of the total vorticity, it may be seen from (6.21) that the value of A turns out to be approximately 3.5. Considering the values of ν reported in Table 2.1, if the fluid is air the time intervals necessary for δ to reach 1 m and 10 m are, respectively, slightly more than 1.5 hours and 6.5 days. For water the same values of δ would be reached after almost 20 hours and 83 days. This example shows that vorticity diffusion is definitely a slow process for many common fluids having a small kinematic viscosity.

6.4.2. *Rotating circular cylinders*

We now consider the evolution of the flow and vorticity fields for a solid circular cylinder of radius r_0, after it has been impulsively set in rotational motion with constant counter-clockwise angular velocity Ω. This is the case analysed in Section 6.3, where it was shown that a global amount $-2\Omega\pi r_0^2$ of vorticity is produced over the cylinder surface at the start of the motion.

It is useful to express the various quantities in cylindrical coordinates (r, θ, z) with origin at the centre of the cylinder and z coinciding with its axis. If we then denote the velocity components as v_r, v_θ and v_z, both in the fluid and within the cylinder, the components of the vorticity vector are given by relations (3.43), which we rewrite here for convenience:

$$\omega_r = \frac{1}{r}\frac{\partial v_z}{\partial \theta} - \frac{\partial v_\theta}{\partial z}, \qquad \omega_\theta = \frac{\partial v_r}{\partial z} - \frac{\partial v_z}{\partial r}, \qquad \omega_z = \frac{1}{r}\frac{\partial(rv_\theta)}{\partial r} - \frac{1}{r}\frac{\partial v_r}{\partial \theta}.$$

The present flow is two-dimensional and thus the velocity component v_z and the derivatives in the z direction are zero. Consequently, the only non-zero component of the vorticity vector is ω_z. Moreover, considering that we also have $v_r = 0$, it is easy to verify that inside the cylinder, where $v_\theta = \Omega r$, we have $\omega_z = 2\Omega$, as expected.

The vorticity generated just after the start of the motion in a thin layer over the cylinder surface will then diffuse in the infinite fluid domain. In effect, as happened for the infinite plate, in this case no further vorticity is produced at the cylinder surface, and convection has the sole role of continuously redistributing, in the fluid field, the same amount of vorticity along each circumference at $r =$ constant. Therefore, diffusion is the only mechanism by which vorticity evolves in the flow.

Once again, the finite amount of vorticity initially generated in the fluid will diffuse in the infinite region outside the cylinder and, as time tends to infinity, the vorticity will locally tend to a value that is infinitesimal, but such that, when integrated over the infinite fluid domain, it still gives a global value $-2\Omega\pi r_0^2$. Therefore, the velocity field that will be produced after an infinite time from the start of the motion is characterized by an effectively zero local value of vorticity.

As previously described in Section 3.6, a flow with circular streamlines in which v_θ varies as $1/r$ is irrotational. Therefore, considering that the boundary condition at the cylinder surface requires that $v_\theta = \Omega r_0$ for $r = r_0$, the asymptotic velocity distribution in the fluid will be given by the expression

$$v_\theta = \frac{\Omega r_0^2}{r}. \tag{6.26}$$

As can be seen, the total velocity field is identical to the one that was seen in Section 3.6 to correspond to a Rankine vortex, with the core coinciding with the solid rotating cylinder. We have already observed that a Rankine vortex is only a simplified model of a real flow in which vorticity is concentrated in a small region. In fact, the discontinuity in vorticity that is present at $r = r_0$ would be immediately cancelled out by the action of the diffusive term in the vorticity dynamics equation. Conversely, it would be possible to produce, in principle, the above flow field if a cylinder were kept in motion with constant angular velocity for a sufficiently long time (theoretically infinite).

Another interesting fact is that the velocity field given by (6.26), which satisfies the no-slip boundary condition at the cylinder surface, is one of the few exact solutions of the Navier–Stokes equations that correspond to irrotational flows. Note, however, that this is not in contradiction with the findings of Section 5.3.3 because, in this case, the solid body is not translating but only rotating in the fluid.

As was the case for an infinite plate, if the cylinder is suddenly stopped after the asymptotic steady flow has been reached, a layer of counter-clockwise (positive) vorticity, globally equal to $2\Omega\pi r_0^2$, will form over its surface, and will then diffuse in the flow, annihilating the previously present vorticity of opposite sign. In a sense, it is as if the vorticity that was present inside the rotating cylinder were suddenly transferred to its surface. Obviously, the diffusion of the newly formed vorticity corresponds to a progressive decrease of the velocity in the fluid, and thus, after a theoretically infinite time, a state of rest is again reached in the fluid.

A slight variation of the above problem is the case in which the cylinder is hollow and has negligible thickness. In this case, at the start of the motion, the no-slip condition will also cause a layer of positive vorticity to form over the inner surface of the cylinder, while the fluid contained within it will remain at rest. The total value of this inner vorticity is $2\Omega\pi r_0^2$, thus balancing the negative vorticity generated over the outer surface. The subsequent phenomenology relating to the evolution of the outer vorticity will be equal to that previously described. Conversely, the positive vorticity generated over the inner surface will diffuse in the limited fluid region inside the cylinder. Therefore, at the end of a (much shorter) transient, one will attain a final condition in which the above global value of positive vorticity is uniformly distributed inside the cylinder. Consequently, the final local value of vorticity will be constant in this region, and equal to the ratio between the global vorticity initially generated at the inner surface and the area of the cylinder cross-section. In other words, the local vorticity will be 2Ω, and all the fluid within the cylinder will finally rotate with the same angular velocity Ω, as if it belonged to a solid body. The situation will thus be exactly equal to the previously examined case. Should the cylinder be suddenly stopped, an amount $-2\Omega\pi r_0^2$ of clockwise vorticity would form over the inner surface, and then, with its diffusion, would finally cancel the pre-existing vorticity and bring the inner fluid to rest.

We have seen that the irrotational flow field (6.26) is a steady-state solution of the equations of motion for a viscous fluid outside a rotating circular cylinder. It is easy, then, to be convinced that the same solution would be also valid when any of the circular streamlines of this flow, at a generic $r = r_1 > r_0$, coincides with the inner surface of another concentric hollow cylinder, provided this new cylinder rotates with an angular velocity with the value $\Omega_1 = \Omega r_0^2 / r_1^2$. In effect, in that case the outer cylinder inner surface would be moving at exactly the velocity given by (6.26) for $r = r_1$.

However, there is a subtle difference between the two cases, and this can be appreciated by analysing again the evolution of the flow after the two cylinders start impulsively from rest, with angular velocities Ω and Ω_1. It is readily seen that the positive vorticity layer initially generated over the inner surface of the outer cylinder would exactly match the negative one produced over the outer surface of the inner cylinder. Therefore, these two amounts of opposite-sign vorticity would annihilate after the transient, thus producing a truly irrotational flow between the cylinders, rather than a flow with infinitesimal local vorticity. As for the region outside the outer cylinder, the situation would be similar to that of the outer region of the previous case, with exactly the same total negative vorticity as before, balancing the positive vorticity contained inside the inner cylinder.

It should be noted that if the cylinders in the above example had non-negligible thickness, the vorticity contained within their solid parts (equal to twice their angular velocity) should have been considered, and it is easy to see that the global conservation law (6.14) would still hold. Finally, as was previously pointed out, all the above phenomenology could also have been described in terms of the progressive action of the viscous tangential stresses on the fluid particles.

6.4.3. *Finite plate*

The simplest situation of a body translating in a flow is the finite flat plate moving impulsively from rest with velocity $-U$ in a direction parallel to its surface, which was considered in Section 6.1. Even if this is still a somewhat idealized case, as the flow is assumed to be two-dimensional, it is certainly closer to a possible real situation, and satisfies the conditions for the validity of the constancy of the total vorticity.

We have already seen that, at the start of the motion, two layers of opposite-sign vorticity form over the upper and lower surfaces of the plate, with global values $-UL$ and UL respectively. After its generation, this vorticity will immediately start diffusing inside the fluid, similar to what happened for the infinite plate. However, the finiteness of the plate introduces a fundamental difference with respect to that case, namely the fact that now convection plays a role. In a reference system fixed to the plate, the diffusing vorticity is transported downstream by particles placed at increasing distance from the plate, while no vorticity is brought over its surface from upstream.

Consequently, as time passes, vorticity will be present not only around the plate, but also in a region of flow of ever increasing length behind it. This region is known as a *wake*, and contains the vorticity of both signs released from the upper and lower surfaces of the plate. The width of the wake also increases progressively moving downstream, due to diffusion. Thus, the flow will be divided in two different parts: the wake and a region around the plate, in which vorticity is present, and the outer region, where vorticity may be neglected.

It is important to point out that the vorticity-containing region is also the one where variations of velocity have occurred due to viscous stresses. In other words, in a reference system fixed to the plate, the diffusion of vorticity implies the progressive decrease of velocity with respect to the upstream asymptotic value U. Thus, in the region around the plate

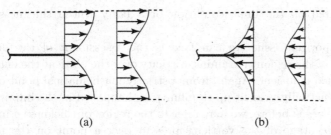

Fig. 6.10. Velocity field at the end of the plate in different reference systems: (a) fixed to the plate, (b) fixed to the fluid.

and in the wake, the momentum will be lower than in the undisturbed stream. Conversely, in a reference system fixed to the fluid, the same flow condition is described by saying that a portion of fluid around and behind the plate is moving in the same direction. In other words, the fluid in the wake is dragged by the moving plate. The flow fields corresponding to these two points of view are qualitatively depicted in Fig. 6.10 for the region comprising the end of the plate and the beginning of the wake.

From what has been described above, it should also be clear that, strictly speaking, exactly steady flow conditions can never be reached in this case, because the length of the wake will continuously increase with time. However, if we are particularly interested in the flow field around the plate, we can say that steady conditions have been reached in that region when the wake has become so long that its further lengthening does not produce any measurable difference in the flow. In practice, we may assume that the transient taking place around the plate is finished when the effects of convection and diffusion balance in such a way that the unsteady term in Eq. (6.12) becomes negligible.

Nonetheless, we may also note that the progressive lengthening of the wake implies that new vorticity of both signs is continuously fed into the fluid, and it is pertinent to enquire where and how it is generated. As we previously observed, if the body forces are conservative, solid surfaces are the only source of new vorticity in incompressible flows; therefore, it is reasonable to infer that this vorticity originates from the surface of the plate. However, this particular example is not adequate for explaining this continuous production of new vorticity, due to the singular geometry of the body we are considering, and in particular to its zero thickness. We therefore postpone this explanation to the following section, where the motion of a finite body with thickness will be analysed, and the flat plate case will be considered as the limit condition of a body whose thickness tends to zero.

An important issue we can face is the assessment of the order of magnitude of the region containing vorticity over the plate at the end of the above-defined transient. Again, from a strictly mathematical point of view, vorticity is actually present up to infinity, as happened in the infinite plate case. However, as before, we may refer to the region of thickness δ in which a large percentage of the vorticity present over a point on the plate is contained. A particularly interesting point is the one at the end of either surface of the plate, which is defined by the streamwise coordinate $x = L$.

Also in this case, the value of δ at that point may be derived from relation (6.25), but with a different value of the coefficient A; in effect, the flow conditions are different, because both convection and a continuous production of new vorticity are now present. For instance, as will be shown in next chapter, A is approximately 5 if δ is chosen so that it contains 99% of the total vorticity present above a point of the plate.

To perform our estimate, we must choose a reasonable typical time to be used in (6.25). To do this, we observe that, in the present flow, U and L represent natural reference values for velocity and length, respectively. Therefore, it is reasonable to choose $t^* = L/U$ as a suitable reference time. Note that t^* corresponds to the time necessary for the plate to move for a distance equal to L or, alternatively, for a fluid particle outside the vorticity-containing region to travel along the whole plate. We then obtain

$$\delta(L) = A\sqrt{\nu t^*} = A\sqrt{\nu \frac{L}{U}}. \tag{6.27}$$

However, what is really interesting is not the value of δ, but rather the ratio between δ and the length of the plate, which is then given by

$$\frac{\delta(L)}{L} = A\sqrt{\frac{\nu}{UL}} = \frac{A}{\sqrt{Re}}, \tag{6.28}$$

where $Re = UL/\nu$ is the Reynolds number, defined in terms of the reference velocity and length.

In our analysis, the result given by expression (6.28) is a fundamental one. In effect, it shows that if the Reynolds number is sufficiently large, we may assume vorticity to remain confined within a small layer adjacent to the plate even after the initial transient of the motion. Outside this layer, the amount of vorticity is so small that completely neglecting its presence is a very good approximation of the real flow condition. Therefore, the situation may be modelled accurately by assuming that the flow is irrotational at distances from the plate larger than δ.

The above-described influence of the Reynolds number on the space evolution of vorticity may become clearer if we estimate the order of magnitude of the ratio between the convective and diffusive terms in the vorticity dynamics equation. We find

$$\frac{[\boldsymbol{V}\cdot\mathrm{grad}\boldsymbol{\omega}]}{[\nu\nabla^2\boldsymbol{\omega}]} = \frac{U(UL^{-1})L^{-1}}{\nu(UL^{-1})L^{-2}} = \frac{UL}{\nu} = Re. \tag{6.29}$$

In Chapter 5, it was shown that the Reynolds number expresses the ratio between the orders of magnitude of the inertial and viscous forces in the momentum balance equation. Now, relation (6.29) provides a further physical interpretation and shows that Re may also be considered as a parameter giving the ratio between the orders of magnitude of the convective and diffusive terms in the vorticity equation. More precisely, the ratio between the velocity of vorticity convection — which is of the order of U — and the velocity of vorticity diffusion — which may be estimated from the value of $\partial\delta/\partial t$ for $t = L/U$ — is easily seen to be proportional to the square root of the Reynolds number, and this result gives a clear justification of the form of relation (6.28).

6.4.4. *Finite body and vorticity sources*

Let us now consider the two-dimensional body whose starting flow, in a reference system fixed to the body, is represented in Fig. 6.4. As happened for the finite plate, the extremely thin layer of vorticity produced around the body surface at the onset of the motion will immediately start diffusing to adjacent particles in the fluid, which will then convect it downstream. If, as we have assumed, the shape of the body is such that its cross-flow dimension is sufficiently smaller (say no more than 20%) than its length L, then the subsequent evolution of the flow will be qualitatively similar to that of the finite plate, although quantitative differences will be present.

Therefore, after a sufficiently long time from the start of the motion, vorticity will be contained in a region around the body and in a wake trailing downstream (see Fig. 6.11). The thickness δ of the vorticity-containing region at the end of the body may again be estimated from (6.28), with a value of the coefficient A that depends on the shape of the body. As a consequence, δ/L will be inversely proportional to the

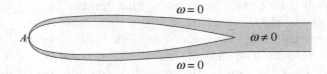

Fig. 6.11.   Flow regions around a finite-thickness body.

square root of the Reynolds number, evaluated using the length of the body and its translational velocity as reference quantities. In the next chapter, we shall see that if the body does not have the previously-cited geometrical characteristics, or if it moves in a different direction, then the above conclusion is not necessarily true, and the vorticity-containing wake trailing behind the body may be much wider.

In any case, we may note that diffusion causes vorticity to be present also in front of the body or, more precisely, along the streamline leading to the stagnation point A. It is of some interest, then, to estimate the thickness δ_0 of the vorticity-containing region upstream of this point.

The order of magnitude of δ_0 may be obtained by observing that this is the distance from the body at which diffusion of vorticity from the surface must be exactly balanced by convection of vorticity in the opposite direction. To obtain the ratio between the orders of magnitude of the vorticity convection and diffusion terms in this region, we may carry out an analysis similar to the one described by (6.29). However, the reference length is now δ instead of L, i.e. we must choose the typical size of the vorticity-containing region, because this is the length within which significant velocity variations occur. Therefore, the above ratio is equal to a Reynolds number defined as $Re_\delta = U\delta/\nu$. Consequently, δ_0 corresponds to the value of δ for which Re_δ becomes equal to one, and thus we get an estimate $\delta_0 \sim \nu/U$. For usual flow conditions, this is a small value indeed, if compared to the characteristic dimension of the body: in effect, we have $\delta_0/L \sim 1/Re$.

We turn now to the problem of discussing the origin and rate of production of the vorticity continuously fed into the lengthening wake behind the body. We have already pointed out that this vorticity must be generated at the surface of the body due to the necessity of always satisfying the viscous no-slip boundary condition. Thus, the situation may be described by saying that the surface is a continuous source of vorticity, which subsequently diffuses into the fluid and is convected downstream.

Therefore, when the vorticity distribution does not vary appreciably in time above the body surface, this actually means that the effects not only of convection and diffusion, but also of generation of new vorticity at the surface must sum up to zero.

Following a suggestion by Lighthill (1963), it is useful to describe this continuous production of vorticity using the concept of a *vorticity flux* at the surface. Considering the analogy between the mechanisms of vorticity diffusion and heat diffusion, it is suitable to express this flux, in a generic three-dimensional case, in the following form:

$$\sigma = -\nu \boldsymbol{n} \cdot \operatorname{grad}\omega, \tag{6.30}$$

where $\boldsymbol{n}$ is the outwards normal unit vector (i.e. pointing towards the fluid) at the considered point on the surface.

In the present two-dimensional case, in which the vorticity vector is directed perpendicular to the plane of the flow, say with component ω, relation (6.30) reduces to

$$\sigma = -\nu \frac{\partial \omega}{\partial n}. \tag{6.31}$$

It should be emphasized that, although the quantity defined in these relations has the dimension of vorticity times velocity, there is no real flux of vorticity through the body surface, where the relative velocity is zero. Thus, it is perhaps more appropriate to use the term *vorticity source strength* to denote σ, as it represents the amount and sign of vorticity that is introduced in the flow at the body contour, per unit surface and unit time.

In any case, expressions (6.30) and (6.31) are definitely useful, because they can provide a quantitative evaluation of the amount of vorticity being continuously produced at a solid wall. To show this, we return to the two-dimensional body whose flow is shown in Fig. 6.11, and consider a curvilinear coordinate system fixed to the body with origin at the stagnation point A, with the x axis along the body surface and the y axis perpendicular to the surface and directed towards the fluid. We denote the velocity components in these two directions by u and v respectively, and we write the component in the x direction of the momentum balance equation, neglecting body forces and assuming the curvature of the wall to be sufficiently small for its effects to be negligible. From (5.35), we obtain

$$\frac{\partial u}{\partial t} + u\frac{\partial u}{\partial x} + v\frac{\partial u}{\partial y} = -\frac{1}{\rho}\frac{\partial p}{\partial x} + \nu\left(\frac{\partial^2 u}{\partial x^2} + \frac{\partial^2 u}{\partial y^2}\right). \tag{6.32}$$

Let us now write Eq. (6.32) for points on the body surface, where, due to the no-slip condition, we have identically $u = v = 0$ for any value of x and t. We then obtain the following equation, where the subscript s indicates quantities evaluated at the surface,

$$\frac{1}{\rho}\frac{\partial p}{\partial x}\bigg|_s = \nu \frac{\partial^2 u}{\partial y^2}\bigg|_s. \tag{6.33}$$

We now recall that the vorticity in the z direction (orthogonal to the flow plane) is given by

$$\omega = \frac{\partial v}{\partial x} - \frac{\partial u}{\partial y}, \tag{6.34}$$

and, consequently, its value at the surface is

$$\omega_s = -\frac{\partial u}{\partial y}\bigg|_s. \tag{6.35}$$

By introducing (6.35) into (6.33), and noting that the y direction coincides with the n direction of relation (6.31), we finally obtain

$$\sigma = -\nu\frac{\partial \omega}{\partial y}\bigg|_s = \frac{1}{\rho}\frac{\partial p}{\partial x}\bigg|_s. \tag{6.36}$$

This relation permits us to associate the strength of the vorticity sources at the body surface with the local value of the component of the pressure gradient along the surface. The utility of this relation in estimating the vorticity being released in the wake in terms of the velocity field will become clearer in next chapter. However, we may now use it to discuss which are the regions where vorticity is continuously being introduced along the surface of the body of Fig. 6.11 during its motion.

We first anticipate that, after sufficient time from the start of the motion, the pressure distribution that can be measured over the body surface will be qualitatively similar to the one shown in Fig. 6.12. From this figure, it is seen that over the upper surface of the body the pressure decreases from the front stagnation point up to point C_u, after which it starts increasing again for the remaining part of the upper surface, from C_u to B. Thanks to the symmetry of the body and of the motion, the situation over the lower surface is similar, with a pressure minimum occurring at point C_l.

If we now consider that, starting from the stagnation point, the coordinate x is increasingly positive over the upper surface and increasingly

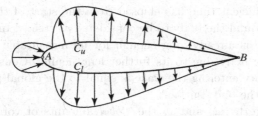

Fig. 6.12.   Pressure distribution around a streamlined body.

negative over the lower one, it is then evident from (6.36) that σ is negative over the upper surface from A to C_u, and positive over the lower surface from A to C_l, while it is positive from C_u to B and negative from C_l to B.

We may thus interpret this result as follows: negative and positive vorticity are continuously introduced into the flow, respectively over the upper surface from A to C_u and over the lower surface from A to C_l. This is in perfect agreement with the fact the vorticity is clockwise over the upper surface and counter-clockwise over the lower surface. Over the remaining parts of the two surfaces, the vorticity sources change sign. As a consequence, positive vorticity is introduced in the flow over the upper surface after point C_u, and tends to decrease, by annihilation, the amount of negative vorticity that has been previously produced upstream of that point and which has been brought by convection over that part of the surface. A similar situation, with changed signs, occurs over the lower surface. Sometimes, the vorticity evolution taking place after points C_u and C_l, over the upper and lower surfaces respectively, is described as being due to the action of 'sinks' of the previously produced vorticity. Even if, from the phenomenological point of view, this description is equivalent to the one based on the change of sign of the vorticity source terms, it is intrinsically non-local, in the sense that the statement that a sink is present implies previous knowledge of the sign of the vorticity produced upstream and present over the considered point on the surface. Therefore, we prefer to always refer to vorticity sources, having different sign according to the sign of the pressure gradient along the surface, and to subsequently analyse what their effect may be according to the sign, amount and evolution of the vorticity that has been produced over the upstream portion of the surface.

If we now want to estimate how much vorticity of both signs is instantaneously introduced in the wake, we first observe that we have

assumed that sufficient time has elapsed from the start of the motion for
the conditions around the body to be considered as steady. In other terms,
the trailing wake is already long enough for no measurable variations to
occur around the body due to its further lengthening. Consequently, the
amount of vorticity entering the wake is equal to the global production of
new vorticity at the body surface.

Let us denote by ω_u and ω_l the *global* amounts of vorticity shed in
the wake in unit time from the upper and lower surfaces respectively; their
values may be found by integrating the vorticity sources present over the
upper and lower surfaces from the stagnation point to point B. Using (6.36),
and noting that, in the flow direction, the x coordinate increases over the
upper surface and decreases over the lower one, we get

$$\omega_u = \frac{1}{\rho}[p(B) - p(A)]; \quad \omega_l = \frac{1}{\rho}[p(A) - p(B)]. \tag{6.37}$$

Relations (6.37) show that, as expected, equal global amounts of
negative and positive vorticity are continuously produced at the surface
and introduced into the wake from the upper and lower surfaces of the
body respectively.

We also note now that Bernoulli's theorem applies along the streamline
coming from upstream and ending at the stagnation point A, because the
flow is irrotational in front of the body. Therefore, if p_∞ is the value of the
undisturbed pressure far upstream, the pressure at point A is equal to

$$p(A) = p_\infty + \frac{1}{2}\rho U^2, \tag{6.38}$$

which is the stagnation pressure and corresponds to the highest value of
the pressure that may be found in an incompressible flow.

We are now able to qualitatively explain the generation of the new
vorticity introduced into the wake for the case of the finite plate considered
in the previous section. To this end, the plate must be considered as the
limit of the body of Fig. 6.12, or simply of a thick plate, as the thickness
is reduced to zero. It may be seen that, with the progressive decrease
of the thickness, the pressure gradients become increasingly strong and
concentrated in the front rounded part of the body, on the two sides
of the stagnation point. Simultaneously, the pressure over the rear part
of the body tends to approach the undisturbed value p_∞, and thus
$p(B) = p_\infty$. From relations (6.37) and (6.38), it may then be deduced
that the amounts of vorticity of the two signs introduced into the wake in

unit time become $-1/2\,U^2$ from the upper side and $1/2\,U^2$ from the lower side.

In a real practical situation, in which the thickness of the plate cannot be exactly zero, this would suffice to explain the continuous production of new vorticity, even if in a very restricted area in the fore part of the plate. On the other hand, in the ideal limit situation in which the thickness is further decreased to zero, even if the pressure gradients would tend to become infinite, they would act over regions on the two sides of the stagnation point that tend to become infinitesimal. Consequently, the amounts of vorticity introduced in the wake would remain the same both during the limit process and at the limit condition, being determined only by the values of the stagnation and undisturbed pressures. For an infinitely thin plate we may thus say that singular vorticity sources exist at the front stagnation point, and that they continuously produce exactly the same amounts of negative and positive vorticity as are instantaneously introduced into the wake from the rear extremity of the plate.

Relation (6.36), giving the strength of the vorticity sources at the contour of a two-dimensional body, may be generalized to the case of three-dimensional flow conditions, and becomes

$$\boldsymbol{\sigma} = -\boldsymbol{n} \times \frac{1}{\rho}\mathrm{grad}\,p. \tag{6.39}$$

It may be seen that a more general expression applies, which is reported, together with its derivation, in Chapter 11. A term connected with the body acceleration also appears explicitly in that expression, and this is reasonable because, from what has been described in the present chapter, it should be clear that the acceleration of a solid surface must be one of the sources of vorticity in the flow. In practice, this directly derives from the fact that the no-slip boundary condition implies that a fluid particle in contact with a surface must have not only the same velocity but also the same acceleration as the surface.

6.5. General Discussion on Vorticity Dynamics

In the previous sections of this chapter the mechanisms at the basis of the generation and evolution of vorticity in incompressible flows have been analysed, using examples relating to some particular and rather simple cases. The results of this analysis will be of great help in understanding many features of generic types of flow, and also for justifying the generation and magnitude of forces acting on bodies moving in a fluid. Furthermore, the mathematical implications of the presence or absence of vorticity, already discussed in Chapter 5, can lead to the development of simplified prediction procedures, which may be used in cases that are precisely defined through a close scrutiny of the behaviour of vorticity in the flow. It might thus be useful to summarize and further discuss the main points arising from our efforts to understand the dynamics of vorticity.

The first important finding is that, in an incompressible flow, vorticity is introduced into the field exclusively from solid surfaces, as a result of the action of the viscous no-slip boundary condition. In an impulsively started motion, the production of vorticity is instantaneous and occurs in a layer of infinitesimal thickness adjacent to the surface. In this layer, the vorticity value is infinite, but the global amount of generated vorticity is finite, and its value is completely determined by the characteristics of the surface and of its motion.

Immediately after generation, vorticity diffuses from the solid surfaces to the neighbouring fluid regions. The action of diffusion is connected with the difference in vorticity content between adjacent fluid particles, thus being extremely large in the initial stages. However, it rapidly decreases as the region where vorticity is present increases in size. Furthermore, as soon as diffusion causes vorticity to spread into the fluid, convection comes into play, and the velocity field carries the vorticity-containing particles downstream. In three-dimensional flow conditions, vorticity tubes may be also stretched and tilted, thus varying the vorticity of the single particles composing them.

In the case of flows produced by the translational motion of finite bodies in the fluid, continuous production of vorticity at their surface occurs so that the no-slip boundary condition is always satisfied while diffusion and convection cause vorticity to be carried downstream in a wake of ever-increasing length. The amount and sign of the vorticity being instantaneously generated at every point of the body surface can be evaluated from the knowledge of the pressure gradients along the surface and of the acceleration of the body, if present. Furthermore, from integration over the closed body surface of the expression giving the strength of the vorticity sources, it is easy to verify that the generated amounts of positive and negative vorticity are always equal.

However, it must be pointed out that this does not mean that the global vorticity being instantaneously introduced in the wake is always zero, as happened in the examples of the previous section. In effect, although we found that the total vorticity present in the whole fluid domain around translating bodies does not vary in time, and is zero if the motion has started from rest, one should not confuse the amount of vorticity being instantaneously produced at a body surface with the amount of vorticity present around it at the same time. Thus, if the flow conditions are not steady, it may happen that the actions of diffusion and convection of the vorticity generated at previous times cause the vorticities of the two signs being released into the wake not to balance. If this occurs, even if the instantaneous generation of vorticity at its surface sums up to zero, the global vorticity present around the body may be different from zero. Nonetheless, we know that its value is always equal to the opposite of the global vorticity that is simultaneously present in the wake, where it has been introduced previously. An example of such a situation will be described in Chapter 8.

More generally, we have seen that the principle of conservation of total vorticity requires that we also include, in the evaluation, the vorticity inside the bodies, which is equal to twice their angular velocity (and is thus zero for translating bodies). This integral conservation law is an important constraint for the vorticity field, and may often be used to gain a deeper insight into the evolution of the flow and on the origin of the forces acting on moving bodies.

Another extremely important result of our previous analysis is the role of the Reynolds number in the evolution of vorticity around a moving body from whose surface it was generated. In effect, we found that high values of *Re*, which actually occur in many (or perhaps most) practical applications, imply that vorticity remains confined in a thin layer around the body, while the flow outside it may be considered as irrotational. In principle, if the interest is in the pressure and in the tangential viscous stresses acting on the body surface, this would not seem a great advantage because the mathematical simplifications that apply for an irrotational flow are not valid inside the vorticity-containing layer separating the outer flow from the solid surface. However, the equations of motion applicable within this layer may be simplified if its thickness is small with respect to a typical dimension of the body, and it can be seen that this leads to important results regarding the pressure field acting on the body surface.

An in-depth mathematical and physical analysis of the characteristics and of the behaviour of the vorticity-containing layer for high Reynolds numbers is the subject of the following chapter. Our approach will essentially be based on the aspects of vorticity dynamics and on the consequences of irrotationality that have been amply described and discussed in the present and previous chapters. In particular, we shall extensively use the observation that the equations of motion of an incompressible irrotational viscous flow coincide with those that would apply if the fluid were non-viscous.

Chapter 7

INCOMPRESSIBLE BOUNDARY LAYERS

7.1. Boundary Layers as Vorticity Layers

In previous chapters, we have stressed the importance of the dynamics of vorticity in the flow. In particular, the interpretation of vorticity as twice the instantaneous angular velocity of a small fluid element was given in Chapter 3, and in Chapter 5 it was shown that the pressure and velocity fields in an incompressible flow can be derived directly from the mass conservation and momentum balance equations. However, and perhaps more importantly as regards our present analysis, it was also highlighted that in the regions of an incompressible flow where vorticity is absent, or in which its value is so small that it can be neglected, the viscous term in the momentum balance equation vanishes. This means that the resultant of the viscous forces over the small surface bounding an elementary volume of fluid is zero, although the viscous forces acting on single surface elements are not necessarily zero.

Finally, the origin of vorticity in incompressible flows due to the viscous no-slip condition at solid surfaces was discussed in Chapter 6, where the equations governing the dynamics of vorticity were also derived and interpreted from the physical point of view. A fundamental result of that analysis is that, if the typical Reynolds number of the flow is large, then vorticity may remain confined in regions of small thickness adjacent to the bodies from whose surface it has been generated and in a thin wake trailing downstream. Nonetheless, it was also pointed out that, for this to occur, certain hitherto unspecified conditions as regards the geometry and the motion of the bodies must be satisfied.

185

By summarizing all the above findings, we may conclude that if the Reynolds number of an incompressible flow is sufficiently high, it is possible that the effects of viscosity be 'felt' only within a thin layer adjacent to the body surfaces, where vorticity is present, while in the outer irrotational flow the particles move as if they belonged to a non-viscous fluid.

The concept of a layer adjacent to solid walls where viscous effects are important, whereas they can be neglected in the outer flow, was first introduced by Prandtl (1904), who called this flow region a *boundary layer* (*Grenzschicht* in German). Starting from the assumption that boundary layers are thin, Prandtl developed a simplified mathematical theory for their treatment and experimentally studied their dynamics as a function of the geometry of the moving bodies and of the general flow features. An account of the origin and development of boundary layer theory is reported, among others, by Tani (1977). As mentioned in Chapter 1, Prandtl's proposal was something of a breakthrough in the historical development of fluid dynamics. It provided a theoretical basis to explain the reason why, in certain cases, a non-viscous flow model may give useful results, and to identify the necessary flow conditions for this to occur. Moreover, it permitted procedures to be devised to estimate the fluid dynamic loads on certain bodies, and this is invaluable in many engineering applications, particularly in the aeronautical field.

Nonetheless, boundary layer theory is more than the basic tool for an approximate treatment of certain particular flow problems. In fact, predicting the dynamics of boundary layers is a fundamental step in understanding fluid flows and mastering their behaviour, in the sense of being able to obtain desired flow features through suitable modifications of the geometry or of the conditions of motion. For example, later in this chapter we shall see that bodies may be divided into different classes according to the behaviour of the boundary layers developing over their surfaces when they move in a fluid. Furthermore, we shall also realize that the type and magnitude of the forces acting on moving bodies are strictly connected with the flow features deriving from the behaviour of the boundary layers.

It must be pointed out that the boundary layer concept applies not only to the regions adjacent to solid surfaces, but also to all zones in the flow where vorticity is confined within small-thickness layers. This may happen in the wakes behind certain bodies or when two streams at different velocities come in contact as, for instance, at the boundary of a jet issuing into a still fluid. These flow regions are sometimes called 'free boundary layers' or, more often, 'free shear layers'. The cross-flow dimension of these regions remains small as long as convection prevails over diffusion, which implies that the typical Reynolds numbers are sufficiently high. Then, we may assume that the flow is everywhere potential except in these thin vorticity layers and use all the consequent mathematical simplifications.

The flow outside a boundary layer is often termed non-viscous (or *inviscid*) flow. We do not follow this practice in this book, in order to stress the fact that in an incompressible flow it is not necessary to neglect fluid viscosity for the viscous term in the momentum balance equation to vanish. In fact, it is sufficient that the motion of the fluid be such that it may be considered, for all practical purposes, as irrotational (or, in other terms, a potential flow). We thus distinguish between flow conditions that strictly refer to the kinematics of the fluid particles only, and flow conditions that allow one physical characteristic of the fluid to be neglected, such as those that were specified in detail in Chapter 5 to be necessary for an incompressible flow treatment. The main reason for this choice is to recall and emphasize that, in a potential flow, viscous stresses are generally not zero and do work during the motion. Consequently, potential flows are usually dissipative, and this fact, together with its possible implications, will be further discussed in Chapter 10. We restrict the use of the term 'non-viscous flow', then, to those cases in which the viscous stresses and their dissipative effects are indeed neglected. For instance, this will occur in Chapter 15 for the analysis of certain flows in which compressibility is the primary physical characteristic influencing the flow features and viscosity may be shown to produce second-order effects only.

In the following sections, we shall describe the simplifications deriving from boundary layer theory and discuss the main aspects and consequences of the dynamics of boundary layers. The two-dimensional case will be mainly considered, as it highlights, with a simpler treatment, all the fundamental features of boundary layer behaviour.

7.2. Boundary Layer Equations

Let us consider a two-dimensional motion past a solid surface where a thin boundary layer is presumed to exist with characteristic thickness δ, defined in an appropriate — albeit still unspecified — way. We further assume that the radius of curvature of the surface is large with respect to the boundary layer thickness, so that terms in the equations deriving from curvature may be neglected. We shall use a Cartesian reference system, fixed to the

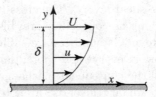

Fig. 7.1. Boundary layer and relevant reference system.

surface, with the x and the y axes parallel and orthogonal to the surface, respectively, and denote by u and v the corresponding velocity components (see Fig. 7.1).

We now assume the thickness, δ, to be small compared to any linear dimension L in which velocity variations occur in the x direction. In other words, δ and L are the typical reference dimensions of the flow, in the y and in the x directions, respectively, and everywhere within the boundary layer we have $\delta/L \ll 1$. Therefore, in a boundary layer the velocity varies rapidly from zero at the surface to a value U that is characteristic of the outer flow. We take U to represent the order of magnitude of the velocity component u within the boundary layer, even if it may actually be closer to the maximum value of u.

Let us now write the equations of motion, neglecting the effect of body forces and recalling that, for an incompressible flow, the energy balance equation is not necessary to derive the velocity and pressure fields. The required equations are, then, those expressing the mass conservation and the momentum balance in the x and y directions; for a two-dimensional flow, they become

$$\frac{\partial u}{\partial x} + \frac{\partial v}{\partial y} = 0, \tag{7.1}$$

$$\frac{\partial u}{\partial t} + u\frac{\partial u}{\partial x} + v\frac{\partial u}{\partial y} = -\frac{1}{\rho}\frac{\partial p}{\partial x} + \nu\left(\frac{\partial^2 u}{\partial x^2} + \frac{\partial^2 u}{\partial y^2}\right), \tag{7.2}$$

$$\frac{\partial v}{\partial t} + u\frac{\partial v}{\partial x} + v\frac{\partial v}{\partial y} = -\frac{1}{\rho}\frac{\partial p}{\partial y} + \nu\left(\frac{\partial^2 v}{\partial x^2} + \frac{\partial^2 v}{\partial y^2}\right). \tag{7.3}$$

Considering that $\delta/L \ll 1$, the velocity varies much more rapidly in the y direction than in the x direction, and so we may use the following approximations:

$$\left|\frac{\partial^2 u}{\partial x^2}\right| \ll \left|\frac{\partial^2 u}{\partial y^2}\right|, \quad \left|\frac{\partial^2 v}{\partial x^2}\right| \ll \left|\frac{\partial^2 v}{\partial y^2}\right|. \tag{7.4}$$

Consequently, the first terms of the Laplacians in Eqs. (7.2) and (7.3) may be neglected with respect to the second ones.

We now proceed to estimate the order of magnitude of the various terms appearing in the above equations, to assess if they may be further simplified thanks to the assumptions characterizing a boundary layer flow. Note, however, that there is no reason why all terms in an equation should be of the same order. Thus, when the order of magnitude of a term is specified, it should be interpreted as the maximum order of magnitude of the absolute value of that term. In other words, it may well happen that, in certain cases, a term is smaller than our estimate (or even zero), but its order of magnitude cannot be larger than that estimate.

Let us start by estimating the order of magnitude of the velocity component v. Even if we already imagine this component to be smaller than u, we must also recall that mass conservation implies a balance between the variations of the two velocity components. For instance, if the boundary layer thickness grows in the x direction, the velocity u at a certain fixed y coordinate will decrease in the x direction; consequently, $\partial u/\partial x$ will be negative and $\partial v/\partial y$ must be equal to the opposite of $\partial u/\partial x$. Then, the following estimate is derived from the continuity equation (7.1) for the absolute value of v:

$$v \sim U\frac{\delta}{L}, \tag{7.5}$$

and the assumption that $\delta/L \ll 1$ leads then to the result that $v \ll u$ within the boundary layer, as might have been expected.

We can now carry out an order of magnitude analysis of the remaining equations. Let us consider the terms appearing on the left-hand side of Eq. (7.2) and observe that the order of magnitude of the time variations in the unsteady term may be assumed to be L/U. Therefore, the orders of magnitude of the various terms are

$$\frac{\partial u}{\partial t} \sim U\frac{U}{L} = \frac{U^2}{L}, \quad u\frac{\partial u}{\partial x} \sim U\frac{U}{L} = \frac{U^2}{L}, \quad v\frac{\partial u}{\partial y} \sim U\frac{\delta}{L}\frac{U}{\delta} = \frac{U^2}{L}.$$

Thus all these terms, which represent the 'inertia' forces per unit mass acting on a particle, are of the same (maximum) order U^2/L. We now observe that the boundary layer is defined as the region in the flow where the force per unit mass due to viscosity is comparable in magnitude with the 'inertia' terms. Therefore, taking the first of (7.4) into account, we can regard the boundary layer as a region where the following order-of-magnitude relation holds:

$$\frac{U^2}{L} \sim \nu \frac{U}{\delta^2},$$ (7.6)

and, consequently,

$$\frac{\delta^2}{L^2} \sim \frac{\nu}{UL} = \frac{1}{Re}.$$ (7.7)

This relation is particularly important. In effect, it connects the boundary layer thickness with the Reynolds number, defined in terms of typical quantities of the flow outside the boundary layer, and shows that Re must be sufficiently high for the assumption $\delta/L \ll 1$ to be valid. This fact, together with the comparison between relations (7.7) and (6.28), confirms that the boundary layer may be seen as the region adjacent to a surface where vorticity (originating from that surface) is confined by the prevalence of convection over diffusion.

From the above estimates, we may conclude that the maximum order of magnitude of the pressure variation along the surface is given by

$$\frac{\partial p}{\partial x} \sim \frac{\rho U^2}{L},$$ (7.8)

and that no further simplifications, besides those given in (7.4), may be introduced in Eq. (7.2), which then becomes

$$\frac{\partial u}{\partial t} + u \frac{\partial u}{\partial x} + v \frac{\partial u}{\partial y} = -\frac{1}{\rho} \frac{\partial p}{\partial x} + \nu \frac{\partial^2 u}{\partial y^2}.$$ (7.9)

Using relations (7.5), (7.6) and (7.7), we may now analyse the order of magnitude of the terms appearing in the momentum equation for the y direction, (7.3),

$$\frac{\partial v}{\partial t} \sim U \frac{\delta}{L} \frac{U}{L} = \frac{U^2}{L} \frac{\delta}{L}, \qquad u \frac{\partial v}{\partial x} \sim U \frac{U\delta}{L} \frac{1}{L} = \frac{U^2}{L} \frac{\delta}{L},$$

$$v \frac{\partial v}{\partial y} \sim U \frac{\delta}{L} U \frac{\delta}{L} \frac{1}{\delta} = \frac{U^2}{L} \frac{\delta}{L}, \qquad \nu \frac{\partial^2 v}{\partial y^2} \sim \nu U \frac{\delta}{L} \frac{1}{\delta^2} = \frac{U^2}{L} \frac{\delta}{L}.$$

Therefore, we see that all the above terms have a maximum order of magnitude that is equal to that of the terms in the momentum equation in the x direction multiplied by (δ/L). Consequently, for the maximum order of magnitude of the pressure variation in the y direction, we find the following estimate:

$$\frac{\partial p}{\partial y} \sim \frac{\rho U^2}{L} \frac{\delta}{L}. \tag{7.10}$$

Now, considering that from the estimate (7.7) and the assumed high value of the Reynolds number the quantity (δ/L) is very small, the momentum equation in the direction normal to the wall reduces, with a very good approximation, to the simple expression

$$\frac{\partial p}{\partial y} = 0. \tag{7.11}$$

If we had taken the curvature κ of the surface into account, in our analysis, the result would have been slightly different:

$$\frac{\partial p}{\partial y} \sim \rho \kappa U^2. \tag{7.12}$$

However, provided the radius of curvature, $1/\kappa$, of the surface is large with respect to the local value of δ (which is normally the case if sharp edges do not occur on the surface), the total variation of pressure across the boundary layer due to curvature effects remains definitely negligible. Therefore, relation (7.11) holds to a very good approximation and becomes an absolutely fundamental *result* of our scrutiny of the possible simplifications deriving from the boundary layer assumptions. In effect, the assumption that the boundary layer is thin does not permit, in principle, the *a priori* assertion that the pressure variations across its thickness are negligible. In fact, this conclusion may only be reached after carrying out the above-described order-of-magnitude analysis of the terms appearing in the equations of motion.

It immediately follows from relation (7.11) that we may put

$$p(x, 0) = p(x, \delta), \tag{7.13}$$

from which we derive that the pressures acting on a moving body over whose surface a thin boundary layer exists can be obtained when the velocity field in the irrotational flow region outside the boundary layer is known. In effect, we have seen that the viscous term in the momentum equation vanishes in a potential flow and, therefore, the pressure variation along the body surface can be recovered from the value of the velocity at the border of the boundary layer. More precisely, if we denote as $U_e(x, t)$ the x component of the velocity in the potential flow at the edge of the boundary layer, we have

$$\frac{\partial U_e}{\partial t} + U_e \frac{\partial U_e}{\partial x} = -\frac{1}{\rho} \frac{\partial p}{\partial x}, \qquad (7.14)$$

which can be obtained from Eq. (7.2) by considering that in the outer potential flow the term $v\partial U_e/\partial y$ can be neglected and the viscous term is zero.

This leads to a reappraisal of the importance and utility of the solution of an outer potential flow problem in which the boundary condition is applied not at the body surface, but at a still undefined contour, whose position is connected with the development of the boundary layer and where the velocity must match the velocity at the outer edge of the boundary layer.

From these observations it should also be immediately clear that the mathematical problems of the outer potential flow and of the boundary layer are strictly coupled. In fact, Eq. (7.9) characterizing the boundary layer flow can be solved, together with Eq. (7.1), only if the pressure term $\partial p/\partial x$ is given. However, this term derives from the solution of the outer potential flow problem which, in turn, can be obtained only if the boundary layer problem has been already solved. Apparently, this seems to be a double problem leading to no solution. Fortunately it is not so, and an iterative procedure may be envisaged which, under certain conditions, converges to a solution that is more than satisfactory from the practical point of view. Nevertheless, before describing this procedure and discussing its limits of application, several issues must still be made clear. One of these is related to the fact that, contrary to what might be deduced from a first reading of the above observations, the edge of the boundary layer is *not* a streamline. For instance, in the case of a flat plate, which will be considered in more detail in Section 7.3 and whose general flow evolution was described in Section 6.4.3, fluid particles coming from upstream at distance y from the plate will progressively be decelerated by the viscous forces when vorticity, by diffusion, reaches the distance y. In other words, they will enter the boundary layer. This leaves the proper formulation of the boundary conditions for the outer potential flow still undefined, but this point will become clearer as we proceed in our analysis.

In any case, it should already be clear that we have to deal with two different mathematical problems for the inner and outer regions, which are ruled by different equations and boundary conditions. Let us return to the equations of motion inside the boundary layer, which we now rewrite for the simplified situation of steady flow. In this case Eq. (7.1) remains unaltered and Eq. (7.9) becomes

$$u\frac{\partial u}{\partial x} + v\frac{\partial u}{\partial y} = -\frac{1}{\rho}\frac{\partial p}{\partial x} + \nu\frac{\partial^2 u}{\partial y^2}. \tag{7.15}$$

This equation has a different mathematical character from the complete momentum equation. In particular, it is *parabolic* instead of *elliptic*. The prototype example of a parabolic equation is the diffusion equation, which we have already encountered, for instance, as Eq. (6.13) describing the evolution of vorticity in a two-dimensional problem. Without going into the details of this classification, the important issue, as regards the parabolic character of Eq. (7.15), is that the solution can be constructed by proceeding from upstream to downstream. In other words, the solution at a certain x depends on the solution for smaller values of x only. This means that a boundary condition must be given for the velocity profile at the coordinate $x = x_0$ at which the calculation is started, but no downstream condition is necessary. Then, the boundary conditions to be added to Eqs. (7.1) and (7.15), so that the boundary layer mathematical problem is correctly formulated, are

$$u(x, y = 0) = v(x, y = 0) = 0, \tag{7.16}$$

$$u(x, y \to \infty) = U_e(x), \tag{7.17}$$

$$u(x_0, y) = u_0(y). \tag{7.18}$$

As already pointed out, we assume that both functions $U_e(x)$ and $u_0(y)$ are known. Consequently, the pressure variation is also known and is given by the relation

$$-\frac{1}{\rho}\frac{\partial p(x)}{\partial x} = U_e(x)\frac{\partial U_e(x)}{\partial x}. \tag{7.19}$$

The boundary condition (7.17) deserves some further comment. It should not be interpreted as a condition applying at infinite distance from the surface, where we have seen that we would go outside the boundary layer and inside the irrotational flow. Conversely, it denotes a condition to be applied at the limit of the region where the boundary layer equations hold, which means that the y coordinate is assumed to refer to a domain that is always inside the boundary layer, whose 'thickness' we know to be extremely small. The y coordinate, then, should be thought of as being scaled by a very small typical length, and we could even have used a subscript or a different symbol, instead of y, to stress the fact that it always refers to locations within the boundary layer. In other words, it is as if we were studying the boundary layer domain with a microscope, so that it is enormously enlarged. Therefore, in terms of this inner y coordinate, the outer potential flow is effectively at infinity, and the boundary condition (7.17) is adequate to obtain a solution of the boundary layer equations that permits good matching with the outer potential flow.

Perhaps some vagueness still remains in this matching procedure but, hopefully, it will be reduced by the analysis in the next sections. For the moment, we may point out that boundary layer theory should be seen primarily as a practical tool to obtain good estimates of the fluid dynamic actions on certain moving bodies, rather than as a rigorous mathematical theory that may be used to accurately describe the flow features in the neighbourhood of their surfaces. Furthermore, we may recall that we have interpreted the boundary layer as the region close to a solid surface where vorticity is concentrated. In Chapter 6, however, we also noted that, immediately after the production of vorticity due to the no-slip condition, diffusion acts in such a way that, from a purely mathematical point of view, vorticity is immediately present up to an infinite distance from the surface. On the other hand, we have also seen that, if the Reynolds number is high, the simultaneous presence of convection confines almost all vorticity in a very thin layer close to the surface. Therefore, from an engineering point of view, we may say that vorticity decreases so rapidly with increasing distance from the surface that, for all practical purposes, it is effectively zero outside this layer. Yet, mathematically, we should say that vorticity is still present up to infinity. We shall see that the boundary layer equations, with the above-cited boundary conditions, are the basic means for obtaining good estimates not only of the thickness of the vorticity-containing layer, but also of the pressures and viscous stresses acting over the surfaces of moving bodies.

Before proceeding in our analysis, it is convenient to illustrate the consequences of the boundary layer simplifications on the connection between the velocity and vorticity fields. Considering that the flow is two-dimensional, the vorticity vector is directed orthogonally to the plane of motion, with a magnitude ω that is rigorously given by

$$\omega = \omega_z = \left(\frac{\partial v}{\partial x} - \frac{\partial u}{\partial y} \right).$$

Now, the boundary layer assumptions imply that the term $\partial v / \partial x$ is definitely negligible with respect to the term $\partial u / \partial y$. In effect, not only

$v \ll u$, as can be derived from (7.5), but also the variations along x are much smaller than those along y. Consequently, inside a boundary layer we have, to a very good approximation,

$$\omega = -\frac{\partial u}{\partial y}. \tag{7.20}$$

From this expression, we readily see that the velocity u at a generic coordinate y within the boundary layer can be written as

$$u = -\int_0^y \omega dy. \tag{7.21}$$

As $U_e(x)$ is, by definition, the velocity in the flow just outside the boundary layer, relation (7.21) shows that using the boundary condition (7.17) is equivalent to asserting that the boundary layer is indeed the region containing all the vorticity present above a point on the surface. This is in perfect agreement with our previous interpretation, although it was not an initial explicit assumption of boundary layer theory, as can be seen from a careful analysis of its derivation.

Nowadays, the boundary layer equations may be solved quite effectively using numerical procedures. However, many ingenious methods, developed when computers were not available, are still useful and capable of rapidly providing good estimates of most quantities of interest. Particularly important is the solution for a flat plate parallel to the flow, which will be described in Section 7.4, after the introduction of the main parameters characterizing a boundary layer.

7.3. Boundary Layer Parameters

7.3.1. *Boundary layer thickness*

It has been already anticipated that the velocity within the boundary layer, from a theoretical point of view, tends to the value of the outer flow only asymptotically. This can be recast in different terms by saying that vorticity tends to zero approaching the edge of the boundary layer, but that only at infinite distances from a solid surface it is actually zero.

Therefore, it is clear that any definition of the boundary layer thickness must necessarily be a conventional one, and the most used is that $\delta(x)$ is the distance at which the velocity u is equal to $0.99U_e$, although other numerical coefficients could be used (for instance, 0.95 or 0.999). This implies that a satisfactory estimate of the outer velocity $U_e(x)$ must be available, and it will be seen that it may be obtained from an iterative procedure for the evaluation of the boundary layer and of the outer potential flow.

However, we have also seen at the end of the previous section that, considering the boundary layer approximations, $U_e(x)$ is equal to the absolute value of the total vorticity present above a point at coordinate x over the surface. Therefore, we may also define the boundary layer thickness as the distance within which, say, 99% of the total vorticity present along a line normal to the surface is contained. Obviously, from the point of view of the boundary layer domain and simplified equations, the two definitions are perfectly equivalent. Nonetheless, it may also happen that, as a result of either experiments or numerical computations, the whole velocity field around a moving body is available, and that it may be necessary or useful to estimate the thickness of the boundary layer, without having to further carry out iterative potential flow and boundary layer calculations. In that case, apart from very particular situations, such as the flat plate problem that will be analysed in the next section, the function $U_e(x)$ will not be known in advance, and will be different from the velocity that actually occurs at infinite distance from the surface. On the other hand, as may be deduced from the discussion in Chapter 6 on the origin and evolution of vorticity in an incompressible flow, we can evaluate the integral of the vorticity along a line that is perpendicular to the surface at a certain point, because it converges to a finite value. In fact, the evaluation of the integral of the vorticity is an unambiguous procedure that is valid for any type of two-dimensional or three-dimensional flow. Moreover, provided the available data are sufficient, the local vorticity might even be determined without any approximation.

Therefore, there are good reasons to assert that the vorticity-based definition of the boundary layer thickness is either equivalent or more satisfactory and more general than the one based on the velocity profile. Furthermore, this definition is consistent with our interpretation of the flow outside the boundary layer as the region where vorticity is so small that it may be neglected and, consequently, viscosity has no effect on the motion of the fluid particles.

7.3.2. *Displacement thickness*

Besides the conventional boundary layer thickness, other quantities, which have the dimension of length and are connected with the boundary layer thickness at a certain fixed x over a solid surface, can be defined. These 'thicknesses' are often very useful from a practical point of view, and also have interesting physical interpretations.

One of these quantities, and perhaps the most important one, is the *displacement thickness*, which is usually denoted by δ^* or δ_1, and is defined by the following relation:

$$\delta^* = \int_0^\infty \left(1 - \frac{u}{U_e}\right) dy. \tag{7.22}$$

It should be pointed out that, in this definition, the upper limit of the integral implies that the displacement thickness is evaluated using the result of the solution of the boundary layer equations in which, as we have seen, U_e is the asymptotic value to which the velocity u tends as the 'inner' coordinate y goes to infinity. In this case, the integrand function tends to zero as the edge of the boundary layer is approached.

However, we have already pointed out that if the velocity $u(y)$ is available, in some way, as a function of the actual distance from the surface, U_e is not generally known in advance, and the integrand does not necessarily vanish as y tends to infinity. Therefore, it is more suitable to use the conventional thickness δ as the upper limit of the integration, and to assume that the velocity at this distance from the surface is a satisfactory estimate of the velocity U_e at the edge of the boundary layer, or of the integral of the vorticity present within it. We then have

$$\delta^* = \int_0^\delta \left(1 - \frac{u}{U_e}\right) dy. \tag{7.23}$$

This alternative definition is also helpful for explaining the physical meaning of the displacement thickness. Considering that the density ρ is constant, we may write

$$\rho U_e \delta^* = \int_0^\delta \rho(U_e - u)dy = \rho U_e \delta - \int_0^\delta \rho u \, dy. \tag{7.24}$$

This relation shows that the quantity $\rho U_e \delta^*$ is equal to the deficit of mass flux caused by the presence of the boundary layer, i.e. to the difference between the mass that would flow per unit time between $y = 0$ and $y = \delta$ if the velocity were equal to U_e, which is equal to $\rho U_e \delta$, and the actual mass flux through the boundary layer, which is given by the integral of ρu. Relation (7.24) can also be recast as follows:

$$\int_0^\delta \rho u \, dy = \rho U_e (\delta - \delta^*), \tag{7.25}$$

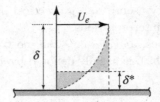

Fig. 7.2.   Interpretation of the displacement thickness.

and we see that the actual mass flux through the boundary layer is equal
to a flux of mass at velocity U_e through a distance $(\delta - \delta^*)$. This is also
illustrated in Fig. 7.2, where the displacement thickness is shown to be the
length such that the two shaded areas are equal.

The displacement thickness may also be interpreted as the distance
through which the streamlines in the external potential flow are displaced
outwards due to the decrease in velocity within the boundary layer.

From the above physical interpretations it should be possible to infer
that the displacement thickness is the adequate quantity to take the global
effect of the boundary layer into account when evaluations of the outer
potential flow are carried out. In effect, relation (7.25) shows that, if a
potential flow is required to be tangent not to the actual surface but to the
edge of the displacement thickness, the mass balance relevant to the real
flow is satisfied with an order of approximation that is consistent with the
boundary layer treatment.

Finally, it may be useful to derive another interesting property of the
displacement thickness, namely that it coincides with the centre of gravity
of the vorticity present within the boundary layer, say $y_{cg\omega}$. In effect, taking
(7.20) into account, we easily obtain

$$y_{cg\omega} = \frac{\int_0^\delta \omega y\,dy}{\int_0^\delta \omega\,dy} = \frac{-\int_0^\delta y\frac{\partial u}{\partial y}\,dy}{-\int_0^\delta \frac{\partial u}{\partial y}\,dy} = \frac{1}{U_e}\left([uy]_0^\delta - \int_0^\delta u\,dy\right)$$

$$= \frac{1}{U_e}\left(U_e\delta - \int_0^\delta u\,dy\right) = \frac{1}{U_e}\int_0^\delta (U_e - u)\,dy = \int_0^\delta \left(1 - \frac{u}{U_e}\right)dy = \delta^*.$$

7.3.3. *Momentum thickness*

The *momentum thickness*, which is denoted by θ or, alternatively, by δ_2,
is connected with the deficit of momentum flux within a boundary layer
with respect to the momentum flux that would occur if all the flow were

potential. It is defined as

$$\theta = \int_0^\delta \frac{u}{U_e}(1 - \frac{u}{U_e})dy. \tag{7.26}$$

Then, from this definition, it is easy to derive the following relation, which highlights the physical meaning of θ,

$$\rho U_e^2 \theta = \int_0^\delta \rho u(U_e - u)dy = U_e \int_0^\delta \rho u\, dy - \int_0^\delta \rho u^2 dy. \tag{7.27}$$

To interpret this relation from the physical point of view, we must note that within the boundary layer the mass flowing per unit time in a streamtube of height dy is given by $\rho u\, dy$. The integral of this quantity between 0 and δ then gives the total flux of mass through the boundary layer. Therefore, relation (7.27) represents the difference between the ideal momentum flux that would correspond to this mass if it were moving with the outer potential flow velocity U_e and the actual momentum flux through the boundary layer.

Note that, again, both in definition (7.26) and in the consequent relation (7.27), the upper limit of integration could have been set to infinity, provided we introduce a velocity $u(y)$ that is the solution of the boundary layer problem in which condition (7.17) has been used. Nonetheless, using δ as the upper limit of integration and recalling the physical meaning of the displacement thickness given by expression (7.25), we may manipulate relation (7.27) to obtain the following:

$$\int_0^\delta \rho u^2 dy = \rho U_e^2(\delta - \delta^*) - \rho U_e^2 \theta = \rho U_e^2[\delta - (\delta^* + \theta)], \tag{7.28}$$

which shows that the actual momentum flux through the boundary layer is equal to the momentum flux of fluid moving at a constant velocity U_e through a distance $[\delta - (\delta^* + \theta)]$.

It should be pointed out that the ratio

$$H = \frac{\delta^*}{\theta}, \tag{7.29}$$

which is called *shape factor*, is an important parameter characterizing the velocity profile within the boundary layer, and is strictly connected with the rate of variation of the pressure along the surface.

Finally, it is important to observe that the momentum thickness provides useful information on the viscous stresses acting over a surface.

This can be shown by deriving the so-called *momentum-integral equation*, also denoted as *von-Karman's integral equation*, which is obtained by integrating the boundary layer equations in the y direction.

To this end, we start from the boundary layer momentum equation (7.15), and recall that we are using quantities relating to the domain where boundary condition (7.17) applies. Therefore, the upper limit in our integration will be $y = \infty$, and the same is true in the definitions of the displacement and momentum thicknesses. By introducing into Eq. (7.15) the expression of the pressure variation given by (7.19), we obtain

$$u\frac{\partial u}{\partial x} + v\frac{\partial u}{\partial y} - U_e\frac{\partial U_e}{\partial x} = \nu\frac{\partial^2 u}{\partial y^2}. \tag{7.30}$$

Let us sum and subtract in (7.30) a term $u\partial U_e/\partial x$ and introduce the term $-v\partial U_e/\partial y$, which, considering that U_e is not a function of y, is equal to zero. By integrating, and changing the sign of the whole equation, we get

$$\int_0^\infty (U_e - u)\frac{\partial U_e}{\partial x}dy + \int_0^\infty u\frac{\partial(U_e - u)}{\partial x}dy$$
$$+ \int_0^\infty v\frac{\partial(U_e - u)}{\partial y}dy = -\nu\int_0^\infty \frac{\partial^2 u}{\partial y^2}dy. \tag{7.31}$$

The last integral on the left-hand side can be integrated by parts, so that, considering the boundary conditions (7.16) and (7.17) and using the continuity equation (7.1), we get

$$\int_0^\infty v\frac{\partial(U_e - u)}{\partial y}dy = \left[v(U_e - u)\right]_0^\infty - \int_0^\infty (U_e - u)\frac{\partial v}{\partial y}dy$$
$$= \int_0^\infty (U_e - u)\frac{\partial u}{\partial x}dy. \tag{7.32}$$

If we now introduce (7.32) into (7.31) and recall that $\partial U_e/\partial x$ is not a function of y, we obtain

$$\frac{\partial U_e}{\partial x}\int_0^\infty (U_e - u)dy + \frac{\partial}{\partial x}\int_0^\infty u(U_e - u)dy = -\nu\left[\frac{\partial u}{\partial y}\right]_0^\infty. \tag{7.33}$$

We now observe that $\partial u/\partial y$ vanishes at the outer edge of the boundary layer, and that, if τ_w is the tangential viscous stress at the surface, the following relation holds:

$$\nu \left.\frac{\partial u}{\partial y}\right|_{y=0} = \frac{\tau_w}{\rho}. \tag{7.34}$$

Therefore, by introducing relation (7.34) into (7.33) and using the definitions of the displacement and momentum thicknesses, we finally obtain the following form of the momentum-integral equation:

$$\delta^* U_e \frac{dU_e}{dx} + \frac{d}{dx}(U_e^2 \theta) = \frac{\tau_w}{\rho}, \tag{7.35}$$

where the ordinary differentiation symbol has been introduced to stress the fact that all the quantities appearing in the equation are functions of x only.

7.3.4. *Energy thickness*

The last boundary layer parameter having the dimension of length that we introduce is the *energy thickness*, δ_e (also denoted in the literature by δ_3 or δ^{**}), which is connected with the global deficit of kinetic energy flux in a boundary layer compared to that corresponding to a completely potential flow. Its definition is

$$\delta_e = \int_0^\delta \frac{u}{U_e}\left(1 - \frac{u^2}{U_e^2}\right) dy, \tag{7.36}$$

which can be recast in the following form

$$\frac{1}{2}\rho U_e^3 \delta_e = \int_0^\delta \frac{1}{2}\rho u(U_e^2 - u^2) dy = \frac{1}{2}U_e^2 \int_0^\delta \rho u\, dy - \int_0^\delta \frac{1}{2}\rho u^3\, dy. \tag{7.37}$$

Obviously, the same observations that were made for the momentum and displacement thicknesses as regards the upper limits of integration in the related integrals are also valid for the energy thickness.

As can be seen, relation (7.37) gives the difference between the ideal flux of kinetic energy, which would correspond to the mass flowing per unit time in the boundary layer if it were moving with velocity U_e, and the actual kinetic energy flux through the boundary layer.

From relations (7.37) and (7.25) it is easy to derive the expression for the kinetic energy flux in the boundary layer as a function of the remaining characteristic thicknesses:

$$\int_0^\delta \frac{1}{2}\rho u^3 \, dy = \frac{1}{2}\rho U_e^3(\delta - \delta^*) - \frac{1}{2}\rho U_e^3 \delta_e = \frac{1}{2}\rho U_e^3[\delta - (\delta^* + \delta_e)]. \quad (7.38)$$

Therefore, the actual kinetic energy flux through the boundary layer is equal to the kinetic energy flux of fluid moving at the outer potential flow velocity U_e through a distance $[\delta - (\delta^* + \delta_e)]$.

We may also derive an *energy-integral equation* showing the relation between the energy thickness and the dissipation within the boundary layer. To this end, we use a procedure that is similar to the one that led to the derivation of the momentum-integral equation, with the sole difference that Eq. (7.30) is first multiplied by u before carrying out its integration. Furthermore, it is again expedient to introduce an additional term whose value is zero, namely

$$-\frac{1}{2}v\frac{\partial U_e^2}{\partial y}.$$

We thus obtain, after simple manipulations,

$$\frac{1}{2}\int_0^\infty u\frac{\partial(u^2 - U_e^2)}{\partial x}dy + \frac{1}{2}\int_0^\infty v\frac{\partial(u^2 - U_e^2)}{\partial y}dy = \nu\int_0^\infty u\frac{\partial^2 u}{\partial y^2}dy. \quad (7.39)$$

The second integral on the left-hand side and the integral on the right-hand side may be integrated by parts, so that, by applying the continuity equation and the boundary conditions for the boundary layer flow, we easily get

$$\frac{1}{2}\int_0^\infty v\frac{\partial(u^2 - U_e^2)}{\partial y}dy = \frac{1}{2}\int_0^\infty (u^2 - U_e^2)\frac{\partial u}{\partial x}dy, \quad (7.40)$$

$$\nu\int_0^\infty u\frac{\partial^2 u}{\partial y^2}dy = -\nu\int_0^\infty \left(\frac{\partial u}{\partial y}\right)^2 dy. \quad (7.41)$$

If we now introduce (7.40) and (7.41) into (7.39), use relation (7.37) and recall the definition of the kinematic viscosity, we finally obtain the following form of the energy-integral equation:

$$\frac{1}{2}\rho\frac{d(U_e^3 \delta_e)}{dx} = \mu\int_0^\infty \left(\frac{\partial u}{\partial y}\right)^2 dy. \quad (7.42)$$

Within the boundary layer approximations, it may be seen that the integral on the right-hand side of (7.42) is linked to the dissipation that takes place inside the boundary layer. Furthermore, it is also equal to the integral of the square of vorticity. We shall use the energy-integral equation in next section and recall it in Chapter 10, where the role of the energy balance equation and of dissipation in incompressible flows will be treated in more detail.

7.4. Boundary Layer Over a Flat Plate

In spite of its apparent simplicity and limited generality, the problem of a two-dimensional boundary layer developing over a negligible-thickness flat plate immersed in a parallel flow is particularly important for applications. In effect, it serves as a standard of comparison for slender bodies, i.e. bodies whose cross-flow dimension is much smaller than their length and which are aligned with the free-stream. Moreover, a solution to this problem was first given by Blasius (1908), and it provides fundamental reference values for the skin friction acting on the plate as a function of the Reynolds number. Finally, the solution is an example of the wider class of so-called *similarity solutions*, and the procedure by which it is derived may thus be of interest in itself.

We have already analysed, in Chapter 6, the generation and dynamics of vorticity over a flat plate of length L placed parallel to an upstream flow with velocity U. In particular, we found that, if the Reynolds number is sufficiently high, convection prevails over diffusion and the vorticity remains confined within a thin layer adjacent to the plate. We now recognize that this layer coincides with the boundary layer introduced by Prandtl and that we can study its evolution through the equations derived in Section 7.2.

It should be noted that, due to the growth of the boundary layer, the streamlines in the outer potential flow are not parallel to the plate, but are deflected laterally. In other words, it is as if the outer potential stream had to flow along a plate whose thickness increases in the downstream direction. We have also seen that a measure of this lateral deflection is given by the displacement thickness of the boundary layer. Therefore, an estimate of the velocity at the edge of the boundary layer might be obtained by solving the potential flow equations for a velocity field tangent to a fictitious body, constructed by adding to the plate a growing thickness equal to $\delta^*(x)$. We shall see that, in principle, this is indeed the correct procedure to obtain the appropriate value of the outer potential velocity to be used as the external condition for the solution of the boundary layer equations. However, in the present case of the boundary layer over a flat plate we shall assume, as a first approximation, that the boundary layer thickness is so small that its effect on the velocity in the potential flow region may be completely neglected. This corresponds to assuming that the velocity along the outer edge of the boundary layer, $U_e(x)$, is constant and equal to the free-stream velocity U. Therefore, from Eq. (7.19) we readily derive that the pressure is likewise uniform just outside the boundary layer, and thus, considering relation (7.11), everywhere throughout the whole boundary layer. This permits further simplification of the momentum equation (7.30), which then reduces to

$$u\frac{\partial u}{\partial x} + v\frac{\partial u}{\partial y} = \nu\frac{\partial^2 u}{\partial y^2}. \tag{7.43}$$

The continuity equation (7.1) must also be fulfilled, and the boundary conditions for the velocity field are

$$u = v = 0 \quad \text{for } y = 0 \quad \text{and} \quad 0 \le x \le L, \tag{7.44}$$

$$u(x; y \to \infty) = U \quad \text{for } 0 \le x \le L, \tag{7.45}$$

$$u(0, y) = U \quad \text{for any } y. \tag{7.46}$$

We now recall that Eq. (7.43) is parabolic; hence, the velocity profile at a certain x is not influenced by the downstream conditions, but only by the upstream ones. Consequently, the length, L, of the plate does not play a role in the flow conditions at a generic x. Considering relation (7.7), the thickness of the boundary layer may then be expected to depend on a Reynolds number evaluated using the outer velocity U and the length x. In other words, δ increases with x in proportion to the quantity

$$\Delta = \sqrt{\frac{\nu x}{U}}. \tag{7.47}$$

Furthermore, due to the constancy of U, the condition (7.45) at the edge of the boundary layer does not depend on x. This suggests that the velocity profiles at varying distances from the leading edge of the plate are similar, i.e. that they may be made identical if the vertical coordinate y and the velocity u are scaled by a length related to the increasing boundary layer thickness and by the outer velocity U, respectively. Therefore, we seek a solution that is a function of a non-dimensional coordinate η only, defined as

$$\eta = \frac{y}{\Delta} = y\sqrt{\frac{U}{\nu x}}. \tag{7.48}$$

To find the appropriate form of the non-dimensional velocity, we first observe that Eq. (7.1), which is valid for any incompressible two-dimensional flow, imposes a constraint on the velocity components that is equivalent to stating that a scalar function $\psi(x, y)$ exists such that

$$u = \frac{\partial \psi}{\partial y}; \quad v = -\frac{\partial \psi}{\partial x}. \tag{7.49}$$

The function ψ is called the *stream function*, a name that may be considered quite appropriate by noting that ψ is constant along a streamline. In effect, the total differential of ψ is given by

$$d\psi = \frac{\partial \psi}{\partial x} dx + \frac{\partial \psi}{\partial y} dy = -v\, dx + u\, dy, \tag{7.50}$$

and is thus zero along a streamline, as may be derived from the special form of Eq. (3.10) that applies for a two-dimensional flow.

Incidentally, there is also an interesting relationship between vorticity and the stream function; in fact, in a two-dimensional flow in the $x-y$ plane we have

$$\omega_z = \left(\frac{\partial v}{\partial x} - \frac{\partial u}{\partial y} \right) = -\nabla^2 \psi. \tag{7.51}$$

Thus, if a two-dimensional flow is irrotational, a Laplace equation holds not only for the velocity potential but also for the stream function.

We now observe that for the flow outside the boundary layer over the flat plate the stream function is equal to

$$\psi = Uy. \tag{7.52}$$

It is reasonable, then, to expect that inside the boundary layer the stream function will be proportional to $U\Delta$; therefore, considering relation (7.47), ψ may tentatively be expressed in terms of a non-dimensional function $f(\eta)$ as

$$\psi(x, y) = \sqrt{\nu U x}\, f(\eta). \tag{7.53}$$

If we insert this form of the stream function into relations (7.49) to express the velocity components, we get

$$u = U f'(\eta), \quad v = \frac{1}{2}\sqrt{\frac{\nu U}{x}}(\eta f' - f), \tag{7.54}$$

where the prime indicates differentiation with respect to η.

Expressions (7.54) can now be inserted into Eq. (7.43) and, after carrying out the various differentiations and some straightforward manipulations, the following ordinary differential equation is obtained:

$$f''' + \frac{1}{2}ff'' = 0. \tag{7.55}$$

The boundary conditions to be applied to this equation are easily derived from (7.44)–(7.46), and are

$$f = f' = 0 \quad \text{for } \eta = 0, \tag{7.56}$$

$$f' \to 1 \quad \text{for } \eta \to \infty. \tag{7.57}$$

Note that both conditions (7.45) and (7.46) collapse to (7.57). Equation (7.55) is non-linear and was first solved by Blasius using series expansion but nowadays it may be solved numerically. The variation of u/U as a function of η is shown in Fig. 7.3, and the function $u(y)$ can thus be found for any value of x and U by using definition (7.48). It should be pointed out that the velocity distribution of Fig. 7.3 is in very good agreement with the available experimental data.

Once the velocity distribution is obtained, the variation of all the characteristic boundary layer thicknesses can be readily derived as a function of x and of a local Reynolds number Re_x, defined as

$$Re_x = \frac{Ux}{\nu}. \tag{7.58}$$

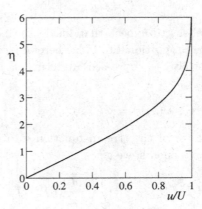

Fig. 7.3.   Velocity distribution in the boundary layer on a flat plate.

We thus get

$$\delta(x) = \frac{4.91x}{\sqrt{Re_x}}, \tag{7.59}$$

$$\delta^*(x) = \frac{1.7208x}{\sqrt{Re_x}}, \tag{7.60}$$

$$\theta(x) = \frac{0.664x}{\sqrt{Re_x}}, \tag{7.61}$$

$$\delta_e(x) = \frac{1.044x}{\sqrt{Re_x}}. \tag{7.62}$$

Therefore, all the various thicknesses increase as the square root of x, and the shape factor $H = \delta^*/\theta$ is slightly less than 2.6.

It should be pointed out that the value of the boundary layer thickness given in (7.59) is the one that corresponds to $u = 0.99U$ (the numerical coefficient would be 6.012 for $u = 0.999U$); conversely, all the other thicknesses are evaluated by extending the upper limit of the relevant defining integrals to infinity. This can be done in this case because the velocity in the outer potential flow is everywhere equal to U and does not vary for increasing distance from the plate. Therefore, the integrands in definitions (7.22), (7.26) and (7.36) tend to zero outside the boundary layer and the various integrals converge to finite values.

From the practical point of view, one of the most important results deriving from the above-described solution is the prediction of the tangential viscous stress τ_w acting over the plate surface; in terms of a *friction coefficient* C_f, obtained by making τ_w non-dimensional through the kinetic pressure of the outer flow, we find

$$C_f(x) = \frac{\tau_w}{\frac{1}{2}\rho U^2} = \frac{0.664}{\sqrt{Re_x}}. \tag{7.63}$$

It is interesting to note that the viscous friction force on a plate is proportional to the outer velocity at the power 3/2. Furthermore, for a given outer velocity, τ_w is inversely proportional to the square root of x, i.e. of the distance from the leading edge of the plate. This is reasonable, because U is constant with x, while the thickness of the boundary layer increases as the square root of x, causing a proportional decrease of the velocity gradient at $y = 0$. This has the interesting implication that, on a flat plate, the regions near the leading edge give a larger contribution to the total friction force than the downstream portions.

The total friction force on a plate of length L (or, more precisely, on one of its faces) can be obtained by integrating the expression given by (7.63) from $x = 0$ to $x = L$. The force, per unit length in the direction perpendicular to the plane of motion, is thus found to be

$$F = \int_0^L \tau_w(x)dx = \rho U^2 \frac{0.664L}{\sqrt{Re_L}} = \rho U^2 \theta(L), \qquad (7.64)$$

where $Re_L = UL/\nu$. The dependence of the global friction force on $\theta(L)$ should not come as a surprise; in fact, it derives directly from the application of the momentum-integral equation (7.35) to the flat plate case, in which U_e is constant and equal to U. The same result can also be obtained by applying the integral momentum balance to a volume of fluid confined between the plate surface and the streamline passing through a point just outside the boundary layer at $x = L$. It is thus seen that, in this case, knowledge of the local value of the momentum thickness at a certain coordinate x implies that the global friction force acting on the plate from its leading edge up to x is also known.

It is convenient to introduce a global friction force coefficient C_F:

$$C_F = \frac{F}{\frac{1}{2}\rho U^2 L} = \frac{1.328}{\sqrt{Re_L}} = \frac{2\theta(L)}{L}. \qquad (7.65)$$

We may now observe that the normal velocity v, although small, is not zero inside the boundary layer, its ratio to u being of order $(Re_x)^{-1/2}$. Its asymptotic value at the outer edge of the boundary layer is

$$\frac{v_\infty(x)}{U} = \frac{0.8604}{\sqrt{Re_x}}. \qquad (7.66)$$

The value given by (7.66) is particularly interesting because it is easily seen that it corresponds to $d\delta^*/dx$. We may thus say that the inclination of the streamline at the outer edge of the boundary layer coincides with the inclination of the streamline of an ideal fully potential flow tangent to the surface of a virtual body obtained by adding the displacement thickness to the plate surface. This confirms that δ^* is indeed the appropriate quantity to be used in an iterative evaluation of the coupling between the boundary layer and the external potential flow. In effect, after having solved the boundary layer equations for the flat plate as just described — i.e. with the outer velocity set equal to U — we may solve the Laplace equation for the velocity potential of a flow around the body obtained by adding $\delta^*(x)$ to both sides of the plate, immersed in an upstream flow with velocity U

and with the sole boundary condition of non-penetration at the surface (as if the fluid were non-viscous). The result of such a calculation will give a second-order estimate of the velocity U_e at the edge of the boundary layer, this time no longer constant, which can be introduced in another evaluation of the boundary layer problem. Therefore, the iterative procedure may be continued as long as the results obtained in two subsequent steps for a quantity of interest (e.g. the thickness of the boundary layer or, perhaps more sensibly, the total friction force on the plate) differ by less than a prescribed maximum value.

In reality, for a flat plate in parallel motion this iterative procedure is hardly necessary, but we shall see that it can be extended to any body characterized by the presence of a thin boundary layer over its entire surface. Furthermore, it is worth recalling that placing the surface of the virtual body, for the outer potential flow evaluation, at a distance $\delta^*(x)$ from the actual surface corresponds to positioning it at the centre of gravity of the total vorticity present inside the boundary layer at that x.

It is interesting to note that the Blasius solution for a flat plate implies that the total amount of vorticity present within the boundary layer over any point of the plate is constant and equal to U. Thus, no further vorticity is introduced along the plate surface, and this is in agreement with the discussion on Chapter 6 on the connection between the generation of new vorticity at a solid surface and the pressure gradient existing along it. We may also evaluate the flux of vorticity through the boundary layer over the upper surface of the plate, and the result is

$$\int_0^\delta \omega u \, dy = - \int_0^\delta u \frac{\partial u}{\partial y} dy = -\frac{1}{2} U_e^2 = -\frac{1}{2} U^2. \qquad (7.67)$$

The negative sign is due to the fact that, with the chosen reference system, the vorticity over the upper surface is negative. Therefore, relation (7.67) gives an evaluation of the vorticity instantaneously shed in the wake from the upper surface of the plate. This confirms the result that was previously derived in Section 6.4.4, and the discussion reported therein on the continuous source of new vorticity being positioned at the leading edge of the plate, which becomes a singular point when the thickness of the plate is reduced to zero. In reality, it may be seen that also in the above-described boundary layer solution the leading edge is a singular point where, for instance, the friction coefficient in (7.63) becomes infinite. This is understandable, because the boundary layer equations are no longer valid at the leading edge of the plate and, in reality, for all values of x such that

Re_x is very small. Thus, in principle, for that region either the full Navier–Stokes equations or higher-order boundary layer approximations should be used. In practice, this is seldom necessary for engineering applications and the corrections for this very small region are usually not taken into account.

Finally, for the flat plate case, the velocity is constant outside the boundary layer and thus, in that region, $\Phi = 0$ and no dissipation occurs. On the other hand, inside an incompressible two-dimensional boundary layer the expression of the dissipation function (4.77) becomes simply

$$\Phi = \mu \left(\frac{\partial u}{\partial y} \right)^2. \tag{7.68}$$

The whole dissipation above a plate of length L is then given by the value of the energy thickness at the end of the plate. In effect, from the energy-integral equation (7.42), we get

$$\int_0^L dx \int_0^\infty \Phi dy = \mu \int_0^L dx \int_0^\infty \left(\frac{\partial u}{\partial y} \right)^2 dy = \frac{1}{2} \rho U^3 \delta_e(L). \tag{7.69}$$

7.5. The Effects of Pressure Gradients

7.5.1. *Generalities*

As already pointed out, the results for the boundary layer over a flat plate parallel to the flow, i.e. with zero pressure gradient, are of great practical importance, as they provide good first-order estimates of the tangential viscous stresses over the plate surface. Furthermore, these results may also be directly used for obtaining the order of magnitude of the friction forces acting on bodies characterized by a prevailing dimension in the direction of motion.

In more general cases, in which the pressure varies along the surface, the boundary layer equations must be solved with the value of $\partial p / \partial x$ given as an input. We shall not go through the description of methods to carry out this calculation, but we shall discuss in some detail the qualitative effects that may be expected from the presence of pressure gradients of different sign along the surface. In any case, we shall keep the flat plate results as a reference, trying to highlight the differences caused by the pressure gradients on the evolution of the boundary layer compared to the zero pressure gradient case.

We first analyse the effect of $\partial p / \partial x$ on the growth of the boundary layer thickness along the surface. To this end, it is expedient to consider the behaviour of the velocity component v normal to the surface. From the continuity equation (7.1), we get

$$v(y) = - \int_0^y \frac{\partial u}{\partial x} dy. \tag{7.70}$$

For the flat plate case, the velocity component u at a fixed value of y decreases with increasing x, i.e. $\partial u / \partial x < 0$. This may be derived by differentiating the first of (7.54) with respect to x, taking the definition (7.48) into account, but should also be quite obvious, considering the increase of the boundary layer thickness with x. In effect, the actual velocity profile $u(y)$ is obtained from the non-dimensional one shown in Fig. 7.3 by continuously stretching its vertical coordinate as x increases; consequently, considering a fixed value of y, the velocity u decreases with increasing values of x. Therefore, in the zero pressure gradient case, the normal velocity v is positive throughout the boundary layer, and its value at the edge is given by relation (7.66).

Let us now consider the case in which $\partial p / \partial x$ is negative; from (7.19), this condition corresponds to an accelerating flow along the edge of the boundary layer. We may then expect this acceleration also to affect the particles moving inside the boundary layer, and in particular those close to its edge. Therefore, compared to the zero pressure gradient case, $\partial u / \partial x$ inside the boundary layer will become less negative, or even positive for sufficiently strong outer accelerations. This implies that the velocity at the edge of the boundary layer, as obtained from the integral (7.70) extended up to $y = \delta$, will be less positive than in the flat plate case, or even negative. Consequently, the thickness of the boundary layer will increase less rapidly along the surface, or even decrease for highly negative values of $\partial p / \partial x$. This lower growth of the boundary layer, together with the increasing outer velocity, may be expected to give rise to an increased friction force at the surface. The effect of a negative pressure gradient on the boundary layer thickness may also be interpreted in terms of vorticity dynamics, by noting that the accelerating outer flow increases the role of convection and, consequently, reduces the action of diffusion in moving vorticity away from the surface.

Obviously, if $\partial p / \partial x$ is positive and the outer velocity decreases, the opposite occurs so that, compared to the zero pressure gradient case, the normal velocity v is more positive and the boundary layer thickness increases more rapidly along the surface. In summary, we may say that, in comparison to what happens for a zero pressure gradient flat plate, an acceleration of the outer flow tends to keep the vorticity within a thinner layer adjacent to the surface, while a deceleration tends to increase the thickening of the boundary layer along the surface.

In addition to their effect on the downstream evolution of the boundary layer thickness, pressure gradients have also a profound influence on the shape of the velocity profile within the boundary layer. This can be readily seen by recalling that the velocity is zero at the surface, and thus the momentum balance equation reduces to (6.33), which we recast as follows:

$$\left.\frac{\partial^2 u}{\partial y^2}\right|_{y=0} = \frac{1}{\mu}\frac{dp}{dx}. \tag{7.71}$$

This expression shows that a direct relation exists between the pressure gradient and the *curvature* of the velocity profile at the surface. In particular, we find that in the zero pressure gradient case the velocity profile is linear at the surface, and this is indeed the result given by the Blasius solution, as can also be noticed from Fig. 7.3. On the other hand, when $\partial p/\partial x < 0$, the curvature is negative over the whole boundary layer thickness, and the slope $\partial u/\partial y$ gradually decreases from its maximum value at $y = 0$ to (practically) zero at $y = \delta$. Finally, when $\partial p/\partial x > 0$, the curvature is positive at the surface, close to which $\partial u/\partial y$ increases for increasing values of y. Therefore, considering that $\partial u/\partial y = 0$ at $y = \delta$, a point of inflection, where the slope reaches a maximum value, must exist within the boundary layer thickness.

Typical velocity profiles for the different cases are shown in Fig. 7.4, where the variation of the vorticity ω is also reported. In fact, we have seen that, in a boundary layer, vorticity is given to a very good approximation by relation (7.20), and is thus immediately linked to the slope of the velocity profile.

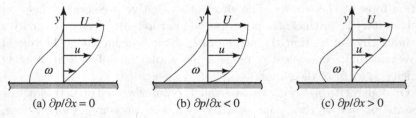

(a) $\partial p/\partial x = 0$ (b) $\partial p/\partial x < 0$ (c) $\partial p/\partial x > 0$

Fig. 7.4. Boundary layer velocity and vorticity profiles for different pressure gradients.

We may interpret the velocity profiles associated with different values of $\partial p/\partial x$ in terms of vorticity generation and dynamics. In Chapter 6 a pressure variation over a solid surface was shown to correspond to a source of vorticity along the surface. Thus, the different velocity profiles in Fig. 7.4 may be explained by noting that, compared to the zero pressure gradient case, further negative (clockwise) vorticity is generated at the surface and enters the flow when $\partial p/\partial x < 0$. Conversely, when $\partial p/\partial x > 0$, there is a production of positive (counter-clockwise) vorticity at the surface, with a consequent decrease of the vorticity of opposite sign that was already present in the neighbourhood of the surface. Therefore, the inflection point that exists in the velocity profile within the boundary layer for $\partial p/\partial x > 0$ coincides with the location where a maximum absolute value of the vorticity is present. Note that the above-mentioned vorticity signs refer to the conditions depicted in Fig. 7.4, where flow is present 'above' a certain surface; the opposite signs would apply if the flow, still coming from the left, were bounded by a surface placed on its upper side.

The velocity profiles of Fig. 7.4 confirm that, for equal velocity outside the boundary layer, a higher friction force acts at the surface when the outer flow is accelerating, while a lower one acts when a deceleration is present. In effect, the value of $\partial u/\partial y$ at $y = 0$ is higher for $\partial p/\partial x < 0$ than for $\partial p/\partial x > 0$. It may also be easily verified that, compared to the zero pressure gradient case, the boundary layer shape factor, $H = \delta^*/\theta$, is larger for $\partial p/\partial x > 0$ and smaller for $\partial p/\partial x < 0$.

In spite of the higher associated friction forces, a flow in which the pressure decreases along a solid surface, in connection with an increasing velocity at the edge of the boundary layer, is generally referred to as a condition in which a *favourable* pressure gradient exists, while an increasing pressure is denoted as an *adverse* pressure gradient. The reason for this notation will become clearer in next section, where the possible consequences of a positive pressure gradient on the evolution of a boundary layer are analysed in more detail.

7.5.2. *Boundary layer separation*

The situation in which the flow outside the boundary layer decelerates is particularly important because it may lead to radical changes in the character of the whole flow field. In particular, the existence of a thin vorticity-containing boundary layer adjacent to a solid surface may be completely hampered, even for very high Reynolds numbers.

Let us consider, then, a portion of a solid surface where a positive pressure gradient exists, after a region or location where $\partial p/\partial x = 0$. The type of velocity profile in the boundary layer will then change from the one represented in Fig. 7.4(a) to that in Fig. 7.4(c). If we follow the evolution of the velocity profile moving downstream inside the region with increasing pressure, we observe a progressive upward displacement of the inflection point. Simultaneously, the thickness of the boundary layer increases and the value of $\partial u/\partial y$ at the surface decreases.

The change in the shape of the velocity profile for $\partial p/\partial x > 0$, caused by the rapid reduction of the velocity close to the surface, may also be ascribed to the invariance of the pressure in the direction perpendicular to the surface. In effect, every particle within the boundary layer experiences the *same* resultant pressure force per unit volume, directed against the motion. Therefore, the relative variation of momentum will be larger for the lower-velocity particles moving close to the surface than for the particles that are nearer to the edge of the boundary layer.

If the positive pressure gradient is strong enough, or if it acts over a sufficiently long distance, a limit condition may be reached such that we have $\partial u/\partial y = 0$ at a certain point S on the surface, where, consequently, the viscous tangential stress also vanishes. Beyond that point, the particles close to the surface move backwards, i.e. in the opposite direction relative to the outer stream. Simultaneously, the vorticity adjacent to the surface changes sign with respect to the one that was present upstream.

This phenomenon, qualitatively depicted in Fig. 7.5, is known as *boundary layer separation*, and point S is the *separation point*. The occurrence of separation leads to a drastic change of the flow features, with significant effects not only on the forces acting on moving bodies, but also on the possibility of devising simplified procedures for their prediction.

From the case sketched in Fig. 7.5 it can be deduced that boundary layer separation may also be interpreted in terms of vorticity generation

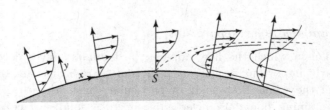

Fig. 7.5.   Sketch of boundary layer separation.

and dynamics. We can say that the sources of positive (counter-clockwise) vorticity connected with the increasing pressure are sufficiently strong or extensive to cause, at least immediately close to the surface, a complete cancellation, by annihilation, of the negative (clockwise) vorticity produced upstream and brought over that region by convection. The region with negative vorticity is then moved outwards from the surface and a completely new flow condition is produced, with a layer of counter-clockwise vorticity immediately over the surface, while clockwise vorticity is still present further outside.

A particularly important feature of flows in which boundary layer separation occurs is that, however high the Reynolds number, vorticity spreads in the field, moving to large distances from the solid surfaces from which it was generated. Thus, beyond the separation point, the irrotational flow is displaced outwards, and the region where vorticity is present is no longer thin and close to the surface. Conversely, vorticity fills the entire downstream separated region, forming a wake where, generally, significant velocity fluctuations are also present. Unsteady conditions are thus typical of separated flows, and this feature adds further difficulty to their prediction.

From the above description of the evolution of the velocity profiles leading to boundary layer separation, it should be clear that the existence of a region where $\partial p/\partial x > 0$, i.e. where the outer flow decelerates, is a necessary condition for separation to occur. Obviously, and fortunately, it is not a sufficient condition, and this means that if the pressure gradient along the surface is positive but not too strong, or if it does not act over a sufficiently long surface extension, then it is possible that the boundary layer continues to be thin and adjacent to the surface (or, as we usually say, that it remains an *attached* boundary layer).

On the other hand, experience shows that it is not exactly true that a positive pressure gradient must necessarily exist for the occurrence of boundary layer separation. The existence of a positive $\partial p/\partial x$ is a strictly necessary condition only in the case of separation along slightly curved surfaces (usually termed *natural separation*). Conversely, if sharp edges (with sufficiently small included angles) are present on the surface, separation always occurs at their location (as shown in the sketch of Fig. 7.6), even when the upstream pressure gradient is negative. This behaviour is plausible because it can be shown that if the boundary layer were able to contour the sharp edge without separating, then the outer potential flow would be characterized by a strong convex curvature of

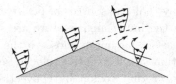

Fig. 7.6. Boundary layer separation at a sharp edge.

the streamlines adjacent to that region, followed by a concave curvature to realign the flow towards the upstream direction. Consequently, a high acceleration, and thus a very low pressure, would occur at the sharp edge, and would then be followed by an abrupt deceleration causing so strong an adverse pressure gradient that it would be incompatible with the existence of an attached boundary layer.

Apparently, the separation of a two-dimensional boundary layer over mildly curved surfaces may be predicted by solving the boundary layer equations up to the point where $\partial u/\partial y = 0$ on the surface. However, such predictions should be considered with extreme caution, because the boundary layer equations progressively lose their validity as the separation point is approached. In fact, some of the assumptions of boundary layer theory become less and less applicable. In particular, it is no longer true that $\partial u/\partial x \ll \partial u/\partial y$ close to the surface; moreover, near separation, the normal velocity component v cannot be assumed to be of small order of magnitude compared to the tangential component u. Finally, the occurrence of separation drastically changes the outer potential flow with respect to the case of an attached boundary layer, due to the above-mentioned outward movement of the vorticity-containing region. Consequently, the pressure distribution over the surface, which is an input for a boundary layer calculation, can no longer be obtained by means of simple procedures. In conclusion, using the classical boundary layer equations one cannot accurately predict separation; at most, one may obtain a first-order assessment of the compatibility of a given pressure distribution with the existence of an attached boundary layer.

It must also be pointed out that, up to now, we have been analysing two-dimensional boundary layers only. As may be easily imagined, when the flow is three-dimensional the situation becomes even more complex as regards both the solution of the boundary layer equations and, especially, the prediction of separation. A detailed analysis of the characteristics and methods of prediction of three-dimensional boundary layers is outside the scope of the present book. Suffice it here to say that, in a general three-dimensional boundary layer, the velocity vector changes direction from the wall to the edge of the boundary layer. In other words, the streamlines just outside the boundary layer generally have a different direction with respect to those close to the surface. Furthermore, one particular and important feature of three-dimensional separation is that it may occur along lines on the surface where the tangential viscous stresses are not necessarily zero.

The study of the evolution, dynamics and separation of boundary layers is a fundamental research topic in fluid dynamics. In effect, the development of methods for predicting and influencing the behaviour of boundary layers is a primary issue for the control of the fluid dynamic loads acting, for instance, on moving bodies or on bodies immersed in the atmospheric wind. Furthermore, boundary layer separation has a considerable influence on the predictability of flow features and loads by means of simplified methods. This leads to a classification of bodies in terms of the behaviour of the boundary layers over their surfaces.

7.5.3. *Aerodynamic bodies and bluff bodies*

In previous chapters we have seen that when a thick body moves in a fluid, the pressures acting over the fore part of its surface, near the stagnation point, are higher than the pressure p_∞ existing in the undisturbed fluid. Moving downstream along the body surface, the pressure rapidly decreases and becomes lower than p_∞, reaching a minimum over a region near the maximum thickness of the body. Subsequently, the pressure increases again over the aft portion of the body surface. However, depending on the shape of the body, we also noticed that significant differences may exist in the pressure distribution, especially as regards the rear part of the surface. For certain bodies, the pressure in this region may approach the undisturbed value and the whole pressure distribution may be analogous, albeit not equal, to the one that can be obtained from an analysis that neglects the action of viscosity (see, e.g., Fig. 5.1). Conversely, for other bodies, such as the circular cylinder in Fig. 5.2, the pressures on the rear part of the surface remain lower than p_∞, and the whole pressure distribution is completely different from the one corresponding to a non-viscous flow.

Having analysed the mechanisms of generation and evolution of vorticity in an incompressible flow, and the results of boundary layer theory as regards both the invariance of the pressure across the boundary layer and the possibility of separation in adverse pressure gradients, we are now able to explain the above observations. In effect, it should be clear that separation of the boundary layer will not occur over the whole surface of a body provided its shape and orientation in the flow are such that the positive pressure gradients over its rear part are sufficiently moderate to permit the boundary layer to remain attached up to the end of the body. In other words, the pressure gradients in the outer potential flow at the edge of the boundary layer must be compatible with the existence of a completely attached boundary layer, and this depends precisely on the shape and type of motion of the body. Therefore, we may proceed to the following classification of bodies according to the type of flow they produce when they are in relative motion in a fluid and, in particular, to the behaviour of the boundary layer — and thus of vorticity — over their surfaces.

Aerodynamic bodies are those characterized by thin boundary layers that are completely attached over their whole surface, and by thin and generally steady vorticity-containing wakes trailing behind them.

Bluff bodies are those characterized by a more or less premature separation of the boundary layer from their surface, and by wakes having significant lateral dimensions and generally unsteady velocity fields.

This classification is, evidently, a purely aerodynamic one, in the sense that it is connected with the flow features produced by the bodies. In particular, it implies that the typical Reynolds number of the flow is sufficiently high, so that it is meaningful to refer to thin boundary layers and to their separation. Furthermore, it is also clear that the above definitions have implications as regards both the geometry and the type of motion of the bodies, as may be deduced from the examples shown in Figs. 7.7 and 7.8. In effect, the main geometrical constraint for bodies to be aerodynamic is that they must be elongated in the direction of motion, i.e. their streamwise dimension must be sufficiently larger than their cross-stream dimension. However, they must also move with their larger dimension at small angles to the free-stream because otherwise the adverse pressure gradients over their upper surface become excessive, and flow separation occurs (see the last example in Fig. 7.8). Therefore, there are no bodies that are aerodynamic for any direction of the flow, whereas bodies characterized by similar

Fig. 7.7.   Example of an aerodynamic body.

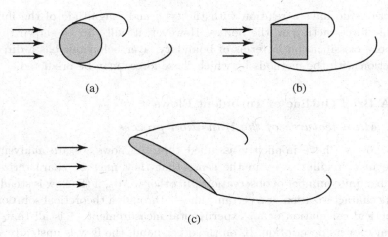

Fig. 7.8. Examples of bluff bodies.

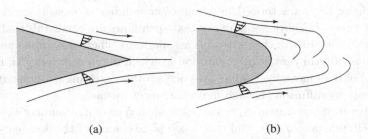

Fig. 7.9. Rear end of: (a) an aerodynamic body, (b) a bluff body.

dimensions in all directions are bluff whatever their orientation to the free-stream.

It must also be noted that for a body to be strictly aerodynamic its rear end must be sharp, as shown in Fig. 7.9(a). In effect, separation inevitably occurs for a rounded rear end, such as that in Fig. 7.9(b), because otherwise the excessively concave curvature of the streamlines in that region would cause decelerations of the flow that, in reality, are not compatible with the existence of an attached boundary layer.

Nonetheless, one may distinguish between bodies with different degrees of 'bluffness', according to the extent of their surface with attached boundary layer or to the width of their wake compared to their cross-flow dimension. As will become clearer in the next chapter, these further distinctions do have a relation with the type and magnitude of the fluid dynamic forces acting on the bodies. However, it will also be shown that the above classification in terms of boundary layer behaviour has a direct connection with the methods by which those forces may be predicted.

7.6. A Brief Outline of Turbulent Flows

7.6.1. *Main features of the transition process*

Up to now we have implicitly assumed that the flows we are analysing behave in a 'regular' way, in the sense that they may be characterized through a finite number of observations. In other words, if the flow is steady, then its characterization at a certain time — through a theoretical solution, a numerical calculation or an experimental measurement — is all that is necessary for its description. If, on the other hand, the flow is unsteady — for instance due to a regular variation of boundary conditions, as happens when a body is accelerating or oscillating in a fluid — then its variation in time may be characterized by a sufficient number of measurements at different times. In short, we may say that such flows are predictable, albeit with some difficulties in the unsteady case, perhaps. These flows are denoted as *laminar*, a term that suggests motion in which different layers ('lamina' in Latin) of fluid move by 'sliding' over one another, exchanging momentum and energy by diffusion through molecular mechanisms.

However, it is common experience that such type of motion may exist in certain circumstances only, and that a complete change of the flow features may occur, with the appearance of highly irregular fluctuations. We then say that the flow has become *turbulent*, or that *turbulence* has appeared. This change of the flow character is a drastic one, and it may have significant effects on the global flow features and on the forces acting on bodies in motion. The process by which a flow changes from the laminar to the turbulent state is called *transition*, and may be different for different flows or conditions.

A fundamental contribution to the study of turbulence and transition was given by Osborne Reynolds, who carried out accurate experiments describing the transition from laminar to turbulent flow in a circular pipe. Furthermore, he was able to identify the fundamental flow parameter influencing transition, which is now known as the Reynolds number, Re (Reynolds, 1883). In effect, turbulence arises only if the typical Reynolds number becomes higher than a certain critical value, which depends on the type of flow.

Turbulence is still one of the unsolved (and fascinating) problems of fluid dynamics. This does not mean that progress has not been made; intense research work has been carried out for more than a century, especially since the advent of the computer era. Nonetheless, a full understanding of the mechanisms of turbulent flows and, in particular, of their onset, is still far from being achieved. A detailed treatment of transition and turbulence is outside the scope of this book, and the interested reader may refer to available specialized texts, for instance, Tennekes and Lumley (1972), Hinze (1975), Pope (2000), Schmid and Henningson (2002), Drazin and Reid (2004), Lesieur (2008). Here, we limit our analysis to a brief description of the main aspects of the transition process, of the general features of turbulent flows, and of the fundamental effects of turbulence on the forces acting on moving bodies.

Let us start by considering the laminar–turbulent transition. Even if our main focus is on boundary layers, it must be pointed out that the transition to turbulence occurs in many other types of flows, such as jets, wakes and, in general, any shear flow. A fundamental common feature of flows that may undergo transition is that they are rotational, i.e. vorticity must be present for turbulence to arise. Several types of perturbations may cause transition, such as velocity fluctuations, acoustic waves, surface roughness and vibration, and temperature variations. Nonetheless, all types of perturbations may probably be expressed, in some way or other, in terms of perturbations of the vorticity field.

One might wonder why an irregular, and largely unpredictable, velocity field can arise from the deterministic equations of fluid motion. A widely accepted answer is that this behaviour is strictly connected with the non-linearity of the Navier–Stokes equations. In fact, it can be seen that non-linear equations may give rise to 'chaotic' solutions when some parameters appearing in the equations reach certain critical values. This is substantially due to an excessive sensitivity to the initial conditions, whose variation, even by a very small amount, may lead to the time evolution of completely different solutions. As perturbations introduced at a certain time and space in a flow may be seen as variations of initial conditions, the hypersensitivity to these variations may explain the outburst of turbulence.

The start of transition may also be viewed as the result of instability of the laminar flow that may occur when a perturbation is introduced. A flow is said to be unstable when an initial perturbation is not damped out in time and space, perhaps after some transient, and, conversely, increases indefinitely. Strictly speaking, instability does not necessarily lead to a turbulent flow, and another laminar flow, albeit generally more complicated, may also arise. However, we are more interested in the cases in which the flow evolves, through different scenarios, to the turbulent condition. In general, transition does not occur abruptly, and a finite region may be recognized in which the unsteadiness and the irregularity of the flow gradually increase.

As already mentioned, the Reynolds number — based on appropriate reference velocities and lengths according to the type of flow — is the fundamental parameter in the transition process; thus, it is quite natural to find that it also determines the stability level of the flow. For increasing Re the flow passes through different conditions, which may be defined in terms of certain characteristic values of the Reynolds number. In particular, if Re is sufficiently low, say below a certain value Re_m, the kinetic energy of any perturbation decreases in time and space, and the flow is said to be *monotonically stable*. For values of Re between Re_m and another typical value Re_g, the perturbation energy may first increase, but will eventually decay, and the flow is termed *globally stable*. For higher Re, the flow is *conditionally stable*, i.e. perturbations are ultimately damped down only provided their initial energy is below a threshold value, which decreases with increasing Re. Finally, above a critical Reynolds number Re_c, any perturbation, even of infinitesimal intensity, is amplified and may thus lead to transition.

Flow stability analyses for the prediction of the critical Reynolds numbers may be carried out by assuming that velocity and pressure disturbances are superposed to a given base flow. The equations governing the disturbances are then derived by subtracting the equations for the base flow from the equations of the total flow. Finally, the disturbance equations, where the base flow components are also present, are linearized by assuming the perturbations to be much smaller than the corresponding quantities in the base flow. The derived equations are then used to study the evolution of perturbations, which are added to different base flows in the form of oscillations in space and time with varying frequencies and wavelengths. Through this procedure, simple two-dimensional flows may be studied, both neglecting and considering the viscous terms in the equations.

Actually, some of the results obtained with this approach may be considered disappointing. For example, certain flows, such as the one caused by two parallel plates in relative motion and the flow driven by a pressure gradient in a circular pipe, respectively known as Couette and Hagen–Poiseuille flows (see Chapter 9), are found to be stable for any Reynolds number. This is in contrast with experimental evidence, which shows that transition may start at Reynolds numbers — based on the semi-distance between the plates or the diameter of the pipe — of the order of 360 and 2000 for the two cases, respectively. In other cases, such as the plane Poiseuille flow, Re_c is theoretically found to be 5772, while transition may actually occur at Re as low as 1000.

Nonetheless, important results may be obtained from linear stability analyses, such as the finding that, if the viscous terms in the perturbation equations are neglected, a necessary condition for instability — although not a sufficient one — is that an inflection point be present in the basic velocity profile. This is known as *Rayleigh's criterion*, but a more specific condition is given by *Fjørtoft's criterion*, which states, in practice, that instability is possible only if the inflection point corresponds to a maximum absolute value of the vorticity (see Fig. 7.10).

The first important consequence of the above criteria is that viscosity must play a fundamental role in causing the instability of flows that show no inflection points, such as boundary layers with zero or negative pressure gradient, or the Poiseuille flow. This is indeed what is found from a linear stability analysis in which, conversely, the viscous terms are taken into account. The destabilizing effect of viscosity may be surprising, as viscosity is usually associated with increased damping of disturbances. However, it can be seen that viscosity may provide contributions to the motion that are out of phase with the perturbation-induced oscillations, and this may lead to the appearance of instability for certain wavelengths, λ, of the perturbations. Therefore, the flow may be unstable above a certain value of Reynolds number, even if it is found to be stable as Re tends to infinity.

This result is exemplified in Fig. 7.11, where qualitative neutral stability curves for a boundary layer are shown as a function of the Reynolds number based on the displacement thickness, $Re_{\delta^*} = U_e \delta^*/\nu$, and of the non-dimensional wave number, $2\pi\delta^*/\lambda$, of the perturbation waves. The curves separate the regions where the perturbation decays from those where it

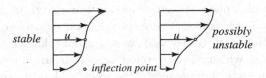

Fig. 7.10. Stable and unstable velocity profiles according to Fjørtoft's criterion.

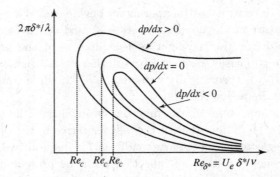

Fig. 7.11.   Qualitative neutral stability curves for a boundary layer.

is amplified, and Re_c is the minimum value of the Reynolds number for which an unstable perturbation wavelength exists. As can be seen from the figure, for zero or favourable pressure gradients the instability region shrinks to zero for Re tending to infinity, in agreement with the non-viscous results. This does not occur for the curve representing a case with an adverse pressure gradient, and this is reasonable, because, as we have seen, in that condition the velocity profile within the boundary layer has an inflection point. It should also be noted that, in practice, the decrease of Re_c due to an adverse pressure gradient is indeed significant. Conversely, a favourable pressure gradient delays the onset of instability.

For a boundary layer, the critical Reynolds numbers derived from the linear stability theory provide an indication of the onset of instability but, in general, largely underestimate the real start of turbulent flow. As an example, for the case of a flat plate with zero pressure gradient, the value of Re_{δ^*} for instability is about 520, which corresponds to a Reynolds number Re_x, based on the distance x along the plate, equal to 91 400. However, a more likely value for the occurrence of transition might be $Re_x \simeq 500\,000$. In fact, for very smooth surface conditions, values above 10^6 may even be found.

The reason for this discrepancy is that transition in a boundary layer is actually a complex phenomenon, involving the growth of perturbations through different mechanisms. The most common scenario comprises the amplification of unstable transversal waves, usually known as *Tollmien–Schlichting waves*, which then develop a non-linear three-dimensional instability, leading to the appearance of the so-called hairpin vortices. These vortices rise up, become inclined with respect to the surface at an angle of approximately 45°, and later break up into ever smaller structures, which may coalesce and cause the complete transition to turbulent flow. Especially in the case of strong local disturbances, the appearance of so-called *turbulent spots*, i.e. space-limited three-dimensional regions with seemingly turbulent flow, may also occur. These structures have the shape of an arrow, are carried with the flow and, simultaneously, may expand in all directions and coalesce further downstream, thus leading to complete transition.

A widely used method for predicting transition in boundary layers, which is often introduced in numerical codes, is to determine the growth of an initial perturbation after the instability region, and then assume that turbulence has been established when an amplification factor equal to e^N has been reached. The value of N is mostly empirical, and different values may be used for different types of boundary layers, but a typical order of magnitude may be $N = 9$ (corresponding to $e^N \simeq 8100$).

Different transition mechanisms come into play for other types of flows, such as free shear layers separating two streams at different velocities. These flows have a point of inflection in their velocity profiles and thus, as already pointed out, are highly unstable. Consequently, a perturbation with a wavelength larger than the thickness of the layer causes the vorticity present within it to concentrate in restricted regions of space or, in other words, in finite *vortices*: this is known as *Kelvin–Helmholtz instability*. The vortices are carried downstream, and merging between subsequent ones may occur, with the appearance of new and larger vortices. Subsequently, three-dimensional instabilities cause a breakdown of the vortices into a large number of smaller ones and, eventually, transition to fully turbulent flow is attained.

From the above brief description, it should be clear that much work is still needed to fully understand the complex phenomena leading to transition. However, the dynamics of vorticity has a primary role in the process; in particular, the appearance in the flow, due to external perturbations, of localized maxima of the absolute vorticity seems to be an important factor promoting instability and transition.

7.6.2. *General characteristics of turbulent flows*

As anticipated, we shall not attempt to give an account of the extensive theoretical and experimental research that has been carried over the years to try to unveil the secrets of turbulence, and we limit ourselves to a general description of some of the main features of turbulent flows.

The first point to be made is that there are many different types of turbulent flows. As is the case for the transition process, the characteristics of turbulence in a boundary layer are not necessarily the same as those in a free shear layer, in a jet or in the wake of a body. Nonetheless, certain features are common to all types of turbulent flows.

In particular, the first and most typical feature of every turbulent flow is the presence of irregular fluctuations, in both time and space, of all three velocity components. Thus, strictly speaking, turbulent flows are always unsteady and three-dimensional. However, it would not be appropriate to say that turbulent fluctuations are completely 'random', since this term is usually ascribed to a process in which no particular pattern may be distinguished. In fact, numerous results in turbulence research indicate that distinct types of 'structures', usually called *eddies*, are present in different classes of turbulent flows, and one of the fundamental problems is then to single out these structures and to characterize their relation to the observed fluctuations.

More specifically, turbulent fluctuations are indeed unpredictable, in the sense that it is not possible to predict, from an observation during a finite time interval, the exact value of the velocity at future times. The only thing that can be done is to evaluate the probability that this value be within a certain range, and this statistical type of characterization is indeed what turbulence has in common with all random processes. Nonetheless, the particular statistical parameters defining the various types of turbulence are probably connected with the complex (and largely unpredictable) dynamics of structures of different shapes and dimensions that interact in a complicated way, both mutually and with the global flow. Understanding the nature, origin and evolution of these structures would be an essential step towards a deeper comprehension of turbulent flows. One possible approach is to observe that, as already pointed out, all turbulent flows are rotational. Therefore, they may also be described as the result of the dynamics of an unsteady and three-dimensional vorticity field. The structures might then be characterized as regions, with different and continuously changing shapes and dimensions, which move with the flow and where vorticity displays a certain degree of 'coherence' in some average sense, at least for a typical time interval.

One significant feature of turbulent flows is that the energy associated with the fluctuating field is actually transferred from structures having dimensions typical of the global turbulent flow — such as the thickness of a boundary layer or the width of a wake — to structures of ever smaller dimensions, in a process that is usually referred to as a *cascade* of kinetic energy from large to small *scales*. The largest-scale structures derive their energy from the outer flow (e.g. the potential flow outside a boundary layer), and their induced velocity field influences the dynamics of the structures of smaller scale. The local velocity gradients increase with decreasing dimension of the structures, causing increased viscous stresses, and the typical size of the smallest scales in the flow is thus dictated by the dissipation of kinetic energy by viscosity. In the energy cascade process, a significant role is played by vorticity dynamics and, in particular, by the stretching and tilting mechanism that was shown in Chapter 6 to act when the flow field is three-dimensional.

Another fundamental feature of turbulent flows is that they are much more *dissipative* than laminar flows. This is a consequence of the sharp velocity gradients associated with the small-scale eddies, which cause large local values of the viscous stresses and thus of dissipation. Therefore, the significant amount of kinetic energy dissipated by the small eddies must be reintroduced into the flow at the large scales, so that turbulence can be sustained through the cascade process. If this does not happen, turbulence decays and the flow returns to the laminar state.

One useful assumption, often made in theoretical descriptions of turbulence, is that the dynamics of the small scales is independent of that of the largest scales. In other terms, it is assumed that the small-scale turbulence has no 'memory' of the large-scale motion and that it is homogeneous and isotropic, i.e. that it does not depend on space and direction. Actually, this hypothesis is strictly valid only in special circumstances, such as in the flow behind a grid of bars or cylinders placed orthogonally to a uniform stream. Conversely, in most shear flows, it represents a simplifying approximation of reality, which may be applied only at large distances from the location where turbulence starts.

The last common characteristic of all turbulent flows is that they are highly *diffusive*. In laminar flows, transport of momentum — and of any other transportable quantity, such as mass and energy — between adjacent layers of fluid takes place due to the molecular thermal motion that is the basis of viscous diffusion. As we have seen in the examples of Chapter 6, this is a slow process for fluids of small kinematic viscosity, such as air and water. Conversely, in turbulent flows transport phenomena are much stronger, because they are caused by the macroscopic fluctuations of the fluid particles, which migrate in different flow regions carrying with them their associated values of the various quantities. The consequence is that, in turbulent flows, diffusion may be two or even three orders of magnitude larger than in laminar flows.

In applications, the higher mixing property of turbulent flows may be positive or negative, depending on the circumstances. In particular, it is a desirable feature when, as often happens, one wants to rapidly mix fluids with different motions or characteristics. For example, a fast mixing between fuel and oxidizer is generally necessary to promote effective combustion, and turbulent flows are then highly advantageous. The same is true when a hot surface must be cooled by a stream of fluid, in which case a turbulent boundary layer over the surface is much more effective than a laminar one. On the other hand, we shall see in next section that a turbulent boundary layer produces increased friction at the surface.

To conclude this brief outline on turbulent flows, it is worth mentioning that the basic approach for their study is to distinguish between the fluctuations and the *mean* flow, which is obtained from the complete velocity field through a proper averaging operator. Several operators of this type may be defined, and all should possess certain mathematical properties; for instance, they must be linear and commutative with the derivative and integral operators. The most common averaging operator — although not necessarily the best one for theoretical analyses — is the time average, and we refer to it for the sake of simplicity. Therefore, the velocity vector $u(t)$ at each point of space is decomposed in the following manner:

$$u(t) = U + u'(t), \tag{7.72}$$

where U is the time-average value of $u(t)$, and $u'(t)$ is the velocity fluctuation which, by definition, has a zero mean value. This is known as *Reynolds decomposition*, because it was first introduced by Osborne Reynolds (see Reynolds, 1895).

If decomposition (7.72) is inserted into the equations of motion, and the averaging operator is applied, the so-called *Reynolds-averaged Navier–Stokes* (RANS) equations are obtained. In these equations, the mean values of the velocity components and pressure appear, together with the so-called *Reynolds stress tensor*, which describes the effect of the fluctuations on the mean flow. The components of this tensor represent the mutual correlations between the fluctuating velocity components, and are additional unknowns in the problem. Consequently, in order to predict the mean field in a turbulent flow using the RANS equations, the Reynolds stresses must be derived from an appropriate *turbulence model*. This is the so-called *closure problem*, and several turbulence models of increasing complexity have been developed for the analysis of different types of flows by means of numerical codes.

7.6.3. *Effects of turbulence on boundary layer characteristics*

We shall now analyse the effects of turbulence on the evolution of boundary layers over solid surfaces, trying to connect these effects with the characteristics of turbulent flows that have been described in the previous section.

The first point that must be made is that the basic assumptions and results of boundary layer theory — and in particular the constancy of pressure across the thickness — are still valid for turbulent boundary layers. However, their growth and their actions on the surface are significantly different from those of laminar boundary layers. These differences may be better understood from Fig. 7.12, where the flow in the turbulent boundary layer over a flat plate is made visible by means of a technique known as laser-induced fluorescence.

It is clear from the figure that several vortical structures are present, that they are mostly inclined in the flow direction, and that they are capable of dragging the outer potential flow — which may be considered to coincide with the dark zones in the figure — towards regions well inside the boundary layer. The figure also provides an immediate visual idea of the strong mixing mechanisms that are typical of turbulent flows, and which justify a stronger transfer of momentum from the outer potential flow to the boundary layer than in the laminar case. The first consequence of this increased mixing is a much faster growth of turbulent boundary layers with respect to the laminar ones. Finally, Fig. 7.12 shows that the edge of a turbulent boundary layer is very irregular in shape; moreover, as may be easily imagined, it continuously changes in time. Therefore, the different thicknesses of the boundary layer and their evolution along the surface may be suitably characterized only

Fig. 7.12. Visualization of a turbulent boundary layer (Photograph courtesy of Professor Mohamed Gad-el-Hak, Virginia Commonwealth University).

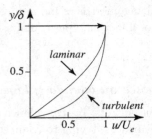

Fig. 7.13.   Velocity profiles of laminar and turbulent boundary layers.

by referring to the mean velocity field. A qualitative comparison between
the mean velocity profiles in a laminar and in a turbulent boundary layer
over a flat plate is given in Fig. 7.13. To allow a direct assessment of the
difference in shape of the velocity profiles, in this figure the coordinates
are made non-dimensional through the outer velocity and the thickness of
the boundary layer, defined with reference to the mean flow.

As can be seen, a peculiar feature of a turbulent velocity profile is that
the velocity is higher near the surface. This is understandable, because the
stronger mixing of the turbulent flow brings higher-velocity particles from
the outer regions towards locations closer to the surface. Therefore, the
mean velocity gradient becomes larger in that zone, causing a significantly
higher viscous stress over the surface.

At variance with the laminar flow condition, no exact solution of the
equations exists for a turbulent boundary layer over a flat plate. Therefore,
reference must be made to available empirical results, such as the following
ones, which refer to a condition in which a completely turbulent boundary
layer is assumed to be present over the plate and x is the distance from the
leading edge:

$$\delta(x) \simeq \frac{0.37x}{Re_x^{0.2}}, \tag{7.73}$$

$$\delta^*(x) \simeq \frac{0.046x}{Re_x^{0.2}}, \tag{7.74}$$

$$\theta(x) \simeq \frac{0.036x}{Re_x^{0.2}}, \tag{7.75}$$

$$C_f(x) = \frac{\tau_w}{\frac{1}{2}\rho U^2} \simeq \frac{0.059}{Re_x^{0.2}}. \tag{7.76}$$

By comparing these results with those of the Blasius theory, it is clear that all the thicknesses of turbulent boundary layers increase along the surface much more rapidly than in the laminar case; in effect, they vary as $x^{0.8}$ rather than as $x^{0.5}$. Furthermore, as anticipated, for the same Reynolds number the turbulent wall friction coefficient is much higher. Another interesting point is that the shape factor $H = \delta^*/\theta$ is about 1.28, i.e. much lower than in the laminar case. In fact, when measurements of the mean velocity profile are carried out, a significant decrease of H may be considered as a clear indication that transition has occurred.

The flat plate results for laminar and turbulent boundary layers can be used to estimate the global friction forces acting over the surface of a plate of length L in which transition from laminar to turbulent flow occurs at a certain given position along the plate. Even if transition generally does not occur at a specific point, but rather over a certain portion of surface, with this procedure we may obtain both a good prediction of the order of magnitude of the forces and an assessment of the sensitivity of the results to the location of the transition region. Thus, in order to estimate the global friction force, we may assume that the flow condition is as depicted in Fig. 7.14, with a sudden increase in the rate of growth of the boundary layer after it becomes turbulent at the point x_T along the plate.

The evaluation of the friction force acting on the portion of the plate with a laminar boundary layer is straightforward: if x is the distance from the leading edge of the plate, it can be carried out by integrating from $x = 0$ to $x = x_T$ the Blasius viscous stress, which may be derived from relation (7.63). Conversely, to estimate the friction stresses over the turbulent portion, a virtual origin of the turbulent boundary layer, say at $x = x_0$, must be defined, so that relation (7.76) can be used and integrated between $(x_T - x_0)$ and $(L - x_0)$. The virtual origin of the turbulent flow can be obtained from a sensible matching between the laminar and turbulent boundary layers. This can be carried out, for instance, by imposing that at $x = x_T$ the thickness of the turbulent boundary layer starting at the

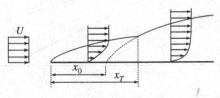

Fig. 7.14. Evolution of the boundary layer for the evaluation of wall friction.

virtual origin $x = x_0$ coincides with that of the laminar one. Actually, the matching could also be based on the momentum thickness, which has a closer connection with the wall friction, but the differences between the various methods are not great. Using such a procedure, the total friction drag F_x acting on one face of a flat plate of length L — per unit length in the direction orthogonal to the plane of motion — can be evaluated as a function of the position of the transition point (i.e. of the ratio x_T/L), and of the Reynolds number based on the length of the plate, $Re_L = UL/\nu$. One then obtains a graph such as the one shown in Fig. 7.15, where the total drag is given in terms of a global friction coefficient, defined as

$$C_F = \frac{F_x}{\frac{1}{2}\rho U^2 L}. \tag{7.77}$$

The flat plate results are particularly interesting because they can be used to obtain a first-order estimate of the friction drag acting on slender bodies with small cross-flow dimension and attached boundary layers, i.e. on aerodynamic bodies. In essence, it is expected that the pressure gradients acting over the surfaces of such bodies are mild enough not to alter the order of magnitude of the friction force compared to the flat plate case. This expectation also originates from the fact that positive and negative pressure gradients have opposite effects on the viscous stresses acting on the surface. Another important outcome deriving from Fig. 7.15 is that it readily provides an estimate of how advantageous it is to keep a boundary

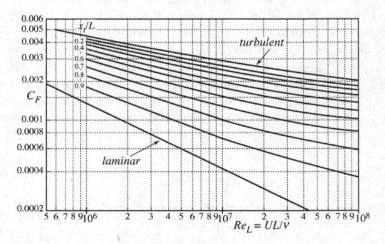

Fig. 7.15.   Friction drag coefficient for a flat plate with different transition location.

layer laminar at high values of the Reynolds number. For instance, at $Re_L = 3 \times 10^6$ the turbulent friction drag on the plate is more than four times higher than the laminar one.

Nonetheless, in spite of their higher friction drag, turbulent boundary layers may also be advantageous. We have already mentioned that turbulence increases heat exchange and that, consequently, it is desirable when a surface must be cooled. However, the greatest advantage of turbulent boundary layers over laminar ones is that they are much more resistant to adverse pressure gradients, and thus produce a delay of separation. This effect is quite understandable, considering that turbulent boundary layers are characterized by a much higher momentum close to the surface (see Fig. 7.13), and are thus further from the limit condition in which the value of $\partial u/\partial y$ at the surface is zero. Consequently, compared to the laminar case, separation occurs only when higher or more extensive adverse pressure gradients have acted on the boundary layer. The stronger resistance to separation of turbulent boundary layers is a favourable feature in many engineering applications. For instance, we shall see in the next chapter that a significant reduction of the total drag of certain bluff bodies may be observed when the boundary layers over their surfaces are turbulent rather than laminar. This feature ensues from the reduction of the portion of drag generated by the pressure forces.

Therefore, in certain circumstances it may be advantageous to promote boundary layer transition, and this is often obtained by adding localized or distributed roughness over the surface of the bodies. However, it must also be highlighted that, while it is commonly recognized that adding random roughness over a surface promotes transition to turbulence, recent experimental results show that, if the roughness has certain geometrical features and is suitably distributed, it may actually delay transition (see Fransson *et al.*, 2006).

Chapter 8

FLUID DYNAMIC LOADS ON BODIES
IN INCOMPRESSIBLE FLOWS

8.1. Generalities

8.1.1. *Fluid dynamic loads*

Up to now, we have been principally concerned with the equations governing the motion of fluids and with the main features of the different types of flow that may occur in various situations. In this chapter, we shall concentrate on one of the most important practical effects of fluid motion, namely the generation of loads on bodies immersed in a stream or moving in an otherwise still fluid.

From the point of view of the fluid, the surface of an immersed finite body is one of the boundaries of the flow. Surface forces act on the fluid in correspondence to this boundary and we have shown that they are the result of the action of pressure and viscous stresses. Obviously, opposite stresses act on the body surface, and their integration over the body contour provides the value of the global fluid dynamic forces.

If we then consider a point over the body surface S, where n is the normal unit vector directed outwards (i.e. towards the fluid), and p and τ_n are the pressure and the viscous stress vector respectively, the resultant fluid dynamic force acting on the body is given by the following expression:

$$\boldsymbol{F} = - \int_S p\boldsymbol{n} dS + \int_S \boldsymbol{\tau}_n dS. \qquad (8.1)$$

Note that relation (8.1) is exactly the same as that which appeared in the integral momentum equation (4.37) to represent the force acting on the surface S bounding a certain volume of fluid. This coincidence is the consequence of the fact that the surface stresses in a fluid change sign with the change of the normal unit vector of the surface on which they act. Therefore, considering that n is directed outwards from the fluid in (4.37) and towards the fluid in (8.1), the two expressions actually correspond to opposite forces, as required.

If we now define a position vector x, connecting a suitably chosen reference point to a generic point over the body surface, the resulting moment acting on the body may be evaluated from the expression

$$M = -\int_S x \times pn dS + \int_S x \times \tau_n dS. \tag{8.2}$$

The force and moment vectors may be decomposed in different ways according to the reference system that is judged to be suitable for the particular problem or application. The most usual one is a reference system connected with the free-stream velocity vector U, which is the velocity of the fluid at a position that is sufficiently upstream of the body not to be influenced by its presence. When a body is immersed in a flowing fluid, U coincides with the velocity that would be present in the absence of the body at the position where it is located. On the other hand, if a body is moving in a still fluid, U is the opposite of the body velocity. The force component in the direction of U is usually called *drag* and, in the case of a body moving at constant velocity, it is always positive, i.e. it is directed as U. Therefore, by definition, drag is the only force component that does work during the motion, and is thus immediately connected with the power that must be spent to keep the body moving at constant velocity. The minimization of fluid dynamic drag is, then, an important issue in the design of any type of vehicle.

The cross-flow force component may be further decomposed in two directions, which may be different for different problems. A particularly important decomposition is the one used for aircraft, in which a plane of symmetry is normally present. The component of the fluid dynamic force, which in this case is more usually referred to as the *aerodynamic* force, lying in that plane and perpendicular to the flight direction — i.e. to the direction of U — is called *lift*. The component of force that is perpendicular to both drag and lift is called *lateral force* or *side force*. Therefore, the lift on a flying aeroplane is not necessarily the vertical force directly opposing the gravity force; for instance, this is not the case when the aeroplane is climbing or turning (see Fig. 8.1).

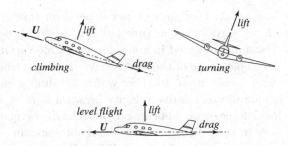

Fig. 8.1.   Lift force on an aeroplane in different flight conditions.

The term 'lift' has become commonplace outside the aeronautical field, to indicate the force perpendicular to the flow lying in a symmetry plane, in all cases in which this exists. It is also widely used to denote the cross-flow force component in two-dimensional flows representing, with sufficient accuracy, the fluid motion around very long cylindrical bodies. Obviously, when we speak of the force acting on a two-dimensional body, we are actually referring to a force per unit value of the length in the direction perpendicular to the plane of motion.

From expression (8.1), it is obvious that the fluid dynamic forces acting on a body are essentially connected with the features of the flow produced by the relative motion between body and fluid. Furthermore, the above relation immediately suggests a classification of the fluid dynamic force according to the origin of the stresses acting over the body surface. In particular, the two terms in (8.1) may be termed the *pressure force* and the *friction force* respectively. In this regard, it must be noted that pressure stresses act perpendicularly to the surface of a body, while the viscous friction stresses are tangential to the surface. Therefore, if a negligible-thickness plate is, for instance, positioned parallel to the free-stream, the drag force will be completely due to the friction stresses, whereas if the plate is set perpendicular to the flow, the drag force is only produced by the pressures acting over its surface.

In general, the absolute and the relative importance of the pressure and friction forces in the various cases will depend on the body geometry and on the type of motion. In particular, we shall see that a striking difference exists between aerodynamic and bluff bodies as regards the magnitude and the relative contribution of the two types of forces.

Depending on the characteristics of the motion and of the flow, the forces acting on moving bodies, or on bodies immersed in a stream, may be unsteady or steady, i.e. they may or not depend on time. An obvious situation in which the forces vary in time is when a body is accelerating, or when it is at rest but immersed in a stream with time-varying velocity. Note that the time dependence of the force derives from the time variation of both velocity and acceleration. In other words, all fluid dynamic forces depend on the instantaneous relative velocity between fluid and body, but there are additional forces that depend on the relative acceleration, and these will then vary in time if the acceleration is not constant.

Incidentally, it must be pointed out that if a body is immersed in a time-varying stream of fluid, the forces acting on it are not the same as would act if the body were moving with the same time-varying velocity. In other words, if a body performs a sinusoidal oscillation in a still fluid, the forces acting on it are not exactly the same as would act if the body were at rest and the fluid were moving with the same oscillating velocity. The difference between the two cases is that, if it is the fluid that is accelerating, then a global force must be acting on each volume of the fluid to cause this acceleration, for instance due to a pressure gradient existing in the flow. Therefore, if the volume of fluid is replaced by a solid body, then the same pressure field would act over its surface and produce a force that is not connected with the relative acceleration or velocity between fluid and body, but with the absolute acceleration of the fluid. This force is known as the *Froude–Krylov force*, and depends on the body's volume and on the acceleration of the fluid. It may be important, for instance, when bodies are immersed inside water waves and thus for marine applications of fluid dynamics. As for the force that depends on the relative acceleration, it is usually known as the *added mass force* and its origin and characteristics will be briefly described in Chapter 10.

However, it is extremely important to point out that unsteady fluid dynamic forces may act on a body even if its motion is a translation at constant velocity or if it is immersed in a perfectly steady free-stream. This is due to the possible unsteadiness of the flow field produced by the interaction between fluid and body, rather than by a time-varying type of motion. For example, at variance with what happens for aerodynamic bodies, the wakes of bluff bodies are almost always characterized by the presence of velocity fluctuations, which then produce fluctuating pressures over the body surface and thus, once integrated, time-varying resultant forces. In general, one may say that the loads experienced by aerodynamic bodies do not vary in time when the free-stream is steady, whereas the opposite is true for bluff bodies.

8.1.2. *Load coefficients*

The components of the resultant loads experienced by a body immersed in a flow are conveniently expressed in non-dimensional form through *force and moment coefficients*, which are usually defined as follows:

$$C_{F_i} = \frac{F_i}{\frac{1}{2}\rho U^2 A}, \tag{8.3}$$

$$C_{M_i} = \frac{M_i}{\frac{1}{2}\rho U^2 A l}. \tag{8.4}$$

In these expressions, F_i and M_i are the components in the x_i direction of the resultant force and moment acting on the body, U is the magnitude of the above-defined free-stream velocity vector, and A and l are a reference area and a reference length of the body respectively. In the case of two-dimensional bodies, the load coefficients are defined by using in the numerator the load per unit distance along the direction perpendicular to the motion, and a reference length — in place of the reference area — in the denominator.

The introduction and definition of the load coefficients deserve further comments as regards their significance and practical utility. For instance, it appears that if a force coefficient for a body with a given shape and moving with a certain orientation is known for one value of the velocity and of the reference area — perhaps as a result of experiments or computations — then relation (8.3) would permit the prediction of the corresponding force component for different dimensions of the body or velocity magnitudes, and even for fluids of different density. However, this is strictly true only if the force coefficient does not vary with varying flow conditions, which may happen only in particular circumstances and if the quantities that were used to define the non-dimensional coefficients have been properly chosen. In particular, one should use the quantities on which the forces *principally* depend.

In general cases, one may expect the load coefficients to depend on all the non-dimensional parameters influencing the flow. Two of these have already been introduced in previous chapters, namely the Mach number and the Reynolds number. As we have already observed, for low values of the Mach number the flow may usually be considered as incompressible and, in that case, the main parameter playing a role is the Reynolds number, defined using suitably chosen reference values for velocity and length. Therefore, knowledge of the variation of the force and moment coefficients with Reynolds number is generally sufficient to predict the loads acting on bodies of similar shape moving in different fluids and with different magnitudes of the velocity.

In certain cases, other parameters may be of interest. In particular, if gravity has a significant effect on the motion, then the force coefficients would also depend on the *Froude number*, defined as

$$Fr = \frac{U}{\sqrt{gl}}. \tag{8.5}$$

This may happen, for instance, for the forces on a boat generating waves during its motion; in effect, the portion of drag due to the generation of the wave system trailing behind a sailing vessel is fundamentally influenced by the value of the Froude number.

Returning to the definition of the load coefficients, in principle any reference area may be used. However, it is more convenient to use a value that refers, in some way, to the surface that most contributes to the generation of the force. Thus, for the non-dimensionalization of the forces on the wing of an aeroplane it is customary to use the wing *planform*, which is usually defined as the area of the wing as seen from above. In practice, it is the area of the projection of the wing on a conventionally defined wing plane, which is not far from a plane containing the direction of the free-stream velocity vector. In general, a similar choice is made for any aerodynamic body for which two dimensions prevail over the third, i.e. bodies that may be seen as a plate with a superposed moderate thickness. Conversely, for a slender three-dimensional body with a single prevailing dimension in the direction of motion, such as a submarine or an airship, it is more usual and appropriate to choose the *wetted area*, i.e. the area of the body surface that is directly exposed to the flow. For bluff bodies, on the other hand, it is more suitable to use the *frontal area*, i.e. the area of the projection of the body on a plane that is perpendicular to the free-stream. For example, this is the usual reference area for road vehicles. Finally, in two-dimensional flow conditions, a length in the direction of the flow or perpendicular to it is appropriate for aerodynamic and bluff bodies respectively.

In all cases, the main feature of a suitable reference area is that the force components with greatest interest for applications should be more or less proportional to that area, so that the corresponding force coefficients turn out to be as constant as possible with the variation of the flow parameters.

The choice is then dictated by an assessment of the contribution of the pressure and friction forces experienced by the considered body. Thus, for instance, the planform area is a suitable reference area for wings because lift is mainly connected with the pressure forces acting on the portion of the surface that is almost parallel to the velocity direction, which is also the one where the friction forces give the greatest contribution to drag. On the other hand, the drag of a bluff body, which is typically the most important force component in engineering designs, is seen to be strictly connected with the cross-flow area of the wake which, in turn, essentially depends on the cross-flow area of the body.

The observation that the load coefficients, for practical reasons, should be preferably defined in such a way that their variation with the flow parameters be as small as possible, also implies that the definitions given in (8.3) and (8.4) are actually adequate only when the typical Reynolds numbers are sufficiently high. In effect, their use is based on the assumption that the forces are mainly caused by dynamical pressure variations, i.e. by variations due to the inertial terms in the equations of motion, whose order of magnitude is determined by the quantity $(1/2\rho U^2)$.

In practice, most applications are characterized by flows at high Reynolds numbers, and thus the above definitions are satisfactory. Nonetheless, there are particular situations in which the Reynolds numbers are very low; an example is the motion of dust particles or small droplets in the air, for which Re may even be significantly smaller than one. In those cases, inertia forces tend to be very small and viscous effects become dominant. Considering that the order of magnitude of the viscous stresses is $(\mu U/l)$, viscous forces may then be expected to vary in proportion to the quantity (μUl), and the same is true, say, for the global drag force acting on a body. Consequently, in order to define a force coefficient that is almost constant with Re (provided it remains in the very low values range), one should use the quantity (μUl) for the non-dimensionalization of the force. Indeed, it is easily seen that, in those conditions, the application of definition (8.3) would give a force coefficient that is inversely proportional to the Reynolds number.

Similarly, when the body or the fluid are accelerating, and a significant force proportional to acceleration is expected, the proper quantity to define a force coefficient would be $(\rho \mathcal{V}a)$, where $\mathcal{V}$ is a reference volume — generally the volume of the body — and a is the magnitude of the acceleration.

8.2. Loads on Aerodynamic Bodies

8.2.1. *Main flow features and drag force*

We have already described some of the flow features of aerodynamic bodies, and we shall now concentrate on their connection with the consequent fluid dynamic loads.

We start by recalling that the characteristic feature distinguishing an aerodynamic body is that the boundary layer over its surface always remains thin and attached. Consequently, the pressures on the body surface coincide with those in the potential flow just outside the boundary layer, which, in turn, are directly related to the velocity in the same region. If we then describe the pressure distribution on the body surface by means of the pressure coefficient defined by relation (5.59), and denote as U_e the velocity at the edge of the boundary layer at a generic point over the body surface, the following relation derives from the application of Bernoulli's theorem for steady flow and negligible or constant body forces:

$$C_p = \frac{p - p_\infty}{\frac{1}{2}\rho U^2} = 1 - \frac{U_e^2}{U^2}. \tag{8.6}$$

As an example, let us consider a two-dimensional symmetrical elongated body, such as the one that was analysed in Chapters 5 and 6 and which, from the definition introduced in Chapter 7, we recognize now to be an aerodynamic body. The relevant flow field and the consequent pressure distribution are shown again in Fig. 8.2.

As anticipated in Chapter 7, the velocity at the edge of an attached boundary layer, such as the one existing around an aerodynamic body, may be derived from the evaluation of a potential flow tangent to a modified body obtained by adding the boundary layer displacement thickness over the surface of the original body. Therefore, a deeper understanding of the connection between the flow and the pressure distribution on the body can be obtained from a thorough analysis of the characteristic features of potential flows. The methods and results of this type of analysis, which is essential for the prediction of the loads on aerodynamic bodies, will be illustrated in Chapter 12. Here, we limit ourselves to a qualitative description of the connection between the evolution of the boundary layer, the outer potential flow and the pressure distribution for the typical aerodynamic body depicted in Fig. 8.2.

As can be seen, the maximum pressure is found at the stagnation point (where the pressure coefficient is 1), and the minimum pressure occurs in

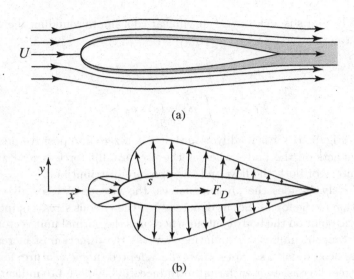

(a)

(b)

Fig. 8.2.   Flow field (a) and pressure distribution (b) around an aerodynamic body. The boundary layers and the wake are schematically indicated by the grey regions.

the regions of maximum thickness, where the velocity outside the upper and lower boundary layers reaches a maximum value that is larger than U and, consequently, the pressure coefficient is negative.

The subsequent decrease of the thickness of the body corresponds to the region where the velocity magnitude decreases, with a consequent increase of the pressure. Simultaneously, we observe a progressive increase of the thickness of the upper and lower boundary layers, which then merge to form the downstream wake. This causes the velocity at the edge of the boundary layers to smoothly align in the direction of the free-stream, and to attain a magnitude that is close to U. Consequently, the value of the pressure coefficient at the end of the body approaches zero, in agreement with relation (8.6). The pressure distribution may then be used to obtain, by integration, the pressure force components.

If we choose a reference system with the x axis in the direction of U and the y axis in the cross-flow direction (see Fig. 8.2(b)), the force components in the x and y directions correspond, respectively, to the drag and lift forces (per unit length in the direction perpendicular to the plane of motion), which we indicate by F_D and F_L. The pressure contribution to these force components is thus given by the following relations, where e_x and e_y are the unit vectors in the x and y directions respectively, and s is a clockwise curvilinear coordinate rounding the whole body contour C:

$$F_{Dp} = - \int_C p(s) \boldsymbol{n}(s) \cdot \boldsymbol{e}_x ds, \qquad (8.7)$$

$$F_{Lp} = - \int_C p(s) \boldsymbol{n}(s) \cdot \boldsymbol{e}_y ds. \qquad (8.8)$$

From Fig. 8.2(b) it is readily seen that a non-zero and positive pressure drag force acts on the body, whereas the pressure lift force is zero due to the symmetry of both the flow and the pressure distribution.

Obviously, besides the pressure forces, the viscous stresses give their contribution to the forces. Let us denote as τ_{ns} the viscous stress component at a generic point on the body contour where $\boldsymbol{n}$ is the normal unit vector and $\boldsymbol{e}_s$ is the tangent unit vector pointing towards the direction of increasing curvilinear coordinate s. Note that the adopted nomenclature for the viscous stress is consistent with that introduced in Chapter 4 to indicate the generic components of the viscous stress tensor. With this notation, the contributions of the viscous stresses to the drag and lift forces are given by

$$F_{D\tau} = \int_C \tau_{ns}(s) \boldsymbol{e}_s(s) \cdot \boldsymbol{e}_x ds, \qquad (8.9)$$

$$F_{L\tau} = \int_C \tau_{ns}(s) \boldsymbol{e}_s(s) \cdot \boldsymbol{e}_y ds. \qquad (8.10)$$

As would be expected, the surface viscous stresses acting over both the upper and lower surfaces of the body give a positive contribution to drag. This derives from the fact that τ_{ns} is positive over the upper surface of the body in Fig. 8.2 and negative over the lower one, but its sign is always the same as that of the scalar product between the unit vectors appearing in (8.9). Due to the symmetry of the flow, the viscous stresses, like the pressures, give no contribution to lift.

Therefore, the resultant force on the body shown in Fig. 8.2 is a drag, and is given by the sum of the pressure and friction parts (8.7) and (8.9). In general, the relative contribution of friction drag for an aerodynamic body is significantly higher than that of the pressure drag, typical values lying between 80% and 90%, with increasing percentage values for decreasing cross-flow dimension of the body.

Actually, the friction drag is never zero, because the viscous stresses always act over the upper and lower surfaces of the body. As a first approximation, the order of magnitude of this portion of drag may be estimated from the values of the friction drag coefficients for a flat plate shown in Fig. 7.15, once the position of the transition point has been tentatively fixed by referring to a local value of the Reynolds number. Indeed, as was anticipated in Chapter 7, we may reasonably predict that in the fore part of the body, where a favourable (i.e. negative) pressure gradient is present, the boundary layer should be thinner and the wall viscous stresses higher than in the case of a zero-pressure-gradient flat plate, whereas the opposite should occur in the rear part, where an adverse pressure gradient exists. Thus, even if the two effects cannot be expected to have equal magnitude, their opposite signs contribute to the fact that an evaluation based on the flat plate results is a sound first-order estimate of the real friction drag.

For an aerodynamic body, the effect of increasing the relative thickness, i.e. the ratio between the cross-flow dimension and the streamwise length of the body, can be analysed on the basis of the above considerations regarding Fig. 8.2 and of what has been learned in previous chapters. The first point to be observed is that increasing the relative thickness implies a decrease of the minimum pressure, which is found in the maximum-thickness regions. Usually, the zones of the surface of a body where the pressure coefficient is negative are called regions of 'suction'. Using this terminology, we may then say that the maximum suction increases with increasing relative thickness. Therefore, the pressure gradients in the fore part of the body become more favourable, whereas in the rear part they become more adverse. The consequence is then a progressive decrease of the thickness of the boundary layer over the fore part of the body and an increase over the rear part. As already pointed out, the global effect on friction drag will be the result of an increase of the viscous stresses on the fore part and a decrease on the rear part, so that its prediction is not immediate. However, we may expect an increase of the pressure drag, whose value is directly connected with the thickness of the boundary layer on the rear part of the body. In effect, it is the presence of a non-zero-thickness boundary layer that gives rise to a finite pressure drag.

Eventually, the maximum thickness of the body will reach a value that corresponds to so strong an adverse pressure gradient that boundary layer separation occurs, together with a much more significant increase of the pressure drag. Strictly speaking, the body can no longer be considered as aerodynamic after the appearance of separation.

Obviously, the above description of the evolution of the flow and the drag force with relative thickness is only qualitative, as many more

variables may play a role. In particular, due to the prevalence of friction
effects, the most important feature influencing the drag of an aerodynamic
body is the occurrence of the boundary layer transition from laminar to
turbulent conditions, with the associated increase of surface viscous stresses.
Consequently, all the geometrical and flow parameters affecting transition
have a significant influence on the drag of aerodynamic bodies. For instance,
apart from the obvious influence of the Reynolds number, other important
features are the degree of surface roughness and the streamwise position
of the maximum thickness, which influences the extent and the magnitude
of the favourable and adverse pressure gradients, and thus the transition
point.

Therefore, once the typical Reynolds number for a certain application
has been given, the basic means for reducing the drag of an aerodynamic
body is, in principle, devising some techniques to delay the boundary
layer transition. However, it must also be recalled that turbulent boundary
layers are much more resistant to separation, and thus a larger maximum
allowable thickness without separation may be expected for a turbulent
boundary layer than for a laminar one. In any case, we shall see later
that the values of the drag coefficients of aerodynamic bodies are very low
when compared to those of bluff bodies with similar cross-flow dimensions.
For instance, for a two-dimensional body like the one shown in Fig. 8.2,
typical values of the drag coefficient — defined using the length in free-
stream direction as the reference length — might range between 0.006
and 0.01.

All the above considerations, regarding the low drag of aerodynamic
bodies and its dependence on geometrical and flow features, are not
restricted to two-dimensional bodies, but apply to slender three-dimensional
bodies as well, such as an airship, a fish or a dolphin, all of them characterized by attached boundary layers over their whole surface.

However, the other fundamental feature of aerodynamic bodies is
their effectiveness in generating lift forces with low drag penalty. This
is characteristic both of two-dimensional bodies and three-dimensional
bodies that have one dimension (in the cross-flow direction) that is much
smaller than the other two, i.e. bodies that are similar to plates with a
small thickness superposed over them, such as aeroplane wings. Due to its
importance in practical applications, we devote the following paragraph to
a deeper analysis of the mechanism of generation of lift on aerodynamic
bodies in incompressible flow.

8.2.2. *The generation of lift on aerodynamic bodies*

The two-dimensional body whose flow and pressure distribution are shown in Fig. 8.2 is an example of a symmetrical *airfoil* (*aerofoil* in British English), i.e. of the cross-section of an infinite wing. The aerodynamic characteristics of different types of airfoils and the methods for predicting the loads acting on them in various flow conditions will be analysed in more detail in Chapter 12. Here, we limit ourselves to the explanation of the physical mechanisms that allow an airfoil to generate significant lift forces.

To this end, it is expedient to provide some definitions that are widely used for the geometrical description of airfoils. In particular, the extreme fore and aft points of the airfoil in Fig. 8.2 are denoted the *leading edge* and the *trailing edge* respectively. The leading edge actually corresponds to the fore point with maximum curvature and, as we have already pointed out, the trailing edge must be sharp to avoid separation of the boundary layer. The segment connecting the leading and trailing edges is known as the *chord* of the airfoil. Finally, the angle α between the free-stream direction and the chord line is the geometrical incidence, or *angle of attack*.

Up to now we have been considering only the case in which this simple type of airfoil is aligned with the free-stream. In particular, we have analysed, in Chapter 6, the flow evolution when we imagine this body to suddenly start moving impulsively from rest in a direction parallel to the chord, and we have seen that the flow that tends to be produced is the potential flow tangent to the surface of the body, satisfying the non-penetration condition at the solid surface and the requirement that no circulation exists around the airfoil. This flow is characterized by two stagnation points (A and B in Fig. 6.4), which coincide with the leading and trailing edges of the airfoil. However, the no-slip boundary condition must also be satisfied at the surface and thus, when the airfoil starts moving, two layers of opposite-sign vorticity are simultaneously produced over its upper and lower surfaces. Then, this vorticity diffuses into the flow and is convected downstream by the velocity field. After a transient, and for sufficiently high Reynolds numbers, the vorticities over the upper and lower surfaces of the airfoil remain confined within two thin boundary layers, which join to produce a thin wake, as shown schematically in Fig. 8.2.

Let us now assume that the body we consider is still a symmetrical airfoil, but that its motion starts with a non-zero (even if small) angle of

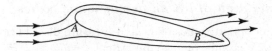

Fig. 8.3.   Starting flow around a symmetrical airfoil at a small angle of attack.

attack. The flow that initially tends to be produced is again the potential flow tangent to the surface, satisfying the non-penetration condition and with zero circulation around the body. However, it may be seen that in this case the solution of the problem is characterized by two stagnation points that no longer coincide with the leading and trailing edges, the front one being positioned on the fore part of the lower surface, while the rear one is on the aft part of the upper surface (see Fig. 8.3).

In this flow, the important feature is the presence of streamlines rounding the sharp trailing edge. Now, the trailing edge may be seen as a convex corner, and one important result of the potential flow mathematical problem is that velocity becomes infinite at a convex corner, whereas it is zero at a concave corner. Incidentally, we may note that points A and B represent concave corners for the flow, and this explains why they must be stagnation points. Therefore, in this starting potential flow, the velocity tangent to the body is ideally infinite at the trailing edge, and then rapidly drops to zero at the rear stagnation point.

However, the above ideal flow is not exactly the one that occurs at the start of the motion because the action of the viscous no-slip condition causes the simultaneous generation of two vorticity layers around the body surface, which tend to develop into thin boundary layers separating the potential flow from the body surface. In any case, we may imagine that, at the trailing edge, the potential flow 'sees' a boundary with extremely high curvature and, consequently, a very high velocity may be expected to occur in that region, followed by a rapid deceleration towards the rear stagnation point. Therefore, the lower boundary layer (which contains counter-clockwise vorticity) will not be able to go around the sharp trailing edge without separating, due to the extremely high adverse pressure gradient that is present from the trailing edge to the rear stagnation point. In fact, we have already pointed out in Chapter 7 that boundary layers cannot remain attached over surfaces with sharp edges. Thus, immediately after the start of the motion of the body (say, in the leftwards direction), the lower boundary layer separates at the sharp trailing edge. After a very short transient — whose details will not be discussed — the counter-clockwise

Fig. 8.4.　Flow around a symmetrical airfoil at a small angle of attack after the shedding of the starting vortex. The grey and black layers (whose thickness is not representative) indicate the zones where, respectively, clockwise and counter-clockwise vorticity are present.

vorticity contained in this separating boundary layer remains in the flow and rapidly concentrates in a small region of space, forming what is known as a *starting vortex*, whose total strength we denote as Γ. If this event is observed in a reference frame fixed to the fluid, the sudden start of the airfoil in the leftward direction immediately produces a counter-clockwise vortex at its trailing edge, which remains practically in that location while the airfoil moves away. Conversely, in a reference frame fixed to the airfoil, the starting vortex is seen to be rapidly carried downstream by the main flow (see Fig. 8.4).

The shedding of the starting vortex from the sharp trailing edge is an extremely rapid event and takes place together with the establishment of a new flow field around the airfoil, with the upper-surface boundary layer remaining attached up to the trailing edge, and joining smoothly with the lower-surface boundary layer to produce a thin wake. In other words, after the initial transient, the airfoil is, again, an aerodynamic body. Nonetheless, we can see that the situation is definitely different from that occurring when the same body was moving at zero angle of attack.

As a matter of fact, if we now analyse the vorticity field, and recall that the total vorticity in the flow must be conserved — and thus it must still be equal to zero as it was before the start of the motion — we conclude that a global surplus of clockwise vorticity $-\Gamma$, equal and opposite to the global counter-clockwise vorticity left in the starting vortex, must have remained in the boundary layers around the airfoil.

The above conclusion also derives from the fact that the global amount of vorticity present in the wake, which links the airfoil to the starting vortex, is zero, at least in the boundary layer approximation. In effect, the integral of the vorticity contained in a generic cross-section of the wake is given by the relation

$$\int_{\delta^-}^{\delta^+} \omega\,dy = -\int_{\delta^-}^{\delta^+} \frac{\partial u}{\partial y}\,dy = -(U_e^+ - U_e^-), \qquad (8.11)$$

where we have denoted by U_e^+ and U_e^-, respectively, the velocities at the upper and lower edges of the wake, which are identified by the coordinates δ^+ and δ^-.

We now note that the boundary layer approximations still hold in the thin wake and that the wake cannot withstand a pressure difference. Therefore, the pressures p^+ and p^- at the upper and lower boundaries of the wake cross-section must be equal. To link pressures with velocities, we cannot use Bernoulli's theorem across the wake, due to the presence of vorticity. However, we may apply the theorem along the streamlines passing through two points at the upper and lower boundaries of the wake, which are in the potential flow region. We call these points 1^+ and 1^-, and consider two further points, 2^+ and 2^-, belonging to the same streamlines, but placed upstream of the airfoil (see Fig. 8.5). In that upstream region, no vorticity is present, and thus the sum of pressure and kinetic energy per unit volume has the same value for all streamlines. We may then write

$$p_{1+} + \frac{1}{2}\rho U_{1+}^2 = p_{2+} + \frac{1}{2}\rho U_{2+}^2 = p_{2-} + \frac{1}{2}\rho U_{2-}^2 = p_{1-} + \frac{1}{2}\rho U_{1-}^2. \quad (8.12)$$

Consequently, the fact that the values of the pressure at any two corresponding points at the upper and lower edges of a wake cross-section are equal implies that the magnitudes of the velocities at those points are also equal. Therefore, from (8.11) we deduce that the integral of the vorticity contained in the considered wake cross-section is zero or, in other words, equal amounts of positive and negative vorticities are present. Due to the conservation of the total vorticity in the flow, this confirms that the global amount of vorticity present in the boundary layers around the airfoil must be equal and opposite to the global vorticity released in the starting vortex. In fact, one may easily infer that the surplus of clockwise vorticity around the airfoil is contained within its upper-surface boundary layer.

The consequence of the above is that a non-zero circulation is now present around the airfoil, as can be determined from the application of Stokes' theorem to the circuits shown in Fig. 8.6. In effect, considering that the vorticity vector is perpendicular to the plane of motion, and thus the

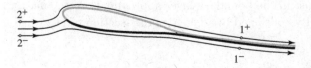

Fig. 8.5.   Streamlines for application of Bernoulli's theorem.

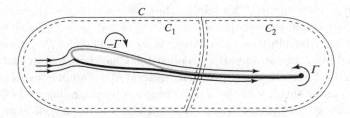

Fig. 8.6.   Application of Stokes' theorem to the flow around an airfoil.

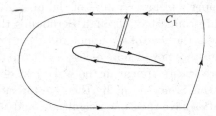

Fig. 8.7.   Application of Stokes' theorem to a circuit rounding the airfoil.

fluxes of vorticity through the surfaces bounded by the circuits C_1 and C_2 in Fig. 8.6 are $-\Gamma$ and Γ respectively, we obtain

$$\oint_C \boldsymbol{V} \cdot dl = \oint_{C_1} \boldsymbol{V} \cdot dl + \oint_{C_2} \boldsymbol{V} \cdot dl = -\Gamma + \Gamma = 0. \qquad (8.13)$$

It should be noted that the application of Stokes' theorem to circuit C_1 must be made with some caution. In this case, the necessary hypothesis that the field be simply connected — or, in other words, that the circuit be *reducible* — is not satisfied. However, as was pointed out in Chapter 1, we may introduce a cut from the outer circuit to the airfoil, and then follow a path comprising the cut, the airfoil contour, and the cut in the opposite direction (see Fig. 8.7). Now, along the airfoil contour the contribution to circulation is zero because the velocity over the airfoil surface is zero, and the path of the cut is followed twice in opposite directions, so that the corresponding integrals cancel out. Therefore, only the integral along the outer curve remains and, from Stokes' theorem, it is equal to the flux of the vorticity contained in the boundary layers that are present around the airfoil.

We may note, now, that the influence of the starting vortex on the flow near the airfoil rapidly fades away with increasing mutual distance between vortex and airfoil, and it becomes negligible when this distance becomes greater than a few chords. This implies that, after a sufficient time from the start of the motion, the conditions around the airfoil, in a reference system fixed to it, may be effectively considered as steady. Therefore, as it can also be deduced from the considerations reported in Chapter 6, and in particular in Section 6.4.4, the amounts of vorticity generated instantaneously over the upper and lower surfaces of the airfoil are given by the integrated effect of the vorticity sources present over those surfaces or, in view of expression (6.36), by the integration of the pressure derivatives along the upper and lower surfaces from the stagnation point to the trailing edge. Thus, they are determined by the upper and lower values of the pressures at the trailing edge; in turn, these are connected, through Bernoulli's theorem, with the values of the velocities at the edge of the upper and lower boundary layers at the same location, which we now denote as U_{tu} and U_{tl} respectively. Unless second-order curvature effects appear in the wake just downstream of the trailing edge — and this is not normally the case for moderate angles of attack — these velocities have the same magnitude, and thus the pressures are equal. Consequently, equal global amounts of opposite-sign vorticity are produced in unit time over the upper and lower surfaces, and coincide with the fluxes of vorticity entering the wake from the two surfaces. In effect, as can also be deduced from relation (7.67), the corresponding fluxes are $-U_{tu}^2/2$ and $U_{tl}^2/2$.

In conclusion, once the initial transient of the motion is over and the starting vortex is far enough from the airfoil for its influence to be negligible, equal amounts of positive and negative vorticity are continuously produced over the airfoil surface and released into the wake, and a surplus of clockwise vorticity remains inside the upper boundary layer of the airfoil. The presence around the airfoil of this vorticity surplus, and of the consequent circulation, is the fundamental cause of the generation and continued existence of a lift force on the airfoil during its motion, because it induces around the airfoil a velocity field that must be added to the one that would be caused by the free-stream if the vorticity surplus were absent. The overall effect is a higher average velocity of the fluid particles that pass alongside the upper surface boundary layer, and a lower one for those passing alongside the lower surface boundary layer. This velocity difference then produces, according to Bernoulli's theorem, significantly lower pressures over the upper surface of the airfoil than over the lower one, and thus a lift force. A qualitative sketch of the typical flow field and of the consequent pressure distribution around a lifting airfoil is shown in Fig. 8.8. Further details on the characteristics of different airfoils will be

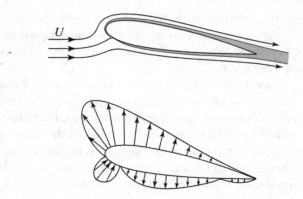

Fig. 8.8.   Flow field and pressure distribution around a lifting airfoil.

given in Chapter 12, but we note here that lift is essentially due to the lower pressures existing over the upper surface with respect to those acting on the lower surface. The prevailing pressures on the lower surface may be higher or lower (as in Fig. 8.8) than the undisturbed pressure, depending on the shape of the airfoil and the angle of attack. In any case, the key role in lift production is played by the higher suction present over the upper surface due to the higher average velocity existing alongside the corresponding boundary layer.

A point that clearly emerges from the above explanation is that the origin of lift is strictly connected with the viscosity of the fluid. No circulation could be produced around the airfoil without the presence of vorticity over its surface which, in turn, is generated by the viscous no-slip boundary condition. Incidentally, it should be pointed out that experimental evidence corroborating the fundamental role of viscosity in lift generation was provided by the tests carried out at Caltech by Craig (1959), who measured the forces acting on an airfoil placed in a flow of superfluid helium II. At temperatures as low as 1.3 K, at which the viscosity of helium II is negligible, no significant lift was found below certain critical fluid velocities above which, in any case, the measured lift was at least two orders of magnitude lower than would be expected for a viscous fluid with the same density.

The importance of the shape of the airfoil should also be emphasized, especially as regards the amount of lift and drag that is produced. The sharp trailing edge immediately triggers boundary layer separation at the start of the motion, and is thus essential for the rapid development of a strong starting vortex. Furthermore, the fact that the airfoil is elongated in the direction of motion guarantees that, after the initial transient, a flow with an attached boundary layer over the whole surface is produced, with a consequent small drag value.

Note that, in general, the drag of a lifting airfoil is higher than that corresponding to zero lift; as will be explained in Chapter 12, this increase is mostly due to the higher pressure drag due to the increased thickness of the upper surface boundary layer which, in turn, is caused by the higher adverse pressure gradients in the aft region of the airfoil. Nonetheless, the values of drag remain rather small, as for all aerodynamic bodies, and airfoils are very effective lift-producing bodies; indeed, the ratio between lift and drag may attain values as high as 100.

It should also be evident that the lift on an airfoil increases with increasing angle of attack. It is easy to see that the strength of the starting vortex is an increasing function of α, and the same is thus true for the opposite vorticity and the consequent circulation remaining around the airfoil. Therefore, the pressure distribution shown in Fig. 8.8 changes in such a way that the pressures over the upper surface further decrease, while those over the lower surface become higher. In particular, the suction peak over the airfoil upper surface increases with the angle of attack. Obviously, the process cannot continue indefinitely because, when α becomes excessive, the adverse pressure gradient over the upper surface will become so strong that it causes separation of the boundary layer, and thus the airfoil can no longer be considered as an aerodynamic body. The lift reaches a maximum value, after which it starts decreasing and, simultaneously, the pressure drag significantly increases. The angle of attack corresponding to this occurrence is known as the *stall angle* and the airfoil is said to be *stalled.*

Returning to conditions of motion at angles of attack that are sufficiently small for the boundary layer to remain attached, we may imagine that, after the end of the above-described transient, either the angle of attack or the velocity of the airfoil is suddenly increased. If this occurs, then a new vortex, sufficiently strong to produce around the body a new flow with completely attached boundary layer, will be shed in the wake. On the other hand, if the airfoil is suddenly stopped, the excess of clockwise vorticity that is present over its surface will rapidly gather at the trailing edge, forming a concentrated vortex that is equal and opposite to the starting vortex (see Fig. 8.9).

In agreement with the vorticity dynamics equation, the subsequent evolution of these vortices will be a displacement in the direction of the

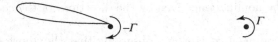

Fig. 8.9. Vortices produced by suddenly starting and stopping an airfoil.

mutually induced velocity field and a progressive diffusion of their vorticity until, in principle, complete mutual annihilation of the two opposite-sign vorticities occurs. Obviously, in practical situations the two vortices might be very far from each other, and each of them would then diffuse in a virtually infinite domain, or perhaps interfere with the vorticity that the vortices themselves produce on a closer boundary as, for instance, the surface of a runway (see Chapter 11 for further details).

Incidentally, it should be noted that the assumption of impulsive motion, which has been made up to now for simplicity, is actually not essential. In all the above cases, the motion could have started with any acceleration law, and the vorticity in the flow would have gradually increased up to the appropriate value corresponding to the final velocity, even though with a different transient. As an example, for an accelerating airfoil the counter-clockwise vorticity would have been shed from the trailing edge in a progressive way, rather than in a concentrated starting vortex. However, for the same final velocity and angle of attack, the final global value of the shed vorticity would be the same as in the impulsive motion case, and the same opposite-sign circulation would remain around the airfoil. Consequently, after a sufficient time from the attainment of the constant final velocity, the conditions in the two cases would coincide.

In practical situations, airfoils are cross-sections not of infinite but of finite wings, and the flow is actually three-dimensional everywhere, with the possible exception of the wing symmetry plane when the flight conditions are symmetrical. Compared to the two-dimensional case, the finiteness of the wing introduces modifications in the vorticity field shed from the wing trailing edge, both at the start of the motion and in constant-velocity conditions. In particular, vorticity components in the free-stream direction will also be present in the wake. These differences, which have consequences both for the value of the lift forces acting on the wing and for the methods through which they can be evaluated, will be described and discussed in some detail in Chapter 13. Nonetheless, the basic physical mechanisms of lift generation that have been described above are still qualitatively valid for each wing cross-section. We shall see that the main difference is simply that the effective angle of attack of each airfoil does not generally coincide with the nominal one given by the directions of the chord and the free-stream.

As a final comment, it is worth observing that the above explanation of the mechanisms through which the lift force is generated on an airfoil requires the use of most of the physical and mathematical concepts that have been introduced in the previous chapters.

8.3. Main Features of the Loads on Bluff Bodies

The separation of the boundary layer that characterizes the flow field around bluff bodies is also the basic feature determining the type and magnitude of the loads acting on them. This fact is particularly evident when these loads are compared with those acting on aerodynamic bodies, and the qualitative and quantitative differences between the two cases are thus highlighted.

Indeed, we have already seen, for instance, that aerodynamic bodies are characterized by small values of the drag coefficients and by a pressure drag that is usually significantly smaller than the friction drag. This is a direct consequence of the small thickness of the attached boundary layers and of the resulting wake, so that the order of magnitude of the drag is not very different, in many situations, from that of a flat plate aligned with the flow.

Conversely, pressure drag is definitely prevalent for bluff bodies, and is normally one order of magnitude higher than the friction drag, whose value remains of the same order as that of aerodynamic bodies. In practice, in many cases the pressure drag is so high that the friction drag corresponds to less than 2% of the total drag, and may even be neglected. The above is true for all bluff bodies in which the portion of surface with attached boundary layer is comparable with that over which the boundary layer is separated. Obviously, different relative contributions apply in the case of bodies that are very elongated in the flow direction, with separation occurring only at their rear end.

A particularly remarkable example of the different magnitudes of the drag acting on aerodynamic and bluff bodies is given in Fig. 8.10,

Fig. 8.10. Comparison of the dimensions of an airfoil and of a circular cylinder experiencing the same drag force for the same value of free-stream velocity.

where the dimensions of an airfoil and of a circular cylinder experiencing the same drag value when immersed in a free-stream at the same velocity are qualitatively shown. Although the effects of Reynolds number and of the chord-to-thickness ratio of the airfoil should be taken into account for an exact comparison, the dimensions of the two bodies shown in the figure are indeed plausible and typical. Incidentally, this example clearly illustrates the advantage of surrounding a bluff body with a *fairing*, i.e. with an aerodynamic-shape cover.

The striking disparity in size of the two bodies shown in Fig. 8.10 is a direct consequence of the fact that the drag coefficient of an airfoil (based on the cross-flow dimension) is 10 to 20 times smaller than that of a circular cylinder. This difference is essentially due to the fact that, contrary to what happens for aerodynamic bodies, the width of the wake of bluff bodies is of the same order as their cross-flow dimension and opposite-sign concentrated vortices are often continuously shed in the wake. As may also be deduced from Fig. 8.11, this implies the introduction in the flow

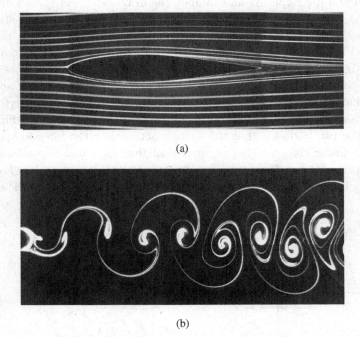

(a)

(b)

Fig. 8.11.   Visualization of the flow around an airfoil (a) and a circular cylinder (b). ((a) Photo copyright H. Werlé/ONERA: The French Aerospace Lab, reprinted with permission; (b) Photo courtesy of Professor Charles Williamson, Cornell University.)

of very considerable velocity perturbations. In other words, the variation of kinetic energy produced by the relative motion between body and fluid is much larger for a bluff body, and corresponds to a much greater work done on the fluid by the opposite of the drag force acting on the body (see Chapter 10 for more details).

In general, the pressure drag on a bluff body may be divided into two contributions, the first given by the *forebody*, i.e. the front part of the body with attached boundary layer, and the second by the *afterbody* (or *base*), i.e. the portion of the body surface lying inside the separated wake. Note, however, that the term *afterbody* is often used to indicate all the portion of surface behind the location of the maximum thickness of the body, where the cross-section decreases, irrespective of whether the flow in that region is separated or not. The afterbody always produces a large pressure drag (also known as the *base drag*), which is determined by the value of the suction acting over that portion of the surface. Conversely, the contribution of the forebody to the total drag of a bluff body may be large or small, depending on its shape. As an example, the drag coefficient (based on the body cross-flow dimension) of a flat plate perpendicular to the free-stream is approximately 2.0 and that of a circular cylinder — in certain ranges of Reynolds number — is about 1.2, in spite of the fact that, as may be seen in Fig. 8.12, the base suctions on the two bodies are similar. Note that boundary layer separation occurs at the lateral edges of the plate, and at approximately 80° from the free-stream direction over the circular cylinder surface. As may also be deduced from Fig. 8.12, the reason for the different drag of the two bodies is that the suction acting over the forebody of the circular cylinder gives a higher negative contribution to drag; consequently, that portion of the surface generates almost half of the pressure drag for the flat plate, but less than 20% for the cylinder. When a forebody is sufficiently

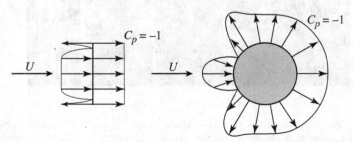

Fig. 8.12.   Comparison of pressure distributions on a flat plate and on a circular cylinder.

elongated — as happens, for instance, for the ogival nose of a missile — its contribution to the total pressure drag of a bluff body may become almost negligible.

The suctions acting on the base are then essential in determining the value of the drag of a bluff body, and it may be seen that their magnitude is primarily connected with the velocity outside the boundary layer at the separation point, say U_s. Particularly in the case of a base with small extension in the free-stream direction, the mean pressures in that region are almost constant, and practically equal to the pressure at the edge of the boundary layer at the separation point. We then derive from relation (8.6) that the pressure coefficient over the base is given by — or strictly connected with — the following expression:

$$C_{ps} = 1 - \frac{U_s^2}{U^2}. \tag{8.14}$$

Consequently, we deduce that the higher the velocity outside the boundary layer in the separation region, the lower the base pressure coefficient and the higher the base drag. In turn, it may be seen that the magnitude of the velocity at separation is a function of the curvature of the streamlines that are adjacent to the separating boundary layers and to the initial boundary of the wake. As will be further discussed in Chapters 11 and 14, this is in agreement with the fact that the drag coefficient of a bluff body is generally an increasing function of the ratio between the width of the wake and the cross-flow dimension of the body.

A consequence of the importance of the relative wake width is that if the boundary layer becomes turbulent before the occurrence of laminar separation then the drag coefficient of a bluff body is usually found to decrease. Indeed, we have already seen in Chapter 7 that turbulent boundary layers are more resistant to adverse pressure gradients, and this generally results in a delayed separation compared to the laminar condition. However, this happens only in the case of natural separation, i.e. if there are no sharp edges forcing the boundary layer to separate, irrespective of its being laminar or turbulent. Therefore, the drag of a bluff body with natural separation depends on all the flow and geometrical parameters, such as Reynolds number and surface roughness, which influence the transition from laminar to turbulent flow.

An aspect that must be pointed out is that two-dimensional bodies (in practice, highly elongated cylindrical bodies in cross-flow) are subject to higher loads than short three-dimensional bodies with analogous cross-section. Thus, for instance, the drag coefficient of an infinite flat plate perpendicular to the free-stream is approximately 2.0, whereas the drag coefficient of a square plate is slightly less than 1.2. Analogously, in a certain range of Reynolds numbers, the drag coefficient of a circular cylinder is 1.2, and the corresponding one of a sphere is 0.5. This difference is reasonable because the disturbance caused by a three-dimensional body is lower, due to the fact that the flow is also free to deflect in an additional direction. However, we shall see in Chapters 11 and 14 that the different behaviour of the drag force is also connected with a different strength and organization of the vorticity structures that are present in the separated wake.

A further typical feature of the flow around bluff bodies is that, at variance with what happens for aerodynamic bodies, their wakes are generally characterized by large velocity fluctuations (see, e.g., Fig. 8.11), which then produce significant fluctuating surface pressures. Consequently, bluff bodies are usually subjected to unsteady loads, whose magnitude and frequency content depend on the shape of the body. Further information on these aspects can be found in Chapter 14, where the aerodynamic features of bluff bodies are analysed in more detail. Here, we just mention that oscillatory phenomena may occur when the unsteady forces acting on a bluff body are characterized by significant energy at frequencies near the natural frequencies of a structural element to which the body belongs. These possible dynamical effects must be carefully taken into account in engineering design.

Finally, we note that, as happens for aerodynamic bodies, bluff bodies are also subject to forces in the cross-flow direction when the pressure distribution over their surface is not symmetrical. For two-dimensional bodies, this force component is often denoted as lift, in analogy to the corresponding force acting on airfoils, and depends both on the body shape and on the orientation of the incoming free-stream. In particular, it may be significant if a two-dimensional body is sufficiently elongated (like an ellipse) and inclined to the flow. Its origin is still due, as for an airfoil, to the shedding of a starting vortex and to the permanence of an average vorticity of opposite sign around the body during motion, but the generation mechanism is not exactly the same as that of an aerodynamic body. The absence of a sharp trailing edge does not permit the boundary layer to remain attached after the end of the initial transient and significantly reduces the strength of the starting vortex compared to that of an airfoil with similar length-to-width ratio. Considering also the higher drag of a bluff body, the consequence is that the lift-to-drag ratio turns out to be much lower than that of an airfoil. Furthermore, bluff bodies with boundary layer separation fixed by sharp corners, such as cylinders with square or rectangular cross-sections, may show non-trivial trends of the loads as a function of wind direction. In particular, they may be subject to mean cross-flow forces that are in the opposite direction to those acting on elongated bodies.

8.4. Some General Comments on Fluid Dynamic Loads

The point of view adopted in this book is that an adequate familiarity with the equations ruling the motion of fluids and with their implications is a necessary prerequisite to properly understanding the origin and the main features of the loads acting on moving bodies. This is why, in previous chapters, particular attention has been devoted to certain topics, such as the conditions of validity and the consequences of the incompressible flow assumption, the origin and dynamics of vorticity, and the development and behaviour of boundary layers. Deep comprehension of all this matter also leads to rapid appreciation of the methods that can be used for the prediction and control of fluid dynamic loads.

We now briefly make some further comments on the connection between the characteristics of the loads acting on different types of bodies, which have been the subject of the present chapter, and the concepts that were previously introduced in our treatment. To this end, let us first return to the discussion at the end of Chapter 5, and in particular to Figs. 5.1 and 5.2, where we highlighted that the comparison between the real pressure distribution and the one that would correspond to a non-viscous flow leads to definitely different conclusions for two bodies with different shapes. From the classification introduced in Chapter 7, it should now be evident that those bodies are examples of an aerodynamic body and a bluff body respectively. We can thus explain why the above-recalled comparisons are so different and, simultaneously, summarize some of the most significant results of the previous chapters.

To do so, we first recall that the incompressible flow equations for a non-viscous fluid coincide with those of the irrotational motion of a viscous fluid, and that in practice they reduce to the Laplace equation for the velocity potential and to Bernoulli's theorem for pressure. However, for a non-viscous fluid the boundary condition to be satisfied at the surface of a moving body is that the velocity must be tangent to the body surface, and a completely irrotational solution for this problem exists. This is not the case for a viscous fluid, because the no-slip boundary condition holds at the body surface, which then becomes a source of vorticity in the flow. From the theorem of total vorticity conservation, we also know that equal amounts of positive and negative vorticity must be produced if the motion has started from rest.

The subsequent evolution of this vorticity depends on the shape of the body and on the type of motion, and this is where the distinction between aerodynamic and bluff bodies comes into play. Thus, if the typical Reynolds number of the flow is sufficiently high and the shape and motion of the body satisfy certain requirements, the vorticity remains confined in a thin boundary layer adjacent to the whole surface, and we say that we are dealing with the motion of an aerodynamic body. One fundamental feature of this flow is that a simplified set of equations may be devised for the boundary layer region, with the remarkable result that the pressure at the edge of the boundary layer effectively coincides with the pressure acting over the body surface.

Consequently, for an aerodynamic body it is reasonable to expect that a first-order approximation of the actual flow may be obtained by first letting a sufficient time from the start of the motion to elapse, so that an effectively steady condition is reached, and then imagining squeezing the boundary layer over the body surface, thus shrinking to zero the thickness of the sole region where vorticity is contained. The result is an approximate model in which a viscous potential flow is tangent to an infinitesimally thin layer containing the global vorticity present in the boundary layer; however, for all practical purposes, this flow may be viewed as being tangent to the body surface. In a sense, in this first-order model the vorticity squeezed over the surface may be considered as the flow mechanism through which the boundary condition changes from no-slip to non-penetration. Thus, from the mathematical point of view, the non-viscous flow problem and the first-order approximated model of the actual viscous flow coincide. This explains why the actual pressure distribution around an airfoil has significant similarities with the one that would be obtained if the flow were considered to be non-viscous.

Furthermore, we may also understand why differences remain near the trailing edge between the pressures obtained from the non-viscous flow solution — which coincides with the solution of the first-order model of the viscous flow — and the real pressures acting on the airfoil. The reason is that the region near the trailing edge is the only one where a large difference exists between the velocity tangent to the body in the non-viscous model and the velocity outside the boundary layer in the actual flow. In effect, the solution of the tangent irrotational flow requires the velocity to vanish at the sharp trailing edge, whereas in the real viscous case the boundary layer causes the outer potential flow to move outwards in that region and the velocity to remain finite, and actually comparable to the free-stream velocity. However, this discrepancy between the actual viscous flow and its first-order approximation may be eliminated by means of subsequent iterative calculations, which include the solution of the boundary layer equations.

In practice, the solution of the first-order problem of the flow around an aerodynamic body, with the boundary layer squeezed over its surface, provides a first estimate of the pressure distribution around the body. This first evaluation then provides the local pressure gradients that are the necessary input for a first solution of the boundary layer equations, from which we obtain a first-order estimate of the tangential viscous forces and of the evolution of the boundary layer displacement thickness over the body surface. We may then proceed to set up the iterative procedure that was repeatedly mentioned in Chapter 7. In particular, a second-order estimate of the pressure distribution may be obtained by solving the Laplace equation for the potential flow tangent to a modified body, derived from the original one by adding the displacement thickness over its surface. This new pressure distribution may again be used to solve the boundary layer equations, and thus to get a second evaluation of the displacement thickness to replace the first one in the construction of another modified body. The calculations can then be repeated, and the iterative procedure converges if the body is aerodynamic, though certain technical precautions may sometimes be necessary to avoid unrealistic separations of the boundary layer that may occur in its first evaluation, due to excessive adverse pressure gradients near the trailing edge.

A similar procedure can also be applied for the evaluation of both drag and lift on an airfoil at a non-zero angle of attack. However, in Chapter 12 we shall see that the first-order flow field obtained by squeezing the boundary layer over the body surface is equivalent not to a generic solution of the non-viscous flow problem, but to the solution of a particular problem in which a further mathematical condition must be imposed. It will be shown that this further condition could never have been derived without a careful analysis of the viscous-flow evolution.

To further appreciate the significance of the first-order approximation of the flow around an aerodynamic body, we may observe that, although the mathematical problem that is solved at the first step of the iterative procedure is apparently identical to the one corresponding to a non-viscous flow, neglecting viscosity is not necessary for its introduction. In fact, there is a profound conceptual difference between neglecting viscosity from the beginning and constructing a simplified, first-order model of the real flow, in which the effect of viscosity is taken into account by imagining the vorticity to be squeezed to a zero-thickness layer over the surface of the body. Furthermore, if the assumption of non-viscous flow had been made from the beginning, no justification could have been found either for the appearance of circulation around the airfoil or for the introduction of the further condition that, as we shall see in Chapter 12, is necessary for the first-order prediction of lift.

In a sense, one might say that reducing the boundary layer thickness to zero is equivalent to progressively increasing the Reynolds number up to infinity. However, this limit procedure must be carried out only *after* the initial transient of the motion, at a finite value of the Reynolds number, is over and the flow has evolved to an almost steady condition. With this proviso, we might then say that an aerodynamic body is such that the limit of the solution of the relevant viscous flow problem for Reynolds number tending to infinity coincides with the solution of the limit of the equations of motion for Reynolds number tending to infinity. Conversely, neither this statement nor the above iterative procedure may be applied for a bluff body, because the spread of vorticity in a wide wake due to boundary layer separation implies that a potential flow tangent to the body surface is not a first-order approximation of the actual flow. Furthermore, boundary layer theory is no longer applicable beyond separation, and even slightly before it. This explains why the prediction of the flow field and of the fluid dynamic loads is much more complex for bluff bodies than for aerodynamic bodies. Therefore, experiments and, nowadays, the numerical solution of the full Navier–Stokes equations are, in practice, the only means to obtain estimates of the main features of bluff body flows and of the consequent fluid dynamic loads. Nonetheless, significant difficulties exist even in using these prediction tools, particularly because the flows at high Reynolds numbers often encountered in practical applications can hardly be reproduced both experimentally and numerically.

In conclusion, the classification of bodies that we have introduced, according to the possible existence of boundary layer separation, implies that aerodynamic bodies are the only ones for which the fluid dynamic loads can be evaluated by means of a simplified procedure, which, in addition to being mathematically accessible, may provide excellent predictions for many interesting flow conditions. Furthermore, we have seen that these bodies are precisely those that are capable of producing significant cross-flow forces with a minimum drag penalty. Therefore, if we consider that the wings of an aeroplane are, perhaps, the most notable example of aerodynamic bodies, the practical importance of these results should be evident. On the other hand, the prediction of the loads on bluff bodies is of great relevance in many non-aeronautical applications, such as the design of road vehicles or of civil structures exposed to the wind. Consequently, careful use of the available information and of previous experience is still fundamental in all these areas of engineering practice.

In all cases, physical insight is fundamental to properly analyse and exploit the results of any type of research activity. Understanding the role of vorticity dynamics may be of significant help, and this is why significant emphasis has been given to this subject in previous chapters. A fruitful idea, which should smoothly derive from the approach adopted in this book, is that a body moving in an otherwise still fluid — or immersed in a uniform stream of fluid — may be seen as a generator of equal amounts of vorticity of opposite sign. A further and consequential concept is that the manner in which these opposite vorticities subsequently develop in the flow determines not only the type and magnitude of the forces that act on the body, but also their predictability. This aspect should become clearer in Chapter 11, which is devoted to some supplementary notions on the connection between the evolution of vorticity and the fluid dynamic loads. It will be seen that the forces experienced by bodies in motion in a fluid may be directly expressed in terms of the dynamics of the vorticity field.

For instance, the fluid dynamic drag of a body will be shown to be dependent on the amount and organization of the vorticity introduced in the wake. In particular, drag generally becomes higher when the distance between the layers of opposite vorticity increases, and this contributes to explaining the different orders of magnitude of the drag of aerodynamic and bluff bodies. On the other hand, the lift on a body is connected with the continuous increase of the distance between the starting vortex and the vorticity of opposite sign remaining around the body during its motion, and is directly proportional to the total strength of these opposite vorticities. Finally, recognizing which wake vorticity structures are more directly connected with the various force components, and understanding the mechanisms through which this connection arises, may be invaluable in devising methods for the control of the type and magnitude of those forces.

If such knowledge is available, one may be able to change the geometry of a body — or introduce external devices — in order to interfere with the development of the main vorticity structures in a favourable way, so that a given objective regarding fluid dynamic loads is achieved. A particularly important field of application of such a procedure is the reduction of the drag or of the unsteady forces acting on bluff bodies.

Part II

DEEPER ANALYSES AND CLASSICAL APPLICATIONS

Chapter 9

EXACT SOLUTIONS OF THE
INCOMPRESSIBLE FLOW EQUATIONS

9.1. Introductory Remarks

As we have already pointed out in previous chapters, the equations of motion of a fluid are extremely complex, especially due to their non-linearity, and thus solutions cannot be derived easily for flow problems characterized by *generic* initial and boundary conditions. However, certain solutions of the equations can be found, particularly for the case of incompressible flows. We shall restrict our attention to this case and briefly discuss some general features of the possible solutions of the relevant equations. We recall that in the incompressible flow case the mass and momentum balance equations are decoupled from the energy equation, and form a set of four scalar equations in which the unknown functions are the three components of the velocity vector and the pressure. The solution of this set of equations is indeed the mathematical problem to which one refers when dealing with incompressible flows. Thus, the energy equation is usually disregarded, and it is only used when it is required to derive the temperature field, once the velocity and the pressure fields have been obtained.

Let us consider again the equations governing an incompressible flow. The mass balance requires the volume of a material element of fluid to remain constant during the motion; thus, as the divergence of velocity represents the time variation of the volume of a fluid element, per unit volume, this requirement implies that the velocity field must be solenoidal, i.e. divergence-free, as expressed by Eq. (5.31).

Therefore, the momentum balance equation (5.35) becomes the basic equation describing the motion, and is now recast as follows:

$$\frac{\partial \boldsymbol{V}}{\partial t} + \boldsymbol{V} \cdot \mathrm{grad}\,\boldsymbol{V} - \nu\nabla^2\boldsymbol{V} = \boldsymbol{f} - \mathrm{grad}\left(\frac{p}{\rho}\right). \qquad (9.1)$$

It is often convenient to use relation (5.41) to derive the following alternative formulation of the equation:

$$\frac{\partial \boldsymbol{V}}{\partial t} + \boldsymbol{\omega}\times\boldsymbol{V} - \nu\nabla^2\boldsymbol{V} = \boldsymbol{f} - \mathrm{grad}\left(\frac{p}{\rho} + \frac{V^2}{2}\right), \qquad (9.2)$$

which, when the body force is conservative ($\boldsymbol{f} = -\mathrm{grad}\,\varPsi$), becomes

$$\frac{\partial \boldsymbol{V}}{\partial t} + \boldsymbol{\omega}\times\boldsymbol{V} - \nu\nabla^2\boldsymbol{V} = -\mathrm{grad}\left(\frac{p}{\rho} + \frac{V^2}{2} + \varPsi\right). \qquad (9.3)$$

Finally, we recall that the viscous term in the above equations may be also expressed in terms of the vorticity vector using relation (5.42), which gives

$$\nu\nabla^2\boldsymbol{V} = -\nu\mathrm{curl}\,\boldsymbol{\omega}. \qquad (9.4)$$

This expression is useful to recall that the viscous term in the momentum equations disappears if the motion is such that $\mathrm{curl}\,\boldsymbol{\omega} = 0$, and in particular when $\boldsymbol{\omega}$ is either zero or a constant.

It is now necessary to explain, in more detail, how we define an 'exact solution' of the equations of motion. In effect, there are several ways in which this definition may be given. In particular, a first possibility is to assert that two fields, $\boldsymbol{V}(\boldsymbol{x},t)$ and $p(\boldsymbol{x},t)$, constitute an exact solution of the flow equations if, besides satisfying Eqs. (5.31) and (9.1) for all $\boldsymbol{x}$ and t and for given values of ρ, ν and $\boldsymbol{f}$, they also fulfil initial and boundary conditions that are relevant to a real (or reasonably approximate) physical problem. This last condition restricts the number of the available exact solutions of the Navier–Stokes equations, as a direct consequence of the already mentioned intrinsic mathematical difficulties of the equations, mainly connected with their non-linearity.

Therefore, a second possibility might be to content ourselves with the requirement that the functions $\boldsymbol{V}(\boldsymbol{x},t)$ and $p(\boldsymbol{x},t)$ be a solution of the equations of motion, irrespective of whether or not they satisfy boundary conditions that may describe, in some way, a realistic physical problem. This is, for instance, the point of view adopted by Berker (1963), and it has several advantages. The first and most obvious one is that the number of known exact solutions increases significantly and a large number of them are listed by Berker (1963). More importantly, these solutions may give useful clues to the nature of the flows corresponding to certain real problems, or provide new ideas on how those problems might be described through simplified models that may be considered as first-order approximations of the real situation. A possible example might be the case in which two exact solutions, not satisfying a given boundary condition along a line or a surface, may be joined through a boundary layer, which is an approximate solution that might be representative of certain flow conditions. Furthermore, it might even happen that, by observing the flow corresponding to an exact solution, one may devise a new physical problem corresponding to that flow in a practical situation. For instance, it is possible that a streamline of a general solution might be replaced by a solid surface whose local shape and velocity coincide with those of that streamline. In this respect, it must be observed that the boundary conditions at solid surfaces may be suitably changed, in practice, by using suction or injection of fluid with different local velocities, and this is indeed a developing and widely studied technique for flow control.

In general, we then consider as valid this wider and more general definition of exact solution, even if the examples described in the following sections mainly refer to cases that may be representative of practical flow conditions. However, it will be immediately appreciated that this applicability to practical cases derives from the fact that the solid surfaces present in various situations are characterized by a shape and a velocity that coincide, in an appropriate reference system, with those of a streamline of a general solution.

Other possible distinctions may be introduced regarding the *form* of the velocity and pressure functions that may be accepted as exact solutions. In particular, one might accept only solutions that are expressed in closed form, by means of elementary functions or special well-known functions (such as the Bessel functions). Infinite series are also generally considered as acceptable, even if this opinion is not shared by all authors (see Wang, 1991). On the other hand, direct numerical simulations cannot be viewed as providing exact solutions because, apart from the approximations due to the numerical schemes used, they only refer to flows at particular values of the Reynolds number, and thus lack the expected generality. Conversely, one significant application of exact solutions is precisely their use to validate numerical solutions.

Nonetheless, when studying the characteristics of exact solutions of the Navier–Stokes equations and investigating their possible utility for describing, either accurately or approximately, flows of practical interest, one must always recall that some solutions may be unstable, or may become so in certain flow conditions. Another note of caution regards steady solutions, which, in order to be physically representative, must be the limit of a feasible unsteady solution for time tending to a finite or, at least, an infinite value. This derives from the fact that, disregarding steady space-uniform flows, which correspond to a condition of zero velocity in an appropriate reference system, all real flows may be considered to start from rest or from a physically realizable initial condition. Finally, an exact solution can never be physically meaningful in the points or regions where it corresponds to infinite values of the velocity. However, if those regions can be excluded by means of boundaries where the velocity matches the one given by the solution, then its representativeness may be recovered.

It is outside the scope of the present book to provide an exhaustive list of exact solutions of the incompressible flow equations, or of the sometime complex mathematical techniques that may be used for their derivation. The interested reader may refer to the available literature on the subject, such as the first and fundamental review by Berker (1963), the subsequent ones by Wang (1989, 1991) and the book by Drazin and Riley (2006). Here we shall only give some further general information on the mathematical aspects of the problem, and describe in more detail some classical particular solutions that may be useful either to describe certain real flows or to obtain useful reasonable approximations of their main features. Nonetheless, we stress again that deriving or studying exact solutions should not be merely seen as a fascinating mathematical challenge, but also as one of the most important means to derive precious clues to the behaviour of real flows and to devise physically and mathematically sound, albeit approximate, simplified flow models.

9.2. Basic Mathematical Considerations

9.2.1. *General exact solutions*

It is useful to recall that, by taking the curl of Eq. (9.2) and considering that both velocity and vorticity are solenoidal vectors, we derived in Chapter 6 the vorticity dynamics equation for incompressible flows, which we rewrite now in the following form:

$$\frac{\partial \boldsymbol{\omega}}{\partial t} + \boldsymbol{V} \cdot \operatorname{grad} \boldsymbol{\omega} - \boldsymbol{\omega} \cdot \operatorname{grad} \boldsymbol{V} - \nu \nabla^2 \boldsymbol{\omega} = \operatorname{curl} \boldsymbol{f}. \tag{9.5}$$

For conservative body forces, the equation becomes

$$\frac{\partial \boldsymbol{\omega}}{\partial t} + \boldsymbol{V} \cdot \operatorname{grad} \boldsymbol{\omega} - \boldsymbol{\omega} \cdot \operatorname{grad} \boldsymbol{V} - \nu \nabla^2 \boldsymbol{\omega} = 0. \tag{9.6}$$

We have extensively discussed the importance of these equations to describe both the dynamics of vorticity and the relevant physical mechanisms. We now further observe that, as vorticity is the curl of velocity and pressure is not present, Eqs. (9.5) and (9.6) can be viewed as *compatibility equations* for the velocity field. This interpretation is due to Berker (1963), and its meaning is that, together with Eq. (5.31), these equations express the necessary and sufficient conditions for a vector function $\boldsymbol{V}(\boldsymbol{x}, t)$ to represent a velocity field that is a solution of the Navier–Stokes equations for non-conservative and conservative body forces respectively. For instance, if a solenoidal velocity field is a solution of Eq. (9.6), then a corresponding pressure field (apart from an additive constant) may be derived from Eq. (9.3), and together they form a solution of the equations governing the incompressible flow of a homogeneous Newtonian viscous fluid in the presence of conservative body forces.

From Eq. (9.6) it is readily seen that irrotational solenoidal velocity fields, i.e. potential flows satisfying the Laplace equation for the velocity potential, are general solutions of the Navier–Stokes equations. Obviously, we have already noticed that, apart from special cases (such as rotating cylinders), these flows cannot satisfy the no-slip boundary condition at solid walls, but only the non-penetration condition. Nonetheless, they are an example (albeit perhaps a trivial one) of a general solution of the equations of motion that can be matched to the real flow existing over a surface (in this case a solid body surface) by means of a boundary layer. Although not an exact solution, this would be a more than satisfactory approximation of the real situation in many circumstances, and in particular if the body is aerodynamic and moves at moderate angles of attack to the free stream. Actually, as we have already seen, and will further discuss in Chapters 12 and 13, in that case even the potential flow tangent to the body surface is a first-order model of the real flow, obtained by squeezing the boundary layers over the body surface while keeping unaltered the global vorticity they contain.

Another immediate result deriving from Eq. (9.6) is that two-dimensional constant-vorticity flows are also a general solution of the equations of motion. By analogy with the previous case, one may infer that regions with constant vorticity separated from potential flows by thin boundary layers could be an approximate model for certain types of two-dimensional flows with practical interest.

An effective way of utilizing the compatibility condition is to first assume that the components of the velocity field have a certain form, usually chosen in order to simplify the problem, and then derive the consequent relations that those components must satisfy to comply with the solenoidality condition and with Eq. (9.6).

As an example, one may assume the flow to be *unidirectional*, i.e. that only one component of the velocity is not zero, say w in direction z. Then, from the continuity equation (5.31) we find that

$$\frac{\partial w}{\partial z} = 0, \tag{9.7}$$

and thus w is constant along z although, obviously, it may vary in the x and y directions. As for the vorticity field, we obtain

$$\omega_x = \frac{\partial w}{\partial y}, \quad \omega_y = -\frac{\partial w}{\partial x}, \quad \omega_z = 0. \tag{9.8}$$

If we now introduce these components into Eq. (9.6), and further assume the motion to be steady, after simple manipulations we obtain

$$\frac{\partial^2 w}{\partial x^2} + \frac{\partial^2 w}{\partial y^2} = K, \tag{9.9}$$

where K is an arbitrary constant. From the momentum equation (9.1) it is easily seen that this constant is related to the pressure field. In effect, the x and y components of that equation show that $\partial p/\partial x = \partial p/\partial y = 0$, while from the equation in the z direction one obtains, for $\boldsymbol{f} = -\text{grad } \Psi$,

$$\frac{\partial}{\partial z}\left(\frac{p}{\rho} + \Psi\right) = \nu\left(\frac{\partial^2 w}{\partial x^2} + \frac{\partial^2 w}{\partial y^2}\right) = \nu K. \tag{9.10}$$

This relation shows that when $\partial \Psi/\partial z$ is a constant (which is almost always the case) the same is also true for the pressure gradient $\partial p/\partial z$, and thus the pressure is either constant (if $K = 0$) or varies linearly along the z direction. If the pressure gradient is given, then the velocity field may be derived from the solution of the following Poisson equation:

$$\frac{\partial^2 w}{\partial x^2} + \frac{\partial^2 w}{\partial y^2} = \frac{1}{\mu}\frac{\partial}{\partial z}(p + \rho\Psi). \tag{9.11}$$

To render this case physically representative, one may require the velocity to be prescribed along certain curves that coincide with solid boundaries, either at rest or in motion. The solution of the equation will then be more or less difficult according to the shape of the bounding curves.

Another interesting example of application of the compatibility equation is the case in which we assume the motion to be characterized by *concentric circular streamlines*, which we choose to be parallel to the x–y plane and with their centre on the z axis. In this case, it is advantageous to use cylindrical coordinates (r, θ, z), and the relevant velocity components may then be written as

$$v_r = 0, \quad v_\theta = v_\theta(r, \theta, z, t), \quad v_z = 0. \tag{9.12}$$

The equations of motion in cylindrical coordinates, together with the expressions of the main differential operators, are reported in Appendix B. In particular, the continuity equation (5.31) is seen to become

$$\frac{\partial v_r}{\partial r} + \frac{1}{r}\frac{\partial v_\theta}{\partial \theta} + \frac{\partial v_z}{\partial z} + \frac{v_r}{r} = 0. \tag{9.13}$$

By introducing the velocity components (9.12) into this equation, we can derive that $\partial v_\theta/\partial\theta = 0$, so that we must have $v_\theta = v_\theta(r, z, t)$.

We may now assume that the body forces are conservative, and thus use the form (9.6) of the compatibility equation. To this end, we first derive the components of the vorticity vector in cylindrical coordinates by introducing the velocity components (9.12) into the relations (3.43), so that we obtain

$$\omega_r = -\frac{\partial v_\theta}{\partial z}, \quad \omega_\theta = 0, \quad \omega_z = \frac{\partial v_\theta}{\partial r} + \frac{v_\theta}{r}. \tag{9.14}$$

Then, we consider the component of Eq. (9.6) in the θ direction. By using the expressions of the various vector operators in cylindrical coordinates reported in Appendix B, after some straightforward algebraic manipulations the following relation is obtained:

$$\frac{\partial v_\theta}{\partial z} = 0, \tag{9.15}$$

from which we deduce that $v_\theta = v_\theta(r,t)$. In other words, the assumption that the streamlines are circular and concentric implies that the motion must be necessarily two-dimensional and occur in the r–θ plane. This result was first derived by Cisotti (1924), using a different approach; however, as pointed out by Berker (1963), it may readily be obtained from the compatibility equation through the above-described procedure.

If we now project the compatibility equation in the r direction, and further assume the motion to be steady, we derive the ordinary differential equation that must be satisfied by the function $v_\theta = v_\theta(r)$:

$$\frac{d^3 v_\theta}{dr^3} + \frac{2}{r}\frac{d^2 v_\theta}{dr^2} - \frac{1}{r^2}\frac{d v_\theta}{dr} + \frac{v_\theta}{r^3} = 0. \tag{9.16}$$

The general solution of this equation is

$$v_\theta = \frac{A}{r} + Br + Cr\ln r, \tag{9.17}$$

where A, B and C are scalar constants. However, as shown by Berker (1963), the requirement that the pressure be single-valued necessarily implies that $C = 0$. Therefore, the final expression for v_θ becomes

$$v_\theta = \frac{A}{r} + Br. \tag{9.18}$$

As a last example, let us consider the simple case of a *parallel axisymmetric flow*, i.e. a flow that is parallel to the z axis and whose components in a cylindrical reference system are

$$v_r = 0, \quad v_\theta = 0, \quad v_z = v_z(r,z,t). \tag{9.19}$$

From the continuity equation (9.13) we find that $v_z = v_z(r,t)$, so that the motion of each particle is rectilinear and uniform. Likewise, if we assume the body force to be conservative and the motion to be steady, the following ordinary differential equation for v_z may be obtained from the compatibility equation (9.6):

$$\frac{d^3 v_z}{dr^3} + \frac{1}{r}\frac{d^2 v_z}{dr^2} - \frac{1}{r^2}\frac{d v_z}{dr} = 0, \tag{9.20}$$

whose general solution is

$$v_z = Ar^2 + B\ln r + C. \tag{9.21}$$

Once again, A, B and C are scalar constants, which may be chosen so that suitable boundary conditions are satisfied.

9.2.2. *Superposable solutions*

As we have shown in the previous section, the necessary and sufficient condition for a solenoidal vector function $V(x, t)$ to represent the velocity field of an exact solution of the Navier–Stokes equations is that it be a solution of the relevant compatibility equation. We may now wonder if a condition exists for two exact solutions V_1 and V_2 to be *superposable*, i.e. to be such that the function $V = V_1 + V_2$ is still an exact solution of the equations of motion. This problem has been considered by several authors, including Ballabh (1940), Strang (1948), Ergun (1949), Bhatnagar and Verma (1957), Remorini (1983a, 1983b).

We first observe that, in general, solutions of the equations of motion are not superposable due to the non-linearity of the momentum equation. To ascertain when, on the contrary, this is possible, let us assume that V_1 and V_2 are the solenoidal velocity fields corresponding to two known exact solutions (V_1, p_1) and (V_2, p_2) of the momentum equation in the presence of a conservative body force $f = -\text{grad}\,\Psi$, and let us denote as ω_1 and ω_2 the relevant corresponding vorticities. Then, the two solutions are superposable if a pressure field p may be determined such that $(V = V_1 + V_2, p)$ is a solution of the momentum equation. To derive the condition that must be satisfied for this to be possible, we write Eq. (9.3) for the three velocity fields V_1, V_2 and V, with the related pressure fields,

$$\frac{\partial V_1}{\partial t} + \omega_1 \times V_1 - \nu\nabla^2 V_1 = -\text{grad}\left(\frac{p_1}{\rho} + \frac{V_1^2}{2} + \Psi\right), \qquad (9.22)$$

$$\frac{\partial V_2}{\partial t} + \omega_2 \times V_2 - \nu\nabla^2 V_2 = -\text{grad}\left(\frac{p_2}{\rho} + \frac{V_2^2}{2} + \Psi\right), \qquad (9.23)$$

$$\frac{\partial(V_1 + V_2)}{\partial t} + (\omega_1 + \omega_2) \times (V_1 + V_2) - \nu\nabla^2(V_1 + V_2)$$

$$= -\text{grad}\left(\frac{p}{\rho} + \frac{V_1^2 + V_2^2}{2} + V_1 \cdot V_2 + \Psi\right). \qquad (9.24)$$

If we now subtract Eqs. (9.22) and (9.23) from Eq. (9.24), we get

$$\omega_1 \times V_2 + \omega_2 \times V_1 = -\text{grad}\left(\frac{p}{\rho} - \frac{p_1}{\rho} - \frac{p_2}{\rho} + V_1 \cdot V_2 - \Psi\right), \qquad (9.25)$$

from which we deduce that the necessary and sufficient condition for the two flows (V_1, p_1) and (V_2, p_2) to be superposable is that

$$\text{curl}(\omega_1 \times V_2 + \omega_2 \times V_1) = 0. \qquad (9.26)$$

In effect, if relation (9.26) is satisfied, then a scalar function χ exists such that we can put

$$\boldsymbol{\omega}_1 \times \boldsymbol{V}_2 + \boldsymbol{\omega}_2 \times \boldsymbol{V}_1 = \operatorname{grad} \chi, \qquad (9.27)$$

and thus the pressure p corresponding to the velocity field $\boldsymbol{V} = \boldsymbol{V}_1 + \boldsymbol{V}_2$ may be readily obtained from relation (9.25), which also shows that, in general, $p \neq p_1 + p_2$. Condition (9.26) applies for both steady and unsteady flows, and an immediate (and well-known) result following from it is that two solenoidal irrotational flows are always superposable; however, even in that case, $p = p_1 + p_2$ only if $\boldsymbol{V}_1 \cdot \boldsymbol{V}_2 = \Psi + constant$.

Thus, superposable solutions are a subset of the exact solutions of the incompressible flow equations, and are definitely attractive from the point of view of the construction of new solutions that may be used for the description, either exactly or in an approximate manner, of flows of practical significance. For instance, two superposable solutions that do not comply with realistic boundary conditions separately, might give rise, when added, to a flow characterized by certain streamlines and velocities representing the motion of a solid surface or the velocity field at the edge of a boundary layer.

We shall not go through a detailed inventory of all the available steady and unsteady superposable solutions, for which the reader is referred to the existing literature; in this respect, an extensive bibliography is provided by Remorini (1983b), and several superposable solutions are also quoted by Wang (1989, 1991). Here we only list a few of these solutions, extracted from the above-mentioned references, and stress again the interest, from both the conceptual and the practical points of view, of the *superposability condition* (9.26).

We start by considering the case in which $\boldsymbol{V}_1$ and $\boldsymbol{V}_2$ are the velocity vectors of *two steady two-dimensional motions parallel to the same plane*, say the x–y plane. The corresponding vorticity vectors are then

$$\boldsymbol{\omega}_1 = \omega_1 \boldsymbol{e}_z, \quad \boldsymbol{\omega}_2 = \omega_2 \boldsymbol{e}_z, \qquad (9.28)$$

from which we derive that

$$\boldsymbol{\omega}_1 \cdot \operatorname{grad} \boldsymbol{V}_2 = \boldsymbol{\omega}_2 \cdot \operatorname{grad} \boldsymbol{V}_1 = 0. \qquad (9.29)$$

We may now use vector identity (6.6) to express Eq. (9.26), so that, considering relations (9.29) and the fact that both the velocity and the vorticity vectors are solenoidal, the superposability condition becomes

$$\boldsymbol{V}_2 \cdot \operatorname{grad} \omega_1 + \boldsymbol{V}_1 \cdot \operatorname{grad} \omega_2 = 0. \qquad (9.30)$$

From this relation we immediately deduce that any two constant-vorticity plane motions are superposable, and their addition is obviously a constant-vorticity motion. Furthermore, we also find that a steady constant-vorticity motion may be obtained from the addition of an irrotational flow and a constant-vorticity unidirectional plane flow (Couette flow) or of an irrotational flow and a constant-vorticity flow with concentric circular streamlines.

Other interesting results that may be obtained for plane flows using condition (9.30) are listed below.

- Let V_1 and V_2 be two unidirectional motions and α the included angle between their directions. Then if $\alpha = 0$ the two motions are always superposable; conversely, if $\alpha \neq 0$ the two motions are superposable if and only if they are both constant-vorticity flows and, in this case, the motion $V = V_1 + V_2$ is still a constant-vorticity flow, but not necessarily a unidirectional motion, unless both motions are uniform.

- A unidirectional motion V_1 and a motion with concentric circular streamlines V_2 are superposable if and only if they are both constant-vorticity motions.

- Two motions V_1 and V_2 with concentric circular streamlines and respective centres in points C_1 and C_2 are always superposable if the two centres coincide. If this is not true, they are superposable if and only if they are constant-vorticity motions.

Let us now assume that V_1 is a *steady parallel axisymmetric flow* which, as previously shown, may be expressed in the form

$$V_1 = v_{z1} e_z, \qquad (9.31)$$

where

$$v_{z1} = Ar^2 + B \ln r + C. \qquad (9.32)$$

The vorticity vector ω_1 is then

$$\omega_1 = \omega_{\theta 1} e_\theta = -\left(2Ar + \frac{B}{r}\right) e_\theta. \qquad (9.33)$$

The superposability condition (9.26) may then be written as

$$\operatorname{curl}(\omega_{\theta 1} e_\theta \times V_2 + \omega_2 \times v_{z1} e_z) = 0. \qquad (9.34)$$

From this relation we find that a generic steady parallel flow (9.31) is superposable to a velocity field V_2 that is parallel to the plane orthogonal to the z axis and is such that

$$V_2 = g(r)e_\theta. \tag{9.35}$$

Therefore, V_2 represents a plane motion with concentric streamlines. Further results and more details on the derivation of superposable solutions are given by Remorini (1983a, 1983b).

9.3. Some Examples of Steady Exact Solutions

9.3.1. *Plane Couette and Poiseuille flows*

Let us consider the steady flow between two parallel infinite flat plates in relative motion, placed a distance d apart. We take a reference system with the y axis perpendicular to the plates and origin on the lower plate, which is assumed to be at rest, whereas the upper plate moves in the x direction with a constant velocity U (Fig. 9.1).

As the plates are infinite and the motion is in the x direction, there are neither velocity components nor velocity variations in the z direction. The motion is then two-dimensional in the x–y plane, and the velocity components in the x and y directions are denoted as u and v respectively. The boundary conditions for the velocity field are

$$u(x,0) = v(x,0) = v(x,d) = 0, \quad u(x,d) = U. \tag{9.36}$$

The boundary conditions do not vary in the x direction and the plates are infinite, so that the motion may be assumed not to depend on x; in particular, $\partial u/\partial x = 0$. Therefore, considering the boundary conditions for the velocity component v and the two-dimensional form of the continuity equation, one easily finds that $v = 0$ in the whole flow, which is thus a unidirectional flow. If we now assume that no pressure variation is present and that we may neglect the presence of a component of the body force in

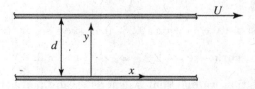

Fig. 9.1.   Coordinate system for plane Couette flow.

the x direction, the momentum equation in the x direction gives

$$\frac{\partial^2 u}{\partial y^2} = 0, \tag{9.37}$$

and thus, by denoting as a and b two scalar constants,

$$u = ay + b. \tag{9.38}$$

By imposing the boundary conditions (9.36), we obtain the velocity field corresponding to this simple problem, which is known as *plane Couette flow*,

$$u = \frac{U}{d} y. \tag{9.39}$$

It is interesting to observe that the velocity field does not depend on the viscosity coefficient. However, as may easily be imagined, the origin of this flow is completely due to viscosity and to the necessity of satisfying the no-slip condition at the two plates. In fact, the following constant tangential stress acts on each fluid layer and on the lower plate:

$$\tau_{yx} = \mu \frac{\partial u}{\partial y} = \mu \frac{U}{d}. \tag{9.40}$$

Consequently, an opposite tangential stress acts on the moving upper plate. We finally note that a constant vorticity vector $\boldsymbol{\omega} = \omega_z \boldsymbol{e}_z$ is present in the whole flow, with

$$\omega_z = -\frac{\partial u}{\partial y} = -\frac{U}{d}. \tag{9.41}$$

The resulting velocity and vorticity fields are shown in Fig. 9.2.

Let us now consider the so-called *plane Poiseuille flow*, in which the two plates are at rest and a pressure gradient $\partial p / \partial x$ is present. It is easy to see that the pressure gradient must be negative for the motion to be in the positive x direction (i.e. rightwards), and thus we put, for convenience, $G = -\partial p / \partial x$.

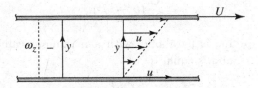

Fig. 9.2.   Velocity and vorticity in plane Couette flow.

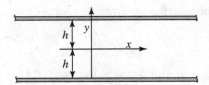

Fig. 9.3.   Coordinate system for plane Poiseuille flow.

We now denote as $2h$ the distance between the plates, and use the new coordinate system shown in Fig. 9.3, in which $y = 0$ along the centreline. The boundary conditions for the flow then become

$$u(x, h) = u(x, -h) = v(x, h) = v(x, -h) = 0. \qquad (9.42)$$

Considering also the continuity equation, the flow is again found to be unidirectional and thus, neglecting the presence of body forces, the equation for the velocity u becomes

$$\frac{\partial^2 u}{\partial y^2} = -\frac{G}{\mu}. \qquad (9.43)$$

Note that, in agreement with our findings in Section 9.2.1 for unidirectional flows, the pressure gradient $-G$ must be a constant. This means that, given a certain pressure difference between two cross-sections at different values of x, the pressure always varies linearly between the two cross-sections.

The general solution of (9.43) is then

$$u = -\frac{G}{2\mu}y^2 + By + C, \qquad (9.44)$$

where B and C are scalar constants to be determined from the boundary conditions (9.42). Then, we can easily find that the velocity field of the plane Poiseuille flow, in the reference system of Fig. 9.3, is given by

$$u = \frac{G}{2\mu}(h^2 - y^2). \qquad (9.45)$$

The velocity profile is parabolic, with a maximum value that occurs along the centreline, and is equal to

$$U_m = \frac{Gh^2}{2\mu}, \qquad (9.46)$$

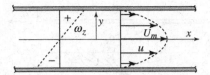

Fig. 9.4. Velocity and vorticity for plane Poiseuille flow.

so that relation (9.45) may also be recast in the form

$$u = U_m \left(1 - \frac{y^2}{h^2} \right).$$ (9.47)

The vorticity is found to be

$$\omega_z = -\frac{\partial u}{\partial y} = \frac{G}{\mu} y,$$ (9.48)

and thus varies linearly across the channel, with positive and negative values above and below the centreline respectively. The velocity and vorticity variations between the plates are depicted in Fig. 9.4.

Note that, in this case, both velocity and vorticity are proportional to the opposite of the pressure gradient and inversely proportional to the viscosity coefficient. It is also easily seen that the tangential stresses acting on the two plate surfaces are equal, and that their common value is given by the relation

$$\tau_w = Gh = 2\mu \frac{U_m}{h}.$$ (9.49)

A quantity that may be of interest is the volume flow rate Q through a cross-section of the channel, per unit length in the z direction. This is readily evaluated from the expression

$$Q = \int_{-h}^{h} u\,dy = \frac{2}{3} \frac{Gh^3}{\mu} = \frac{4}{3} U_m h,$$ (9.50)

from which we derive that the average velocity through the channel is

$$U_a = \frac{Q}{2h} = \frac{2}{3} U_m.$$ (9.51)

We observe now that the velocity fields given by relations (9.38) and (9.44) correspond to two unidirectional flows with common direction, and thus, from one of the results described in Section 9.2.2, they are superposable. Therefore, as the boundary conditions of the plane Couette

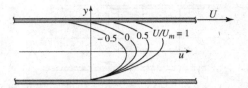

Fig. 9.5.   Velocity profiles for Couette–Poiseuille flow and different values of U/U_m.

and Poiseuille flows are imposed on the same points, i.e. on the surfaces of the two plates, the two problems are indeed superposable, and the solution obtained by adding the corresponding velocity fields is the solution of the composite problem, which is called *Couette–Poiseuille flow*. Actually, this is readily seen by imposing on the velocity expression (9.44), which is the general solution of equation (9.43) for a unidirectional flow in presence of a pressure gradient, the boundary conditions corresponding to the addition of the two problems. If we use the reference system of Fig. 9.3, the new boundary conditions are

$$u(x, -h) = v(x, h) = v(x, -h) = 0, \quad u(x, h) = U. \tag{9.52}$$

The solution of this problem may be easily shown to correspond to the following velocity field:

$$u = \frac{G}{2\mu}(h^2 - y^2) + \frac{U}{2}\left(1 + \frac{y}{h}\right) = U_m\left(1 - \frac{y^2}{h^2}\right) + \frac{U}{2}\left(1 + \frac{y}{h}\right), \tag{9.53}$$

which is indeed the addition of the solutions of the Poiseuille flow and of the Couette flow in the new reference system. The resulting velocity profiles are reported in Fig. 9.5 for different values of the ratio U/U_m.

9.3.2. *Hagen–Poiseuille flow*

We now consider the flow inside an infinite-length pipe with circular cross-section, radius equal to a and axis in the z direction. This flow is known as *Hagen–Poiseuille flow* and, for its analysis, it is obviously convenient to use cylindrical coordinates (r, θ, z).

We now assume that neither swirl nor circumferential variation are present, so that $v_\theta = \partial/\partial\theta = 0$. The boundary conditions for the other two velocity components are that both v_r and v_z must be zero at $r = a$, for any value of z. The continuity equation (9.13) may then be written as follows:

$$\frac{1}{r}\frac{\partial(rv_r)}{\partial r} + \frac{\partial v_z}{\partial z} = 0. \tag{9.54}$$

However, the problem is such that, again, we may assume the motion not to depend on z, and thus the solution of Eq. (9.54) is

$$(rv_r) = constant, \tag{9.55}$$

which, considering the boundary condition for v_r, implies that $v_r = 0$ everywhere. Therefore, the motion is unidirectional and $v_z = v_z(r)$ is the only non-zero velocity component.

We consider the case in which the motion is driven by a pressure gradient along the pipe axis, and the body forces may be neglected. Once more, the pressure may vary only in the z direction, and the quantity $G = -\partial p/\partial z$ is found to be a constant.

From the momentum balance in the z direction we derive the equation that corresponds to Eq. (9.11) for this flow; by using the expressions of the various operators in cylindrical coordinates (see Appendix B), we get

$$\frac{\partial^2 v_z}{\partial r^2} + \frac{1}{r}\frac{\partial v_z}{\partial r} = -\frac{G}{\mu}. \tag{9.56}$$

The solution of this equation is

$$v_z = -\frac{G}{4\mu}r^2 + B\ln r + C, \tag{9.57}$$

which agrees with the general form (9.21) that has been introduced in Section 9.2.1. We now observe that the velocity must be finite at $r = 0$, which implies that $B = 0$. By imposing the boundary condition at the pipe inner surface, we finally obtain the solution of this problem:

$$v_z = \frac{G}{4\mu}(a^2 - r^2). \tag{9.58}$$

We thus see that the velocity variation is described by a paraboloid of revolution, with a maximum velocity on the pipe axis, equal to

$$U_m = \frac{Ga^2}{4\mu}. \tag{9.59}$$

The flow rate through the pipe cross-section is readily obtained by integration:

$$Q = \int_0^a 2\pi r v_z dr = \frac{\pi G a^4}{8\mu}, \tag{9.60}$$

and thus the average velocity is

$$U_a = \frac{Q}{\pi a^2} = \frac{G a^2}{8\mu} = \frac{U_m}{2}. \tag{9.61}$$

The viscous stress acting on a generic point of the pipe surface, which is the opposite of the stress acting on the fluid, is

$$\tau_w = -\left(\mu \frac{\partial v_z}{\partial r}\right)_{r=a} = \frac{G a}{2}. \tag{9.62}$$

Considering that the pressure gradient is constant, the total frictional force acting on a length L of the pipe at whose upstream and downstream extremities the pressures are p_1 and p_2 respectively, is then

$$F_z = 2\pi a L \tau_w = \pi a^2 G L = \pi a^2 (p_1 - p_2). \tag{9.63}$$

This expression should not come as a surprise, since the fluid within the considered portion of tube is in steady motion, and the pressure is constant in each cross-section. Therefore, the global pressure force on that volume of fluid, which is given by the pressure difference between the two end faces times the area of the cross-section, is balanced by the total frictional force exerted by the pipe walls on the fluid.

We finally observe that the vorticity vector is easily seen to be equal to $\boldsymbol{\omega} = \omega_\theta \boldsymbol{e}_\theta$, with

$$\omega_\theta = -\frac{\partial v_z}{\partial r} = \frac{G r}{2\mu}, \tag{9.64}$$

which shows that the sign of the vorticity coincides with that of G and that the field lines of the vorticity field, denoted as *vortex lines*, are directed as $\boldsymbol{e}_\theta$, i.e. they form circular loops concentric to the z axis.

9.3.3. *Flow in pipes of various cross-sections*

Besides the Hagen–Poiseuille flow, other exact solutions of the Navier–Stokes equations may be found for unidirectional flows in pipes with different cross-sections. As already found in Section 9.2.1, for such flows the pressure is uniform on any cross-section and the pressure gradient along the pipe is a constant.

We now choose the pipe walls to be parallel to the z direction, and denote as w the velocity component in the same direction; furthermore, we neglect the body force and still use the notation $G = -\partial p/\partial z$. The velocity field is then the solution of the resulting form of Eq. (9.11):

$$\frac{\partial^2 w}{\partial x^2} + \frac{\partial^2 w}{\partial y^2} = -\frac{G}{\mu}, \tag{9.65}$$

with the boundary condition $w(x, y) = 0$ on the contour C of the pipe cross-section.

The above mathematical problem is analogous to the one that must be solved to obtain the stress field in an elastic cylinder subjected to torsion, and solutions are available for several cross-section shapes. A significant result that derives from theoretical considerations is that if $G > 0$, then $w(x, y)$ is a *superharmonic function*, and its minimum is at the boundary of the domain. Consequently, considering the boundary condition for w, in that case we have $w > 0$, which means that inside the pipe the velocity is in the positive z direction everywhere. Obviously, the opposite occurs if $G < 0$, as could be expected. An extensive bibliography regarding the solutions of this problem and a detailed description of some of them are reported by Berker (1963); here, we consider only some of the most interesting cases.

9.3.3.1. *Elliptic cross-section*

Let the pipe cross-section be described by the curve

$$\frac{x^2}{a^2} + \frac{y^2}{b^2} = 1. \tag{9.66}$$

The velocity distribution inside the pipe is then found to be

$$w(x, y) = \frac{G}{2\mu} \left(\frac{a^2 b^2}{a^2 + b^2} \right) \left(1 - \frac{x^2}{a^2} - \frac{y^2}{b^2} \right). \tag{9.67}$$

By putting $b = a$ in this expression, it is easy to check that it reduces to (9.58) for a circular cross-section. The flow rate through the pipe cross-section turns out to be

$$Q = \frac{\pi G}{4\mu} \left(\frac{a^3 b^3}{a^2 + b^2} \right), \tag{9.68}$$

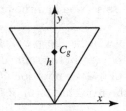

Fig. 9.6.   Reference system for pipe with equilateral triangular cross-section.

and the average velocity is

$$U_a = \frac{Q}{\pi ab} = \frac{G}{4\mu}\left(\frac{a^2 b^2}{a^2 + b^2}\right) = \frac{U_m}{2}, \tag{9.69}$$

where U_m is the maximum velocity, which occurs at the centre of the ellipse, i.e. for $x = y = 0$.

9.3.3.2. *Equilateral triangular cross-section*

Considering the reference system of Fig. 9.6, let the pipe cross-section contour be limited by the straight lines $y = h$ and $y = \pm x\sqrt{3}$. The velocity, the flow rate and the average velocity are then given by the following expressions:

$$w(x, y) = \frac{G}{4\mu h}(y - h)(3x^2 - y^2), \tag{9.70}$$

$$Q = \frac{\sqrt{3}}{180}\frac{Gh^4}{\mu}, \tag{9.71}$$

$$U_a = \frac{\sqrt{3}Q}{h^2} = \frac{Gh^2}{60\mu} = \frac{9}{20}U_m, \tag{9.72}$$

where U_m is the maximum velocity, which occurs at the centroid of the triangle, i.e. for $x = 0$, $y = (2/3)h$.

9.3.3.3. *Rectangular cross-section*

We assume the contour of the cross-section to be defined by the straight lines $x = \pm a$, $y = \pm b$, with $a \geq b$. In this case there is no closed-form solution of Eq. (9.65), but the velocity field may be expressed in terms of a converging infinite series.

By introducing, for convenience, the quantity

$$\beta_n = \frac{(2n+1)\pi}{2b}, \tag{9.73}$$

the velocity may be written as follows:

$$w(x,y) = \frac{G}{2\mu} \left[b^2 - y^2 - \frac{32b^2}{\pi^3} \sum_{n=0}^{\infty} \frac{(-1)^n}{(2n+1)^3} \frac{\cosh(\beta_n x)}{\cosh(\beta_n a)} \cos(\beta_n y) \right]. \tag{9.74}$$

The volume flow rate is then found to be

$$Q = \frac{4}{3} \frac{Gab^3}{\mu} \left[1 - \frac{192}{\pi^5} \frac{b}{a} \sum_{n=0}^{\infty} \frac{\tanh(\beta_n a)}{(2n+1)^5} \right], \tag{9.75}$$

and thus the expression of the average velocity is

$$U_a = \frac{Q}{4ab} = \frac{Gb^2}{3\mu} \left[1 - \frac{192}{\pi^5} \frac{b}{a} \sum_{n=0}^{\infty} \frac{\tanh(\beta_n a)}{(2n+1)^5} \right]. \tag{9.76}$$

Again, the maximum velocity is found at the centre of the rectangular section, and may thus be obtained by putting $x = y = 0$ in relation (9.74). An approximate formula for the flow rate, giving a relative error smaller than 1.6%, is the following (Berker, 1963):

$$Q = \frac{4}{3} \frac{Gab^3}{\mu} \left[1 - \frac{192}{\pi^5} \frac{b}{a} \tanh\left(\frac{\pi a}{2b} \right) \right]. \tag{9.77}$$

9.3.4. *Rotating circular cylinders*

We consider now the steady flow between two concentric infinite circular cylinders of radius r_0 and r_1, with $r_1 > r_0$, which rotate with constant angular velocities Ω_0 and Ω_1 respectively. We assume that the common axis of the cylinders coincides with the z axis of a cylindrical reference system (r, θ, z), and seek a solution with concentric streamlines around that axis.

As already found in Section 9.2.1, the general solution for the only non-zero velocity component $v_\theta = v_\theta(r)$ is given, with the above assumptions, by expression (9.18). The scalar constants A and B appearing in that relation can be obtained by imposing the boundary conditions of the problem which, in this case, are

$$v_\theta(r_0) = \Omega_0 r_0, \quad v_\theta(r_1) = \Omega_1 r_1. \tag{9.78}$$

By introducing these conditions into (9.18), we readily find

$$A = \frac{(\Omega_0 - \Omega_1)r_0^2 r_1^2}{(r_1^2 - r_0^2)}, \tag{9.79}$$

$$B = \frac{\Omega_1 r_1^2 - \Omega_0 r_0^2}{r_1^2 - r_0^2}. \tag{9.80}$$

Therefore, the expression for the velocity between the two cylinders becomes

$$v_\theta = \frac{(\Omega_0 - \Omega_1)r_0^2 r_1^2}{r(r_1^2 - r_0^2)} + \left(\frac{\Omega_1 r_1^2 - \Omega_0 r_0^2}{r_1^2 - r_0^2}\right)r. \tag{9.81}$$

It is now interesting to note that, using the same line of reasoning that was introduced in Chapter 6, this result could also have been obtained by considering the evolution of the vorticity from a condition of impulsive rotation from rest of the two cylinders, and allowing a sufficient time to elapse for a steady condition to be attained. In effect, if we assume the two rotations to be counter-clockwise, at the start of the motion of the two cylinders a total quantity of vorticity $\Gamma_1 = 2\Omega_1 \pi r_1^2$ would be generated over the inner surface of the outer cylinder, whereas the total vorticity generated over the outer surface of the inner cylinder would be $\Gamma_0 = -2\Omega_0 \pi r_0^2$. These vorticities would then diffuse within the available space between the cylinders, and partially annihilate, as they are of opposite sign. At the end of the transient, the attained steady condition would then be such that the remaining vorticity in the annulus between the cylinders would be constant, and its local value would be equal to the algebraic sum of the vorticities produced over the two surfaces divided by the area of the annulus, which is $\pi(r_1^2 - r_0^2)$. In other words, in the final steady condition, the vorticity component in the z direction — the only non-zero one in this flow — would be, in each point,

$$\omega_z = 2\frac{\Omega_1 r_1^2 - \Omega_0 r_0^2}{r_1^2 - r_0^2}. \tag{9.82}$$

The velocity field may then be found by considering that it must be composed of a rotational flow with the above value of the vorticity plus, at most, an irrotational flow in order to satisfy the appropriate boundary conditions. As a matter of fact, the equation to be solved is now the one corresponding to the definition of ω_z, which, considering that $v_r = 0$, is

$$\frac{\partial v_\theta}{\partial r} + \frac{v_\theta}{r} = \omega_z, \tag{9.83}$$

and the general solution of this equation, for constant value of ω_z, is readily seen to be

$$v_\theta = \frac{A}{r} + \frac{\omega_z}{2}r. \tag{9.84}$$

It is now easy to check that A may be determined by imposing any of the boundary conditions at the surfaces of the two cylinders, and that the obtained value is the same as that given by relation (9.79). Therefore, by introducing this value into (9.84), together with expression (9.82) for ω_z, the velocity field given by (9.81) is recovered.

From the above solution, the velocity distributions of specific steady flows may be easily obtained. In particular, if we put $r_0 = 0$, and Ω_0 is finite, we get the velocity inside a rotating hollow cylinder:

$$v_\theta = \Omega_1 r, \tag{9.85}$$

which is a solid-body rotation, with the same angular velocity as the cylinder with radius r_1 and with a local value of the vorticity $\omega_z = 2\Omega_1$.

Conversely, we may recover the flow field outside a single rotating cylinder by letting r_1 tend to infinity and putting $\Omega_1 = 0$. The result is

$$v_\theta = \frac{\Omega_0 r_0^2}{r}, \tag{9.86}$$

which is the irrotational flow satisfying the no-slip condition at the cylinder surface that has already been derived in Chapter 6 from the analysis of the generation and subsequent dynamics of vorticity.

For the present problem, the pressure distribution in the flow may be readily obtained from the equation of motion, as a function of the value of the pressure acting on a boundary. In effect, from the projection of the momentum equation in the r direction, expressed in cylindrical coordinates (see Appendix B), it may be easily verified that

$$\frac{\partial p}{\partial r} = \rho\frac{v_\theta^2}{r}, \tag{9.87}$$

which shows that the radial variation of pressure only balances the centrifugal force acting on any elementary volume of fluid. Note that this is understandable, because solution (9.18), corresponding to a steady motion with circular concentric streamlines, is the superposition of an irrotational flow and a flow with constant vorticity $\omega_z = 2B$, and thus the viscous term in the momentum equation, $-\nu \text{curl}\,\omega$, is identically zero everywhere. Obviously, as repeatedly pointed out in previous chapters, this does not mean that the viscous stresses are zero, but that their resultant over the surface of any elementary volume is zero.

Let us analyse, then, the pressure field for the case of a hollow cylinder with negligible thickness and radius r_0, which rotates with angular velocity Ω_0 in an infinite fluid domain. The velocity field outside the cylinder is given by relation (9.86) and thus, if we denote as p_∞ the pressure at infinity, the integration of (9.87) for $r > r_0$ gives

$$p(r) = p_\infty - \int_r^\infty \frac{\rho \Omega_0^2 r_0^4}{r^3}\,dr = p_\infty - \frac{\rho \Omega_0^2 r_0^4}{2r^2} = p_\infty - \frac{\rho v_\theta^2}{2}, \qquad (9.88)$$

as could be expected, considering that the flow is irrotational and thus Bernoulli's theorem applies. In particular, the pressure on the outer surface of the cylinder is

$$p(r_0) = p_{r_0}^+ = p_\infty - \frac{\rho}{2}\Omega_0^2 r_0^2. \qquad (9.89)$$

As for the velocity inside the cylinder, its expression is $v_\theta = \Omega_0 r$, and thus we easily obtain

$$p(r) = p(r_0) - \frac{\rho}{2}\Omega_0^2(r_0^2 - r^2) = p_{r_0}^- - \frac{\rho}{2}\Omega_0^2(r_0^2 - r^2). \qquad (9.90)$$

In the particular case in which the pressures on the two faces of the cylinder surface are the same, i.e. when $p_{r_0}^+ = p_{r_0}^- = p_{r_0}$, then the pressure for $r < r_0$ becomes

$$p(r) = p_\infty - \frac{\rho}{2}\Omega_0^2(2r_0^2 - r^2), \qquad (9.91)$$

and, in particular, the pressure at the centre of the cylinder is

$$p_0 = p(r = 0) = p_\infty - \rho \Omega_0^2 r_0^2. \qquad (9.92)$$

We note now that, in this particular case in which the pressure is continuous on the two sides of the cylinder surface, both the velocity and pressure fields are those that would correspond to a Rankine vortex whose core radius is r_0, which we have seen in Chapter 3 to be a model for a real vortex. As can be appreciated from the above formulas and from Fig. 9.7, the pressure drop from $r = \infty$ to $r = r_0$ is equal to that from $r = r_0$

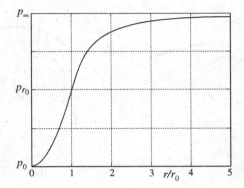

Fig. 9.7.   Pressure distribution for a Rankine vortex.

to $r = 0$, and may be significant if the angular velocity of the core is high. However, it is more interesting to analyse the pressure drop as a function of the core radius of a Rankine vortex always containing the same global amount of vorticity $\Gamma_0 = 2\Omega_0\pi r_0^2$. In effect, from relation (9.92) we find

$$p_0 = p_\infty - \frac{\rho\Gamma_0^2}{4\pi^2 r_0^2}, \tag{9.93}$$

which shows that the smaller the core radius, the higher the pressure drop.

Furthermore, the maximum velocity at the core edge also increases with decreasing r_0 because its expression can be written as

$$v_\theta(r_0) = \Omega_0 r_0 = \frac{\Gamma_0}{2\pi r_0}. \tag{9.94}$$

Therefore, the more a certain quantity of vorticity is concentrated in a small region, the higher will be its effect upon the velocity and pressure fields, a result that is also true for real vortices. As we saw in Chapter 6, the cross-section of vortex tubes may be reduced by the stretching action of a velocity gradient component existing in the direction of the vorticity, as may occur, for instance, in hurricanes or tornados.

Finally, it is interesting to consider the fluid particles that, at a certain time, are positioned outside the core along a radius $\theta = 0$, and to visualize their positions after the edge of the core has rotated by a given angle θ_0. It may be readily seen that, in the corresponding time interval, the particles at a generic $r > r_0$ rotate by an angle $\theta = \theta_0(r_0/r)^2$, and the curves representing the locus of their positions for rotations $\theta_0 = 180°$ and

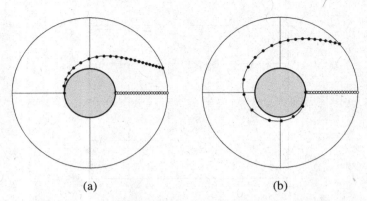

Fig. 9.8. Position of fluid particles initially positioned at $\theta = 0$ after a rotation of the core of 180° (a) or 360° (b). White dots: initial position of the particles.

$\theta_0 = 360°$ are shown in Fig. 9.8. As can be appreciated, the particles are placed along spiralling curves, and one may observe that this is a common situation not only for clouds around the core of a storm, but also for any rotating motion such that the moment of momentum remains constant, like the one given by (9.86).

9.4. Some Examples of Unsteady Exact Solutions

9.4.1. *Parallel motion of an infinite plate*

Considering now a two-dimensional unsteady flow in the x–y plane, we assume that no body forces are present and that the motion is unidirectional in the x direction, so that the velocity components are given by

$$u = u(x, y, t), \quad v = 0.$$

Application of the continuity equation shows that u does not depend on x, whereas the compatibility equation (9.6) becomes

$$\frac{\partial^2 u}{\partial y \partial t} - \nu \frac{\partial^3 u}{\partial y^3} = 0, \tag{9.95}$$

which may be integrated once with respect to y to give

$$\frac{\partial u}{\partial t} - \nu \frac{\partial^2 u}{\partial y^2} = f(t), \tag{9.96}$$

where $f(t)$ is an arbitrary function of time.

By comparison with the x component of the momentum equation, we find that

$$f(t) = -\frac{1}{\rho}\frac{\partial p}{\partial x}, \tag{9.97}$$

which shows that $\partial p/\partial x$ is, at most, a function of time, and that the general solution for the pressure is

$$p = -\rho[f(t)x + g(t)], \tag{9.98}$$

where $g(t)$ is another arbitrary function of time.

The compatibility equation, representing the dynamics of vorticity, may also be rewritten in terms of ω_z, which is the only non-zero component of vorticity,

$$\frac{\partial \omega_z}{\partial t} - \nu\frac{\partial^2 \omega_z}{\partial y^2} = 0. \tag{9.99}$$

Both Eq. (9.96) and Eq. (9.99) are diffusion equations, and their solutions may be obtained once appropriate initial and boundary conditions are specified.

In Chapter 6, we have already considered the particular case of an infinite plate, placed at $y = 0$, impulsively set in motion parallel to its surface, with the fluid initially at rest in the region $y \geq 0$, where the pressure is constant. If we take the impulsive motion to be in the positive x direction with velocity U (opposite to what was assumed in Section 6.4.1), we have $f(t) = 0$ in Eq. (9.96) and the initial and boundary conditions are

$$u(y,0) = 0 \quad \text{for } y > 0, \quad u(0,t) = U \quad \text{for } t > 0. \tag{9.100}$$

The solution is then

$$u = U\left[1 - \text{erf}\left(\frac{y}{2\sqrt{\nu t}}\right)\right], \tag{9.101}$$

where the error function $\text{erf}(\eta)$ is defined by relation (6.19).

By recalling that

$$\frac{\partial}{\partial \eta}\text{erf}(\eta) = \frac{2}{\sqrt{\pi}}e^{-\eta^2}, \tag{9.102}$$

we immediately find that the friction stress at the plate surface is

$$\tau_w = \mu\left(\frac{\partial u}{\partial y}\right)_{y=0} = -\rho U\sqrt{\frac{\nu}{\pi t}}, \tag{9.103}$$

and we note that it is infinite at the initial time and then progressively decreases, tending to zero when t goes to infinity.

We now consider the case in which the plate is oscillating with a velocity $U\sin(\omega t)$, and seek a solution for a time in which any transient flow due to the start of the motion has died out. It may be seen that this solution exists and corresponds to a periodic motion described by the expression

$$u = Ue^{-\alpha y}\sin(\omega t - \alpha y), \qquad (9.104)$$

where we have put

$$\alpha = \sqrt{\frac{\omega}{2\nu}}. \qquad (9.105)$$

Relation (9.104) shows that the motion of the wall is propagated into the fluid as a damped transverse wave, with a *wavelength*

$$\lambda = \frac{2\pi}{\alpha} = 2\pi\sqrt{\frac{2\nu}{\omega}}, \qquad (9.106)$$

and with a *phase velocity* (i.e. the velocity at which the phase of the wave travels in space) given by

$$v_{ph} = \frac{\lambda\omega}{2\pi} = \sqrt{2\nu\omega}. \qquad (9.107)$$

It is interesting to note, from (9.104), that viscosity has a strong effect on the damping of the oscillation amplitude with increasing distance from the plane. In effect, at a distance from the wall of one wavelength, the ratio between the local amplitude of the velocity oscillation and that of the plane is $e^{-2\pi} \simeq 0.002$. Therefore, the motion in the fluid is effectively confined to a very thin region adjacent to the plate surface, of the order of $(\nu/\omega)^{1/2}$, which is usually called the *Stokes layer*.

The viscous stress acting on the plate surface is readily found to be

$$\tau_w = \mu\left(\frac{\partial u}{\partial y}\right)_{y=0} = -U\sqrt{\rho\mu\omega}\sin\left(\omega t + \frac{\pi}{4}\right). \qquad (9.108)$$

The stress is thus not in phase with the plate velocity, and it opposes the motion for $3/4$ of the period $T = 2\pi/\omega$. The work done on the fluid in one period by a portion of plate of length dx and unit length in the z direction may thus be evaluated, and is

$$W = dx\int_0^T -\tau_w u\,dt = \pi\sqrt{\frac{\nu}{2\omega}}\rho U^2 dx. \qquad (9.109)$$

9.4.2. *Unsteady plane Couette and Poiseuille flows*

Let us reconsider the situation and reference system described by Fig. 9.1, with two parallel infinite plates in relative motion, and let us analyse the transient leading to the steady Couette flow (9.39), assuming that the upper plate moves impulsively with velocity U, while the lower one remains at rest. The problem is then given by the solution of Eq. (9.96) with $f(t) = 0$ and the following initial and boundary conditions:

$$u(y,0) = 0 \quad \text{for } 0 \leq y < d,$$
$$u(0,t) = 0, \quad u(d,t) = U \quad \text{for } t > 0. \tag{9.110}$$

Using the solution for the final steady condition, we may express the time-varying solution in the form

$$u(y,t) = \frac{U}{d}y - f(y,t), \tag{9.111}$$

where $f(y,t)$ is a function to be determined, which satisfies the equation

$$\frac{\partial f}{\partial t} - \nu \frac{\partial^2 f}{\partial y^2} = 0, \tag{9.112}$$

with the initial and boundary conditions

$$f(y,0) = \frac{U}{d}y, \quad \text{for } 0 \leq y < d,$$
$$f(0,t) = 0, \quad f(d,t) = 0 \quad \text{for } t > 0. \tag{9.113}$$

The homogeneous boundary conditions at $y = 0$ and $y = d$ suggest a solution in the following form:

$$f(y,t) = \sum_{n=1}^{\infty} A_n e^{-n^2 \pi^2 \nu t/d^2} \sin \frac{n\pi y}{d}, \tag{9.114}$$

where the constants A_n can be obtained from the initial condition. The final velocity distribution is then found to be (see Erdogan, 2002)

$$u(y,t) = \frac{U}{d}y + \frac{2U}{\pi} \sum_{n=1}^{\infty} \frac{(-1)^n}{n} e^{-n^2 \pi^2 \nu t/d^2} \sin \frac{n\pi y}{d}. \tag{9.115}$$

The evolution of the velocity distribution is shown in Fig. 9.9, from which it can be deduced that the steady-state condition is practically attained for $\nu t/d^2 = 1$; actually, it may be considered to be already reached, with an approximation of the order of 1%, for $\nu t/d^2 = 0.5$.

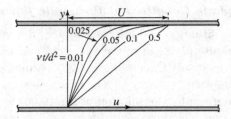

Fig. 9.9.   Velocity distribution of unsteady Couette flow for different values of $\nu t/d^2$.

The friction stress acting on the upper wall turns out to be

$$\tau_w = -\mu \left(\frac{\partial u}{\partial y} \right)_{y=d} = -\frac{\mu U}{d} \left(1 + 2 \sum_{n=1}^{\infty} e^{-n^2 \pi^2 \nu t/d^2} \right). \tag{9.116}$$

Therefore, when t goes to infinity the viscous stress tends to the value that corresponds to the steady state. Conversely, at the initial time, the stress is infinite due to the production of an infinitely thin layer of vorticity over the upper plate surface. In fact, the situation is similar to the case of the flat plate in impulsive motion, and this is understandable, because the effect on the flow of the presence of the lower plate is negligible for small values of $\nu t/d^2$.

The vorticity $\omega_z = -\partial u/\partial y$ can be readily obtained from the velocity distribution, and it may be interesting to note that its integral over the whole width of the channel is, at any time, always equal to $-U$, which is the value of the vorticity generated over each point of the upper plate surface at the start of the motion. Therefore, the final steady-state solution corresponds to the condition in which this vorticity has filled up the whole available space, attaining a constant local value $\omega_z = -U/d$.

Let us now consider the case of an unsteady plane Poiseuille flow, i.e. the motion of a fluid between two parallel plates at rest caused by the sudden application of a constant pressure gradient in direction x. For convenience, we use the reference system defined by Fig. 9.3, in which the distance between the plates is $2h$ and the origin of the y axis is on the centreline of the channel, and we put $G = -\partial p/\partial x$.

The governing equation is then

$$\frac{\partial u}{\partial t} - \nu \frac{\partial^2 u}{\partial y^2} = \frac{G}{\rho}, \tag{9.117}$$

with the initial and boundary conditions

$$u(y, 0) = 0 \quad \text{for } -h \le y \le h,$$
$$u(\pm h, t) = 0 \quad \text{for all } t. \tag{9.118}$$

Once more, we use the steady-state solution (9.45), and express the general form of the solution as

$$u(y, t) = \frac{G}{2\mu}[h^2 - y^2 - f(y, t)], \tag{9.119}$$

where the function $f(y, t)$, to be determined, satisfies the differential equation (9.112) and the appropriate initial and boundary conditions deriving from relations (9.118):

$$f(y, 0) = h^2 - y^2 \text{ for } -h \le y \le h,$$
$$f(\pm h, t) = 0 \qquad \text{for all } t. \tag{9.120}$$

The solution of Eq. (9.112) satisfying the boundary conditions is expressed as

$$f(y, t) = \sum_{n=0}^{\infty} A_n e^{-(2n+1)^2 \pi^2 \nu t/(4h^2)} \cos \frac{(2n+1)\pi y}{2h}, \tag{9.121}$$

and the coefficients A_n are obtained by imposing the initial condition.

The final result for the complete velocity field is (Erdogan, 2002)

$$u(y, t) = \frac{G}{2\mu}(h^2 - y^2)$$
$$- \frac{16Gh^2}{\pi^3 \mu} \sum_{n=0}^{\infty} \frac{(-1)^n}{(2n+1)^3} e^{-(2n+1)^2 \pi^2 \nu t/(4h^2)} \cos \frac{(2n+1)\pi y}{2h}, \tag{9.122}$$

and is plotted in Fig. 9.10 as a function of $\nu t/h^2$.

The viscous stress acting on the two walls may be evaluated from

$$\tau_w = \mu \left(\frac{\partial u}{\partial y} \right)_{y=-h} = -\mu \left(\frac{\partial u}{\partial y} \right)_{y=h}$$
$$= Gh \left[1 - \frac{8}{\pi^2} \sum_{n=0}^{\infty} \frac{e^{-(2n+1)^2 \pi^2 \nu t/(4h^2)}}{(2n+1)^2} \right]. \tag{9.123}$$

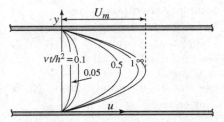

Fig. 9.10. Velocity distribution of unsteady Poiseuille flow for different values of $\nu t/h^2$.

As can be seen, when t goes to infinity the viscous stress tends to the steady-state value. On the other hand, since

$$\sum_{n=0}^{\infty} \frac{1}{(2n+1)^2} = \frac{\pi^2}{8}, \qquad (9.124)$$

the viscous stress is zero for $t = 0$, as could be expected.

In this flow a continuous introduction of vorticity occurs due to the presence of the constant pressure gradient. In particular, for $G > 0$, equal amounts of negative and positive vorticity are introduced at the lower and upper surfaces respectively. Therefore, the integral of the vorticity between the two plates is always zero. However, the absolute value of the vorticity increases with time, and in the final steady state it reaches its maximum, equal to Gh/μ, at the plates. At any time, there is a continuous annihilation of vorticities of opposite sign that penetrate inside the flow due to diffusion. During the transient, the vorticity of each sign that is annihilated in unit time is less than the one that is introduced at the walls, whereas, in the final state, a complete balance is reached, and the amount of newly introduced vorticity is always equal to the amount that is annihilated.

9.4.3. *Unsteady pipe flow*

We now consider the unsteady motion in a pipe when a constant pressure gradient is suddenly applied in the direction of the pipe axis, which we assume to be the z axis. We shall only analyse the cases in which the cross-section of the pipe is circular or rectangular.

For a pipe with circular cross-section and radius equal to a, we use cylindrical coordinates (r, θ, z), and assume that no circumferential velocity or variations are present. Consequently, from the continuity equation we find the motion to be unidirectional, with $v_z = v_z(r, t)$, and the pressure gradient $\partial p/\partial z = -G$ to be, at most, a function of time. From the compatibility condition and the momentum equation we can derive the generalization to the unsteady case of Eq. (9.56), which is

$$\frac{\partial v_z}{\partial t} - \nu \left(\frac{\partial^2 v_z}{\partial r^2} + \frac{1}{r} \frac{\partial v_z}{\partial r} \right) = \frac{G}{\rho}. \tag{9.125}$$

The initial and boundary conditions are

$$v_z(r,0) = 0 \quad \text{for } 0 \leq r \leq a,$$
$$v_z(a,t) = 0 \quad \text{for all } t. \tag{9.126}$$

As we assume G to be a constant, the asymptotic solution is the velocity field of the Hagen–Poiseuille flow, given by relation (9.58). We then seek a solution for our problem that is the addition of the steady-state solution and a function that vanishes for t tending to infinity and satisfies appropriate boundary and initial conditions. The result of this procedure is the following velocity field (see, e.g., Batchelor, 1967)

$$v_z(r,t) = \frac{G}{4\mu}(a^2 - r^2) - \frac{2Ga^2}{\mu} \sum_{n=1}^{\infty} \frac{J_0(\lambda_n r/a)}{\lambda_n^3 J_1(\lambda_n)} e^{-\lambda_n^2 \nu t/a^2}, \tag{9.127}$$

where $J_n(\lambda)$ are the Bessel functions of the first kind of order n, and λ_n are the positive roots of the equation $J_0(\lambda) = 0$.

The unsteady part of the velocity distribution (9.127) depends on the parameter $\nu t/a^2$, and the first term is the one that dominates the series. The order of magnitude of the decay time of the transient contribution is then $a^2/(\nu \lambda_1^2)$, where $1/\lambda_1^2 \simeq 0.1729$.

The evolution of the flow rate through the pipe is found to be given by the following expression

$$Q = \frac{\pi G a^4}{8\mu} \left(1 - 32 \sum_{n=1}^{\infty} \frac{1}{\lambda_n^4} e^{-\lambda_n^2 \nu t/a^2} \right). \tag{9.128}$$

Considering that

$$\sum_{n=1}^{\infty} \frac{1}{\lambda_n^4} = \frac{1}{32}, \tag{9.129}$$

it is seen that the flow rate, as expected, is zero at the initial time whereas it attains virtually the steady-state value (9.60) for $\nu t/a^2 = 1$.

For a pipe with rectangular cross-section, we turn again to a Cartesian (x, y, z) reference system, denote as w the velocity in the z direction, parallel to the pipe axis, and assume the contour of the cross-section to be defined by the straight lines $x = \pm a$, $y = \pm b$, with $a \geq b$. The governing equation is then

$$\frac{\partial w}{\partial t} - \nu \left(\frac{\partial^2 w}{\partial x^2} + \frac{\partial^2 w}{\partial y^2} \right) = \frac{G}{\rho}, \tag{9.130}$$

with initial and boundary conditions

$$w(x, y, 0) = 0 \quad \text{for} \ -a \leq x \leq a \ \text{and} \ -b \leq y \leq b$$
$$w(\pm a, y, t) = 0 \quad \text{and} \quad w(x, \pm b, t) = 0 \ \text{for all} \ t. \tag{9.131}$$

Once more, the solution may be obtained as a superposition of the steady-state solution (9.74) and a transient velocity field, and it is found that it may be expressed in the following form (Erdogan, 2003):

$$w(x, y) = \frac{G}{2\mu} \left[b^2 - y^2 - \frac{32b^2}{\pi^3} \sum_{n=0}^{\infty} \frac{(-1)^n}{(2n+1)^3} \frac{\cosh(\beta_n x)}{\cosh(\beta_n a)} \cos(\beta_n y) \right]$$
$$- \frac{Gb^2}{2\mu} \sum_{p=0}^{\infty} \sum_{q=0}^{\infty} A_{pq} e^{-(\alpha_p^2 + \beta_q^2)\nu t} \cos(\alpha_p x) \cos(\beta_q y), \tag{9.132}$$

where β_n is given by (9.73) and

$$\alpha_p = \frac{(2p+1)\pi}{2a}, \quad \beta_q = \frac{(2q+1)\pi}{2b},$$
$$A_{pq} = \frac{128(-1)^{p+q}}{(2p+1)(2q+1)^3 \pi^4} \left[1 - \frac{b^2/a^2}{(2q+1)^2/(2p+1)^2 + b^2/a^2} \right]. \tag{9.133}$$

The velocity depends on two parameters, $\nu t/b^2$ and b/a. For the particular case of a square cross-section, the order of magnitude of the decay time of the transients is $a^2/(\nu \pi^2/2)$.

9.4.4. *Flows with concentric circular streamlines*

We have already shown in Section 9.2.1 that a motion with concentric circular streamlines is necessarily plane and such that its components, in cylindrical coordinates, are

$$v_r = 0, \quad v_\theta = v_\theta(r, t), \quad v_z = 0. \tag{9.134}$$

Considering the compatibility equation and the necessity that the pressure be single-valued, the equation that must be satisfied by the circumferential velocity component in the unsteady case, which represents the momentum balance in the θ direction, is seen to be

$$\frac{\partial v_\theta}{\partial t} - \nu \left(\frac{\partial^2 v_\theta}{\partial r^2} + \frac{1}{r} \frac{\partial v_\theta}{\partial r} - \frac{v_\theta}{r^2} \right) = 0. \tag{9.135}$$

The particular problem considered is defined by appropriate initial and boundary conditions. It may also be useful to write explicitly the compatibility equation, which expresses the dynamics of vorticity; considering that the only non-zero vorticity component is $\omega_z = \omega_z(r,t)$, in this case that equation becomes simply

$$\frac{\partial \omega_z}{\partial t} - \nu \nabla^2 \omega_z = \frac{\partial \omega_z}{\partial t} - \nu \left(\frac{\partial^2 \omega_z}{\partial r^2} + \frac{1}{r} \frac{\partial \omega_z}{\partial r} \right) = 0. \tag{9.136}$$

Therefore, we note that, as expected, the vorticity satisfies a diffusion equation. It may be also observed that, in view of the expression (9.14) for ω_z, we may recast Eq. (9.135) in the form

$$\frac{\partial v_\theta}{\partial t} - \nu \frac{\partial \omega_z}{\partial r} = 0. \tag{9.137}$$

As for the pressure field, it may be obtained from Eq. (9.87), which is also valid for an unsteady flow, and thus we have, in general,

$$p(r,t) = \rho \int \frac{v_\theta^2}{r} dr + f(t), \tag{9.138}$$

where $f(t)$ is an arbitrary function of time.

We now analyse some flows that are characterized by circular streamlines and may thus be analysed through the above equations.

9.4.4.1. *Rotating circular cylinder*

As a first problem, we consider the flow inside a circular cylinder of radius a that is set impulsively in motion with angular velocity Ω_0. The initial and boundary conditions complementing Eq. (9.135) are

$$\begin{aligned}
v_\theta(r,0) &= 0 && \text{for } 0 \leq r < a, \\
v_\theta(a,t) &= \Omega_0 a && \text{for } t > 0, \\
v_\theta(0,t) &= 0 && \text{for all } t,
\end{aligned} \tag{9.139}$$

and the solution is again found by adding a transient function, vanishing for t tending to infinity, to the steady-state solution $v_\theta = \Omega_0 r$, which corresponds to a rigid rotation. The result is the following time-varying velocity distribution (Batchelor, 1967):

$$v_\theta(r,t) = \Omega_0 r + 2\Omega_0 a \sum_{n=1}^{\infty} \frac{J_1(\lambda_n r/a)}{\lambda_n J_0(\lambda_n)} e^{-\lambda_n^2 \nu t/a^2}, \tag{9.140}$$

where $J_n(\lambda)$ are the Bessel functions of the first kind of order n, and λ_n are the positive roots of the equation $J_1(\lambda) = 0$.

The first term in the series in (9.140) is, again, the one that survives longest and, considering that $\lambda_1 \simeq 3.83$, in practice the transient can be assumed to be over for $\nu t/a^2 = 0.5$.

More complicated solutions, involving integrals of the Bessel functions of both kinds, may also be derived for the flow outside a circular cylinder set in impulsive rotation (Mallick, 1957).

9.4.4.2. *Evolution of vorticity distributions*

Let us now consider another situation, still characterized by concentric circular streamlines, in which we study the evolution in time of a given initial vorticity distribution that is constant along θ and may, in general, vary along r. In other words, if $f(r)$ is a known prescribed function, we must find a solution of Eq. (9.136) such that

$$\omega_z(r,0) = f(r), \tag{9.141}$$

and also satisfying Eq. (9.137).

The procedure to solve this problem is described in detail by Berker (1963), and is based on expressing the solution in the form

$$\omega_z(r,t) = F(t)G(r), \tag{9.142}$$

and further requiring that the function $F(t)$ must vanish for $t \to \infty$.

The solution for a general function $f(r)$ may be expressed in terms of integrals of Bessel functions, and one of the possible forms is

$$\omega_z(r,t) = \frac{1}{2\nu t} e^{-\frac{r^2}{4\nu t}} \int_0^{\infty} f(\xi) e^{-\frac{\xi^2}{4\nu t}} I_0\left(\frac{\xi r}{2\nu t}\right) \xi d\xi, \tag{9.143}$$

where ξ is an auxiliary integration variable and I_0 is the modified Bessel function of the first kind of order zero.

An interesting particular case is when the vorticity is initially constant within a circle of radius a and is zero outside. In other words,

$$\omega_z(r,0) = \omega_0 \quad \text{for } 0 \le r < a,$$
$$\omega_z(r,0) = 0 \quad \text{for } r > a,$$

(9.144)

which corresponds to the vorticity distribution of a Rankine vortex whose core radius is a. In this case, relation (9.143) becomes

$$\omega_z(r,t) = \frac{\omega_0}{2\nu t} e^{-\frac{r^2}{4\nu t}} \int_0^a e^{-\frac{\xi^2}{4\nu t}} I_0\left(\frac{\xi r}{2\nu t}\right) \xi d\xi.$$

(9.145)

This expression may be also used to evaluate the vorticity evolution in the limit case in which the radius a of the initial vorticity core tends to zero while, simultaneously, the value ω_0 tends to infinity in such a way that the total vorticity in the core (and the circulation around it),

$$\Gamma_0 = \pi a^2 \omega_0,$$

(9.146)

remains constant. This limit of the Rankine vortex for vanishing core radius and constant total circulation is usually called an *ideal point vortex* (or, more simply, a *point vortex*) and corresponds to a completely irrotational flow with circulation around any curve enclosing the centre, where a singularity is present. It will be seen in Chapter 12 that this flow is one of the elementary solutions of the Laplace equation that are used to solve the problem of finding the potential flow tangent to a body.

Actually, the problem of the time evolution of an ideal point vortex may be solved in different ways, without recourse to (9.145), in terms either of vorticity or directly of the velocity field, by solving Eq. (9.135) with the initial condition

$$v_\theta(r,0) = \frac{\Gamma_0}{2\pi r}.$$

(9.147)

The solution is found to depend on the similarity variable $r^2/\nu t$, and corresponds to the following vorticity and velocity distributions:

$$\omega_z(r,t) = \frac{\Gamma_0}{4\pi\nu t} e^{-\frac{r^2}{4\nu t}},$$

(9.148)

$$v_\theta(r,t) = \frac{\Gamma_0}{2\pi r}\left(1 - e^{-\frac{r^2}{4\nu t}}\right).$$

(9.149)

This solution is termed an Oseen vortex, and the velocity and vorticity distributions corresponding to different values of νt are shown in Fig. 9.11.

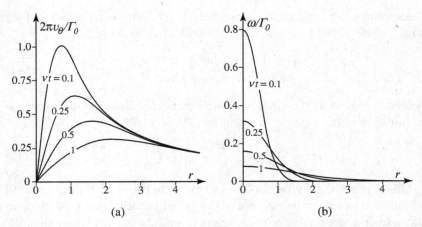

Fig. 9.11. Velocity (a) and vorticity (b) profiles for an Oseen vortex for various values of νt.

As can be seen, the vorticity profile coincides with a Gaussian curve with decreasing maximum for increasing time. However, it may be easily verified that the total amount of vorticity in the field remains constant in time and equal to Γ_0. Furthermore, the velocity has a maximum for $r_1 \simeq 2.24\sqrt{\nu t}$, and for smaller values of r the motion is similar to a solid-body rotation; in effect, in that region the velocity profile may be approximated by the relation

$$v_\theta(r, t) \simeq \frac{\Gamma_0}{8\pi\nu t} r. \qquad (9.150)$$

Finally, we note that, for all the above cases of flows with circular concentric streamlines with no dependence on θ, the velocity field may be obtained from the vorticity field through the relation

$$v_\theta(r, t) = \frac{1}{r} \int_0^r \omega_z(r, t) r \, dr, \qquad (9.151)$$

which immediately derives either from (9.83) or from the application to the velocity vector of Stokes' theorem (1.15), considering a closed curve coinciding with a generic streamline of radius r.

9.5. Final Comments on Exact Solutions and their Application

It may be useful to add some further comments on the derivation and on the possible applications of the exact solutions of the Navier–Stokes equations and, in particular, of those that have been analysed in more detail in the present chapter.

The first point to be observed is that most of the unsteady solutions we have described were obtained by a technique in which time-varying functions are added to the steady-state solution, with appropriate initial and boundary conditions derived from the particular problem. The reason of this procedure is that, in all the cases in which it was used, the only non-linear term in the equations of motion, namely the convective term, vanished identically; as a result, the mathematical problems to be solved to obtain the global unsteady flow were relatively simple. The situation is much more complex when the non-linearity cannot be avoided, and in that case using superposable solutions might be helpful.

Another important issue concerns the practical utility of the exact solutions that have been previously described, and in particular of the steady solutions in which the flows have been assumed to be unidirectional. In effect, this assumption allows exact solutions to be found for flows bounded by infinite surfaces having the same direction as the velocity, and we have described solutions for both plane walls and ducts of different cross-sections. However, as no infinite plates or ducts exist in reality, it is quite reasonable to enquire in which conditions these solutions may have any resemblance to real flows of practical interest.

To answer this question, one may refer, for instance, to a duct connecting a reservoir to another one where a lower pressure is present. For simplicity, we may imagine that at the entrance of the duct the velocity is constant over the whole cross-section. However, the no-slip condition will immediately cause the development of a lower-velocity boundary layer over the inner surface of the duct, which grows with increasing distance from the entrance, and the mass balance will imply a consequent velocity increase in the inner regions of the duct cross-section. We may also describe the same flow development in terms of vorticity, by stating that vorticity is produced at the duct walls, diffuses inwards and is convected downstream, occupying an increasing portion of the duct cross-section. Therefore, in this developing entrance region the velocity will not be exactly unidirectional, velocity gradients along the duct axis will be present, and the axial pressure gradient will not be constant. Nonetheless, the boundary layers will subsequently merge and the flow will progressively approach the one corresponding to the exact solution for steady pipe flow. As an example, for a circular cross-section the pressure gradient will tend to become constant, and the velocity field will tend to be described, with higher and higher accuracy, by the Hagen–Poiseuille profile (9.58). This new condition is reached after a distance from the pipe inlet that is termed *entry length*, downstream of which the flow is usually said to be *fully developed*.

The entry length, say L_e, depends on several factors, among which is the detailed duct inlet geometry, but the most important one is certainly the typical flow Reynolds number, defined, for instance, as $Re_d = Ud/\nu$, where U is the average velocity in the duct (i.e. the flow rate divided by the cross-section area) and d is the duct diameter. It should be clear that with increasing Reynolds number the boundary layer growth rate decreases, as a consequence of the increased relative importance of convection with respect to diffusion. A useful formula for the order of magnitude of the entry length is

$$\frac{L_e}{d} \simeq 0.06 Re_d. \qquad (9.152)$$

However, it should be recalled that this relation, as well as all the above exact solutions of the equations of motion, refers to laminar flows only. In fact, we have already seen that when the Reynolds number exceeds a certain value, which is dependent on the considered situation (being of the order of $Re_d = 2300$ for a circular duct), the flow becomes turbulent, with the appearance of irregular velocity fluctuations superposed on a mean velocity profile. Now, one of the characteristic features of turbulent flows is their higher diffusivity with respect to laminar flows, and thus turbulent boundary layers grow much more rapidly than laminar ones. Therefore, if we still define a fully developed flow as the condition in which the boundary layers have filled all the duct cross-section and the velocity profile no longer depends on the streamwise distance, it may be expected that for a completely turbulent flow condition the entry length will be definitely smaller than in the laminar case. In effect, for a turbulent flow L_e may be estimated from the new relation

$$\frac{L_e}{d} \simeq 4.4 Re_d^{1/6}. \qquad (9.153)$$

In conclusion, it is clear that the above-described exact solutions of the incompressible flow equations may be applied directly to cases of practical interest only when the typical Reynolds numbers of the real flows they represent (albeit perhaps in an approximate way) are sufficiently low. Nonetheless, we have already seen in Section 9.1 that exact solutions are also useful for many other reasons; in particular, they can be used as basic reference flows for the theoretical and numerical stability analyses that are essential for studying the complex process of transition.

Chapter 10

THE ROLE OF THE ENERGY BALANCE
IN INCOMPRESSIBLE FLOWS

10.1. Introduction

We have already pointed out, in Chapters 5 and 9, that the energy balance
equation is generally disregarded in incompressible flow theory, particularly
if the object of the analysis is the prediction of the loads on moving bodies.
However, there are situations in which the energy equation is useful even
in incompressible flows, for both theoretical and practical reasons.

From the theoretical point of view, the energy balance may help in
understanding flow behaviour in all those circumstances in which sudden
discontinuities appear in a flow, as a consequence of small variations
of certain geometrical or dynamical parameters. The obvious example
of such a situation is the transition from laminar to turbulent flow,
whose full comprehension is still far from being attained. But it may
also happen that a laminar steady flow becomes unstable and evolves
to a new flow that is still regular and spatially ordered but essentially
periodic, and thus unsteady. A possible approach to all these problems
related to flow instability is analysing in detail the energy balance and
trying to understand the behaviour of the various terms appearing in
the relevant equation, expressed either in differential or in integral form.
For example, the energy balance may be seen as a constraint on the
possible solutions satisfying the momentum and mass balances. This might
be particularly useful when simplified models are developed, in which
certain terms are either disregarded or described in an approximate way.
Alternatively, in order to obtain a flow satisfying given conditions or
to describe its time evolution, one may assume the validity of certain

'principles', which imply that some of the quantities appearing in the energy balance, such as dissipation or entropy production, should be a minimum or a maximum. All these aspects are the object of continuous research, but the reader should be warned that the use of these procedures or criteria raises non-trivial mathematical and physical issues. These involve for instance, the unicity of the solutions of the equations of motion, or the adequacy of neglecting mathematical terms representing physical mechanisms whose effect is small enough to be safely disregarded in normal conditions, but might become important in certain critical regimes or flow regions.

From a more practical point of view, the energy balance may help in understanding the origin and the different values of the drag force acting on moving bodies. To this end, it is useful to write the integral form of the energy balance equation (4.50) again, neglecting the effects of the body forces. Considering a volume v of fluid, bounded by a surface S, we get, with a slightly different organization of the terms,

$$\int_v \frac{\partial \rho(e + V^2/2)}{\partial t} dv + \int_S \rho(e + V^2/2) \boldsymbol{V} \cdot \boldsymbol{n} dS + \int_S \boldsymbol{q} \cdot \boldsymbol{n} dS$$
$$= - \int_S p \boldsymbol{V} \cdot \boldsymbol{n} dS + \int_S \boldsymbol{\tau}_n \cdot \boldsymbol{V} dS. \tag{10.1}$$

We now note that the sum of the two integrals on the right-hand side represents the work done on the fluid, in unit time, by the surface forces acting on the volume boundary. Therefore, Eq. (10.1) shows that, if the heat flux through the boundary is zero (i.e. the flow is globally adiabatic), this work is equal to the time variation of the total energy in the fluid. If we then consider, for instance, a body moving in an unlimited fluid domain, the boundary comprises an outer boundary tending to infinity, where the velocity may be assumed to become negligible, and an inner boundary coinciding with the body surface. Consequently, the work done on the fluid in unit time is equal to the opposite of the work done in unit time by the resultant drag force acting on the body. This provides an energetic interpretation of the drag experienced by moving bodies, and, for instance, an immediate explanation of the larger drag of bluff bodies with respect to that of aerodynamic bodies. In effect, as already anticipated in the comments to Fig. 8.11, the two types of bodies are characterized by different orders of magnitude of the perturbation energy introduced in the flow, and in particular in their wakes, and thus by significantly different works exerted on the fluid. Although this energetic interpretation of drag might seem, to a certain extent, fairly obvious, it opens the way to a deeper understanding of the origin of the fluid dynamic loads, and also to a rationale for the development of procedures aimed at drag reduction. In effect, it suggests the utility of identifying the geometrical and fluid dynamical features influencing the amount of energy being introduced in the wake of a body, so that one can devise means to modify the flow in a favourable way. These aspects will be further considered in Section 10.4 of this chapter and also in Chapter 11, where the connection between the perturbation energy caused by a moving body and the amount and organization of the vorticity released in its wake will also be discussed.

Further insight may be gained by writing the integral equation of the kinetic energy balance for an incompressible flow in a particular form, derived by subtracting the equation of the internal energy balance from the one expressing the total energy balance. To this end, we may first integrate Eq. (5.36) over the considered volume, and subtract it from Eq. (4.50), after neglecting the work of the body forces. We thus get

$$
\int_v \frac{\partial \rho(V^2/2)}{\partial t} dv + \int_S \rho(V^2/2)\boldsymbol{V} \cdot \boldsymbol{n} dS + \int_v \Phi dv
$$
$$
= -\int_S p\boldsymbol{V} \cdot \boldsymbol{n} dS + \int_S \boldsymbol{\tau}_n \cdot \boldsymbol{V} dS.
$$
(10.2)

This equation is particularly expressive and shows, for instance, that if the motion is steady and the global flux of kinetic energy through the boundaries is zero, then the work done in unit time by the pressure and viscous forces acting on the boundaries is equal to the dissipation within the fluid. On the other hand, when the work on the boundaries is globally zero, then the dissipation is directly connected with the decay of the kinetic energy. Therefore, the evaluation of the global and local dissipation may be helpful both to understand the origin of the fluid dynamic forces and to achieve a deeper comprehension of the flow behaviour. For instance, one may try to identify the regions where dissipation is lower, and which may then be more subject to the progressive enhancement of the kinetic energy of perturbations, and, thus, more prone to instability.

It is not within the scope of the present chapter to go through a detailed analysis of all the possible uses of the energy balance in the various areas of research on incompressible flows. We shall limit ourselves to providing some theoretical tools that may be helpful for the evaluation of dissipation, and to describing a few examples of situations in which the energy balance may be used either to solve a certain flow problem or, more simply, to face it from a different point of view.

10.2. Dissipation and the Bobyleff–Forsyth Formula

We now derive a formula that may be very useful for the evaluation of dissipation. We start by recalling that the dissipation function Φ for an incompressible flow is given by

$$\Phi = \tau : \operatorname{grad} V = 2\mu \mathbf{E} : \operatorname{grad} V, \tag{10.3}$$

where $\mathbf{E}$ is the rate of strain tensor, i.e. the symmetric part of the tensor $\operatorname{grad} V$. Therefore, considering that, as may be deduced from definition (1.9), the double dot product of the symmetric and the anti-symmetric parts of a tensor is zero, we get also

$$\Phi = 2\mu \mathbf{E} : \mathbf{E}. \tag{10.4}$$

We now recall that the Eulerian expression of the acceleration of a fluid particle in a generic point of a flow is

$$a = \frac{\partial V}{\partial t} + V \cdot \operatorname{grad} V, \tag{10.5}$$

and thus the divergence of the acceleration vector is given by

$$\operatorname{div} a = \frac{\partial (\operatorname{div} V)}{\partial t} + \operatorname{div}(V \cdot \operatorname{grad} V). \tag{10.6}$$

We introduce the following identity:

$$\operatorname{div}(V \cdot \operatorname{grad} V) = V \cdot \operatorname{grad}(\operatorname{div} V) + \mathbf{E} : \mathbf{E} - \frac{1}{2}\omega^2, \tag{10.7}$$

where ω is the modulus of the vorticity vector $\boldsymbol{\omega}$. The validity of this identity may be easily verified by using the definition of the various tensors and operators. If we introduce relation (10.7) into (10.6), we get

$$\begin{aligned}
\operatorname{div} a &= \frac{\partial(\operatorname{div} V)}{\partial t} + V \cdot \operatorname{grad}(\operatorname{div} V) + \mathbf{E} : \mathbf{E} - \frac{1}{2}\omega^2 \\
&= \frac{D(\operatorname{div} V)}{Dt} + \mathbf{E} : \mathbf{E} - \frac{1}{2}\omega^2.
\end{aligned} \tag{10.8}$$

For an incompressible flow the velocity vector is solenoidal, and thus (10.8) may be written as

$$\mathbf{E} : \mathbf{E} = \frac{1}{2}\omega^2 + \mathrm{div}\mathbf{a}. \tag{10.9}$$

By introducing this relation into (10.4), we finally derive the following expression for the dissipation function:

$$\Phi = \mu\omega^2 + 2\mu\mathrm{div}\mathbf{a}. \tag{10.10}$$

The square of the vorticity modulus appearing in (10.10) is here denoted as the local value of the *enstrophy*, but the reader is warned that, in the scientific literature, this term is often used for the quantity $\frac{1}{2}\omega^2$, or, at times, even for the integral of $\frac{1}{2}\omega^2$ over a given volume of fluid. Apart from these inessential nomenclature issues, relation (10.10) shows clearly that the appearance of vorticity in the flow, whatever its sign, is connected with dissipation. From the principle of total vorticity conservation, we already know that when a body starts moving in an unlimited fluid, initially at rest, equal amounts of positive and negative vorticity are produced at its surface (and thus the global vorticity in the flow remains zero). However, we also realize, now, that this process causes the appearance of a finite positive amount of enstrophy, and thus of the associated viscous dissipation.

In addition, we observe that the second term on the right-hand side of (10.10) is independent of the presence of vorticity, and may thus be different from zero also in an irrotational flow. This is not surprising because we have already pointed out that, even if the resultant viscous force on a fluid element vanishes in an incompressible flow when $\boldsymbol{\omega} = 0$, this is not necessarily true, in general, for the viscous stresses, which may then do work and cause dissipation. Thus, both terms in (10.10) should always be taken into account when evaluating the local dissipation; for instance, we shall show, later in this chapter, that they mutually cancel if the flow corresponds to a rigid rotation, in agreement with the fact that the viscous stresses are zero in such a motion.

If we now consider a volume v of fluid bounded by a surface S with outward unit normal $\boldsymbol{n}$, using relation (10.10) and the divergence theorem we obtain the following expression for the dissipation taking place within the whole volume in unit time:

$$\int_v \Phi dv = \mu \int_v \omega^2 dv + 2\mu \int_S \boldsymbol{a} \cdot \boldsymbol{n} dS. \tag{10.11}$$

This is known as the *Bobyleff–Forsyth formula* (see Bobyleff, 1873, and Forsyth, 1880[a]), and its utility seems to have been largely underestimated in the literature. It often allows the dissipation in a flow to be computed without great difficulty from knowledge of the vorticity field and the acceleration on the boundary. In particular, it shows that when the acceleration on the boundary vanishes, the total dissipation is directly given by the integral of the enstrophy present within the fluid. This, however, does not mean that dissipation is taking place only where the flow is rotational, but simply that the global dissipation (including the part taking place in the irrotational region, if present), may be evaluated from an integral extending only to the regions where vorticity is present, and thus not necessarily to the whole fluid domain. The formula also shows that the dissipation in an irrotational flow (or in the irrotational part of a flow) may easily be obtained from the surface integral of the component of the acceleration normal to the boundary of the considered region.

The surface integral in (10.11) may also be expressed in a different form, in terms of the pressure gradient, by considering the momentum equation of an incompressible flow for the particular case of conservative body forces. In that case the acceleration may be written as

$$a = -\mathrm{grad}\left(\frac{p}{\rho} + \varPsi\right) - \nu\,\mathrm{curl}\,\omega, \tag{10.12}$$

from which, by recalling that the divergence of a curl is always zero and that the divergence of the gradient of a scalar function is equal to its Laplacian, we derive that

$$\mathrm{div}\,a = -\nabla^2\left(\frac{p}{\rho} + \varPsi\right). \tag{10.13}$$

Note that when the body forces are absent, or when $\nabla^2\varPsi = 0$ (as happens for the gravitational field), we have simply

$$\mathrm{div}\,a = -\nabla^2\left(\frac{p}{\rho}\right), \tag{10.14}$$

which can also be used for any type of conservative body forces, if one adopts a notation in which the term p denotes, in place of the pressure, the sum of the pressure and of the quantity $\rho\varPsi$.

---

[a] This article was the first one published by the eminent Scottish mathematician Andrew Russell Forsyth (1858–1942), when he was still an undergraduate at Trinity College in Cambridge. However, apparently this was also the only work in which he studied a problem with direct relevance to fluid mechanics.

If we now compare relations (10.14) with (10.9), we get the following form of the Poisson equation for the pressure:

$$\nabla^2\left(\frac{p}{\rho}\right) = \frac{1}{2}\omega^2 - \mathbf{E} : \mathbf{E}. \tag{10.15}$$

On the other hand, if we introduce Eq. (10.14) into relation (10.10), the expression for the dissipation function becomes

$$\Phi = \mu\omega^2 - 2\mu\nabla^2\left(\frac{p}{\rho}\right), \tag{10.16}$$

and thus, by integration over a fluid volume v, the following alternative form of the Bobyleff–Forsyth formula — in which $\partial p/\partial n$ is the directional derivative along n — is readily obtained:

$$\int_v \Phi dv = \mu \int_v \omega^2 dv - 2\nu \int_S \frac{\partial p}{\partial n} dS. \tag{10.17}$$

In the next section, we shall describe the application of the energy balance to some simple examples of incompressible flows, in which the utility of the Bobyleff–Forsyth formula will also become evident.

10.3. Examples of Application of the Energy Balance

10.3.1. *Plane Couette and Poiseuille flows*

Let us consider the plane Couette flow analysed in Section 9.3.1, i.e. the flow between two parallel infinite flat plates placed a distance d apart, with the upper one in parallel motion with constant velocity U and the lower one at rest. With the reference system of Fig. 9.1, the velocity profile is the linear one given by relation (9.39), i.e. $u = Uy/d$, and the vorticity is $\boldsymbol{\omega} = \omega_z \mathbf{e}_z$, with the constant value of ω_z equal to $-U/d$.

We now consider a volume of fluid with unitary extension in the direction perpendicular to the plane of motion and delimited by the two plates and by two cross-flow planes placed a distance dx apart. As the flow is steady and with no variation in the x direction, the first two integrals in Eq. (10.2) are zero and the energy balance implies that the dissipation in the considered volume in unit time is equal to the power introduced by the surface forces on the boundaries.

The dissipation may be evaluated through the Bobyleff–Forsyth formula (10.11); by observing that, in this case, the surface integral in the formula is zero, we get

$$\int_v \Phi dv = \mu \int_v \omega^2 dv = dx\mu \int_0^d \omega^2 dy = \mu\frac{U^2}{d}dx. \tag{10.18}$$

As for the work of the surface forces on the boundary, the work of the pressure forces is zero, because the pressure is constant over the whole volume boundary, and the work of the viscous stresses reduces to that of the tangential force acting on the upper boundary of the fluid volume, which is the opposite of the force $\tau_p^+ dx$ acting on the moving upper plate. We thus get, in agreement with the energy balance,

$$\int_S \boldsymbol{\tau}_n \cdot \boldsymbol{V} dS = dxU(-\tau_p^+) = dxU\mu\left(\frac{\partial u}{\partial y}\right)_{y=d} = \mu\frac{U^2}{d}dx. \tag{10.19}$$

We now consider the plane Poiseuille flow between two parallel infinite plates at rest placed at a mutual distance $2h$. The flow is induced by a pressure gradient $G = -\partial p/\partial x > 0$ and is described through the reference system of Fig. 9.3; as we have seen, the resulting velocity field is described by the following parabolic profile:

$$u = \frac{G}{2\mu}(h^2 - y^2) = U_m\left(1 - \frac{y^2}{h^2}\right), \tag{10.20}$$

where U_m is the maximum velocity, which occurs at the centreline. Consequently, the vorticity profile is the following linear one:

$$\omega_z = -\frac{\partial u}{\partial y} = \frac{G}{\mu}y = 2\frac{U_m}{h^2}y. \tag{10.21}$$

We consider the same volume already analysed for the Couette flow, and observe that, again, the integrals in the energy balance connected with the variation of kinetic energy vanish, and thus the dissipation is equal to the work of the surface forces on the boundary. The dissipation is obtained through the Bobyleff–Forsyth formula, in which only the volume integral appears because the acceleration is zero. We have, then,

$$\begin{aligned} \int_v \Phi dv &= \mu \int_v \omega^2 dv = dx\mu \int_{-h}^h \omega^2 dy \\ &= dx\frac{G^2}{\mu}\int_{-h}^h y^2 dy = \frac{2}{3}\frac{G^2 h^3}{\mu}dx = \frac{8}{3}\mu\frac{U_m^2}{h}dx. \end{aligned} \tag{10.22}$$

We now observe that, with the chosen reference system, the plates are at rest, and thus the corresponding surface forces do no work on the fluid. Conversely, the work done in unit time by the pressure forces is not zero, due to the presence of the pressure variation in the flow direction, and is given by

$$
-\int_S p\boldsymbol{V} \cdot \boldsymbol{n} dS = -\int_{-h}^{h} \frac{\partial p}{\partial x} dx u dy = dx G \int_{-h}^{h} u dy
$$

$$
= dx \frac{G^2}{2\mu} \int_{-h}^{h} (h^2 - y^2) dy = \frac{2}{3} \frac{G^2 h^3}{\mu} dx = \frac{8}{3} \mu \frac{U_m^2}{h} dx,
$$

$$(10.23)$$

which, as expected, is equal to the total dissipation in unit time.

We may now wonder if it is possible to write a form of the energy balance in which the tangential viscous stresses acting on the plates appear. This can indeed be done by analysing the same problem in a different reference system, and in particular using a system that moves in the x direction with velocity U_m, so that the velocity becomes

$$
u' = u - U_m = -\frac{G}{2\mu} y^2 = -\frac{U_m}{h^2} y^2. \tag{10.24}
$$

The velocity at the centreline is, then, zero and the velocity of the two plates is $-U_m$, i.e. they are moving leftwards; the velocity profile in the new reference system is shown in Fig. 10.1.

We now apply Eq. (10.2) to the same previously considered volume of fluid, and observe that the kinetic energy integrals still vanish and that the dissipation does not change, as it is Galilean invariant. On the other hand, the work does change, and we now have contributions from both the pressures and the viscous stresses.

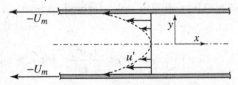

Fig. 10.1. Velocity profile for plane Poiseuille flow in a reference system moving with velocity U_m.

In particular, the work of the pressure forces becomes

$$-\int_S p\boldsymbol{V}\cdot\boldsymbol{n}dS = -\int_{-h}^{h}\frac{\partial p}{\partial x}dx u'dy = dxG\int_{-h}^{h}u'dy$$

$$= -dx\frac{G^2}{2\mu}\int_{-h}^{h}y^2dy = -\frac{1}{3}\frac{G^2h^3}{\mu}dx = -\frac{4}{3}\mu\frac{U_m^2}{h}dx. \tag{10.25}$$

Therefore, from Eq. (10.2) we get

$$\int_S \boldsymbol{\tau}_n\cdot\boldsymbol{V}dS = \int_S p\boldsymbol{V}\cdot\boldsymbol{n}dS + \int_v \Phi dv$$

$$= \frac{4}{3}\mu\frac{U_m^2}{h}dx + \frac{8}{3}\mu\frac{U_m^2}{h}dx = 4\mu\frac{U_m^2}{h}dx, \tag{10.26}$$

which is indeed the value of the global work of the (leftward) tangential viscous forces acting on the upper and lower boundaries of the fluid volume; these are the opposite of the (rightward) forces $\tau_p^+ dx$ and $\tau_p^- dx$ acting on the upper and lower plates respectively. In effect, as the outer normal to the lower boundary of the fluid volume is $-\boldsymbol{e}_y$, we have

$$(-U_m)(-\tau_p^+)dx + (-U_m)(-\tau_p^-)dx$$

$$= -dxU_m\mu\left(\frac{\partial u'}{\partial y}\right)_{y=h} + dxU_m\mu\left(\frac{\partial u'}{\partial y}\right)_{y=-h} \tag{10.27}$$

$$= 2\mu\frac{U_m^2}{h}dx + 2\mu\frac{U_m^2}{h}dx = 4\mu\frac{U_m^2}{h}dx.$$

10.3.2. *Rotating cylinders*

Let us now consider the case of the flow inside and outside a hollow circular cylinder of radius r_0 and negligible thickness, which rotates with constant counter-clockwise angular velocity Ω_0. We further assume that the transient from the start of the motion is over, so that the flow is steady and is characterized by the velocity field derived in Section 9.3.4, in which $v_r = v_z = 0$ and

$$v_\theta = \Omega_0 r \quad \text{for } r \le r_0, \quad v_\theta = \frac{\Omega_0 r_0^2}{r} \quad \text{for } r > r_0. \tag{10.28}$$

We now apply the energy balance to the flow in a domain v contained inside a circumference of varying radius R and unit length in the direction perpendicular to the plane of motion. For $0 < R \le r_0$, i.e. inside the cylinder, it is easy to see that all terms appearing in Eq. (10.2) are identically zero. In effect, the kinetic energy remains constant in the domain, and the integral involving pressure vanishes because there is no radial velocity component. Furthermore, the motion is a solid-body rotation and, according to what we already know, this implies that the viscous stresses are identically zero; consequently, the two integrals representing the work of the viscous forces and the global dissipation are both zero.

Nevertheless, it may be instructive to verify that the same result for dissipation is also obtained from the Bobyleff–Forsyth formula. In this case we have

$$\omega = 2\Omega_0, \quad \boldsymbol{a} \cdot \boldsymbol{n} = -\frac{v_\theta^2}{r} \tag{10.29}$$

and thus the dissipation becomes

$$\int_v \Phi dv = \mu \int_0^{2\pi} \int_0^R 4\Omega_0^2 r d\theta dr - 2\mu \int_0^{2\pi} \Omega_0^2 R^2 d\theta \tag{10.30}$$
$$= 4\pi\mu\Omega_0^2 R^2 - 4\pi\mu\Omega_0^2 R^2 = 0.$$

As previously anticipated, this is an example of a situation in which a cancellation of the two terms in the Bobyleff–Forsyth formula occurs, and this shows that dissipation does not necessarily take place in the flow region where vorticity — and thus enstrophy — is present.

For $r_0 \leq r \leq R$, i.e. in the flow outside the cylinder, the velocity field is irrotational, and, considering the opposite directions of the outer normals for $r = R$ and for $r = r_0$, the Bobyleff–Forsyth formula gives

$$\int_v \Phi dv = 2\mu \int_0^{2\pi} \Omega_0^2 r_0^2 d\theta - 2\mu \int_0^{2\pi} \frac{\Omega_0^2 r_0^4}{R^2} d\theta \tag{10.31}$$
$$= 4\pi\mu\Omega_0^2 r_0^2 - 4\pi\mu\frac{\Omega_0^2 r_0^4}{R^2} = 4\pi\mu\Omega_0^2 r_0^2 \left(1 - \frac{r_0^2}{R^2}\right).$$

The surface integral connected with the pressure in Eq. (10.2) is still zero, and we thus analyse the integral representing the work of the viscous stresses. To this end, we observe first that the only non-zero component of the stress tensor along the two circumferences at $r = r_0$ and at $r = R$ is $\tau_{r\theta} = 2\mu E_{r\theta}$, where (see Appendix B)

$$E_{r\theta} = \frac{1}{2}\left[r\frac{\partial}{\partial r}\left(\frac{v_\theta}{r}\right) + \frac{1}{r}\frac{\partial v_r}{\partial \theta}\right], \tag{10.32}$$

and thus we get, in this case,

$$\tau_{r\theta} = -2\mu\frac{\Omega_0 r_0^2}{r^2}. \tag{10.33}$$

Therefore, taking the different direction of the normal unit vectors into account, we have

$$
\int_S \boldsymbol{\tau}_n \cdot \boldsymbol{V} dS = -2\pi r_0 \Omega_0 r_0 (\tau_{r\theta})_{r_0} + 2\pi R\frac{\Omega_0 r_0^2}{R}(\tau_{r\theta})_R
$$
$$
= 4\pi\mu\Omega_0^2 r_0^2 - 4\pi\mu\frac{\Omega_0^2 r_0^4}{R^2} = 4\pi\mu\Omega_0^2 r_0^2 \left(1 - \frac{r_0^2}{R^2}\right), \tag{10.34}
$$

which, as expected, coincides with the value of the dissipation given by expression (10.31). Therefore, by letting R go to infinity, we find that the total dissipation and the work done on the whole fluid in unit time are

$$\int_v \Phi dv = \int_S \boldsymbol{\tau}_n \cdot \boldsymbol{V} dS = 4\pi\mu\Omega_0^2 r_0^2. \tag{10.35}$$

The quantity in relation (10.35) is the power that must be introduced in the flow to keep the motion steady, and thus also the power to be applied to the cylinder to avoid its rotation being slowed down by the tangential stresses acting over its outer surface, which produce a couple around its axis equal to $M = -4\pi\mu\Omega_0 r_0^2$ per unit length along the cylinder axis. An interesting feature of this flow is that all dissipation takes place in the region outside the cylinder, where the velocity field is irrotational, whereas no dissipation is present in the inner rotational region. Actually, this is a very peculiar situation, and in general dissipation is present both in the rotational and in the irrotational parts of a flow. Nevertheless, it is useful to point out again that we should not confuse rotationality, which is a kinematical feature of a flow, and dissipativity, which is a physical characteristic of the fluid, connected with the fact that it is viscous.

An issue that may cause some concern regarding the physical representativeness of the velocity field outside the cylinder, given by (10.28), is that, if we consider the whole fluid domain extending to infinity, we easily find that it corresponds to an infinite kinetic energy. Indeed, for a domain with unit length along the cylinder axis and $r_0 \leq r \leq R$, we have

$$\int_v \frac{1}{2}\rho V^2 dv = \frac{1}{2}\rho \int_0^{2\pi}\int_{r_0}^R \frac{\Omega_0^2 r_0^4}{r^2}r d\theta dr = \pi\rho\Omega_0^2 r_0^4 \ln\left(\frac{R}{r_0}\right), \tag{10.36}$$

which diverges for R tending to infinity. However, this result is less puzzling than it might seem, and actually shows that it is often useful, and even necessary, to take the transient leading to a certain steady solution into account. As a matter of fact, we have already seen in Chapter 6 that the irrotational flow around a rotating cylinder is the result of a process that comprises first the production — at the start of the motion — of a layer of vorticity over the surface of the cylinder, and then its diffusion into the outer fluid domain. A flow with infinitesimal local vorticity is thus eventually attained, but only after a theoretically infinite time has elapsed. During the transient, finite work is continuously being done on the fluid in unit time by the rotating cylinder, and the final irrotational flow should be seen as the result of finite power having being introduced into the fluid for an infinite time. Therefore, infinite work would actually be necessary for the attainment of the above-discussed velocity field with infinite kinetic energy. Obviously, in practice no infinite domains exist and the probable presence of outer solid boundaries would render the kinetic energy finite in a practical situation. Nonetheless, if these boundaries are sufficiently far from the cylinder, the final local vorticity in the outer flow would still be small enough for the velocity field of (10.28) to be highly representative of the real flow condition.

We consider now the case in which another concentric cylinder is present, with a radius $r_1 > r_0$ and a constant angular velocity Ω_1. The steady solution for the flow between the two cylinders was derived in Chapter 9, and corresponds to the velocity field

$$v_\theta = \frac{A}{r} + Br, \tag{10.37}$$

where

$$A = \frac{(\Omega_0 - \Omega_1)r_0^2 r_1^2}{(r_1^2 - r_0^2)}, \quad B = \frac{\Omega_1 r_1^2 - \Omega_0 r_0^2}{r_1^2 - r_0^2}. \tag{10.38}$$

Once more, the energy balance may be easily verified by considering the volume of fluid with unit length in the direction of the axis of the cylinders and with $r_0 \leq r \leq r_1$. The work done in unit time on the fluid is due to the tangential stresses acting along the surfaces of the cylinders, whose general expression is easily derived to be

$$\tau_{r\theta} = -2\mu \frac{A}{r^2}. \tag{10.39}$$

Therefore, taking the direction of the normal unit vectors into account, we get

$$\int_S \boldsymbol{\tau}_n \cdot \boldsymbol{V} dS = -2\pi r_0 \Omega_0 r_0 (\tau_{r\theta})_{r_0} + 2\pi r_1 \Omega_1 r_1 (\tau_{r\theta})_{r_1}$$
$$= 4\pi\mu A(\Omega_0 - \Omega_1). \tag{10.40}$$

As for the dissipation, it may be derived from the Bobyleff–Forsyth formula; considering the different directions of $\boldsymbol{n}$, we have

$$\omega = 2B, \quad (\boldsymbol{a} \cdot \boldsymbol{n})_{r_0} = \left(\frac{v_\theta^2}{r}\right)_{r_0}, \quad (\boldsymbol{a} \cdot \boldsymbol{n})_{r_1} = -\left(\frac{v_\theta^2}{r}\right)_{r_1}, \quad (10.41)$$

and we thus obtain, after some manipulations,

$$\int_v \Phi dv = 4\mu B^2 \pi (r_1^2 - r_0^2) + 2\mu 2\pi (\Omega_0^2 r_0^2 - \Omega_1^2 r_1^2). \quad (10.42)$$

By introducing the expressions of A and B from (10.38) and rearranging terms, we finally get

$$\int_v \Phi dv = 4\pi\mu \frac{(\Omega_0 - \Omega_1)^2 r_0^2 r_1^2}{(r_1^2 - r_0^2)} = \int_S \boldsymbol{\tau}_n \cdot \boldsymbol{V} dS. \quad (10.43)$$

It is interesting to observe that the velocity field corresponding to a single rotating cylinder, given by relations (10.28), coincides with that of a Rankine vortex having a core with radius r_0 and constant vorticity equal to $2\Omega_0$. Therefore, for such a model vortex, the dissipation within a circular domain bounded by a circumference of radius R and unit length in the direction perpendicular to the motion is still given by the previously obtained formulas (10.30) and (10.31), for the regions inside and outside the core respectively. Actually, for a Rankine vortex, relation (10.31) could have also been derived by applying the Bobyleff–Forsyth formula to the domain $0 \le r \le R$, with $R > r_0$; in effect, considering that vorticity is present only within the core, we have

$$\int_v \Phi dv = \mu \int_0^{2\pi} \int_0^{r_0} 4\Omega_0^2 r d\theta dr - 2\mu \int_0^{2\pi} \frac{\Omega_0^2 r_0^4}{R^2} d\theta$$

$$= 4\pi\mu\Omega_0^2 r_0^2 - 4\pi\mu \frac{\Omega_0^2 r_0^4}{R^2} = 4\pi\mu\Omega_0^2 r_0^2 \left(1 - \frac{r_0^2}{R^2}\right). \quad (10.44)$$

Obviously, in this case the energy balance shows immediately that the dissipation is not equal to the work done in unit time on the boundary of the domain. Therefore, the considered flow field cannot correspond to a steady solution, but at most to an initial flow condition, as already seen in Chapter 9, where it was also shown that vorticity would subsequently diffuse into the flow with the time law given by relation (9.145). Nonetheless, it is instructive to express the dissipation in the whole infinite flow domain outside a Rankine vortex, given by (10.35), as a function of the total vorticity present in the core, $\Gamma_0 = 2\Omega_0\pi r_0^2$:

$$\int_v \Phi dv = 4\pi\mu\Omega_0^2 r_0^2 = \frac{\mu\Gamma_0^2}{\pi r_0^2}. \tag{10.45}$$

From this relation we see that, for constant total vorticity contained in the core, the dissipation is inversely proportional to the square of the radius of the core. Then, if we consider a Rankine vortex as a first-order model of a real vortex, this means that the more a certain quantity of vorticity is concentrated in a small region of space, the higher its associated dissipation will be. It may be seen that the same conclusion can be drawn if a more realistic vortex model is used. For instance, let us consider the vorticity distribution of an Oseen vortex at a generic time t of its evolution, which we have seen in Chapter 9 to be

$$\omega = \frac{\Gamma_0}{4\pi\nu t}e^{-\frac{r^2}{4\nu t}}, \tag{10.46}$$

where Γ_0 is the global vorticity in the flow, initially contained in a core of vanishing radius. By applying the Bobyleff–Forsyth formula we can readily find that the dissipation in unit time, and for unit length in the direction of the axis of the vortex, is given by

$$\int_v \Phi dv = \mu \int_0^{2\pi} \int_0^\infty \omega^2 r d\theta dr = \frac{\mu\Gamma_0^2}{8\pi\nu t}. \tag{10.47}$$

Therefore, we may conclude that the dissipation of an Oseen vortex at time t is equal to the dissipation of a Rankine vortex of equal total strength Γ_0 and having a core radius $r_0 \simeq 2.83\sqrt{\nu t}$. A comparison of the velocity fields corresponding to the two vortices is shown in Fig. 10.2.

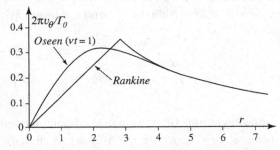

Fig. 10.2. Profiles of $2\pi v_\theta/\Gamma_0$ for Rankine and Oseen vortices with the same strength and producing the same dissipation ($\nu t = 1$ for the Oseen vortex).

10.3.3. *Boundary layer over a flat plate*

Let us now consider the two-dimensional boundary layer over a semi-infinite flat plate parallel to a free-stream U directed in the positive (rightwards) x direction. We first use a reference system fixed to the plate, with $x = 0$ at its leading edge, and y normal to it. We also assume that all the simplifications of boundary layer theory are applicable and that the transient from the start of the motion is over, so that the flow in the chosen reference system is steady.

In order to apply the energy balance, we choose a volume of fluid comprising a portion of the plate of length L with unit extension in the direction perpendicular to the plane of motion, and thus delimited by the coordinates $-\infty \leq x \leq L, 0 \leq y \leq +\infty$. We now note that the velocity over the plate is zero, and that, within the boundary layer approximations, there is no contribution to the last two integrals in Eq. (10.2) from the upstream and downstream boundaries of the chosen volume. Therefore, the work in unit time of the pressure and viscous forces may be neglected, and, considering the steadiness of the flow, we simply have

$$\int_S \rho(V^2/2)\boldsymbol{V} \cdot \boldsymbol{n}dS + \int_v \Phi dv = 0, \qquad (10.48)$$

which means that the net flux of kinetic energy is equal to the opposite of the dissipation.

To evaluate the kinetic energy flux, we note first that, in reality, the left boundary of the considered volume of fluid may be positioned at any coordinate within the interval $-\infty \leq x \leq 0$, because no variation of velocity occurs in that region. In practice, the only variation in kinetic energy within the chosen volume, and the consequent global flux, is due to the reduction of kinetic energy of the fluid within the boundary layer. We must then consider the quantity

$$Q = \int_0^{\delta(L)} \rho u(L,y)dy, \qquad (10.49)$$

which is the mass flowing in unit time through the boundary layer thickness $\delta(L)$ at the coordinate $x = L$, where the velocity is $u(L,y)$. This same quantity of mass enters the upstream boundary of the volume (positioned at any coordinate $x \leq 0$) with velocity U and thus, considering the direction of the normal unit vectors, the first integral in Eq. (10.48) becomes

$$-\frac{1}{2}\rho U^2 \int_0^{\delta(L)} u(L,y)dy + \frac{1}{2}\rho \int_0^{\delta(L)} u^3(L,y)dy = -\frac{1}{2}\rho U^3 \delta_e(L), \quad (10.50)$$

where we have introduced the energy thickness δ_e, defined by relation (7.36), with the velocity outside the boundary layer equal to U.

As for the dissipation, we have already noted in Chapter 7 that, with the boundary layer approximations, the dissipation is zero outside the boundary layer over a flat plate. In effect, from the Bobyleff–Forsyth formula, with the help of the energy-integral equation, (7.42), and considering that the acceleration flux is negligible at the upstream and downstream boundaries, we get

$$\int_0^L dx \int_0^\infty \Phi dy = \mu \int_0^L dx \int_0^\infty \omega^2 dy$$

$$= \mu \int_0^L dx \int_0^\infty \left(\frac{\partial u}{\partial y}\right)^2 dy = \frac{1}{2}\rho U^3 \delta_e(L), \quad (10.51)$$

which shows that the kinetic energy balance is satisfied.

As we have already done for the plane Poiseuille flow, we now express the energy balance in a reference system moving with the free-stream velocity, so that the work done by the friction forces appears. We denote the x component of velocity in the new reference system by $u' = u - U$, and consider the same volume of fluid as before; however, for reasons that will soon be clear, while the downstream boundary is kept at $x = L$, the upstream boundary is taken either at $x = -\infty$ or at a sufficiently negative value of x. Note that the velocity now is zero everywhere, except within the boundary layer, where u' is directed in the negative x direction. The evident disadvantage of the new reference system is that the flow has become unsteady; in effect, in a time interval dt the plate moves by a quantity $dx = Udt$ in the negative x direction, and thus the extent of the portion of lower boundary (at $y = 0$) where the no-slip condition applies is increased by the same amount.

Consequently, the only integral that vanishes in Eq. (10.2) is the one expressing the work of the pressure forces, which are constant around the whole boundary. As for the remaining integrals, we start by observing again that, due to its Galilean invariance, the global dissipation does not change in the new reference system, and is thus still given by expression (10.51). On the other hand, the unsteady term may be evaluated by observing that in a time interval dt the motion of the plate causes a consequent introduction of kinetic energy through the boundary at $x = L$, which may be evaluated, neglecting second-order terms, from the amount of kinetic energy that is present at a generic time t at the coordinate $x = L$ multiplied by $dx = Udt$. The variation of kinetic energy within the domain in unit time may thus be estimated through the following relation:

$$\int_v \frac{\partial \rho(V^2/2)}{\partial t} dv = U \int_0^\infty \frac{1}{2}\rho u'^2 dy = \frac{1}{2}\rho U \int_0^\infty (u^2 + U^2 - 2uU)dy$$

$$= \frac{1}{2}\rho U^3[\delta^*(L) - \theta(L)],$$

$$(10.52)$$

where the values of the velocities appearing in the line integrals are those corresponding to $x = L$, and $\delta^*(L)$ and $\theta(L)$ are the displacement and momentum thicknesses of the boundary layer at $x = L$. In the above derivation we have assumed that at time $t + dt$ the leading edge of the plate is still within the considered fluid volume, and this is why the upstream boundary of the volume must be located at $x \leq -Udt$.

The flux of kinetic energy is non-zero only at $x = L$ and, with the help of simple manipulations, we obtain

$$\int_S \rho(V^2/2)\boldsymbol{V} \cdot \boldsymbol{n}dS = \int_0^\infty \frac{1}{2}\rho u'^3 dy$$

$$(10.53)$$

$$= \frac{1}{2}\rho U^3[-\delta^*(L) + 3\theta(L) - \delta_e(L)].$$

Therefore, by introducing expressions (10.51), (10.52) and (10.53) into Eq. (10.2), we finally obtain the following relation for the work of the viscous forces

$$\int_S \boldsymbol{\tau}_n \cdot \boldsymbol{V}dS = \rho U^3\theta(L) = \boldsymbol{F} \cdot \boldsymbol{U},$$

$$(10.54)$$

where $\boldsymbol{F}$ is the total force acting on the fluid through the plate surface, and $\boldsymbol{U} = -U\boldsymbol{e}_x$ is the velocity of the plate. We thus find that the global force acting on the portion of plate of length L is given by

$$\boldsymbol{F}_x = -\boldsymbol{F} = \rho U^2\theta(L)\boldsymbol{e}_x,$$

$$(10.55)$$

and we recover the result of expression (7.64), which, in turn, is obtained from the application of the momentum-integral Eq. (7.35) to the case of a boundary layer over a flat plate in parallel motion.

10.4. Energetic Interpretation of Drag

As already anticipated in the introduction to this chapter, the energy balance may help in interpreting the different values of the drag experienced by moving bodies. Let us assume that a body is moving with velocity $U = U e_u$ in an unlimited domain of fluid, where the velocity was zero before the start of the motion of the body. We are using a reference system fixed to the fluid, and thus the motion is not steady, as the body occupies different locations in space at different times. The outer boundary may be taken at a very large distance from the body, where the velocity is assumed to have become negligible; this may be shown to be true if the motion has started from rest. In practice, we may assume all the fluxes at the outer boundary to be zero and extend the remaining integrals to a boundary tending to infinity.

With the above assumptions, the work done on the fluid in unit time by the surface forces acting on the body contour is equal to the opposite of the work done by the component along the direction of motion of the total fluid dynamic force acting on the body, say $F_D = F_D e_u$. Note that, with this notation, when F_D is negative the force on the body is in the direction of $-U$, i.e. it opposes the motion and corresponds to drag in the common sense of the word. Conversely, when F_D is positive the body experiences a force that favours its motion, i.e. traction. We thus have

$$-F_D \cdot U = -\int_S pV \cdot n dS + \int_S \tau_n \cdot V dS, \qquad (10.56)$$

where n is directed inside the body. As a result, from Eq. (10.2) we get

$$-F_D \cdot U = -F_D U = \int_v \frac{\partial \rho(V^2/2)}{\partial t} dv + \int_v \Phi dv. \qquad (10.57)$$

We note now that, from our assumptions, the volume of integration is delimited by boundaries through which the kinetic energy flux is globally zero, and the velocity vanishes at the outer boundary in such a way that all the volume integrals are finite. Consequently, from the transport theorem in the form (3.41), we find that the chosen fluid domain behaves like a material volume, and we may write

$$\int_v \frac{\partial \rho(V^2/2)}{\partial t} dv = \frac{d}{dt} \int_v \frac{1}{2} \rho V^2 dv, \tag{10.58}$$

where the ordinary differentiation symbol is used to highlight that the integral on the right-hand side is a quantity that depends on time only.

Therefore, we may recast relation (10.57) in the following form:

$$-F_D U = \frac{d}{dt} \int_v \frac{1}{2} \rho V^2 dv + \int_v \Phi dv = \frac{d}{dt} E_T(t) + D_T(t), \tag{10.59}$$

where we have denoted the total kinetic energy and the total instantaneous dissipation in the fluid at time t by $E_T(t)$ and $D_T(t)$ respectively. From this expression, we readily see that the value of the drag force is strictly connected with the magnitude of the global dissipation and with the rate of variation of the kinetic energy in the whole flow. Incidentally, we note that relation (10.59) also holds when the outer boundary is at rest, i.e. if the body is moving inside a closed solid vessel.

Returning to the case in which the body is moving in an effectively unlimited domain of fluid, let us now assume that it moves with a velocity that is constant in time, i.e. its motion is a uniform translation. The work done on the fluid in an interval of time dt may thus be written, to first order in dt,

$$-F_D U dt = E_T(t + dt) - E_T(t) + \int_t^{t+dt} D_T(t) dt. \tag{10.60}$$

The first observation that follows from this relation is that it provides a mathematical and physical justification of d'Alembert's paradox, i.e. that no drag force acts on a body moving with constant velocity in a non-viscous fluid. In effect, in such an ideal fluid, no dissipation is present, and thus the last integral in (10.60) is identically zero. Furthermore, we have already seen that no vorticity may be generated in an incompressible non-viscous flow starting from rest, and thus the whole velocity field remains irrotational. Consequently, the velocity vector must be the gradient of a potential scalar function which, to comply with the mass balance, satisfies the Laplace equation (5.53). The boundary conditions to be imposed are that the velocity must vanish at infinite distance from the body and must be tangent to the body surface, as enforced by condition (5.56). For our present discussion, the important point is that the solution of this mathematical problem depends on the instantaneous value of the velocity of the body only, and the velocity in any point of the flow domain is linearly dependent on that velocity and is a function of the position of the point with respect to the body. Therefore, although the kinetic energy of the single particles varies in time as the body moves to different positions in space, the total kinetic energy in the infinite domain of fluid does not depend on the position of the body, but on its instantaneous velocity only. Consequently, as this velocity is constant, we have $E_T(t + dt) = E_T(t)$ and, from (10.60), we deduce that the drag force is zero. We also realize that, when a body moves with a time-varying velocity, the proportionality between the perturbation velocities in the fluid and the instantaneous velocity of the body implies that the kinetic energy also varies in time, and thus work must be done on the fluid to account for this kinetic energy variation. Therefore, even a body moving in a non-viscous fluid may experience a fluid dynamic force in the direction of motion, but only if its motion is not uniform in time. We shall return to this fluid dynamic force linked to the body acceleration in the next section, where we introduce the notion of *added mass*.

Let us then apply relation (10.60) to a viscous flow. The fundamental difference with respect to the non-viscous fluid case is that, now, both pressures and viscous stresses act on the particles, and the no-slip condition must be enforced over the body surface in addition to the non-penetration condition. This implies that the fluid particles adjacent to the surface always move with its velocity, and that neighbouring particles must move not only to leave the space necessary for the passing body, but also as a consequence of the viscous stresses generated by the mutual velocity differences. We have already seen this process in action when describing the evolution of a boundary layer over a solid surface. The result is that an ever-lengthening wake is generated behind the body, and thus, even when it is moving at constant velocity, new kinetic energy is continuously introduced in the flow due to the action of viscosity.

The presence in (10.60) of a term containing dissipation deserves some further comments. In effect, this term, which is always positive, tends to decrease the kinetic energy, and is equal to the net increase in internal energy in the fluid. This may be easily understood by imagining a situation in which no work is done on the fluid after time t. In that case, the final kinetic energy $E_T(t + dt)$ would be lower than $E_T(t)$ by the amount given by the dissipation term, which corresponds to the increase in internal energy necessary to keep the total energy of the fluid constant. Therefore, Eq. (10.60) may be interpreted by saying that the work done on the fluid by the opposite of the drag force in the time interval dt is equal to the sum of the final kinetic energy and the dissipation occurring in the time interval *dt* minus the initial kinetic energy. Alternatively, as already pointed out, we may say that it is equal to the increase of the total energy in the considered time interval. In practice, the fact that the kinetic energy continuously introduced into the flow is, afterwards, converted into another — and less valuable — form of energy has no bearing on the drag experienced by the moving body.

We may try to use a simplified example to describe the different behaviour of the fluid particles when a body is moving in an ideal non-viscous fluid or in a viscous fluid. Let us imagine that we have a car which starts moving in an extremely large and very crowded square. Obviously, we assume that the velocity of the car and the reactivity of the people are such that no one is run over by the car. People will start moving to let the car pass and to fill the space left free, as a consequence of the pressure exerted upon them by the surface of the car or by their neighbours. It is easy to be convinced that the motion of the crowd will be in all directions, but will progressively die out at increasing distances from the car. Furthermore, it is apparent that the kinetic energy associated with single individuals varies in time, but the sum of the kinetic energies of all the people in the square only changes when the car is accelerating or decelerating, and remains invariant when it is moving at constant velocity. In that condition, the car is subjected to the retarding pressure of the persons around its front part, and also to a balancing pressure of the people pushing its rear part, so that the total drag acting on it is zero. Therefore, this case represents the flow field produced in a non-viscous fluid.

If we want to use the same example to visualize the flow in a viscous fluid, we must first imagine that all the people are linked together through peculiar linkages providing mutual forces that are proportional to the velocity difference between neighbours. Furthermore, each person initially adjacent to the car surface is compelled to remain attached to it. Therefore, when the car starts moving, people will acquire motion that depends not only on the necessity of letting the car pass, but also on the forces induced by the motion of their neighbours. A procession will then develop behind the car, and its length will continue to increase, with more and more people involved in the motion, even when the car has reached a constant velocity. Correspondingly, in that condition the car will be retarded not only by the pressure of the persons in front of it, but also by the forces caused by those who are attached all around its surface (and in particular on its rear part) and by their following neighbours. Obviously, there would be no mechanism, in this imaginative example, causing dissipation of the kinetic energy induced by the motion of the body. However, the idea of the consequences produced on the fluid motion by the presence of viscosity should be quite clear.

Returning to real fluids, it is seen that the viscous no-slip boundary condition at the body surface is the primary source of a continuous introduction of new kinetic energy in the flow due to a body moving with uniform translational velocity. On the other hand, the dissipation caused by the work of the viscous stresses takes place in the whole flow due to the velocity differences existing between the fluid particles, irrespective of the fact that these velocity differences are caused by the no-slip or by the non-penetration condition. The reader may recognize that this is equivalent to saying that, in a viscous fluid, even an irrotational flow tangent to a moving body would cause dissipation.

The energetic interpretation of drag shows that the higher the perturbation kinetic energy produced by a moving body the higher its drag. Consequently, a good indicator of the drag of a body is the amount of kinetic energy that is present in its wake. Indeed, we have already pointed out, in Chapter 8, that flow visualizations give a clear idea of the very different kinetic energies that are present in the wakes of bluff bodies and aerodynamic bodies, thus justifying their very different drags. The cross-flow dimension of the wake of a body is certainly a primary factor influencing the level of perturbation it causes with its motion. However, in the next chapter it will be shown that the kinetic energy introduced by a body moving in a fluid also depends strongly on the characteristics of the wake vorticity field. In particular, we shall see that it is connected both with the amount of vorticity that is continuously shed into the wake and with its subsequent organization in space.

10.5. The Concept of Added Mass

In the last section we observed that when a body accelerates in a fluid, a force in the direction of motion must act on the body as a consequence of the work that is necessary to account for the variation of the global kinetic energy in the fluid, caused by the change of the body's velocity. This force arises both in viscous and in non-viscous fluids, but its evaluation is definitely easier for the case of non-viscous flows. We thus concentrate on this case first and, subsequently, discuss the practical utility of the results obtained and the consequences of the introduction of viscosity in the analysis.

We assume that the body has started moving from rest in an unlimited domain of fluid and, for the sake of simplicity, that its motion is unidirectional with a velocity U, which may vary in time. We are using a reference system fixed in space with coordinates x, y and z, along which the respective components of the velocity vector V are u, v and w. As the fluid is non-viscous, no vorticity is produced at the body surface and thus the flow remains irrotational for all subsequent times after the start of the motion. Then, the velocity field may be expressed as $V = \mathrm{grad}\varphi$, and the velocity potential φ may be derived from the condition of solenoidality of the velocity field, which reduces to the Laplace equation

$$\nabla^2 \varphi = \frac{\partial^2 \varphi}{\partial x^2} + \frac{\partial^2 \varphi}{\partial y^2} + \frac{\partial^2 \varphi}{\partial z^2} = 0. \tag{10.61}$$

The boundary condition to be applied over the body surface is

$$V \cdot n = \frac{\partial \varphi}{\partial n} = U \cdot n, \tag{10.62}$$

where n is the normal unit vector at the considered point on the body surface and n is the scalar coordinate along the direction of n.

The total kinetic energy present in the fluid at a generic time t may thus be written as follows:

$$E_T(t) = \frac{1}{2}\rho \int_v (u^2 + v^2 + w^2)dv = \frac{1}{2}\rho U^2 k \mathcal{V}, \tag{10.63}$$

where the integrals are extended over the whole fluid domain, U is the modulus of U and $\mathcal{V}$ is the volume of the body. We have also introduced the coefficient k, defined as

$$k = \frac{1}{\mathcal{V}} \int_v \left(\frac{u^2 + v^2 + w^2}{U^2} \right) dv. \tag{10.64}$$

We now observe that, having chosen the fluid domain to extend to infinite distance from the body, k does not depend on time. In effect, Eq. (10.61) is linear and it may be easily seen that its solution only depends on the shape of the body and on the instantaneous value of U. More precisely, all the components of the velocity vector in a generic point of the flow are directly proportional to $U(t)$ and a function of the position of the point with respect to the body surface. Therefore, the ratios between the velocity components and U in fixed points of space (say $u' = u/U, v' = v/U, w' = w/U$) depend on time only because the body occupies different positions for different values of t. However, the fluid domain is infinite and thus the variation of the body's location does not actually change the mathematical problem to be solved. Therefore, the velocity components u', v' and w' do not change in a reference system moving with the body, while in a space-fixed reference system the same values of the non-dimensional velocity components correspond to different points in space. Consequently, in both cases, the integral in (10.64) does not depend on time and the total kinetic energy in the flow changes only if the velocity of the body varies in time. In other words, if the body moves with a constant translational velocity, the velocities of the single particles do vary in time, but the integral of the kinetic energy of all the particles in the infinite fluid domain remains constant.

With the above definitions and observations in mind, the variation of kinetic energy in the flow in unit time may be obtained as follows:

$$\frac{dE_T}{dt} = \frac{d}{dt}\left(\frac{1}{2}\rho U^2 k \mathcal{V}\right) = \rho k \mathcal{V} U \frac{dU}{dt}. \tag{10.65}$$

By recalling that there is no dissipation in a non-viscous fluid, from relation (10.59) we determine that the above quantity is equal to the work done on the fluid by the moving body in unit time, and thus the opposite work done on the body by the fluid is equal to

$$F_D U = -\rho k \mathcal{V} U \frac{dU}{dt}. \tag{10.66}$$

Therefore, the force component acting on the body in the direction of its velocity U is such that its magnitude is given by

$$F_D = -\rho k \mathcal{V} \frac{dU}{dt}. \tag{10.67}$$

We have thus found that a fluid dynamic force acts on a moving body when it is accelerating, and that the direction of the force is opposite to that of the acceleration. In other words, if the velocity of the body increases, then the force opposes the motion, i.e. it acts as a drag; conversely, if the body decelerates, the force is effectively a traction. Due to the obvious analogy between this force and the inertia force acting on any accelerating body, the quantity

$$m_a = \rho k \mathcal{V} \tag{10.68}$$

is commonly denoted as the *added mass* of the body, and k (often also indicated with the symbol C_a) as the *added mass coefficient*. In effect, if a body with inertial mass m is immersed in a fluid with density ρ, the force expressed by (10.67) will cause its dynamics to be analogous to that of a body having a *virtual mass* equal to $(m + m_a)$.

Nonetheless, from the above derivation, it should be clear that the added mass does not represent the mass of a specific finite amount of fluid being 'dragged' by a body moving in a fluid. More correctly, we may say that it is simply the coefficient of proportionality between the fluid dynamic force acting on an a accelerating body, due to the consequent variation of the kinetic energy of the whole fluid, and the opposite of the acceleration of the body.

Obviously, the added mass reveals its presence with significant effects only when the density of the fluid is not negligible with respect to the density of the moving body and, thus, unless we are dealing with the design of an airship, it is usually much more important in water than in air. Actually, it is a common experience that if a certain force is applied to a body in air, it will acquire a much greater acceleration than if an equal force were applied to the same body submerged in water. Another example is that the natural frequency of a submarine structure may be significantly lower than the one evaluated or measured in air. Therefore, the presence of the added mass must be always carefully taken into account when studying the dynamics of bodies immersed in water.

Up to now, we have only considered the simple case of a body accelerating in a certain direction and of the consequent force acting on it in the same direction. Actually, for a body of generic shape accelerating in an arbitrary direction with respect to a chosen reference system, the resulting forces may have components in all directions, and the added mass becomes a tensor. With the assumption of non-viscous fluid, and thus of potential flow, the components of this tensor can be evaluated without great difficulty. Let us now denote as x_1, x_2, x_3 the axes of the chosen reference system and let u_1, u_2, u_3 be the corresponding velocity components. If $\boldsymbol{U} \equiv (U_1, U_2, U_3)$ is the velocity of the body, the velocity in any point of the fluid may be written in the form

$$
\begin{aligned}
u_1 &= u_{11}U_1 + u_{12}U_2 + u_{13}U_3, \\
u_2 &= u_{21}U_1 + u_{22}U_2 + u_{23}U_3, \\
u_3 &= u_{31}U_1 + u_{32}U_2 + u_{33}U_3,
\end{aligned}
\tag{10.69}
$$

where a generic term u_{ij} is the velocity component in the i direction due to a unit component of the body velocity in the j direction, and may be obtained by solving the relevant potential flow problem.

Then, it may be seen that the kinetic energy in the whole fluid domain may be expressed in the form

$$E_T = \frac{1}{2} \sum_{i,j} m_{ij} U_i U_j, \tag{10.70}$$

where the coefficients m_{ij} are the components of the added mass tensor, and are defined as follows:

$$m_{ij} = \rho \int_v \sum_k u_{ki} u_{kj} dv. \tag{10.71}$$

Furthermore, the components of the fluid dynamic force acting on the body due to its acceleration can be written as

$$F_i = -\sum_j m_{ij} \frac{dU_j}{dt}, \tag{10.72}$$

so that we may interpret the generic component m_{ij} of the added mass tensor as the added mass associated with a force component acting on the body in the i direction due to a unit component of the body acceleration in the j direction. Having assumed the flow to be potential, the added mass tensor is found to be symmetric, and thus it may be diagonalized, i.e. a reference system exists such that the corresponding matrix of the coefficients is diagonal.

The added masses may also be easily evaluated as a function of the velocity potential functions associated with the various motions. In effect, if φ_i is the potential corresponding to the steady motion of a body with surface S moving with unit velocity in the i direction, then it may be shown that the following relation applies:

$$m_{ij} = -\rho \int_S \varphi_i \frac{d\varphi_j}{dn} dS, \tag{10.73}$$

where the normal unit vector n is directed outwards from the body.

In order to check rapidly the validity of the above formulas, let us consider a two-dimensional case and a reference system (x_1, x_2) fixed to the still fluid in which a body is moving with velocity $U \equiv (U_1, U_2)$. The flow is assumed to be potential, so that the velocity components in generic points of the flow domain may be written

$$u_1 = u_{11}U_1 + u_{12}U_2,$$
$$u_2 = u_{21}U_1 + u_{22}U_2. \tag{10.74}$$

Hence, the total kinetic energy in the flow, with some rearrangement of the various terms, becomes

$$
\begin{aligned}
E_T &= \frac{1}{2}\rho \iint (u_1^2 + u_2^2)dx_1dx_2 \\
&= \frac{1}{2}\rho \iint (u_{11}^2 + u_{21}^2)U_1^2 dx_1 dx_2 + \frac{1}{2}\rho \iint (u_{12}^2 + u_{22}^2)U_2^2 dx_1 dx_2 \\
&\quad + \frac{1}{2}\rho \iint (u_{11}u_{12} + u_{21}u_{22})U_1 U_2 dx_1 dx_2 \\
&\quad + \frac{1}{2}\rho \iint (u_{12}u_{11} + u_{22}u_{21})U_2 U_1 dx_1 dx_2,
\end{aligned} \tag{10.75}
$$

which is seen to correspond to expression (10.70) provided the added mass components are defined through relation (10.71). We also verify immediately that the added mass matrix is indeed symmetric. Therefore, we may write

$$E_T = \frac{1}{2}(m_{11}U_1^2 + m_{12}U_1U_2 + m_{21}U_2U_1 + m_{22}U_2^2), \tag{10.76}$$

from which, after some manipulations and using the symmetry of the added masses, we obtain

$$\frac{dE_T}{dt} = U_1 \left(m_{11}\frac{dU_1}{dt} + m_{12}\frac{dU_2}{dt} \right) + U_2 \left(m_{21}\frac{dU_1}{dt} + m_{22}\frac{dU_2}{dt} \right). \tag{10.77}$$

This quantity must be equal to the work done on the fluid in unit time by the force $-\boldsymbol{F}$, which is the opposite of the force acting on the body. Therefore, we have

$$\frac{dE_T}{dt} = -\boldsymbol{F} \cdot \boldsymbol{U} = -F_1U_1 - F_2U_2. \tag{10.78}$$

By comparing (10.78) with (10.77), we finally obtain the expressions for the components of the fluid dynamic force acting on the body due to its acceleration:

$$F_1 = -\left(m_{11}\frac{dU_1}{dt} + m_{12}\frac{dU_2}{dt}\right), \quad F_2 = -\left(m_{21}\frac{dU_1}{dt} + m_{22}\frac{dU_2}{dt}\right),$$

$$(10.79)$$

which are immediately seen to coincide with those that may be derived from relation (10.72).

If a body is moving with three components of linear acceleration, we have seen that the added mass is a symmetric tensor with, at most, six distinct components. However, if the shape of the body has certain symmetries, some of these terms may coincide or be zero. The extreme case is, obviously, a sphere, for which the matrix representing the added mass tensor is diagonal, with all the relevant terms equal to $0.5\rho\mathcal{V}$, i.e. to half of the displaced mass of fluid. In other words, the added mass coefficient for a sphere is 0.5. For more general shapes, the components of the added mass tensor depend on the chosen reference system and on the direction of the acceleration. On the other hand, for a two-dimensional body and flow, the added mass coefficients k_{ij} are defined by using its cross-sectional area, rather than the volume of the body. As an example, the two-dimensional body corresponding to a sphere is a circular cylinder which, if its radius is a, has an added mass equal to $\rho\pi a^2$, i.e. its added mass coefficient is equal to 1. Actually, any reference area may be used, and, indeed, this is necessary in the degenerate case of vanishing area of the body cross-section; for instance, the chosen reference area for a flat plate is, usually, that of the cross-section of a circular cylinder whose diameter coincides with the plate length.

Nonetheless, we must point out that the most general situation for a moving body is rather more complex than what we have seen above, because a generic motion comprises three linear accelerations and three angular accelerations of the body, and thus the added mass is expressed by a 6×6 matrix with 21 distinct terms, although this number may be reduced when geometric symmetries are present. Furthermore, when a body is near a solid boundary or a free-surface (such as the boundary between the sea and the atmosphere), then its added mass coefficients are different from those that correspond to an infinite fluid domain, and their evaluation may become significantly more elaborate.

We now observe that all the previous analyses and the consequent results are based on the fact that the ratio between the kinetic energy in the flow and the square of the instantaneous velocity of the body, expressed by the coefficient k defined in (10.64), was kept constant in the

derivation carried out in (10.65). This could be done because the fluid
was assumed to be non-viscous and this, in turn, implied that the flow
was irrotational and dependent on the instantaneous body velocity only.
Obviously, in a real situation in which the fluid is viscous and vorticity is
continuously introduced in the flow due to the no-slip boundary condition,
one may expect the situation to be quite different. In particular, a wake
will generally be present behind the body, with different dimensions and
vorticity organization depending on the shape of the body and on the
orientation of its motion. Therefore, there is no reason to suppose that the
irrotational values of the added mass coefficients may still be used without a
deep scrutiny of the physical conditions. A clear example is the motion of a
circular cylinder which, from the previous potential flow analysis, we found
to be characterized by an added mass coefficient equal to 1 for any direction
of its acceleration. As will be seen in more detail in Chapter 14, in a real flow
condition a wide wake with complex features is generally present behind a
moving cylinder immersed in a steady stream of fluid. Consequently, one
cannot expect the added mass to be the same, for instance, when a cylinder
oscillates in a still fluid or in a steady current and, in the latter case, in
the same direction of the free-stream or in a direction perpendicular to
it. In fact, even for a stream-wise oscillation, the situation would not be
exactly the same for increasing or decreasing velocity of the body. In other
words, the 'history' of the motion, i.e. what happened in previous times,
determines the configuration of the vorticity field at a given time, and may
be expected to affect the variations of kinetic energy of the flow caused by
changes of the body velocity, and thus to influence the values of the added
masses.

From the above considerations it should be evident that prediction
of the added masses in real and general flow conditions is an extremely
hard task: even the identification of all the features and parameters
influencing the fluid dynamic forces caused by the acceleration of a body
poses non-trivial problems. One may then wonder what the practical utility
of all the results derived from the above potential-flow analysis can be
and, more specifically, whether conditions in which they can give at least
a reasonable approximation of the real added mass coefficients do exist. In
practice, it should be quite clear that for this to be true a potential velocity
field tangent to the body must be a reasonable first-order approximation
of the actual flow field, and we have seen in previous chapters that this is
indeed the case immediately after the start of a body from rest in a still fluid.
Experience has also shown that when a body undergoes small-amplitude

oscillations in a fluid otherwise at rest, the potential-flow values of the added mass may provide good predictions, for instance, of the natural frequency of oscillation of the body, which is an important piece of information in the design of underwater structures. Obviously, the above is true as long as the separation of the boundary layer remains restricted to small regions in the rear part of the body, so that no significant wake has time to develop. In practice, this means that the amplitude of the oscillation should be small with respect to a typical dimension of the body (such as the diameter for a circular cylinder). For larger amplitudes of oscillation, or when a body is immersed in currents or waves, separation becomes important and reference should be made to experimental data obtained in flow conditions that are comparable to those of interest. Otherwise, the potential-flow values should be used with extreme caution and, in any engineering design, the influence of possible significant variations of the added masses should be carefully taken into account.

In order to illustrate, by means of a simple example, the importance of the added mass in the description of the dynamics of a body, let us consider the problem of evaluating the initial acceleration of a sphere with density ρ_s and volume $\mathcal{V}$ suddenly released in a fluid of density ρ_f. To this end, we choose a reference system with the z axis in the upwards vertical direction and we denote as w the corresponding velocity of the sphere. In addition to the forces due to inertia, gravity and buoyancy, the added mass force must also be taken into account, and thus the equation describing the initial motion of the sphere becomes

$$-\rho_s \mathcal{V} \frac{dw}{dt} - \rho_s \mathcal{V} g + \rho_f \mathcal{V} g - k\rho_f \mathcal{V} \frac{dw}{dt} = 0, \qquad (10.80)$$

where g is the acceleration of gravity. By introducing the potential-flow value of the added mass coefficient k for a sphere (which is equal to 0.5 and we have seen to be adequate for a starting motion), we then get

$$\frac{dw}{dt} = \frac{\rho_f - \rho_s}{\rho_s + 0.5\rho_f} g. \qquad (10.81)$$

If we now imagine the sphere to be a balloon filled with air initially tied to the bottom of a swimming pool by a string, which is suddenly cut, we readily see that, considering the densities of air and water given in Table 2.1, the balloon starts its motion with an upwards acceleration only slightly lower than $2g$. On the other hand, had we 'forgotten' the presence of the added mass, we would have estimated the upwards acceleration to be more than $800g$. It is also interesting to see from relation (10.81) that even a steel sphere released in air starts its motion with an acceleration that is not exactly equal to $-g$, even if the difference is practically almost imperceptible. Obviously, the above equation only applies for the initial motion, as in subsequent times not only will the added mass coefficient probably change for the previously explained reasons, but also a fluid dynamic drag proportional to the square of velocity will appear, and the body acceleration will progressively decrease, until a constant *terminal velocity* is reached.

It is beyond the scope of the present book to examine all the possible physical features and flow parameters that may influence the unsteady motion of a body immersed in a fluid. The interested reader may refer to the specialized literature on the subject, and in particular to books devoted to applications concerning the design of marine structures, where theoretical and experimental data on added mass are provided, together with deeper critical analyses on the prediction of the loads acting on bodies immersed in currents and waves (see, for instance, Sarpkaya and Isaacson, 1981, and Sarpkaya, 2010).

Chapter 11

SUPPLEMENTARY ISSUES ON VORTICITY IN INCOMPRESSIBLE FLOWS

11.1. General Expression of the Strength of Vorticity Sources

As reported in Chapter 6, the strength of the vorticity source at a solid surface in a generic three-dimensional flow is defined by the relation

$$\sigma = -\nu n \cdot \mathrm{grad}\omega, \tag{11.1}$$

where n is the unit normal vector pointing outwards from the surface, and thus inside the fluid.

We now derive the general relation expressing σ in terms of other quantities evaluated at the surface. We start by rewriting the momentum equation of an incompressible flow (5.35) in a slightly different form in which we use relation (5.42) for the viscous term and denote with a the material derivative of the velocity, which expresses the acceleration of a fluid particle,

$$a = f - \frac{1}{\rho}\mathrm{grad}p - \nu\mathrm{curl}\omega. \tag{11.2}$$

We then take the cross-product of $-n$ and Eq. (11.2), and we get

$$-n \times \left(a - f + \frac{1}{\rho}\mathrm{grad}p \right) = \nu n \times \mathrm{curl}\omega. \tag{11.3}$$

We now use the following vector identity, which is valid for any two vectors b and c,

$$b \times \mathrm{curl}c = (\mathrm{grad}c) \cdot b - b \cdot (\mathrm{grad}c). \tag{11.4}$$

This identity may be derived from relations (1.4)–(1.6) and, incidentally, it recalls that if $\text{curl}\,c = 0$ the tensor $\text{grad}\,c$ is symmetric.

If we now put $b = n$ and $c = \omega$, we have

$$n \times \text{curl}\omega = (\text{grad}\omega) \cdot n - n \cdot (\text{grad}\omega), \qquad (11.5)$$

which can be introduced in (11.3) to obtain

$$-n \times \left(a - f + \frac{1}{\rho}\text{grad}p \right) = \nu\text{grad}\omega \cdot n - \nu n \cdot \text{grad}\omega. \qquad (11.6)$$

This equation is valid for any surface inside the flow field, with the proper choice of the normal unit vector n. If we then apply it at the solid surface of a moving body, introduce (11.1) and observe that, thanks to the no-slip condition, the acceleration a of a fluid particle on the surface is equal to the acceleration a_b of the body, we arrive at the following general expression for the vorticity source over the surface:

$$\sigma = -n \times \left(a_b - f + \frac{1}{\rho}\text{grad}p \right) - \nu\text{grad}\omega \cdot n. \qquad (11.7)$$

This relation shows that the sources of vorticity at a solid surface are connected not only with the pressure gradient along the surface, as already pointed out in Chapter 6, but also with the acceleration of the body and with the body forces, whose action is similar. Note also that, in a given problem, both a_b and f are known, and that only their components tangential to the surface give a contribution to the production of new vorticity.

By using the definitions (1.6) and (1.8) of the involved operators, it may be easily verified that the last term in (11.7) is zero for a two-dimensional flow. In a three-dimensional flow, Wu and Wu (1993) showed that this term may be further divided, so that the role of different quantities may be highlighted (see also Wu *et al.*, 2006). In particular, in the case of motion of a closed body, it is also connected with the curvature of the body's surface. It must be noted, however, that in the presence of significant pressure gradients along the surface, this term is usually negligible in flows at high Reynolds numbers; nonetheless, it may be important in certain cases, such as in turbulent flows.

An important point arising from (11.7) is that, when a body starts moving from rest in a still fluid, the tangential component of the acceleration at the points over its surface is the primary source of vorticity in the flow field, together with the possible action of a body force. The pressure and viscous terms of the vorticity sources originate from the features of the flow field that is then produced around the body by its motion. It should be noted that the pressure field is also significantly connected with the non-penetration boundary condition at solid surfaces, and this is particularly evident for the motion of aerodynamic bodies with thin boundary layers. However, in all cases, the physical mechanism causing vorticity generation is the necessity of satisfying the no-slip condition. When the body reaches a uniform velocity, and assuming that body forces do not act, the continuous production of vorticity in the flow is only due to the pressure gradient along the surface and, to a lesser extent, to the viscous source term.

Further expressions for the vorticity source strength are reported by Wu and Wu (1993), who also consider the more complex case of compressible flow with non-constant viscous coefficients.

In order to gain a deeper insight on the roles of the various terms appearing in relation (11.7), let us consider the simple case of flow over a flat plate lying on the $x-y$ plane, so that $\boldsymbol{n} = \boldsymbol{e}_z \equiv (0, 0, 1)$. We assume the flow to be three-dimensional, and the components of the velocity vector to be denoted as u, v, w. Due to the no-slip condition, all these components are zero over the plate surface, and so are their derivatives with respect to x and y. However, it should be noted that the derivatives with respect to z need not be zero for $z = 0$.

Hence, by using definitions (1.5) and (1.8) for the operators, we find that in this case the vorticity source vector at the surface is equal to

$$\boldsymbol{\sigma} = -\nu \boldsymbol{n} \cdot grad\boldsymbol{\omega} = -\nu \frac{\partial \omega_x}{\partial z} \boldsymbol{e}_x - \nu \frac{\partial \omega_y}{\partial z} \boldsymbol{e}_y - \nu \frac{\partial \omega_z}{\partial z} \boldsymbol{e}_z, \tag{11.8}$$

where ω_x, ω_y and ω_z are the components of the vorticity vector $\boldsymbol{\omega}$ and all the derivatives are evaluated at $z = 0$.

Let us now derive the expression of the last term on the right-hand side of (11.7), which we may denote as the viscous contribution to the vorticity source. By recalling relation (1.6), we have

$$-\nu grad\boldsymbol{\omega} \cdot \boldsymbol{n} = -\nu \frac{\partial \omega_z}{\partial x} \boldsymbol{e}_x - \nu \frac{\partial \omega_z}{\partial y} \boldsymbol{e}_y - \nu \frac{\partial \omega_z}{\partial z} \boldsymbol{e}_z. \tag{11.9}$$

However, ω_z is zero over the surface, and the same is thus true for its derivatives with respect to x and y. Therefore, we obtain

$$-\nu grad\boldsymbol{\omega} \cdot \boldsymbol{n} = -\nu \frac{\partial \omega_z}{\partial z} \boldsymbol{e}_z. \tag{11.10}$$

By comparing (11.10) with (11.8), we note that the viscous contribution accounts for the whole normal component of the vorticity source vector $\boldsymbol{\sigma}$. In more general cases, in which a curvature of the surface is present, the viscous term may also contribute to the tangential components of $\boldsymbol{\sigma}$ (see Wu *et al.*, 2006).

As for the first term on the right-hand side of (11.7), we assume, for the sake of simplicity, that both the acceleration of the surface and the body force are zero. Then, we readily get

$$-\frac{1}{\rho}\boldsymbol{n} \times \mathrm{grad}p = \frac{1}{\rho}\frac{\partial p}{\partial y}\boldsymbol{e}_x - \frac{1}{\rho}\frac{\partial p}{\partial x}\boldsymbol{e}_y. \tag{11.11}$$

We must now show that the two terms on the right-hand side of (11.11) coincide with the corresponding ones of (11.8). To this end, we start by observing that, taking the no-slip condition into account, the components of the momentum balance equation in the x and y directions become simply

$$\frac{1}{\rho}\frac{\partial p}{\partial x} = \nu\frac{\partial^2 u}{\partial z^2}, \tag{11.12}$$

$$\frac{1}{\rho}\frac{\partial p}{\partial y} = \nu\frac{\partial^2 v}{\partial z^2}. \tag{11.13}$$

However, for points over the surface we also have

$$\frac{\partial \omega_y}{\partial z} = \frac{\partial}{\partial z}\left(\frac{\partial u}{\partial z} - \frac{\partial w}{\partial x}\right) = \frac{\partial^2 u}{\partial z^2}, \tag{11.14}$$

$$\frac{\partial \omega_x}{\partial z} = \frac{\partial}{\partial z}\left(\frac{\partial w}{\partial y} - \frac{\partial v}{\partial z}\right) = -\frac{\partial^2 v}{\partial z^2}, \tag{11.15}$$

and thus

$$\frac{1}{\rho}\frac{\partial p}{\partial x} = \nu\frac{\partial \omega_y}{\partial z}, \tag{11.16}$$

$$\frac{1}{\rho}\frac{\partial p}{\partial y} = -\nu\frac{\partial \omega_x}{\partial z}. \tag{11.17}$$

By introducing these relations into vector (11.11), we see that it indeed coincides with the tangential vector component of the vorticity source given by (11.8).

We end this section with some further comments on the normal component of the vorticity source strength, given by (11.10), which we now denote as $\sigma_n \mathbf{e}_z$. We have already pointed out that for points over the solid surface we have

$$\omega_z = \left(\frac{\partial v}{\partial x} - \frac{\partial u}{\partial y}\right) = 0, \tag{11.18}$$

and thus the no-slip condition, which is the vorticity-creating physical mechanism, only produces tangential vorticity at the surface. Therefore, a non-zero value of σ_n at the surface denotes the appearance of a tilting towards the normal direction of the vorticity lines originally parallel to the surface. This mechanism may be seen in action in three-dimensional boundary layer separations or at the surface ends of hairpin vortices present in turbulent boundary layers.

For the previously considered flat-plate flow, we may recast σ_n in terms of the tangential viscous stresses acting over the surface, say τ_{wx} and τ_{wy}. In effect, we have

$$\sigma_n = -\nu\frac{\partial \omega_z}{\partial z} = -\nu\frac{\partial}{\partial z}\left(\frac{\partial v}{\partial x} - \frac{\partial u}{\partial y}\right) = -\nu\frac{\partial}{\partial x}\left(\frac{\partial v}{\partial z}\right) + \nu\frac{\partial}{\partial y}\left(\frac{\partial u}{\partial z}\right). \tag{11.19}$$

Therefore, considering that

$$\tau_{wx} = \tau_{zx}|_{z=0} = \mu\left(\frac{\partial w}{\partial x} + \frac{\partial u}{\partial z}\right)_{z=0} = \mu\left(\frac{\partial u}{\partial z}\right), \tag{11.20}$$

$$\tau_{wy} = \tau_{zy}|_{z=0} = \mu\left(\frac{\partial w}{\partial y} + \frac{\partial v}{\partial z}\right)_{z=0} = \mu\left(\frac{\partial v}{\partial z}\right), \tag{11.21}$$

we finally get

$$\sigma_n = -\frac{1}{\rho}\frac{\partial \tau_{wy}}{\partial x} + \frac{1}{\rho}\frac{\partial \tau_{wx}}{\partial y}. \tag{11.22}$$

This relation shows that an outwards tilting of the vorticity lines at the surface may occur only if the motion is three-dimensional and the two terms on the right-hand side of (11.22) do not mutually cancel.

11.2. Mathematical Aspects of Total Vorticity Conservation

In this section we consider, in more detail, the principle of total vorticity conservation introduced in Chapter 6, which is expressed in mathematical terms by the relation

$$\frac{d}{dt}\int_{R_\infty}\omega dR = 0,\qquad\qquad(11.23)$$

where R_∞ is the two-dimensional or three-dimensional region jointly occupied by the fluid and the finite bodies contained in it. We recall that, for the validity of this relation, we have assumed that a certain number of closed solid bodies, placed at finite mutual distances, are initially at rest in an infinite domain of still fluid, and that the subsequent velocity field in the fluid is caused solely by the motion of the bodies. Thanks to the no-slip condition, this velocity field may be extended inside the region occupied collectively by the bodies, say R_s, whose joint boundary is denoted as B_s and where the vorticity vector is defined as twice the angular velocity of the bodies. On the other hand, the region occupied by the fluid is denoted as R_f, and is bounded internally by B_s.

This extension of the velocity and vorticity fields allows the fluid and the contained bodies to be treated as a single kinematic system and, as discussed by Wu (1981, 2005), it is a key point in the derivation of relation (11.23). Note that the admissibility of using this joint kinematical system derives from the behaviour of velocity and vorticity at the boundaries of the solid bodies. The velocity therein is continuous, and this also implies that the vorticity component normal to the solid boundary is continuous, being derived from velocity derivatives in the tangential directions of the tangential velocity components. In particular, at the solid boundary this component is zero if the body does not rotate, and it is equal to twice the normal component of the body angular velocity if the body is rotating. Conversely, the tangential components of vorticity are generally discontinuous, as a consequence of the fact that the normal gradients of velocity are discontinuous. To check that this is indeed the case, reference may be made to a body that translates in a fluid without rotating. In that situation, all the velocity gradients are obviously zero inside the body, where the velocity is constant, but are generally non-zero on the fluid side of its surface, where the velocity components do change in the normal direction. In summary, we may say that, in the joint system, the velocity is everywhere continuous, whereas the vorticity is piecewise continuous only, because it is generally discontinuous at the surfaces of the solid bodies. However, the kinematical description of the system implies characterizing the whole velocity field; as will be shown in more detail in next section, this may be obtained through a process of integration of a given vorticity field, and thus only a piecewise continuity of the vorticity is required.

The other fundamental point concerns the behaviour of the vorticity field at infinity. It may be demonstrated that (see Wu, 1981, and Wu *et al.*, 2006) if the flow starts from rest in an infinite domain of fluid due to the motion of finite solid bodies (so that at $t = 0$ the vorticity is confined within a finite region around their surface), then, *for any finite value of t*, the vorticity decays exponentially in the far field, i.e. for distances from the bodies tending to infinity. This result is a consequence of the fact that the fundamental solution of the diffusion equation, for any domain dimension and for any finite time, decays as $\exp(-r^2)$, where r is the distance from the finite region in which vorticity is contained at the initial time. Now, the vorticity is everywhere zero in the fluid before the motion of the bodies starts and thus immediately after the onset of the motion vorticity is confined within a finite region near the boundaries of the bodies. Subsequently, vorticity diffuses and is carried away by convection and, moreover, continues to be produced at the boundaries by the no-slip condition. However, convection is a finite-rate process, and thus does not alter the above result on the qualitative decay of vorticity at infinity.

Let us now consider a three-dimensional flow and observe that vorticity is the curl of velocity; thus, by definition, it is solenoidal. The application of the Gauss divergence theorem implies that the global flux of vorticity through the surface bounding a closed volume is zero. Therefore, if a certain amount of vorticity crosses the bounding surface locally, it must be balanced by an equal amount of vorticity crossing the surface in the opposite direction. Considering also the first Helmholtz theorem, this means that all vorticity lines in the combined solid–fluid system form closed loops. However, as just recalled, in a flow produced by the motion of a finite number of bodies initially at rest, vorticity decays exponentially, and thus more rapidly than $(1/r^2)$. Therefore, as r tends to infinity, not only the global but also the local flux of vorticity through the bounding surface will be zero. In other words, if the volume becomes large enough, it will contain all the vorticity loops present in the flow; consequently, one has, at any finite time,

$$\int_{R_\infty} \omega dR = 0, \tag{11.24}$$

which is even more stringent than (11.23). Note that relation (11.24) is actually a direct consequence of the solenoidality of the vorticity field, and a more rigorous mathematical proof of its validity may be derived. To this end, we first observe that a generalized form of Gauss' theorem (1.14) applies for any continuous tensor field $\mathbf{F}$, and may be written

$$\int_v \text{div}\mathbf{F}dv = \int_S \mathbf{n} \cdot \mathbf{F}dS, \tag{11.25}$$

where v is the volume, with boundary S, where $\mathbf{F}$ is defined, and the definitions of the relevant operators, given by (1.5) and (1.7), have been used. We now apply (11.25) to the tensor $\boldsymbol{\omega}\boldsymbol{x}$, defined using (1.10), where $\boldsymbol{x}$ is the position vector with components (x, y, z), and we consider the region occupied by the fluid, say R_f, with outer boundary Σ and inner boundary B_s, coinciding with the surface of the internal solid bodies. We then get

$$\int_{R_f} \text{div}(\boldsymbol{\omega}\boldsymbol{x})dv = \int_\Sigma \mathbf{n} \cdot (\boldsymbol{\omega}\boldsymbol{x})dS + \int_{B_s} \mathbf{n} \cdot (\boldsymbol{\omega}\boldsymbol{x})dS, \tag{11.26}$$

where the normal unit vector $\mathbf{n}$ is directed outwards from the fluid.

However, it may be readily seen, using (1.11) and (1.8), that

$$\text{div}(\boldsymbol{\omega}\boldsymbol{x}) = \boldsymbol{x}\text{div}\boldsymbol{\omega} + \boldsymbol{\omega} \cdot \text{grad}\boldsymbol{x} = \boldsymbol{\omega} \cdot \mathbf{I} = \boldsymbol{\omega}, \tag{11.27}$$

where $\mathbf{I}$ is the identity tensor and we have used the solenoidality of the vorticity vector. Therefore, we obtain

$$\int_{R_f} \boldsymbol{\omega}dv = \int_\Sigma \mathbf{n} \cdot (\boldsymbol{\omega}\boldsymbol{x})dS + \int_{B_s} \mathbf{n} \cdot (\boldsymbol{\omega}\boldsymbol{x})dS. \tag{11.28}$$

We now observe that the following identity may be easily derived from (1.5) and (1.10):

$$\mathbf{n} \cdot (\boldsymbol{\omega}\boldsymbol{x}) = (\boldsymbol{\omega} \cdot \mathbf{n})\boldsymbol{x}, \tag{11.29}$$

and thus (11.28) may be recast as follows:

$$\int_{R_f} \boldsymbol{\omega}dv = \int_\Sigma (\boldsymbol{\omega} \cdot \mathbf{n})\boldsymbol{x}dS + \int_{B_s} (\boldsymbol{\omega} \cdot \mathbf{n})\boldsymbol{x}dS. \tag{11.30}$$

An analogous derivation may be carried out for the region inside the solid bodies, say R_s, with outer boundary B_s, and thus, by noting that the normal unit vector at the boundary is now $-\mathbf{n}$, directed inside the fluid, we obtain

$$\int_{R_s} \boldsymbol{\omega}dv = -\int_{B_s} (\boldsymbol{\omega} \cdot \mathbf{n})\boldsymbol{x}dS. \tag{11.31}$$

If we now add (11.30) and (11.31), let Σ go to infinity, where the vorticity decays exponentially, and use the continuity of the normal vorticity component at the solid boundary, we obtain relation (11.24).

In the two-dimensional case, the vorticity lines do not form closed loops because they extend to infinity in the direction perpendicular to the plane of the flow, and relation (11.26) does not apply. The proof of the constancy of total vorticity is then somewhat more involved. We first rewrite the vorticity dynamics equation for a two-dimensional flow, (6.13), in a slightly modified form, in which identity (5.40) is applied to the vorticity vector and its solenoidality is taken into account:

$$\frac{D\boldsymbol{\omega}}{Dt} = -\nu\text{curl}(\text{curl}\boldsymbol{\omega}). \tag{11.32}$$

We then recall another form of Gauss' theorem, which may be derived by using some vector identities and by applying relation (1.14) to a vector $\boldsymbol{c} \times \boldsymbol{b}$, where $\boldsymbol{c}$ is a generic constant vector:

$$\int_{\mathcal{D}} \text{curl}\boldsymbol{b}d\mathcal{D} = -\int_{B_{\mathcal{D}}} \boldsymbol{b} \times \boldsymbol{n}dB_{\mathcal{D}}. \tag{11.33}$$

In this relation $\mathcal{D}$ is a closed three-dimensional or two-dimensional domain where vector $\boldsymbol{b}$ is defined, $B_{\mathcal{D}}$ is its boundary and $\boldsymbol{n}$ is the unit normal vector directed outwards from $\mathcal{D}$. It may be seen that in the two-dimensional case relation (11.33) reduces to Stokes' theorem (1.15) and, in that case, the methods for its application to non-simply connected domains described in Section 1.2 should be used.

Let us integrate Eq. (11.32) over the fluid region R_f. We use the Reynolds transport theorem, taking the constancy of density into account, and apply theorem (11.33) to the vector $\boldsymbol{b} = \text{curl}\boldsymbol{\omega}$. We thus get

$$\frac{d}{dt}\int_{R_f} \boldsymbol{\omega}dR_f = \nu\int_{B} (\text{curl}\boldsymbol{\omega}) \times \boldsymbol{n}dB, \tag{11.34}$$

where the boundary B comprises the boundary B_s of the solid bodies and a boundary at infinity, B_∞. However, the contribution of B_∞ to the integral is zero because, as we have seen, with the present assumptions $\boldsymbol{\omega}$ approaches zero exponentially with increasing distance r from the origin. The boundary B in Eq. (11.34) may then be replaced by the solid inner boundary B_s.

We now write the momentum balance equation for an incompressible flow in a form that is derived from Eq. (5.35) by introducing the material derivative operator, the expression $\boldsymbol{f} = -\text{grad}\,\Psi$ for the conservative body force and identity (5.42):

$$\frac{D\boldsymbol{V}}{Dt} = -\text{grad}\left(\frac{p}{\rho} + \Psi\right) - \nu\text{curl}\boldsymbol{\omega}. \tag{11.35}$$

If we take the cross product between Eq. (11.35) and $\boldsymbol{n}$, and integrate the result over the boundary B_s, we obtain

$$\int_{B_s} \frac{D\boldsymbol{V}}{Dt} \times \boldsymbol{n}dB_s = -\int_{B_s}\left[\text{grad}\left(\frac{p}{\rho} + \Psi\right)\right] \times \boldsymbol{n}dB_s$$
$$-\nu\int_{B_s}(\text{curl}\boldsymbol{\omega}) \times \boldsymbol{n}dB_s. \tag{11.36}$$

The first integral on the right-hand side of (11.36) can be seen to be zero by applying (11.33) and recalling that the curl of a gradient is always zero. Therefore, Eq. (11.36) and Eq. (11.34) can be combined to get

$$\frac{d}{dt}\int_{R_f}\boldsymbol{\omega}dR_f = -\int_{B_s}\frac{D\boldsymbol{V}}{Dt} \times \boldsymbol{n}dB_s. \tag{11.37}$$

Note that the unit vector $\boldsymbol{n}$ points towards the inside of the solid bodies. We now observe that B_s coincides with the global external boundary of the region R_s occupied by the solid bodies (which is given by the sum of all the regions R_j occupied by the single bodies). The important point to note is that, thanks to the no-slip condition, the material acceleration is continuous on B_s, i.e. it is identical for the solid bodies and the fluid. If we then apply relation (11.33) to the material derivative of velocity and to the domain R_s, also considering that the external normal to the boundary of this domain is $-\boldsymbol{n}$, we obtain

$$\int_{B_s}\frac{D\boldsymbol{V}}{Dt} \times \boldsymbol{n}dB_s = \int_{R_s}\text{curl}\left(\frac{D\boldsymbol{V}}{Dt}\right)dR_s = \frac{d}{dt}\int_{R_s}\boldsymbol{\omega}dR_s. \tag{11.38}$$

By introducing (11.38) into (11.37) we obtain (11.23), i.e. the relation stating that the total vorticity in the flow is constant in time. As already pointed out, this relation is less stringent than (11.24), which is valid for a three-dimensional flow. However, since the motion starts from rest, we initially have $\boldsymbol{\omega} = 0$ everywhere, and thus the integral of vorticity remains equal to zero also for subsequent times.

As we have seen, the viscous no-slip condition plays an essential role in the proof of total vorticity conservation, but we may also note that at least as important is the behaviour of vorticity at infinity, namely the fact that it decays exponentially. Once again, it should be recalled that, for this to be true, the motion must have started from rest (so that vorticity is initially confined to a finite region) and a *finite* time must have elapsed from the start of the flow. Therefore, one should not expect, for instance, that the total vorticity necessarily be zero for a steady flow which, even being a solution of the Navier–Stokes equations, can be reached only after an infinite time from the start of the motion.

We have actually seen a case of this type, in Chapter 6, as regards the irrotational flow outside a circular cylinder of radius r_0 impulsively set in rotational motion with constant angular velocity Ω. Were we to consider the situation after an infinite time from the start of the motion, then we would conclude that the final steady solution is characterized by a vorticity field that is zero outside the cylinder and globally equal to $2\Omega\pi r_0^2$ inside the cylinder, in apparent contrast with (11.23), considering that the initial vorticity was zero. However, the total vorticity is conserved, and equal to zero, for any $t < \infty$ because we have seen that, at the start of the motion, a quantity of vorticity $-2\Omega\pi r_0^2$ is generated over the outer surface of the cylinder and then diffuses in the outer infinite domain, where it becomes locally infinitesimal only after an infinite time. As also discussed in Chapter 10, this final steady solution corresponds to infinite kinetic energy of the flow, and also to an infinite power having been introduced into the fluid, and thus it cannot be considered as strictly realistic from a physical point of view, but only as an approximation for sufficiently large values of t.

Another interesting case, which was discussed in Chapter 8, is the flow around an airfoil over which a lift force acts. We have noted that, after a sufficiently long (albeit finite) time from the onset of the motion, the starting vortex initially shed from the airfoil trailing edge is far enough away not to affect the velocity field around the airfoil, where the flow may thus be considered as steady in a reference system fixed to the body. However, the integral of the vorticity contained in the boundary layers around the airfoil is not zero, and its value is equal and opposite to the strength on the starting vortex. Therefore, to apply (11.23), we must consider the actual unsteady flow that is present for any finite time from the start of the motion, and thus the whole flow domain, including the starting vortex.

11.3. Evaluation of the Velocity Field from the Vorticity Field

11.3.1. *General vorticity distributions*

We consider now the problem of determining the velocity field that corresponds to a given vorticity field. We assume that at a certain time the value of the vorticity vector $\boldsymbol{\omega}$ is known in all the domain of interest, and derive the formulas for the velocity at a generic point of the domain from the definition of vorticity and from the incompressible flow assumption, which implies that the velocity field is solenoidal. More general expressions, including the possible presence of non-zero values of the quantity $\vartheta = \operatorname{div} \boldsymbol{V}$ in certain parts of the flow, may be found, for instance, in Batchelor (1967) or in Wu *et al.* (2006).

The velocity field we seek, say $\boldsymbol{u}_v$, is then such that $\operatorname{div} \boldsymbol{u}_v = 0$ and

$$\operatorname{curl} \boldsymbol{u}_v = \boldsymbol{\omega}. \tag{11.39}$$

Given that the divergence of a curl is zero, it is then natural to introduce the *vector potential* $\boldsymbol{A}$ such that

$$\boldsymbol{u}_v = \operatorname{curl} \boldsymbol{A}. \tag{11.40}$$

Consequently, from (11.39) and (5.40) we get

$$\boldsymbol{\omega} = \operatorname{curl}(\operatorname{curl} \boldsymbol{A}) = \operatorname{grad}(\operatorname{div} \boldsymbol{A}) - \nabla^2 \boldsymbol{A}. \tag{11.41}$$

Let us assume, for the moment, that $\operatorname{div} \boldsymbol{A} = 0$. We then derive from (11.41) that the vector potential satisfies the following Poisson equation:

$$\nabla^2 \boldsymbol{A} = -\boldsymbol{\omega}. \tag{11.42}$$

The general solution of this equation is

$$\boldsymbol{A}(\boldsymbol{x}) = -\int_R G(\boldsymbol{x} - \boldsymbol{x}') \boldsymbol{\omega}(\boldsymbol{x}') dR', \tag{11.43}$$

where R is the region occupied by the fluid, $\boldsymbol{x}$ is the vector defining the position where the solution is evaluated, $\boldsymbol{x}'$ is the integration position vector, $dR' = dR(\boldsymbol{x}')$, and $G(\boldsymbol{x} - \boldsymbol{x}')$ is the Green's function providing the fundamental solution of the Poisson equation, which is given by

$$G(\boldsymbol{x} - \boldsymbol{x}') = \frac{1}{2\pi} \ln |\boldsymbol{x} - \boldsymbol{x}'| \quad \text{in a two-dimensional space,} \tag{11.44}$$

$$G(\boldsymbol{x} - \boldsymbol{x}') = -\frac{1}{4\pi |\boldsymbol{x} - \boldsymbol{x}'|} \quad \text{in a three-dimensional space.} \tag{11.45}$$

For instance, if we use the notation $\boldsymbol{\omega}' = \boldsymbol{\omega}(\boldsymbol{x}')$, $\boldsymbol{r} = \boldsymbol{x} - \boldsymbol{x}'$ and $r = |\boldsymbol{x} - \boldsymbol{x}'|$, we get for a three-dimensional domain

$$\boldsymbol{A}(\boldsymbol{x}) = \frac{1}{4\pi} \int_R \frac{\boldsymbol{\omega}'}{r} dR'. \tag{11.46}$$

We may now enquire into the conditions for the function given by (11.46) to comply with our assumption that div $A = 0$. It may be shown (see, e.g., Batchelor, 1967) that A is indeed solenoidal provided that at the boundary of the considered domain, with normal unit vector n, we have $\omega \cdot n = 0$. We have already seen that this condition is satisfied for an externally unbounded fluid domain, as well as for the flow around a non-rotating body. When $\omega \cdot n \neq 0$ at some points on the boundary, the domain may be extended beyond the actual flow boundary, keeping the normal vorticity component continuous there, towards a new boundary where $\omega \cdot n = 0$. In the case of bodies moving in an unlimited fluid domain, we have already seen that the vorticity and velocity fields may be continued inside the bodies, at whose surface $\omega \cdot n$ is continuous. We then assume that the above condition, $\omega \cdot n = 0$, is satisfied at the actual or virtual boundary of the fluid domain.

The velocity field u_v may then be derived from relation (11.40) by introducing the expression of A derived from (11.43), with the function G given by (11.44) or (11.45), according to whether the domain is two-dimensional or three-dimensional. The result is (see, e.g., Batchelor, 1967, or Wu *et al.*, 2006)

$$u_v(x) = \frac{1}{2(d-1)\pi} \int_R \frac{\omega(x') \times r}{r^d} dR', \qquad (11.47)$$

where d is the domain dimension ($d = 2$ and $d = 3$ in two-dimensional and three-dimensional domains respectively).

In particular, the formula for a three-dimensional domain is

$$u_v(x) = \frac{1}{4\pi} \int_R \frac{\omega' \times r}{r^3} dR'. \qquad (11.48)$$

Due to its analogy with a formula in electromagnetic theory, relation (11.48) is known as the *Biot–Savart law*, and is usually expressed by saying that a certain vorticity field *induces* a corresponding velocity field given by (11.48). Actually, this expression may lead to a certain misinterpretation, giving the impression that vorticity causes the velocity field; more correctly, one should say that (11.48) gives the solenoidal velocity field whose curl has the specified value $\omega(x)$ in the domain, or, in other words, which is associated with the given distribution of vorticity.

It must now be observed that the velocity field (11.40) is not the only one for which the curl is the given distribution of vorticity. In effect, any velocity field V obtained by the addition of u_v and of an irrotational field with potential ϕ is still characterized by the same vorticity field. Therefore, a more general expression for the required velocity field is

$$V = u_v + \text{grad}\phi. \tag{11.49}$$

However, if we want V to represent a possible incompressible flow, it must be solenoidal, and thus, considering that the divergence of a gradient is the Laplacian, the potential ϕ must be *harmonic*, i.e. it must satisfy the Laplace equation

$$\nabla^2 \phi = 0. \tag{11.50}$$

Nonetheless, as an infinite number of harmonic functions exist, it is obvious that (11.49), with u_v given by (11.47) and ϕ satisfying (11.50), is an expression representing an infinite number of functions that are solutions of the problem of finding a solenoidal velocity field with prescribed vorticity values. Therefore, boundary conditions on the velocity field are necessary to characterize a specific solution of the problem that may have a practical interest. For instance, one may consider u_v, namely the function given by the Biot–Savart law, as the particular solution of the problem that corresponds to grad$\phi = 0$ and that is compatible with a vanishing velocity on the boundary B of the considered region and with a non-zero value of vorticity only in a finite region that is at infinite distance from the boundary. In other words, it corresponds to the solution for an infinite space domain with vanishing velocity at infinity and prescribed vorticity values inside the domain.

A particularly important point arising from the mathematical problem we are considering is that it may be shown (see, e.g., Wu *et al.*, 2006) that, for the problem of a finite flow field delimited by a generic boundary, the solution (11.49) becomes unique if one prescribes on the boundary *either* the value of the normal velocity, $V \cdot n$, *or* the value of the tangent vector component, which may be expressed through $V \times n$. Conversely, if both values are given, the problem is over-specified.

As an example, let us assume that the vorticity field is prescribed in the whole domain (which implies that u_v is also known) and $V \cdot n$ is given on the boundary. From (11.49) one gets the following boundary condition:

$$\frac{\partial \phi}{\partial n} = V \cdot n - u_v \cdot n, \tag{11.51}$$

which determines a unique potential function ϕ satisfying the Laplace equation (11.50) in the domain; therefore, the complete velocity field is obtained. However, this solution implies a consequent value of $V \times n$ on the boundary, which might not be in agreement with prescribed conditions on the flow, such as the no-slip condition at the solid surface of a moving body. Analogously, it may be seen that by prescribing the value of $V \times n$ on the boundary, the solution is uniquely determined, together with the relevant boundary value of $V \cdot n$.

As a consequence of the above, it should be clear that not all vorticity distributions are compatible with a given incompressible flow problem in which, for instance, the value of the whole velocity vector is prescribed at a boundary of the fluid domain, as may be the surface of a moving solid body. However, this should not come as a surprise, as the above treatment is based solely on kinematical relations, namely the definition of vorticity and the condition of solenoidality of the velocity deriving from the assumption of incompressible flow. In other words, the dynamical equations of motion have not been considered in any way.

Therefore, expressions (11.47), (11.49) and (11.50) define the velocity field associated, at a certain instant of time, with *any* prescribed vorticity field but, from the dynamical point of view, not all vorticity fields may represent a physically realistic flow problem. Considering, as an example, the flow produced by a moving body starting from rest in an indefinite domain of fluid, we have seen that vorticity is continuously generated at the solid surface of the body in order to satisfy the no-slip condition, and then evolves in the flow as prescribed by the vorticity dynamics equation. Thus, the amount and distribution of vorticity at a certain time is the consequence of the initial condition, of the shape of the moving body, and of the whole history of its previous motion. Incidentally, this is why one may say that *the instantaneous vorticity field contains the memory of the flow*. Therefore, if we want to use relation (11.49) in a procedure for the evaluation of a physically representative velocity field at a certain time, then the vorticity field at that same time must be the one corresponding to the whole previous history of the motion.

A realistic vorticity field may be considered to be known, at least with a satisfactory approximation, when vorticity is concentrated in thin boundary layers and wakes, as happens for the flow around aerodynamic bodies in uniform translation (see Chapters 12 and 13). In more general situations, the solution of a flow problem may also be carried out using a velocity–vorticity approach, in which the kinematical relations are used together with the simultaneous or iterative solution of the dynamical equations. To this end, the following relation, known as the *generalized Biot–Savart law*, may be useful (for its derivation see, e.g., Wu, 2005)

$$V(x) = \frac{1}{2(d-1)\pi} \int_R \frac{\omega' \times r}{r^d} dR'$$

$$+ \frac{1}{2(d-1)\pi} \left(\int_B \frac{(V' \times n') \times r}{r^d} dB' - \int_B \frac{(V' \cdot n')r}{r^d} dB' \right),$$

$$(11.52)$$

where B is the boundary of the fluid domain R, whose dimension is d, $dB' = dB(x')$, $V' = V(x')$ and $n' = n(x')$ is the normal unit vector directed outwards from the domain R. It may be shown that the sum of the two integrals over the boundary B may be expressed as gradϕ, where ϕ is a function satisfying the Laplace equation. Therefore, relation (11.52) is effectively equivalent to (11.49), with the particular feature that the values of both the normal and the tangential components of the velocity over the boundary appear explicitly.

It should be pointed out that (11.52) is not in contradiction with the fact that one cannot specify the value of both the normal and tangential components of velocity on the boundary. Given a certain vorticity field, it may be seen that by prescribing any of the two above quantities, the other one can consequently be derived from appropriate manipulations of the same relation (11.52). However, this relation may also be used to obtain the vorticity field (or, more precisely, the vorticity layer produced instantaneously over the solid boundary) if the velocity on the boundary is known, as is the case when the non-penetration and no-slip conditions are both applied to a body moving from prescribed initial conditions in an unlimited domain of fluid.

The rationale of this approach to the solution of a flow problem is that, if the velocity and vorticity fields are completely known at a certain time (including the one at which the initial conditions are prescribed), the time derivative of vorticity may be derived from the vorticity dynamics equation (6.9), and used to obtain new values of the vorticity in the domain. The first integral in relation (11.52) is then subdivided into two terms, one containing the new field values of the vorticity and the other the unknown values of new vorticity on the boundary that must be produced to comply with the velocity boundary condition, which gives the velocity on the boundary at all times. Relation (11.52) may then be suitably recast as an integral equation for the intensity of a layer of new vorticity over the boundary. Once this equation is solved, new values of the velocity in the field may be obtained from the original form of (11.52), and the procedure may be continued for a subsequent time step. Finally, the values of the pressure in the field may be derived from a Poisson equation obtained by applying the divergence operator to the momentum equation. However, if interest is only in the pressures acting over the boundary of a moving body, as is often the case, this step may be significantly simplified. Furthermore, as will be seen in Section 11.4, the global forces acting on a body may be directly derived from the knowledge of the dynamics of the vorticity field.

This type of approach leads to methods of solution of flow problems that are alternative to those using as unknown functions the so-called primitive variables, namely, velocity and pressure, and may be shown to have some advantages for particular cases (see, e.g., Wu, 2005, and Wu *et al.*, 2006).

In the following sections, we shall assume that the whole vorticity field is known at a certain time in an unbounded domain. This means that, if a finite number of solid bodies are moving in a fluid, the velocity and vorticity fields have been properly extended inside the bodies, so that the fluid and the bodies may be considered as a joint continuous domain with external boundary at infinity, where the velocity and vorticity properly vanish. The original Biot–Savart law (11.47) may then be employed to derive the velocity corresponding to a known vorticity field, and we shall derive the kinematic expressions that apply when the spatial distributions of vorticity are such that certain simplified treatments may be used.

11.3.2. *Vortex sheets*

As has been repeatedly pointed out in previous chapters, when a flow is characterized by high values of the Reynolds number, the convection terms in the vorticity dynamics equation prevail over the diffusion terms. Therefore, in many circumstances, vorticity remains confined within thin layers, either contouring the surface of moving bodies or trailing downstream in their wakes. It is then reasonable to investigate the influence of these thin layers of vorticity in the limit of Reynolds number tending to infinity, which we can consider as a first approximation of the real situation that may occur when the Reynolds number is very high.

The first point to be noted is that, with increasing values of Re, the thickness of a vorticity layer, say δ, will progressively decrease, and we may consider the vorticity to be contained between two surfaces that approach progressively and tend to become parallel. Let us then consider a point P lying on an intermediate surface with normal unit vector $n(P)$, and denote as $\delta(P)$ the thickness of the layer at the same point. We then introduce the following vector, which may be defined for finite values of the Reynolds number,

$$\boldsymbol{\gamma}(P) = \int_{\delta(P)} \boldsymbol{\omega} dx_n = \bar{\boldsymbol{\omega}}(P)\delta(P), \qquad (11.53)$$

where x_n is the distance along $\boldsymbol{n}(P)$, the integral is taken over the thickness of the layer, and we denoted as $\bar{\boldsymbol{\omega}}(P)$ the corresponding mean value of the vorticity.

We may now construct a model of the vorticity layer by letting the Reynolds number tend to infinity, while keeping constant the global amount of vorticity present within the layer. Therefore, $\delta(P)$ becomes zero, and $\bar{\boldsymbol{\omega}}(P)$ becomes infinite, but in such a way that their product, i.e. the quantity $\boldsymbol{\gamma}$ given by (11.53), remains constant. The vorticity layer then becomes a surface S, which we might naturally call a *vorticity sheet*, although it is better known as a *vortex sheet*. The finite global amount of vorticity present at each point of a vortex sheet is given by $\boldsymbol{\gamma}(P)$, which is then denoted as the local *strength* of the vortex sheet. Note that the dimensions of the vector $\boldsymbol{\gamma}$ are vorticity times length (i.e. the same as velocity), and that it is parallel to the vortex sheet, i.e. we have on S

$$\boldsymbol{\gamma} \cdot \boldsymbol{n} = 0. \qquad (11.54)$$

Let us now evaluate the velocity field associated with (or, with the usual notation, induced by) a vortex sheet, assuming vorticity to be zero elsewhere. To this end, we use the Biot–Savart law in the form (11.48), and consider that the integral must now be evaluated only over the domain where vorticity is present, namely the vortex sheet surface S. For the velocity at a generic point, defined by the position vector $\boldsymbol{x}$, we get

$$\boldsymbol{u}_v(\boldsymbol{x}) = \frac{1}{4\pi} \int_S \frac{\boldsymbol{\gamma}(\boldsymbol{x}') \times \boldsymbol{r}}{r^3} dS', \qquad (11.55)$$

where $\boldsymbol{x}'$ is the vector defining the position of the integration points, which now vary over S, $dS' = dS(\boldsymbol{x}')$, and, again, $\boldsymbol{r} = \boldsymbol{x} - \boldsymbol{x}'$.

The velocity given by expression (11.55) is finite when the point $\boldsymbol{x}$ is outside the vortex sheet, whereas if $\boldsymbol{x}$ lies on S then a singular integral arises in the evaluation. It may be shown (see, e.g., Saffman, 1992) that when the evaluation point crosses the vortex sheet, the velocity has a jump that concerns its tangential component only, while the normal component is continuous. More precisely, the velocity is different if we approach a certain

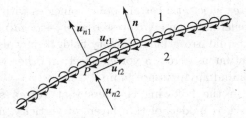

Fig. 11.1.   Vortex sheet and relevant nomenclature.

point P on the sheet from one side or the other. In particular, with reference to Fig. 11.1, if we call 1 and 2 the domains on the two sides of the vortex sheet, n the normal unit vector directed from 2 to 1, $u_1(P)$ and $u_2(P)$ the velocities on the two sides of the sheet in correspondence of point P, and $u_\gamma(P)$ the velocity of the vortex sheet at the same point, we have

$$u_1(P) = \frac{1}{2}\gamma(P) \times n(P) + u_\gamma(P),$$
$$u_2(P) = -\frac{1}{2}\gamma(P) \times n(P) + u_\gamma(P).$$
(11.56)

We thus see that the velocity u_γ of a generic point of the vortex sheet, which may be evaluated from the principal value of a singular integral (see Saffman, 1992), is equal to the arithmetic mean of the velocities that are present on the two sides of the sheet. Furthermore, the velocity jump $\Delta u = u_1 - u_2$ is given by the expression

$$\Delta u = \gamma \times n,$$
(11.57)

which obviously also implies

$$\gamma = n \times \Delta u.$$
(11.58)

In summary, we may say that across a vortex sheet the normal velocity component remains continuous, whereas the tangential velocity component undergoes a discontinuity, whose magnitude is equal to the strength of the vortex sheet.

So far we have considered a vortex sheet as a kinematic concept, which models the distribution of vorticity at a certain instant of time. However, it may also be used as a dynamical model, to represent the time evolution of a thin layer of vorticity, provided its thickness remains always small enough to be modelled through a vortex sheet. In effect, it can be shown that the velocity of the vortex sheet at a certain point coincides with the velocity of the flow, and, in general, it can be expressed as $\boldsymbol{u}_\gamma + grad\phi$, where ϕ is the potential of the external irrotational velocity field. It may also be seen that the pressure is continuous across a vortex sheet; in other words, a vortex sheet cannot withstand any pressure jump.

We shall see in the following chapters that vortex sheets may be advantageously used as models of thin layers of vorticity, particularly for the evaluation of the flow field around aerodynamic bodies. In effect, there are many situations in which vortex sheets may be considered as first-order models of real flows in which vorticity is confined within thin layers, as we have already seen for the case of boundary layers attached over the surface of moving bodies. However, vortex sheets may also be used to model free shear layers, either separating two potential flow regions, or contouring regions of flow containing distributed vorticity. This last case may happen, for instance, when a boundary layer first separates from a solid surface and then reattaches to it, forming a closed region where vorticity tends to become constant, and which is separated from the outer irrotational flow by a thin layer of concentrated vorticity.

Finally, we point out that vortex sheets may also be encountered in flows that we model as being two-dimensional, in which case their shape is defined by a curve and the associated vector $\boldsymbol{\gamma}$ is directed as the normal to the plane of motion, and does not vary in that direction. It is then advantageous to use the two-dimensional form of the Biot–Savart formula, which may be derived from (11.47) by putting $d = 2$.

11.3.3. *Line vortices*

When the flow Reynolds numbers are high, thin vortex layers originating from the separation of boundary layers may rapidly roll up to form tight spirals with many turns. This process is essentially due to convection, but the simultaneous action of diffusion between the closely spaced rolled-up sheets often leads to a situation in which vorticity is confined within extremely limited regions of space. We commonly refer to this condition by saying that a *vortex* has formed, but the reader should be warned that the exact definition of a vortex is still a largely open problem (see, e.g., Wu *et al.*, 2006, and the references therein for a discussion on this topic). As concerns our present analysis, suffice it to say that, in many cases, vorticity becomes concentrated within tubes of very small cross-section, whereas one may safely assume the outer flow to be irrotational.

As we have already seen in Chapter 6, the solenoidality of the vorticity field ensures that the strength Γ of a vortex tube remains constant. Again, we may construct a simple model of the above-described flow by imagining that we progressively contract the cross-section of the vortex tube while keeping Γ constant; therefore, the average vorticity through the tube cross-section tends to become infinite as the cross-section area vanishes, but their product remains constant. The vortex tube then becomes a so-called *line vortex* (not to be confused with a vortex-line), which is characterized by its strength Γ and by the direction of a local unit vector tangent to the line, say t. If we denote as s the curvilinear coordinate along the line vortex, the quantity $\Gamma t ds$ represents the global amount of vorticity contained in an elementary length ds of the line vortex or, more precisely, of the vortex tube whose model is the line vortex. In effect, it must be remembered that the dimensions of Γ are vorticity times surface.

Therefore, the velocity due to the line vortex may be evaluated through the three-dimensional Biot–Savart law (11.48), which becomes

$$u_v(x) = \frac{\Gamma}{4\pi} \int_C \frac{t \times r}{r^3} ds, \qquad (11.59)$$

where the integration is carried out along the whole length of the curve C representing the line vortex. We note that, in order for the velocity field represented by (11.59) to be solenoidal, C must be a closed curve. Nonetheless, we may apply this relation to an infinitely long line vortex, imagining curve C to be closed by a large semi-circle with radius tending to infinity. In effect, it is easy to see that the contribution of this far portion of the line vortex vanishes.

Contrary to the case of a vortex sheet, a *curved* line vortex is not a suitable dynamical model. This is because the velocity induced at a point on a curved line vortex by its remaining part becomes infinite, unless its curvature is zero (see, e.g., Saffman, 1992). This singularity may be avoided by considering a curved vortex tube with a small but finite core, whose self-induced velocity may be shown to be finite. As an example, with this expedient one could analyse the motion of a circular vorticity tube modelling a smoke ring, and find that it is self-propelled, i.e. it moves in space due to its self-induced velocity. However, as this velocity depends on the chosen shape and dimension of the finite core, to obtain a reliable dynamical model one should adopt a procedure to model a realistic evolution of the core, which remains an open and complex problem. On the other hand, a straight line vortex does not have the difficulty of infinite

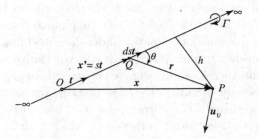

Fig. 11.2.   Velocity induced by a straight line vortex.

self-induced velocity, and may thus be used as a dynamical model; in the following we shall discuss its associated vorticity field in some more detail.

We first consider an infinite straight line vortex, which, as we have seen, may be imagined as a line vortex whose two 'ends' at infinity are joined by a semi-circle with effectively infinite radius. We then choose the origin of the coordinates for the position vectors at a point O on the line vortex, and consider the velocity induced on a generic point $P(\boldsymbol{x})$, while the inducing line vortex element is placed at the variable point $Q(\boldsymbol{x'})$, with $\boldsymbol{x'} = st$, as shown in Fig. 11.2, where the vector $\boldsymbol{r} = \boldsymbol{x} - \boldsymbol{x'}$ is also displayed.

The velocity at P is then given by (11.59), which becomes

$$\boldsymbol{u}_v(\boldsymbol{x}) = \frac{\Gamma}{4\pi} \int_{-\infty}^{+\infty} \frac{\boldsymbol{t} \times \boldsymbol{r}}{r^3} ds. \qquad (11.60)$$

It can then be seen that the induced velocity is directed perpendicularly to the plane defined by $\boldsymbol{t}$ and $\boldsymbol{r}$, as indicated in Fig. 11.2 for a given orientation of the line vortex. Considering that $|\boldsymbol{t} \times \boldsymbol{r}| = r \sin \theta$, with θ defined in Fig. 11.2, the velocity modulus, say $u_v(\boldsymbol{x})$, is then equal to

$$u_v(\boldsymbol{x}) = \frac{\Gamma}{4\pi} \int_{-\infty}^{+\infty} \frac{\sin \theta}{r^2} ds. \qquad (11.61)$$

However, if h denotes the normal distance from P to the line vortex, and s_p the coordinate of the intersection between the normal from P and the line vortex, the following relations hold

$$r = \frac{h}{\sin \theta}, \qquad (11.62)$$

$$s_p - s = h \cot \theta, \qquad (11.63)$$

$$ds = \frac{h}{\sin^2 \theta} d\theta, \qquad (11.64)$$

and thus we get

$$u_v(\boldsymbol{x}) = \frac{\Gamma}{4\pi h} \int_0^\pi \sin\theta\, d\theta = \frac{\Gamma}{2\pi h}. \tag{11.65}$$

Hence, we find that the velocity is inversely proportional to the normal distance of the considered point from the line vortex, and does not depend on the position of the point along the line vortex. If we then consider the motion in any plane perpendicular to the line vortex, the velocity field is effectively two-dimensional, with streamlines that are concentric circles around the intersection of the infinite-length line vortex with the plane of motion, which we may denote as a *point vortex* with strength Γ.

If we then use a cylindrical coordinate system (r, θ, z), with the z axis coinciding with the infinite straight line vortex, the components of the velocity field are $v_r = 0$, $v_z = 0$ and

$$v_\theta = \frac{\Gamma}{2\pi r}. \tag{11.66}$$

This same relation may be easily obtained by using the two-dimensional version of the Biot–Savart formula, and by assuming all the vorticity Γ to be concentrated in the point vortex. The corresponding velocity field is shown in Fig. 11.3, and, by comparison with Fig. 3.6, it is easy to see that it coincides with the velocity field of a Rankine vortex whose core radius has been reduced to zero, while keeping the global contained vorticity constant.

We have already pointed out that a Rankine vortex is just a model of a realistic situation, because constant-vorticity cores cannot actually exist in an evolving flow. Now we see that a point vortex may be considered as an even more simplified model, which leads to a velocity field that is singular at the point of application of the vortex, where v_θ becomes infinite. In spite of this singularity, ideal point vortices may be used to model the

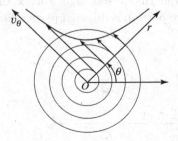

Fig. 11.3. Velocity induced by a point vortex.

effect on the flow of regions with concentrated vorticity, which, as already pointed out, may often occur when convection effects prevail over those of viscous diffusion. We have already seen that both Rankine vortices and more realistic ones, such as the Oseen vortices, are characterized by associated velocity fields that approach the irrotational distribution given by (11.66) as the distance from the centre of the vortex increases. Thus, their effect in regions that are not close to their cores of concentrated vorticity may be appropriately modelled using point vortices.

Furthermore, point vortices may also be used to follow the variation in time of the position of concentrated-vorticity regions, because their application points can be displaced with the velocity corresponding to existing external flow fields, subtracting the singular velocity deriving from relation (11.66). And this is indeed reasonable, because we have seen that the velocity is zero at the centre of realistic vortices placed in an otherwise still fluid. The external velocity field causing the motion of a vortex centre may be due, for instance, to the motion of bodies in the fluid, but it may also be the one induced by other vortices present in the flow. As an example, two point vortices of opposite strengths Γ and $-\Gamma$, placed at a mutual distance h, will both move with velocity $\Gamma/(2\pi h)$ in a direction perpendicular to the segment joining their application points. This motion will closely represent the actual dynamics of two real vortices, at least as long as diffusion does not cause the radius of their vorticity cores to become comparable with $h/2$, and provided an additional velocity field due to adjacent solid boundaries is not present.

Let us now return to the three-dimensional case and analyse the velocity field induced by a finite portion of a straight line vortex. The geometrical configuration will then be as depicted in Fig. 11.4, where θ_1 and θ_2 denote the angles between the vortex unit vector $\boldsymbol{t}$ and the vectors joining the extremities of the segment with the considered field point.

If Γ is, again, the strength of the straight vortex, the velocity associated with the considered finite portion is directed perpendicularly to the plane of

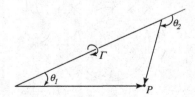

Fig. 11.4.   Velocity induced by a segment of straight line vortex.

the figure, with the orientation deriving from that of vector t; its modulus may be easily obtained with the same procedure used for an infinite vortex, and we thus get the following relation:

$$u_v(\boldsymbol{x}) = \frac{\Gamma}{4\pi h} \int_{\theta_1}^{\theta_2} \sin\theta d\theta = \frac{\Gamma}{4\pi h}(\cos\theta_1 - \cos\theta_2). \qquad (11.67)$$

We have already pointed out that a segment of line vortex cannot exist by itself, but it may be considered as a portion of a closed loop of concentrated vorticity, which can be modelled as being composed of straight segments. We shall see in Chapter 13 that this type of model may be extremely useful for the first-order evaluation of the lift force acting on aeroplane wings.

The formula for a semi-infinite straight line vortex may also be useful, and can be readily obtained by putting $\theta_1 = \pi/2$ and $\theta_2 = \pi$ in relation (11.67). We thus get

$$u_v(\boldsymbol{x}) = \frac{\Gamma}{4\pi h}. \qquad (11.68)$$

11.4. Loads on Moving Bodies in Terms of Vorticity Dynamics

Relations (8.1) and (8.2), introduced in Chapter 8, show that the global fluid dynamic forces and moments acting on moving bodies are derivable from knowledge of the distribution of the stresses (pressures and viscous stresses) acting over the surface of the bodies. In experiments, these global loads may be directly obtained by using suitable balances, whereas the integration of the surface stresses gives good predictions, in practice, only for the pressure component of the loads. This is because the measurement of viscous stresses acting over wind tunnel models, even if feasible, is generally considerably more difficult than the measurement of surface pressures. In fact, even when the fluid dynamic loads are obtained by numerical means, a reliable prediction of the viscous stresses is usually quite problematic when the typical Reynolds number of the flow is high. Therefore, alternative methods to obtain the fluid dynamic loads in terms of quantities evaluated in the flow field, rather than on the surface, are desirable, especially considering the increasingly reliable predictions of these quantities provided by modern experimental and numerical tools.

The global force acting on a solid body immersed in a fluid may be evaluated from a straightforward application of the momentum balance, provided the pressure and velocity fields on a chosen control surface are known. If we neglect the action of body forces, and use the Reynolds transport theorem in the form (3.41) for the variation of the quantity ρV, the integral momentum balance for a generic volume v, bounded by a surface S with outer normal unit vector n, becomes

$$\int_v \frac{\partial \rho V}{\partial t} dv + \int_S n \cdot (\rho VV) dS = - \int_S pn dS + \int_S \tau_n dS. \qquad (11.69)$$

If we then choose a volume of fluid bounded internally by the body surface B and externally by a suitable fixed surface Σ, and recall that

$$n \cdot (VV) = V(V \cdot n), \qquad (11.70)$$

we may recast Eq. (11.69) as follows:

$$\int_v \frac{\partial \rho V}{\partial t} dv + \int_{B+\Sigma} \rho V(V \cdot n) dS = - \int_{B+\Sigma} pn dS + \int_{B+\Sigma} \tau_n dS. \qquad (11.71)$$

If we denote by F the fluid dynamic force acting on the body, and consider that along B the normal n is directed inside the body, we have

$$-F = - \int_B pn dS + \int_B \tau_n dS. \qquad (11.72)$$

We now note that, if the body is at rest in a moving fluid or if it moves in translatory motion, the second integral on the left-hand side of (11.69) vanishes for $S = B$. This derives from the no-slip boundary condition and the application of the divergence theorem to the constant-velocity domain inside the body. We thus get from (11.71)

$$F = - \int_v \frac{\partial \rho V}{\partial t} dv + \int_\Sigma [-pn + \tau_n - \rho V(V \cdot n)] dS. \qquad (11.73)$$

This relation is particularly useful when the flow in the considered volume is steady, so that the first term on the right-hand side of (11.73) vanishes. It could be advantageously used, for instance, to evaluate the drag of an aerodynamic body in a wind tunnel by choosing Σ to be sufficiently far from the body that the flow is practically uniform on its upstream and lateral sides. In that case, only measurements over a downstream cross-plane are necessary. Obviously, the same is not true for unsteady flow conditions, as might be the case when the body oscillates, or when significant fluctuations are present in its wake. For instance, as we shall see in more detail in Chapter 14, bluff bodies are often characterized by a periodic alternate shedding of concentrated vorticity structures in their wakes, which may produce velocity fluctuations of the same order of magnitude as the free-stream velocity.

Nonetheless, load formulas in terms of flow-field variables and, in particular, solely in terms of velocity and vorticity, are useful for many important reasons. Pressures and viscous stresses are the *consequence* of the perturbation flow produced by a body, and it is thus of great importance to achieve a deeper understanding of the connection between surface stresses and flow features. A deep physical insight into the flow structures that contribute most to the various components of the fluid dynamic loads opens the way to the development of simplified procedures for their evaluation. For instance, one may model the essential physical features accurately, and neglect or model at a lower level of accuracy the remaining unessential ones. Furthermore, and perhaps more importantly, knowledge of the physical mechanisms generating the fluid dynamic loads may also provide invaluable clues for modifying the shape of a body in order to obtain certain desired design objectives. In other words, it may be a fundamental tool in load control and optimization.

It has already been pointed out in Chapter 8 that an approximate procedure of load prediction may be devised for aerodynamic bodies moving at low angles of attack due to the particular features of their associated flow field, namely the existence, over their upper and lower surfaces, of thin attached boundary layers which join at the sharp trailing edge, forming a thin wake containing opposite-sign vorticities. As will be better described in Chapter 12, this procedure provides good first-order estimates of the lift and drag acting on an airfoil without having to solve the complete Navier–Stokes equations, and has been the basis of aeronautical design for most of the 20th century. This is thus an example of the construction of an effective simplified mathematical model on the basis of the recognition of fundamental physical aspects of the flow generating the forces. However, the present availability of advanced numerical and experimental methods allows more complex flow fields to be described in detail, even in the presence of boundary layer separation and finite regions containing steady or unsteady vorticity. It is thus very useful to have at one's disposal relations providing the contribution of these flow features to the loads experienced by a moving body. Indeed, one might consider the possibility, also for these more complex flow fields, of constructing simplified models describing, in an accurate manner, only the essential flow features connected directly with the generation of the fluid dynamic loads. Furthermore, one might try to devise procedures for the control of the characteristics and of the magnitude of the loads.

We then consider the special, but fundamental, case of a solid body moving in an infinite domain of fluid R_f, delimited internally by the boundary B of the body; the region occupied by the body is denoted by R_b. The motion is assumed to have started from rest, so that, at any finite time, the vorticity present in the flow decays exponentially with increasing distance from the body, over whose surface it has been generated. In these conditions, we have already seen that the vorticity field may be continued inside the body, where it coincides with twice its angular velocity. We may then denote by R_∞ the whole region jointly occupied by the fluid and the body. With the above assumptions and definitions, it may be shown that the following formula applies for the fluid dynamic force acting on the body:

$$\boldsymbol{F} = -\frac{\rho}{d-1}\frac{d}{dt}\int_{R_\infty} \boldsymbol{x} \times \boldsymbol{\omega} dR + \rho\frac{d}{dt}\int_{R_b} \boldsymbol{V}_b dR, \qquad (11.74)$$

where d is the domain dimension, $\boldsymbol{x}$ is a position vector with respect to a chosen origin, and $\boldsymbol{V}_b$ is the velocity inside the body. Note that if the body is rotating, then the corresponding vorticity value must be included in the first integral, and the velocity $\boldsymbol{V}_b$ is not space-constant inside the body. The proof of relation (11.74) is beyond the scope of the present book, and the reader is referred to Wu (1981, 2005).

Let us now discuss the implications of expression (11.74). The first point to note is that the second integral, representing the force that would be needed to accelerate a mass of fluid filling the region occupied by the body and moving with the same instantaneous velocity, is known once the time history of the body velocity has been given; obviously, this term is zero if the body does not accelerate. On the other hand, the first term on the right-hand side of (11.74) is connected with the rate of variation of the first moment of vorticity in the whole fluid–solid domain, and is generally non-zero even if the body moves with a constant velocity and is subjected to a steady force. Therefore, this is the fundamental term directly connecting flow-field features with the forces acting over the body boundary. In particular, we see that the global force on the body is linked to the evolution in time of vorticity. More precisely, we find that only if the moment of vorticity varies in time does a force act on the body, and, therefore, the magnitude of the force is dependent on the variation of both the amount and the space distribution of vorticity.

To further investigate the physical insight we may gain from expression (11.74), we refer to the simple case of a body moving with uniform translational velocity. As the motion has been assumed to have started from rest, we already know that equal amounts of positive and negative vorticity are present in the fluid domain at any time. These opposite-sign vorticities have been generated over the body surface by the no-slip boundary condition, and have been subsequently introduced and redistributed in the flow by diffusion and convection. Furthermore, we also know that all these processes are continuously taking place as long as a relative motion between body and fluid exists.

Now, we learn from (11.74) that the direction and magnitude of the force acting on the body depend on the time variation of the moment of the opposite-sign vorticities generated by the body's motion. This means, for instance, that different bodies having similar typical dimensions and moving at the same velocity may be subjected to very different forces according to the characteristics of the vorticity field they produce continuously and to its evolution in time. It is easy to be convinced that the different features of the generated vorticity field depend on the shape and type of motion of the body.

As a simple application of the above concepts, we can now explain, not only from a phenomenological point of view, but also in mathematical and physical terms, the significant difference between the drag experienced by aerodynamic and bluff bodies. Let us consider a two-dimensional aerodynamic body, such as an airfoil, and analyse the velocity and vorticity field it produces when it moves in uniform translation, say in the $-x$ direction with velocity magnitude U. As repeatedly pointed out on previous occasions, if sufficient time has elapsed from the start of the motion, the flow field immediately around the body will not appreciably vary in time in a reference system fixed to the body, i.e. in this reference system it will be approximately steady and similar to the one illustrated in Fig. 6.11. However, the flow is never exactly steady if we consider a global domain extending to infinity, whatever the reference system used, because the wake produced by the body lengthens continuously. As a matter of fact, new amounts of vorticity of opposite sign are released constantly in the wake from the boundary layers leaving the upper and lower surfaces of the body and joining at its trailing edge.

The order of magnitude of the drag may then be estimated from an assessment of the variation of the x component of the moment of vorticity — which is the component connected with the drag force — caused by the continuous introduction of vorticities of opposite sign from the upper and lower surfaces of the body. Now, the amount of vorticity of each sign released into the wake in unit time is linked to the flux of vorticity in each boundary layer at the trailing edge, which, as can be deduced from relation (7.67), is of the order of $U_t^2/2$, where U_t is the velocity outside the boundary layer at the same position. If the pressures on the two sides of the trailing edge are equal, as normally happens for aerodynamic bodies, the value of U_t is also the same on the upper and lower surfaces, and thus equal amounts of vorticity of opposite sign are released into the wake. We also note that U_t is of the same order as the velocity of the body, so that, if we put $U_t = kU$, we have $k \simeq 1$. Furthermore, the centre of gravity of the released vorticity is positioned at a distance from the surface of the order of the displacement thickness of each boundary layer. Consequently, the order of magnitude of the x component of the vorticity moment introduced instantaneously into the wake may be estimated to be $k^2 U^2 (\delta_u^* + \delta_l^*)/2$, where δ_u^* and δ_l^* are the displacement thicknesses, at the trailing edge, of the upper and lower boundary layers respectively. Furthermore, as the vorticity from the upper surface is clockwise — and thus negative — whereas that from the lower surface is counter-clockwise, the contribution to the time variation of the x component of the vorticity moment is negative; therefore, due to the negative sign in (11.74), the corresponding force is in the positive x direction, and thus opposes the motion, as expected.

Although this is only a first-order estimate of the actual time variation of the x component of the moment of vorticity, it clearly indicates that the drag of an aerodynamic body must not be very large, considering that, if the Reynolds number of the motion is high, the thickness of the boundary layers is very small compared to the typical dimension of the body. Conversely, if we consider a bluff body, boundary layer separation occurs before the end of the body, and the distance between the separating boundary layers, which release opposite-sign vorticities into the wake, is of the order of the width of the wake, which is clearly much larger than any thickness of the boundary layers. Furthermore, if we denote by $U_s = kU$ the velocity outside the boundary layer at each separation point, we generally have $k > 1$. Therefore, this explains both the much larger drags of bluff bodies, compared to those of aerodynamic bodies of similar typical dimensions, and the dependence of the drag of a bluff body on the width of its wake and on the magnitude of the velocity outside the boundary layer at the separation points. We thus find a confirmation, based on a specific formula for the fluid dynamic forces, of the conclusions we have drawn in Chapter 8 from a phenomenological analysis of the experimentally observed pressure distributions.

We now show that relation (11.74) also allows us to explain that the release of a starting vortex from an airfoil moving at a non-zero angle of attack produces a cross-flow force, i.e. a lift, and also to estimate its magnitude. In effect, with reference to the discussion in Chapter 8 and to Fig. 8.6, after a sufficient time from the start of the motion we may assume that the starting vortex — containing a total amount Γ of positive vorticity — moves away from the airfoil with a velocity U, and thus the instantaneous increase in the y component of the vorticity moment is equal to $-U\Gamma$. Consequently, the cross-flow force acting on the airfoil is positive, i.e. directed upwards, and its magnitude is $\rho U \Gamma$. The same result was anticipated in Chapter 5 with formula (5.57), where Γ was the (negative) circulation around a two-dimensional body immersed in a non-viscous irrotational flow, and will be shown in Chapter 12 to be a fundamental output of the first-order procedure for the evaluation of lift on an airfoil. This result may be put in more general form by writing the following formula for the *cross-flow force* acting on a two-dimensional body immersed in a uniform free-stream with velocity U, namely

$$\boldsymbol{F}_{cf} = \rho \boldsymbol{U} \times \boldsymbol{\Gamma}, \tag{11.75}$$

where $\boldsymbol{\Gamma}$ is a vector representing the resultant of the vorticity present over the body surface, and whose magnitude is then equal to the circulation around the body. From (11.74) and from the conservation of the total vorticity, we determine that formula (11.75) may be applied for any body if $\boldsymbol{\Gamma}$ denotes the resultant of the vorticity present around the body and moving with it, and if we can assume that the global value of vorticity released in previous times from the body, $-\boldsymbol{\Gamma}$, is being carried away with velocity $\boldsymbol{U}$. Therefore, the formula may also be applied, for instance, in the presence of separation bubbles over the body's surface, and for any previous time history of the body motion, provided a sufficiently long time has elapsed from the instant when the motion became a uniform translation.

However, relation (11.74) has a wider range of application, as from it we can derive, perhaps from the outputs of numerical simulations, the fluctuating forces that act on a body due to a periodic shedding of vorticity in their wakes, which we shall see in Chapter 14 to occur for practically all two-dimensional bluff bodies. More generally, it may be applied to any unsteady flow condition, and can provide important clues to the relative contribution of the dynamics of the various vorticity structures released in the wake of any type of body.

We finally observe that expression (11.74) may also be recast in another form, in which the volume integral is solely extended to the fluid domain and only the velocity on the body contour must be known (see, e.g., Wu *et al.*, 2006):

$$\boldsymbol{F} = -\frac{\rho}{d-1}\frac{d}{dt}\int_{R_f} \boldsymbol{x} \times \boldsymbol{\omega} dR - \frac{\rho}{d-1}\frac{d}{dt}\int_B \boldsymbol{x} \times (\boldsymbol{n} \times \boldsymbol{V}_b)dB. \qquad (11.76)$$

We recall that in this relation B is the body surface and the normal unit vector $\boldsymbol{n}$ is directed outwards from the body and towards the fluid.

It is also possible to derive expressions similar to (11.74) and (11.76) for the moment acting on a body moving in an unlimited fluid domain; the corresponding formulas, valid for both two-dimensional and three-dimensional domains, are (see Wu, 1981, Wu *et al.*, 2006)

$$\boldsymbol{M} = \frac{\rho}{2}\frac{d}{dt}\int_{R_\infty} x^2 \boldsymbol{\omega} dR + \rho\frac{d}{dt}\int_{R_b} \boldsymbol{x} \times \boldsymbol{V}_b dR, \qquad (11.77)$$

$$\boldsymbol{M} = \frac{\rho}{2}\frac{d}{dt}\int_{R_f} x^2 \boldsymbol{\omega} dR + \frac{\rho}{2}\frac{d}{dt}\int_B x^2 \boldsymbol{n} \times \boldsymbol{V}_b dB. \qquad (11.78)$$

It should be noted that many other non-standard formulas are available to obtain the loads on bodies in relative motion in a fluid. In particular, different expressions exist for the total fluid dynamic force as a function of various field quantities, or even solely in terms of integrals of velocity and of its derivatives over arbitrary finite control-boundaries. Such formulas are described by, among others, Noca *et al.* (1999), Graziani and Bassanini (2002), Zhu *et al.* (2002), Wu *et al.* (2005). On the other hand, Wu and Wu (1993, 1996) developed a theory to evaluate forces and moments in terms of boundary-vorticity fluxes, which may be applied also for compressible flow and deformable body surfaces. These diverse theoretical approaches may be advantageously used in different circumstances, particularly when data deriving from experimental or numerical investigations are available. However, thanks to the deeper understanding they may provide of the main flow features contributing to fluid dynamic loads, their most promising application fields are probably load control and body shape optimization. For further details on the derivation and possible use of the various force expressions, reference may be made to Wu *et al.* (2006), and Wu *et al.* (2007).

11.5. Wake Vorticity Organization and Drag

In the previous section we have derived alternative mathematical expressions for the evaluation of the fluid dynamic force acting on a body moving in a fluid. In Chapter 10 we introduced an energetic interpretation of drag, connecting its value with the variation of total energy in the fluid caused by the work done by the opposite of the drag force. Furthermore, we have seen that, for a body in uniform translation, this variation — and thus drag — is essentially due to the effect of viscosity, and in particular to the no-slip boundary condition on the body surface, which causes a continuous introduction of new kinetic energy into the flow. However, we have seen that the no-slip condition is also responsible for the generation of vorticity, and that a body moving in an infinite domain of fluid initially at rest introduces in the flow, and particularly in its wake, equal amounts of vorticity of opposite sign. Now we want to analyse, albeit only in a qualitative manner, if and how the organization of this vorticity is connected with the variation of total energy in the flow, and thus with the drag experienced by the moving body generating the vorticity.

Let us consider, for the sake of simplicity, a two-dimensional flow in which two equal amounts of vorticity of opposite sign are present, such as might have been released in the wake of a body in a certain time interval, and try to estimate the kinetic energy of their associated velocity field. To this end, we construct an extremely simplified flow model, in which we assume the relevant vorticities to be concentrated in two circular regions and constant therein. In practice, we are considering the flow around two Rankine vortices with equal core radius a and containing the same total amount of constant vorticity of opposite sign, say Γ and $-\Gamma$. We assume that the two vortices are placed in an infinite fluid domain with a distance $2h$ between their centres, with $h \geq a$; moreover, we neglect, at least for the moment, the presence of convection due to the flow produced by any body or boundary. In other words, we estimate the kinetic energy of the two vortices as if contained in an infinite domain of fluid otherwise at rest.

We observed in Chapter 10, when analysing the flow around a rotating circular cylinder, that the kinetic energy associated with a single Rankine vortex in an unlimited domain is infinite. However, this is not the case for two counter-rotating vortices, and the reason is that the velocity field associated with a couple of opposite-sign vortices decays much more rapidly with increasing distance from their cores. In effect, as can be seen in Fig. 11.5, the velocity vector of the combined field outside the core of the two

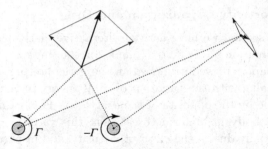

Fig. 11.5. Velocities induced at different points by two opposite-sign Rankine vortices.

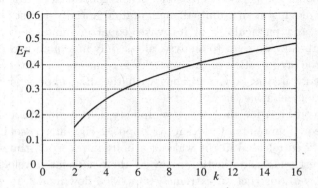

Fig. 11.6. Kinetic energy of two opposite-sign Rankine vortices.

vortices is the sum of two vectors that tend to become equal and opposite for large distances from the two cores.

The evaluation of the kinetic energy E of an isolated pair of Rankine vortices implies both closed-form and numerical integrations, and was carried out by Buresti (2008); the result is shown in Fig. 11.6, where the quantity $E_\Gamma = E/(\rho\Gamma^2)$ is reported as a function of the parameter $k = 2h/a$, i.e. of the ratio between the mutual distance and the core radius of the two vortices. The curve in the figure, which applies for non-superposing cores of the vortices, i.e. for $k \geq 2$, may be closely approximated by the relation

$$E = \rho\Gamma^2(0.0398 + 0.1591 \ln k). \qquad (11.79)$$

We thus see that the energy associated with two Rankine vortices is proportional to the square of their strength, increases with increasing distance between their centres, and decreases with increasing radius of their cores. Obviously, we recover the expected results that the energy diverges if the distance between the vortices goes to infinity and if the radius of their cores tends to zero, in which case the vortices become ideal point vortices with infinite velocity at their centre.

The general conclusion we can draw from the above simplified analysis is that the introduction into the flow of concentrated vorticity structures requires work to be done on the fluid, work that increases with the strength of the structures and with the distance between the regions in which the vorticities of opposite sign are contained. This is in perfect agreement with formula (11.74), which indeed shows that the drag acting on a moving body is proportional to the time variation of the component of the vorticity moment in the direction of motion. The fact that the kinetic energy of two opposite vortices of given strength decreases as their cores increase in size, whereas, as may easily be verified, their moment remains unchanged, does not affect the above conclusion. As a matter of fact, the process of diffusion of vorticity due to the action of viscosity coincides with the dissipation of kinetic energy. Thus, as already pointed out in Chapter 10, what one has to 'pay for' with the drag force acting on the body is the production of kinetic energy associated with newly introduced vorticity structures and the sole part of dissipation connected directly with their production, and not the portion linked to their subsequent diffusion through viscosity, which simply converts kinetic energy into internal energy without changing their sum, i.e. the total energy. Incidentally, from relation (10.45) we found that the dissipation connected with a Rankine vortex is proportional to the square of its strength and to the inverse of the square of its core radius. Furthermore, we may deduce from the Bobyleff–Forsyth formula that the dissipation of two Rankine vortices of opposite sign with non-superposing cores in an infinite domain of fluid is equal to the sum of the dissipations the two vortices would produce if they were isolated. Therefore, we may plausibly infer that the amount of dissipation in the time interval necessary for the production of two new vorticity structures is higher if these structures are concentrated in small regions of space.

At variance with the above described simplified example, in the actual flow around a moving body the dynamics of the vorticity released in the wake from the boundary layers depends on the Reynolds number and on the shape of the body. For high values of the Reynolds number, this dynamics is primarily dominated by convection and, in particular, by the velocity field immediately downstream of the body, i.e. in the region that is usually denoted as the near-wake. For instance, the width of the wake is directly linked to the shape of the body, on which the boundary layer separation points depend, and to the action of confinement exerted by the outer potential flow. Thus, for high Reynolds numbers, the vorticity is convected for a significant extent before starting to diffuse. In fact, the picture may be more complex, and different situations may occur even for similar widths of the wake according to the type of convective action caused by the near-wake velocity field. As will be seen in Chapter 14, for many bluff bodies the vorticity released from the separating boundary layers may undergo, essentially by convection, a process of rapid concentration in small regions of space, thus producing structures not very dissimilar from the Rankine vortices we have just analysed. Obviously, from what we have seen above, this corresponds to high values of the introduced kinetic and internal energies, and thus of the drag force. If this process is hampered in some way, and the concentration of the released vorticities of opposite sign in small and distant regions is avoided, then the total energy variation in the near-wake decreases and, consequently, the same happens for the drag. Note that this does not contradict the connection of drag with the moment of vorticity. In effect, the width of the wake — being mainly determined by the above-cited confinement effect of the outer flow — is not significantly affected by an action to avoid the concentration of vorticity. However, such an action would cause a redistribution of the vorticities of opposite sign in a larger portion of the wake width, with a consequent decrease of the mutual distance between their centres of gravity; therefore, a reduction of the time variation of the vorticity moment would occur.

In conclusion, from all the above analyses we may derive a rationale for decreasing the drag of bodies having a given cross-section in a plane perpendicular to the free-stream. The first point is, obviously, to reduce the width of their wakes by delaying the separation of the boundary layers developing over their surfaces. For fixed wake width, we still have the possibility of trying to reduce the velocity outside the boundary layers at separation, thus decreasing the amount of vorticity introduced instantaneously into the wake. But we may also attempt to change, by passive or active means, the distribution of the vorticity that is shed into the wake. Indeed, we have seen that any action aimed at avoiding the concentration of vorticity of opposite sign in small and distant regions reduces the global introduction of total energy in the flow, and thus provides a reduction of the drag experienced by the body.

11.6. Two Vortices in a Cup

We now propose conducting a simple experiment, which can be easily carried out by the reader, and to interpret its results using as a guide all that has been learned as regards vorticity and its dynamics in the present and previous chapters. Let us take a cup of tea or coffee, perhaps with some milk, and perform the following action: partially immerse a teaspoon into the liquid, in the centre of the cup, move it quickly with a rectilinear motion and then withdraw it rapidly. The result will be the production of two dimples on the downstream side of the two edges of the spoon, around which the fluid moves in opposite directions, as depicted in Fig. 11.7(a). The circulatory motion around the dimples may be made more visible if small floating particles are sprinkled over the surface of the liquid. Subsequently, we observe that the dimples start moving parallel to each other, and also that their separation increases when they approach the surface of the cup. Finally, when they get really close to the cup surface, they start moving parallel to it in opposite directions, further separating from each other, and then rapidly disappear before 'climbing' a significant portion of the cup surface. The trajectory of the dimples is thus as qualitatively shown in Fig. 11.7(b). Obviously, the details of the above phenomenology depend on the initial movement of the spoon, which should be as perpendicular as possible to the spoon concavity in order to get symmetrical dimples of equal intensity; furthermore, higher velocities correspond to deeper dimples.

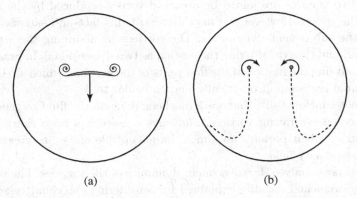

(a)   (b)

Fig. 11.7.   Generation (a) and trajectories (b) of two dimples in a cup of liquid.

To analyse the outcome of this experiment, we start by observing that we have every reason to interpret the dimples as the signatures of two regions with concentrated vorticity or, in other words, of two counter-rotating vortices. In fact, we know that the initial flow caused by the impulsive motion of the spoon is effectively irrotational, but also that thin layers of vorticity of opposite sign are simultaneously generated over the spoon surface due to the no-slip boundary condition. However, the streamlines of the outer potential flow contour the sharp edges of the spoon, where the velocity is then extremely high, with a subsequent rapid deceleration. The consequent adverse pressure gradient present just downstream of the two edges is thus so strong that it cannot be negotiated by the vorticity layers. Therefore, they immediately separate, and the convection of the generated unsteady velocity field is such that their opposite vorticities are almost instantaneously concentrated in two very tight cores of approximately circular shape. We have thus produced two structures not very dissimilar from a couple of Rankine vortices, and this also explains why dimples are produced over the surface of the liquid. In effect, as shown by relation (9.93) and Fig. 9.7, the pressure decreases inside the core of a Rankine vortex, by an amount that is proportional to the square of the vortex strength and inversely proportional to the square of the core radius. Thus a depression — the dimple — forms over the liquid surface, with a shape that depends on the equilibrium between pressure forces, surface tension forces, centrifugal forces and gravity forces; as a result, the depth of the dimple increases with the strength of the vortex. In reality, the two structures that were produced are not exactly Rankine vortices, because the spoon is finite, and thus the flow is three-dimensional and the two vortices are joined by a curved vortex produced by the lower part of the spoon. However, we may disregard this fact and satisfactorily explain the subsequent dynamics of the vortices by assuming them to be rectilinear, and thus considering the motion as two-dimensional. In practice, we are analysing the motion of the first parts of the vortices, which are closer to the liquid surface and practically perpendicular to it.

We note, incidentally, that work has been done on the fluid to generate the two counter-rotating vortices, and thus a force has been exerted on the spoon — albeit perhaps an almost imperceptible one — to create the associated velocity field.

Let us now analyse the subsequent dynamics of the vortices. The initial parallel movement is readily explained by considering the equal velocities induced on each vortex core by the opposite one. If we denote by Γ the absolute value of the strength of each vortex, and by d the distance between the centres of the two vortices, the magnitude of this velocity is equal to $\Gamma/(2\pi d)$. Note that this value is correct not only for a point vortex or a Rankine vortex, but also, with excellent approximation, for any more realistic vortex, provided d is sufficiently larger than the radius of the cores or, more precisely, of the regions where the two opposite vorticities are effectively concentrated; in our experiment we may safely assume this to be practically true.

In order to justify the progressive increase of the mutual distance between the vortices as they approach the inner surface of the cup, we must first analyse the effect of the velocity field induced by the two vortices over that surface. Due to the necessity of satisfying the non-penetration condition, the streamlines of the induced field tend to become tangent to the inner cup surface; as a result, the no-slip condition causes the generation of a layer of vorticity over the whole surface. As shown in Fig. 11.8, the sign of this vorticity is opposite on the two sides of the inner cup surface, and on each side it is opposite to that of the vortex present in the same half of the cup.

It should then be easily understood that the velocity field associated with this additional vorticity will increasingly influence the motion of the two vortices as they get closer to the cup inner surface. In fact, there are two concurrent causes for this increased effect: the first is that the velocity induced by the surface vorticity is inversely proportional to the distance of the vortices, and the other is that the same is also true for the velocity induced by the vortices on the surface. Thus, as the vortices get closer to the cup surface, the strength of the surface vorticity increases,

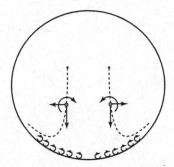

Fig. 11.8. Vorticity produced over the cup surface by the induced velocity of the vortices.

being proportional to the intensity of the approaching vortex and inversely proportional to its distance from the surface. We may then deduce, as a first-order estimate, that each vortex moves laterally with a velocity that is inversely proportional to the square of its distance from the inner cup surface.

We may now also explain why the vortices disappear quite rapidly as they get close to the surface or, in other words, why they are not able to travel round a significant portion of the cup contour before fading away. When each vortex reaches a position very close to the surface, and thus to the opposite-sign vorticity that is present therein, the gradient of vorticity in the direction perpendicular to the surface becomes very high, and this causes strong vorticity diffusion. As a result, each vortex is quickly annihilated by the opposite-sign vorticity that itself has generated over the cup surface. Therefore, we may say that, even if it is indeed diffusion that eventually causes the annihilation of vorticity, the role of the previously acting convection is essential in enhancing this phenomenon.

Chapter 12

AIRFOILS IN INCOMPRESSIBLE FLOWS

12.1. Geometry and Nomenclature

In this chapter we shall consider the incompressible two-dimensional flow around airfoils, which may be seen as cross-sections of hypothetical infinite wings, as we have already pointed out in Chapter 8. In other words, we are dealing with infinite cylinders placed in a stream of fluid, whose cross-sections have shapes with certain specific features. The analysis of this particular situation will be a first fundamental step towards the development of methods for the evaluation of the loads acting on more realistic finite wings of generic shape, provided their motion is such that the flow around them remains subsonic everywhere, i.e. the velocity at every point in the flow is lower than the local velocity of sound. To begin with, let us analyse the main characteristics of the shape of a typical airfoil, and introduce, with the help of Fig. 12.1, the notation that is generally used for its description.

As described in Chapter 8, the fore and aft extremities of an airfoil are termed the *leading edge* and the *trailing edge*, respectively. The former may also be assumed to coincide with the fore point with maximum curvature, and is generally rounded for a subsonic airfoil. Conversely, the trailing edge is always sharp, so that, as already pointed out in Chapter 7, the boundary layers may remain attached over the whole airfoil contour for appropriate directions of the upstream velocity, i.e. for sufficiently small angles of attack.

The segment connecting the leading and trailing edges is called the *chord* (usually denoted by c), and is normally used as the reference length of an airfoil. Furthermore, the leading and trailing edges divide the airfoil contour in two parts, which are generally called the upper and lower *surfaces*

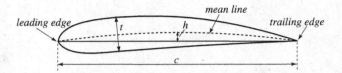

Fig. 12.1.   Airfoil geometry and notation.

of the airfoil, even though they are actually curved lines. In practice, with this notation we are referring to the surfaces of the infinite wing whose cross-section is the considered airfoil.

The line positioned midway between the upper and lower surfaces is called the *mean line* of the airfoil. When this line is curved, we say that the airfoil is *cambered*, and thus the shape of the mean line defines the distribution of *camber*. The important parameters of a cambered airfoil are the value of the maximum camber, say h, and its position along the chord from the leading edge; both these quantities are often given as a fraction or percentage of the chord length. Obviously, when the mean line is straight, it coincides with the chord, and we say that the airfoil is *symmetrical*. In fact, only symmetrical airfoils were considered in Chapter 8. The distance between the upper and lower surfaces of the airfoil identifies the *thickness distribution*. Again, the maximum thickness value, t, and its position along the chord are significant parameters as regards the aerodynamic behaviour of an airfoil.

The geometry of an airfoil may thus be thought as being constructed by starting from a segment representing the chord, adding a curve whose extremities coincide with those of the chord — which corresponds to the mean line — and finally superposing, above and below the mean line and perpendicularly to it, the curve giving the thickness distribution.

The analysis of the loads acting on an airfoil is usually carried out by using a reference system fixed to the airfoil, which is thus assumed to be at rest and immersed in a stream of fluid with velocity U, as shown in Fig. 12.2.

As already pointed out in Chapter 8, the angle α between the direction of the vector U and the chord is called the *angle of attack* of the airfoil, and is usually defined as positive when the airfoil orientation, with respect to the free-stream, is as depicted in Fig. 12.2. The resultant aerodynamic force acting on the airfoil — per unit length in the direction perpendicular to the plane of motion — is decomposed into a component in the direction of U and a component perpendicular to it, respectively known as the *drag*

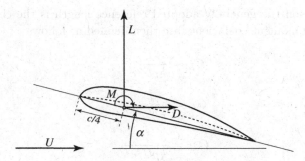

Fig. 12.2. Airfoil geometry and notation.

and the *lift* of the airfoil, which will now be denoted by D and L. If the free-stream is directed rightwards, as in Fig. 12.2, the positive direction of the lift is that obtained through a counter-clockwise rotation from the positive direction of the drag. The lift force is also associated with the definition of an important value of the angle of attack, namely the angle that corresponds to a zero value of the airfoil lift. The zero-lift angle of attack is usually denoted by α_0, and is obviously zero for a symmetrical airfoil. Conversely, it is negative for an airfoil with positive camber, such as the one shown in Fig. 12.2.

For the complete characterization of the load on an airfoil, it is necessary to specify the moment M around a reference point along the chord, where we choose to place the resultant force vector. In fact, the direction of the resultant force intersects the chord at a point around which the value of M is zero; however, this *application point* of the force may also be identified by giving the values of the resultant force and of the moment around a chosen reference point. The choice of this reference point is arbitrary, and thus one might use, for instance, the leading or the trailing edges. However, for reasons that will become clearer in the following, it is usual to adopt the point at a quarter of the chord length, measured from the leading edge. This choice is connected with the definition of another important point, namely the *aerodynamic centre* of the airfoil, which is the point around which the moment does not change with variations of α. Considering that, as shown in Chapter 8, the lift on an airfoil increases with the increase of angle of attack, we may also define the aerodynamic centre as the point of application of the *variations* of lift with α.

The performance of an airfoil is generally analysed through non-dimensional coefficients corresponding to the various load components, for whose definition the generally adopted reference length is the chord. The lift, drag and moment coefficients are then defined as follows:

$$C_L = \frac{L}{\frac{1}{2}\rho U^2 c}, \tag{12.1}$$

$$C_D = \frac{D}{\frac{1}{2}\rho U^2 c}, \tag{12.2}$$

$$C_M = \frac{M}{\frac{1}{2}\rho U^2 c^2}, \tag{12.3}$$

where U is the modulus of the free-stream velocity.

For a given angle of attack, all these coefficients depend on the geometry of the airfoil, and in particular on the distributions of camber and thickness. The effects of the variation of the different geometrical parameters defining the airfoil shape will be discussed briefly in the last section of the present chapter, after the description of the available methods for the prediction, at different levels of approximation, of the loads acting on an airfoil. Nonetheless, we may anticipate that the main objective in the design of an airfoil is usually to obtain, for the most common conditions of utilization, a high value of its *efficiency*, defined as $E = L/D$, i.e. as the ratio between the lift produced and the drag experienced by the airfoil.

12.2. Load Evaluation Procedure

12.2.1. *First-order potential-flow model*

In previous chapters we have already pointed out that a simplified load prediction procedure may be used for aerodynamic bodies, thanks to the fact that the boundary layers remain attached and thin over the whole surface. In that procedure, the equations that apply in the outer potential flow and in the boundary layers are solved iteratively. However, for the sake of simplicity, only the case of a symmetrical airfoil moving parallel to its chord, and thus producing no lift, has been considered so far. We now describe the main features of the application of the iterative method to the case of an airfoil producing lift.

We start by considering a symmetrical airfoil moving with constant velocity and a small positive angle of attack. In Chapter 8, and in particular in Section 8.2.2, we described in detail the physical mechanisms causing the shedding of a starting vortex, say of strength Γ, during the initial transient of the motion, and the production of an equal and opposite circulation

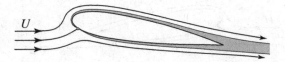

Fig. 12.3. Flow around a symmetrical airfoil at a small angle of attack.

around the airfoil. We now assume that sufficient time has elapsed from the start of the motion, so that the starting vortex is sufficiently far from the airfoil not to affect the velocity field around its contour. Consequently, in a reference system fixed to the airfoil, the situation will be as schematically shown in Fig. 12.3, with two attached boundary layers developing over the upper and lower surfaces of the airfoil and smoothly joining at its trailing ,edge to form a thin wake. We also note that, in the considered reference system, the flow may be assumed to be steady because the continuous lengthening of the wake does not cause any measurable variation of the velocity field around the airfoil.

The flow field depicted in Fig. 12.3 is the one we must predict in order to evaluate the forces acting on the airfoil. In principle, this might be obtained by solving the full equations of motion numerically but, in practice, we can take advantage of the fact that, thanks to the shape of the body and the small value of the angle of attack, the boundary layers are attached over the whole airfoil contour and the pressure is constant across them. Furthermore, in practical applications the Reynolds numbers are generally larger than 10^6 and thus, from the analyses in Chapter 7, we may expect the thickness of the boundary layers at the trailing edge to be, at most, of the order of a few hundredths of the chord length. We also recall that, with excellent approximation, the boundary layers and the wake they produce are the only regions where vorticity is present, while the outer flow is vorticity-free, and thus a potential flow.

Therefore, we may construct a first-order model of the flow around the airfoil by imagining squeezing the boundary layers to zero thickness, keeping the global amount of vorticity present within them unaltered. In this model, the potential velocity field becomes effectively tangent to the airfoil contour although, more precisely, we should say it is tangent to infinitesimally thin vortex sheets. Consistently, we also squeeze the thin wake; however, we recall from Section 8.2.2 that, after the end of the initial transient, equal amounts of vorticity of opposite sign are instantaneously released in the wake from the upper and lower boundary layers. Therefore, the ideal process of squeezing the wake to zero thickness implies that

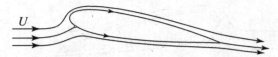

Fig. 12.4.   First-order model flow around a symmetrical airfoil at a small angle of attack.

the vorticity it contains is annihilated. In practice, we neglect the very weak velocity field induced by the adjacent opposite-sign vorticities present in the wake, which then becomes a streamline of the resulting potential flow. The velocity field corresponding to the first-order model may thus be schematically represented as in Fig. 12.4.

Our task now is to develop a consistent mathematical model describing this first-order physical model of the real flow; in other words, we must write down a mathematical problem allowing the evaluation of the potential velocity field, say $V = \mathrm{grad}\varphi$, which corresponds to the flow condition of Fig. 12.4. The flow has been assumed to be incompressible, and this implies that the velocity field must be solenoidal and, consequently, that the Laplacian of the velocity potential must be zero. Furthermore, at infinite distance from the airfoil the velocity must be equal to the free-stream velocity U, while at the airfoil surface the normal velocity component must vanish. Thus, the appropriate mathematical problem is seemingly the following:

$$\begin{cases} \nabla^2\varphi = 0 & \text{in the whole flow field,} \\ \dfrac{\partial\varphi}{\partial n} = 0 & \text{over the airfoil surface,} \\ \mathrm{grad}\varphi = U & \text{at infinity.} \end{cases} \qquad (12.4)$$

However, the flow field is two-dimensional, and thus the domain where the velocity field is defined is not simply connected. It may then be shown that, in this case, problem (12.4) admits an infinite number of solutions, namely one for each value of the circulation around the airfoil, which must then be prescribed to obtain a unique solution of the problem. In effect, we have already learnt that, around the airfoil, a circulation equal and opposite to the circulation around the starting vortex is present, and thus our objective is to complete the above mathematical problem by requiring the value of the circulation to be sufficiently close to the actual value of the flow field in Fig. 12.3. Fortunately, we do not need to do so, but only to specify that, in the model flow of Fig. 12.4, the wake is represented by a streamline starting from the trailing edge and going downstream to infinity.

Fig. 12.5. Streamlines of the potential flow around an airfoil for different values of the circulation.

In order to better explain the above statement, we must describe, at least qualitatively, the main features of the solutions of problem (12.4) for different values of the circulation around the airfoil. Our interest is in the shape of the relevant streamlines and, in particular, in the positions of the front and rear stagnation points over the airfoil surface.

We assume the circulation to be negative, i.e. clockwise, and, for convenience, we put $\Gamma = -\Gamma_0$. As can be seen in Fig. 12.5, each value of the circulation corresponds to different positions of the stagnation points, and only for one value, say $\Gamma_0 = \Gamma_K$, does the rear stagnation point coincide with the trailing edge. In fact, the rear stagnation point is located on the upper surface for $\Gamma_0 < \Gamma_K$, whereas it is located on the lower surface for $\Gamma_0 > \Gamma_K$. Furthermore, we see immediately that this particular value of the circulation is exactly the one that produces, around the airfoil, the flow condition shown in Fig. 12.4. Therefore, in order to obtain the unique particular solution corresponding to the first-order physical model we have developed, we must complete problem (12.4) by imposing that the circulation around the airfoil be equal to $-\Gamma_K$. However, this constraint on circulation may also be expressed through a condition on the velocity field, namely that *there must be no streamline rounding the trailing edge of the airfoil*. In effect, Fig. 12.5 shows that this is equivalent to prescribing the correct value $\Gamma = -\Gamma_K$ for the circulation.

The above condition is known as the *Kutta condition*, after the German mathematician Martin Wilhelm Kutta, and plays an essential role in the potential-flow theory for the prediction of the aerodynamic lift acting on an airfoil. This condition may be implemented in various equivalent ways. In particular, if the trailing edge angle is finite, then the velocity at the trailing edge must be zero. This is a consequence of the fact that the velocity vector must be continuous at the trailing edge, and this can be obtained only if the trailing edge is a stagnation point. In effect, this point may be thought as belonging to both the upper and the lower surfaces of the airfoil, to which the velocity must be tangent. Therefore, if the trailing edge angle is finite, the direction of the velocity vectors is not the same on the upper and lower surfaces, and thus the continuity of the velocity vector at the trailing edge may be obtained only if the two vectors vanish. On the other hand, if the trailing edge is cusped, i.e. with a zero included angle, the velocities are in the same direction on either side of the trailing edge, which then does not need to be a stagnation point. Therefore, the Kutta condition becomes that the velocity must be finite at the trailing edge and have the same magnitude on either side of the airfoil. In other words, there should be no jump in the velocity vector (in magnitude and direction) at the trailing edge. This is in fact a more general statement of the Kutta condition, which is widely used in the computation of the potential flow around airfoils.

The addition of the Kutta condition to problem (12.4) allows us to obtain the particular potential flow corresponding to the first-order model of the actual flow around the airfoil, shown in Fig. 12.4. A brief outline of some methods for the solution of this mathematical problem will be given in Sections 12.3 and 12.5. Here we only point out that, once the velocity potential φ is available in the whole domain, the velocity vector components may be obtained by differentiation, and the pressure field may be derived subsequently, through the application of Bernoulli's theorem. In particular, we obtain the pressure distribution around the airfoil contour, whose integration provides a resultant pressure force. By projection of this force vector in the directions parallel and perpendicular to the free-stream velocity, we then get a first prediction of the drag and lift forces. As anticipated in Chapter 5, the result of this evaluation procedure is

$$D = 0, \tag{12.5}$$

$$L = -\rho U \Gamma, \tag{12.6}$$

where Γ is the circulation around the airfoil, and is negative if clockwise. Note that the value of Γ is an output of the evaluation procedure, which derives from the enforcement of the Kutta condition, and may be obtained, using expression (5.58), by integrating the component of the resulting velocity vector not only along the airfoil surface, but also along any closed circuit rounding the airfoil. Formula (12.6) is known as the *Kutta–Joukowski theorem*.

The result that the drag is zero coincides with what we have found, in Chapter 5, for the flow of a non-viscous fluid around a closed body. This should not come as a surprise, considering that the mathematical problem describing our first-order potential-flow model of the actual viscous flow is the same as one would obtain by assuming the flow to be non-viscous,

albeit with a further constraint, namely the Kutta condition, which could
not have been derived without observing first the evolution of the real
viscous flow and the process of generation of the circulation around the
airfoil. Consequently, the *pressure* drag corresponding to the potential-flow
first-order model is zero, in agreement with d'Alembert's paradox. The
process of squeezing the boundary layer to zero thickness prevents us from
obtaining any estimate of the friction drag, which we know to depend on
the actual value of the finite normal velocity gradient existing on the airfoil
surface, i.e. within the actual boundary layer. Therefore, the first-order
model provides a prediction of the lift force acting on the airfoil, but is not
capable of giving any estimate, not even an approximate one, for the drag
force.

At first sight, the above result might seem disappointing; in reality,
this is far from true. We shall see that the potential-flow model, supple-
mented with the Kutta condition, gives good estimates, perhaps beyond
expectations, of the lift force on an airfoil and of its variation as a function
of angle of attack and geometry. Considering that the main application of
airfoils is in the design of bodies that must provide significant lift, this
is certainly a very important practical result. Furthermore, the effects of
the presence of the boundary layers over the airfoil surface, as regards
a correction to the first-order lift evaluation and, more importantly, the
prediction of both the pressure and friction contributions to the drag force,
may be obtained through an iterative procedure, which starts with the
above-described first-order model. The objective of this procedure is to
evaluate the effects of the boundary layers in terms of a correction to
the basic potential-flow satisfying both (12.4) and the Kutta condition.
In reality, the problem may be put in a more general form, namely the
derivation of appropriate methods to describe and predict the coupling
between the irrotational outer flow and the boundary layers. Considering the coincidence between the irrotational and the non-viscous flow
problems, this is often referred to as the analysis of the 'viscous–inviscid
interaction'.

12.2.2. *Prediction of the 'viscous–inviscid interaction'*

The basic procedure for the prediction of the coupling between outer
potential flow and boundary layers has been already outlined in Chapters 7
and 8. We have indeed seen that the main effect of the presence of the
boundary layers over the upper and lower surfaces of an airfoil is to

displace the streamlines of the potential flow outwards by an amount that is quantitatively equal to the displacement thickness of the boundary layers. Therefore, the pressure distribution that acts on the body surface may be predicted if one can determine the potential flow tangent to the fictitious body that is obtained by adding the displacement thickness to both the real body and the wake. The latter is formed by the smooth joining of the boundary layers emanating from the airfoil trailing edge. Obviously, the actual distribution of the boundary layer displacement thickness is not known from the beginning but, in most conditions, it can be obtained by solving the equations of the outer potential flow and of the boundary layer iteratively.

As already described in Chapter 8, the pressure distribution found from the solution of the first-order potential flow, tangent to the body surface and satisfying the Kutta condition, can be used as an input for the evaluation of the two boundary layers starting from the front stagnation point and developing over the upper and lower surfaces of the airfoil. The solution of the associated mathematical problem permits us to obtain a first-order estimate of the evolution of the boundary layer displacement thickness over the airfoil upper and lower surfaces. Furthermore, a first prediction of the tangential viscous stresses, and then of the relevant contribution to drag, is also derived. Subsequently, the displacement thicknesses obtained from the first calculation of the boundary layers are added to the original airfoil upper and lower surfaces. In this process, the wake must be also represented by appropriately continuing the boundary layer displacement thicknesses on the two sides of the streamline emanating from the trailing edge, derived from the first-order potential flow calculation (see Fig. 12.6). We can then calculate a new potential flow tangent to the 'virtual' body derived from the above procedure, and thus obtain a second-order estimate of the velocity and pressure distributions around the airfoil.

It should be pointed out that some care must be taken in imposing the Kutta condition in the second potential-flow calculation; in practice, this can be done with acceptable accuracy by requiring the velocities

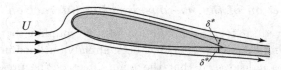

Fig. 12.6.   Virtual body for the calculation of the pressure distribution around the airfoil.

(or the pressures) to coincide on the points of the upper and lower surfaces of the virtual body whose projection coincides with the trailing edge.

The pressure distribution resulting from the second potential-flow calculation around the virtual body also provides a first prediction of the pressure drag on the airfoil. Indeed, the pressures around the virtual body must be projected onto the surface of the real airfoil, and it is easy to see that their integration over that surface will no longer give a zero value of the force component along the direction of the free-stream flow. This is a consequence of the different values of the velocities tangent to the virtual body over the trailing edge region with respect to those of the potential flow tangent to the original airfoil. We have indeed seen that, in the first potential-flow calculation, the Kutta condition implies that a finite-angle trailing edge is a stagnation point, and this produces a pressure recovery that is much higher than that present over the region of the thickened body corresponding to the trailing edge. In fact, the velocities derived from the second potential-flow calculation in that region will be nearer to those that are present over the trailing edge of the real airfoil, which are of the order of the free-stream velocity. The overestimation of the pressure recovery in the first potential-flow calculation may occur even for airfoils with cusped trailing edges, due to an excessive concave curvature of the streamlines over the trailing edge region, which produces lower velocities and higher pressures compared to the real situation. Incidentally, it may happen that the pressure recovery over the trailing edge region deriving from the first potential-flow solution be excessively high even for moderate angles of attack, and be incompatible with the existence of a completely attached boundary layer. This would introduce an unrealistic difficulty in the first boundary layer calculation, which may be easily overcome by modifying the pressure distribution properly. For instance, the pressure obtained in the first potential-flow at the trailing edge may be changed to the free-stream value, and joined smoothly with the values obtained slightly more upstream. This ensures a lower adverse pressure gradient in that region and the possibility of obtaining a more realistic first estimate of the boundary layer behaviour.

Returning to the second potential-flow solution, the corresponding pressure distribution may be used for a new prediction of the boundary layers developing over the upper and lower surfaces of the original airfoil, and thus for a second estimate of their displacement thicknesses. These may be superposed on the *original* airfoil surface, i.e. replacing the displacement thickness of the first estimate, to obtain a new virtual body for another potential-flow calculation. The iterative procedure may then be continued until two consecutive estimates of the desired quantities differ by less than a prescribed threshold, usually in terms of percentage variation. The convergence will obviously be more rapid if the control quantity is global, such as the lift or drag force on the airfoil, and slower if more critical local values are considered, such as the boundary layer thickness in the trailing edge region.

In principle, the above iterative procedure is the simplest, but has the disadvantage that the successive solutions of the potential-flow problem must be carried out around bodies of different geometry because, at each step, different displacement thicknesses are added to the original airfoil surface. In numerical computations, this may be impractical and time-consuming. However, an equivalent alternative approach exists in which the airfoil shape is left unchanged, and the boundary condition is varied in the consecutive steps of the procedure. The idea, first suggested by Lighthill (1958), is to imagine that, in all the various subsequent evaluations, the ideal irrotational flow extends to the airfoil surface, but a fictitious outflow from the surface is present such that a streamline of the ideal flow is tangent, as required, to the 'displaced surface' obtained by the addition of the displacement thickness. In other words, the boundary condition is always imposed at the original airfoil surface; however, the condition is no longer that the normal velocity be zero, but that its magnitude be such that one streamline coincides with the displacement surface. Therefore, the problem is to derive the value of the 'transpiration' velocity that is adequate to fulfil this condition.

Let us then introduce a local reference system such that x and y are the coordinates parallel and normal to the surface respectively. We also assume that the curvature of the surface may be neglected, i.e. that the surface radius of curvature is much larger than the thickness of the boundary layer. The velocity components in the x and y directions in the *real* flow within the boundary layer are denoted as u and v respectively, while the corresponding ones in the *ideal* completely irrotational flow extending to the airfoil surface are denoted as u_i and v_i. Even if not strictly necessary (see Lock and Williams, 1987 for higher-order approximations), the component u_i of the ideal potential-flow within the region actually occupied by the boundary layer is assumed to be equal to the corresponding real velocity component U_e at the edge of the boundary layer; in other words, we have $u_i(y \leq \delta) = u(\delta) = U_e$. Hence, it may be shown (see,

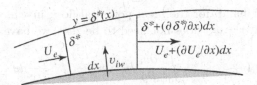

Fig. 12.7. Evaluation of the transpiration velocity at the airfoil surface.

e.g., Lighthill, 1958, Cousteix, 1988) that the value of the ideal normal velocity v_{iw} at the airfoil surface that produces the potential-flow with one streamline coinciding with the curve $y = \delta^*(x)$ and the velocity vector at the edge of the boundary layer equal to that of the real flow is given by

$$v_{iw} = \frac{d}{dx}(U_e \delta^*). \tag{12.7}$$

The validity of relation (12.7) may be intuitively verified with reference to Fig. 12.7, where the ideal flow is shown within a volume (per unit length in the direction normal to the plane of motion), delimited by the airfoil surface, by a curve $y = \delta^*(x)$ and by two normal cross-sections placed at a small mutual distance dx. We recall from Chapter 7 that the variation along the surface of the quantity $\rho U_e \delta^*$ is equal to the evolution of the defect of mass flux caused by the presence of the boundary layer. Now, from Fig. 12.7 it is easily seen that, to first order, the condition for the curve $y = \delta^*(x)$ to be a streamline of the ideal potential-flow, i.e. for no flux to occur along that curve, is that a transpiration velocity v_{iw} be present at the airfoil surface providing a virtual mass flux equal to

$$\rho v_{iw} dx = \frac{d(\rho U_e \delta^*)}{dx} dx, \tag{12.8}$$

from which, considering that ρ is constant due to the assumption of incompressible flow, relation (12.7) immediately derives.

The transpiration velocity of the ideal flow may also be derived through a more rigorous procedure. Considering the no-slip condition at the wall, the continuity equation and the Leibniz rule for differentiation under the integral sign, the normal velocity of the real flow at the edge of the boundary layer, say $v(\delta) = v_e$, may be expressed as follows:

$$v_e = \int_0^\delta \frac{\partial v}{\partial y} dy = -\int_0^\delta \frac{\partial u}{\partial x} dy = -\frac{d}{dx} \int_0^\delta u \, dy + U_e \frac{d\delta}{dx}. \tag{12.9}$$

On the other hand, for the ideal potential-flow, in which the velocity at the surface $v_i(0) = v_{iw}$ does not need to be zero, we have

$$v_i(\delta) = v_{iw} + \int_0^\delta \frac{\partial v_i}{\partial y} dy = v_{iw} - \int_0^\delta \frac{\partial u_i}{\partial x} dy$$

$$= v_{iw} - \frac{d}{dx} \int_0^\delta u_i dy + U_e \frac{d\delta}{dx}.$$

$$(12.10)$$

If we now put $u_i = U_e$ and impose the condition that the velocity vectors of the real and ideal flows be equal at the edge of the boundary layer, i.e. that $v_i(\delta) = v_e$, we obtain the desired expression for v_{iw}

$$v_{iw} = \frac{d}{dx} \int_0^\delta (U_e - u) dy = \frac{d}{dx} (U_e \delta^*). \tag{12.11}$$

It is also easy to evaluate the vertical velocity of the ideal potential-flow along the displacement curve. By considering the interval between $y = \delta^*(x)$ and $y = \delta(x)$, we obtain

$$v_i(\delta) = v_i(\delta^*) + \int_{\delta^*}^\delta \frac{\partial v_i}{\partial y} dy = v_i(\delta^*) - \int_{\delta^*}^\delta \frac{\partial u_i}{\partial x} dy$$

$$= v_i(\delta^*) - \frac{d}{dx} \int_{\delta^*}^\delta U_e dy + U_e \frac{d\delta}{dx} - U_e \frac{d\delta^*}{dx}.$$

$$(12.12)$$

Again, if we impose that $v_i(\delta) = v_e$, introduce (12.9) and recall relation (7.25), which shows that the integral of u over the whole boundary layer thickness may be expressed as $U_e(\delta - \delta^*)$, we easily obtain

$$v_i(\delta^*) = U_e \frac{d\delta^*}{dx}. \tag{12.13}$$

We thus find that the value of the transpiration velocity at the surface given by (12.11) implies that the curve $y = \delta^*(x)$ is a streamline of the ideal potential-flow, and this confirms that the two above-described methods of evaluation of the ideal potential-flow are indeed equivalent, and thus correspond to the same pressure distribution. Incidentally, by using the expression (7.25) giving the mass flux through the boundary layer, we may readily recast relation (12.9) for the outer normal velocity component as follows:

$$v_e = U_e \frac{d\delta^*}{dx} - (\delta - \delta^*) \frac{dU_e}{dx}. \tag{12.14}$$

This relation shows that when the outer velocity gradient is zero, the inclination of the velocity vector at the edge of the boundary layer is equal to that of the displacement surface, in agreement with the result found in Chapter 7 for a flat plate at zero angle of attack. Moreover, we find confirmation of the influence of the outer velocity gradient — and thus of the corresponding pressure variation — on the evolution of the boundary layer thickness. Indeed, from (12.14) it is immediately clear that, considering that $\delta > \delta^*$, an outer accelerating flow implies a decrease of the normal velocity at the edge of the boundary layer, and thus a lower rate of increase of the boundary layer thickness. Conversely, when the outer flow decelerates (and, consequently, pressure increases) v_e is larger than in the zero pressure gradient case, and a faster growth of the boundary layer occurs.

In summary, whichever method is used to determine the ideal potential-flow, the classical approach for evaluation of the effects of viscosity is based on successive solutions of the potential-flow problem for given geometry and of the boundary layer equations for given outer velocity. This type of iterative procedure is denoted the *direct* method of estimation of the viscous–inviscid interaction, and provides good predictions of the pressure and friction forces acting on an airfoil or, more in general, on any aerodynamic body. However, this is only true as long as the boundary layer does not separate due to the presence of excessive adverse pressure gradients, as may happen on the rear part of the airfoil upper surface at high angles of attack. In effect, we have seen that even an airfoil becomes a bluff body when the boundary layer separation causes spreading of vorticity over an extensive flow region and the appearance of a wide and generally unsteady wake. Consequently, in that case the iterative approach based on the coupling of successive solutions of the potential-flow and boundary layer equations can no longer be used for the prediction of the forces on the body.

Nonetheless, there are flow conditions in which boundary layer separation remains restricted to small regions, and does not strongly affect the global character of the flow. This may happen, for instance, when small separation bubbles appear, or when the boundary layer separates exclusively in the very rear part of an airfoil upper surface. In these situations, vorticity may remain confined to limited regions, which are still thin with respect to the typical reference length of the body — such as the chord of an airfoil — and where the flow may be considered as practically steady. If this occurs, it is not unreasonable to hope that an approach based on the coupling between potential-flow and boundary layer calculations may still be useful, provided it is possible to extend its validity beyond the occurrence of an incipient boundary layer separation. However, it is easy to see that this cannot be achieved through the direct method of evaluation of the boundary layer, due to a mathematical singularity that prevents the continuation of the solution of the boundary layer equations beyond the separation point, irrespective of the magnitude of the effects of separation on the global features of the flow. Different techniques have been developed to cope with this problem, and also to devise new procedures that take into account higher-order effects, such as the non-constancy of the pressure across the boundary layers when a high curvature of the streamlines is locally present.

A detailed description of all these methods and of the numerous research activities dedicated to their validation is beyond the scope of the present book, and the interested reader may refer, for instance, to the comprehensive review by Lock and Williams (1987). Suffice it here to say that a basic ingredient of these methods is the idea that the boundary layer equations may also be solved through an *inverse* procedure, i.e. by prescribing a tentative displacement thickness and deriving the corresponding outer velocity distribution. The same is also true for the potential flow calculations, for which it is possible to develop inverse methods of solution that give, as an output, the body geometry producing a certain prescribed velocity distribution. Several methods have been devised by combining in different ways direct and inverse calculations, in an attempt to overcome the shortcomings of the direct approach. These methods are generally classified as *fully inverse* and *semi-inverse*. More advanced procedures have also been developed, in which the potential and boundary layer equations are simultaneously solved, either in their complete form (*fully simultaneous* methods) or using a simplified version of the equations for one of the flows and an appropriate interaction law (*quasi-simultaneous* methods). With different convergence properties and computational efforts, all these various methods can provide satisfactory predictions even when flow separations of limited extent are present and/or the curvature effects become significant. Finally, the coupling between viscous and inviscid flows may also be extended to three-dimensional conditions. In that case, the quasi-simultaneous approach seems to provide the best compromise between predictive performance and computational efficiency (see Coenen, 2001). For details on boundary layer prediction methods, reference may also be made to Schlichting and Gersten (2000).

12.3. Potential Flow Analysis

12.3.1. *Generalities on the mathematical problem and its solution*

We now analyse more closely the mathematical problem of deriving the potential flow that corresponds to the first-order model of the flow around airfoils moving at angles of attack that are small enough for the boundary layers to remain attached over their whole surface.

In body-fixed coordinates, the problem is the one given in (12.4), with the addition of the Kutta condition to specify the value of the circulation around the airfoil. A slightly different formulation may be obtained by expressing the velocity field in the following form:

$$V = \text{grad}\varphi = U + \text{grad}\phi, \qquad (12.15)$$

where the scalar function ϕ is the *perturbation potential*; this notation expresses the fact that the vector gradϕ represents the velocity field induced on a uniform stream with velocity U by the perturbation caused by the presence of the airfoil. Therefore, if we choose, for instance, a reference system with the x *axis* in the direction of a free-stream velocity with magnitude U, the potential of the global flow is

$$\varphi = Ux + \phi, \qquad (12.16)$$

and we immediately recognize that the perturbation potential must also satisfy the Laplace equation. Then the mathematical problem for the perturbation potential is readily derived from (12.4), and becomes

$$\begin{cases} \nabla^2\phi = 0 & \text{in the whole flow field,} \\ \dfrac{\partial\phi}{\partial n} = -U \cdot n & \text{over the airfoil surface,} \\ \text{grad}\phi = 0 & \text{at infinity,} \end{cases} \qquad (12.17)$$

where n is the unit vector normal to the airfoil surface and directed towards the fluid. Obviously, the Kutta condition must also be imposed at the trailing edge. Incidentally, we note that problem (12.17) is the same that would apply if we had considered a reference system fixed to the still fluid in which the airfoil is moving with velocity $-U$.

The above mathematical problem has been widely studied, and the corresponding velocity fields have been shown to possess certain important properties. One of these is that the velocity maximum must occur at the boundary of the fluid field. In our case, this means that the maximum value of the velocity will occur at some point(s) on the airfoil contour and that, thanks to Bernoulli's theorem, a minimum of the pressure will be present at the same location. Another important result (known as Kelvin's minimum energy theorem) is that, among all incompressible flows having a prescribed normal velocity component at the boundary, the irrotational flow has the minimum value of the total kinetic energy. For the proof of these properties, as well as for a deeper analysis of all the characteristics of irrotational incompressible flows, the interested reader may refer, for instance, to Lamb (1932), Serrin (1959), Karamcheti (1966) and Batchelor (1967).

We now observe that the equation that must be satisfied by the velocity potential is a linear one, and this greatly simplifies the task of finding solutions to specific problems, identified by the equation and by appropriate boundary conditions. In particular, we can take advantage of the fundamental property of linear differential equations, namely that a linear combination of solutions is still a solution of the equation. In other words, if the functions $\phi_1, \phi_2, \ldots, \phi_n$ are known to be solutions of the Laplace equation, then the function

$$\phi = A_1\phi_1 + A_2\phi_2 + \cdots + A_n\phi_n \qquad (12.18)$$

is also a solution of the same equation for any value of the scalar coefficients $A_i(i = 1, n)$. The fact that irrotational incompressible flows are superposable has already been noted in Chapter 9, where it was also pointed out that the pressure fields are obviously not superposable, as they depend non-linearly on velocity through Bernoulli's theorem.

Therefore, the solution of a particular irrotational flow problem can be found by expressing the relevant velocity potential in the form (12.18) and deriving the values of the coefficients A_i by imposing a sufficient number of boundary conditions describing the given problem. It is important to note that, in principle, different basic solutions might be used for solving the same problem. However, as already pointed out, it may be proved that the solution of the Laplace equation with Neumann boundary conditions, i.e. with prescribed normal derivatives along a closed boundary around which the circulation is given, not only exists but is also unique. Therefore, different choices of the basic functions will correspond to different values of the coefficients A_i, but the resulting linear combination (12.18) will give rise to the same function ϕ. Nonetheless, the choice is not immaterial and, in practice, it is usually convenient to use a small number of basic solutions or to choose them in such a way that the coefficients are all of the same order. In reality, we shall see that the flow past bodies with particularly simple shapes may even be described by means of a very limited number of basic solutions.

In a Cartesian $x-y$ reference system the Laplace equation in two dimensions becomes

$$\frac{\partial^2 \phi}{\partial x^2} + \frac{\partial^2 \phi}{\partial y^2} = 0, \qquad (12.19)$$

whereas if we use $r - \theta$ cylindrical coordinates, we have

$$\frac{\partial^2 \phi}{\partial r^2} + \frac{1}{r}\frac{\partial \phi}{\partial r} + \frac{1}{r^2}\frac{\partial^2 \phi}{\partial \theta^2} = 0. \qquad (12.20)$$

As already noted in Chapter 11, functions whose Laplacian is zero are denoted as *harmonic* functions and, in the next section, we shall list some of the most important basic (or elementary) solutions of the Laplace equation that may be useful for the evaluation of the potential flow around airfoils. In order to check that these solutions represent irrotational flows, it may be useful to recall here the expressions, in Cartesian and cylindrical coordinates, of the only non-zero vorticity component in a two-dimensional flow, i.e. the one in the direction perpendicular to the plane of motion,

$$\omega_z = \frac{\partial v}{\partial x} - \frac{\partial u}{\partial y}, \qquad (12.21)$$

$$\omega_z = \frac{1}{r}\frac{\partial(rv_\theta)}{\partial r} - \frac{1}{r}\frac{\partial v_r}{\partial \theta}. \qquad (12.22)$$

12.3.2. *Elementary solutions of the Laplace equation*

12.3.2.1. *Uniform stream*

The simplest two-dimensional flow is a uniform velocity field whose components in the x and y directions are U and V respectively. The velocity potential ϕ and the stream function ψ (see Chapter 7) may then be derived from the relations

$$\frac{\partial \phi}{\partial x} = \frac{\partial \psi}{\partial y} = U, \qquad (12.23)$$

$$\frac{\partial \phi}{\partial y} = -\frac{\partial \psi}{\partial x} = V. \qquad (12.24)$$

Apart from integration constants, we obtain

$$\phi(x,y) = Ux + Vy, \tag{12.25}$$

$$\psi(x,y) = -Vx + Uy. \tag{12.26}$$

As already pointed out when discussing relation (7.50), the stream function is constant along streamlines. In effect, the streamlines of a uniform flow are straight lines which, as is evident from expressions (12.23) and (12.24), correspond to $\psi = constant$. For this flow the equipotential lines are also straight lines, which are orthogonal to the streamlines. In reality, streamlines and equipotential lines are always orthogonal because we have

$$\mathrm{grad}\phi \cdot \mathrm{grad}\psi = \frac{\partial\phi}{\partial x}\frac{\partial\psi}{\partial x} + \frac{\partial\phi}{\partial y}\frac{\partial\psi}{\partial y} = -uv + uv = 0. \tag{12.27}$$

12.3.2.2. *Source*

Let us now consider a flow with the following components in a cylindrical coordinate system with origin at a certain point O:

$$v_r = \frac{\partial\phi}{\partial r} = \frac{1}{r}\frac{\partial\psi}{\partial\theta} = \frac{\sigma}{2\pi r}; \quad v_\theta = \frac{1}{r}\frac{\partial\phi}{\partial\theta} = -\frac{\partial\psi}{\partial r} = 0. \tag{12.28}$$

Apart from a constant, the velocity potential and the stream function associated with this flow are then

$$\phi(r,\theta) = \frac{\sigma}{2\pi}\ln r; \quad \psi(r,\theta) = \frac{\sigma}{2\pi}\theta. \tag{12.29}$$

This flow, which is depicted in Fig. 12.8, is termed a *source*, and represents a flux of fluid from the application point O. It is indeed easy

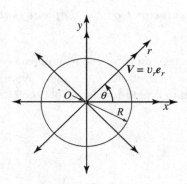

Fig. 12.8. Source flow.

to see that the flux through a circumference of radius R around O does not depend on the value of R, and is given by

$$2\pi R v_r(R) = 2\pi R \frac{\sigma}{2\pi R} = \sigma. \tag{12.30}$$

Therefore, the quantity σ, which is termed the *strength* of the source, represents the volume flux (per unit length in the direction normal to the plane of motion) through any circuit enclosing point O.

By using (12.20) and (12.22) it is simple to check that a source flow is irrotational and that the Laplacian of its velocity potential is zero. However, the velocity becomes infinite when $r = 0$. Therefore, point O is a singular point, and this is why the source flow, like other solutions of the Laplace equation with the same behaviour, is usually described as a *singularity* with *point of application* in O. This point, then, must be excluded from the region where we want to describe the irrotational flow. In practice, this poses no particular problem because, in a potential flow, any streamline may represent a boundary of the flow and thus, for instance, the surface of a body to which the flow is tangent. Therefore, the solution may be considered as representative of an irrotational flow outside the surface of a body provided no singular point is contained in the region of interest; for instance, it may be placed inside the body.

As an example, let us consider the superposition of a uniform flow in the x direction, with velocity magnitude U, and a source of strength σ placed at the origin of a Cartesian reference system. We then have

$$r^2 = x^2 + y^2, \tag{12.31}$$

$$u = U + v_r \cos\theta = U + v_r \frac{x}{\sqrt{x^2 + y^2}} = U + \frac{\sigma}{2\pi} \frac{x}{x^2 + y^2}, \tag{12.32}$$

$$v = v_r \sin\theta = v_r \frac{y}{\sqrt{x^2 + y^2}} = \frac{\sigma}{2\pi} \frac{y}{x^2 + y^2}. \tag{12.33}$$

We readily see that a stagnation point is present on the x axis, and it coincides with the point where the velocity of the free stream is exactly balanced by the velocity in the opposite direction originating from the source. As might be expected, and as should be evident from Fig. 12.9, this happens at a coordinate x_s that is more and more negative as the strength of the source increases, namely

$$x_s = -\frac{\sigma}{2\pi U}. \tag{12.34}$$

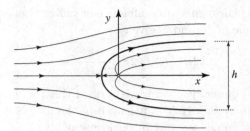

Fig. 12.9.   Flow due to the superposition of a source and a uniform stream.

It is also evident from Fig. 12.9 that the streamline passing through the stagnation point divides the flow into two parts, an outer region originating from the upstream uniform flow and an internal region in which the fluid emanates entirely from the source. This streamline may then be considered as the surface of a two-dimensional semi-infinite body to which the flow is tangent. At very large downstream distances, the velocity tends to become parallel to the x direction and to attain the magnitude U of the free-stream. Consequently, the width of the body becomes equal to $h = \sigma/U$, so that the downstream flux of the inner flow is equal to the strength of the source. Therefore, if h and U are given, the strength of the source generating the semi-infinite body is defined. We note that, if the interest is in the pressures created over the surface of such a body by an ideal irrotational flow with upstream velocity $U e_x$, then the inner flow in Fig. 12.9 may be disregarded, and the pressures may be evaluated from the velocities along the dividing streamline using Bernoulli's theorem. We thus see that the singularity existing at the application point of the source is outside the flow domain of interest, in which the velocity is continuous and finite everywhere.

Let us now add to the above flow another source with a strength that is equal and opposite to the previous one. This negative type of source is usually called a *sink* and, indeed, the relevant velocity vectors are directed towards its application point. In other words, a negative flux is present through any closed circuit enclosing the sink. In the present case, as the source and the sink have the same absolute strength, the positive and negative fluxes balance mutually. We choose the origin of the coordinates midway between the source and the sink, which are then positioned at $x = -a$ and $x = +a$ respectively. The resulting flow is schematically shown in Fig. 12.10, and is characterized by the presence of a closed streamline

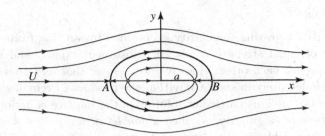

Fig. 12.10. Superposition of a uniform flow, a source and a sink.

with two stagnation points, A and B, whose coordinates are

$$x_A = -a\sqrt{1 + \frac{\sigma}{\pi U a}}, \quad x_B = +a\sqrt{1 + \frac{\sigma}{\pi U a}}. \qquad (12.35)$$

The closed streamline passing through the stagnation points may be seen as the surface of a two-dimensional body, representing an infinite cylinder with a cross-section shape that is known as a 'Rankine oval'. Again, this streamline divides the field into two parts, and it is readily seen that all the streamlines of the inner flow originate at the source and are directed towards the sink. On the other hand, the outer flow represents the potential flow with upstream velocity equal to $U e_x$ and tangent to the Rankine oval defined by the strength and position of the singularities.

It should now be well understood that closed streamlines will be generated whenever sources and sinks with equal global strength are present in the flow. More generally, referring to sources with positive or negative sign, one can distribute either a finite number or even a continuous distribution of sources along a segment of the x axis, and the superposition with a uniform flow will generate a closed streamline provided the integral strength of the sources is zero. By this procedure, one may construct two-dimensional bodies with different cross-sectional shapes by using different distributions of sources. But it should be clear that one may also choose the shape of the body and find the corresponding distribution of sources by imposing that the resulting velocity field be tangent to the given body contour. We shall see later on that this is indeed the rationale behind the evaluation of the potential flow tangent to generic bodies through the superposition of a uniform flow and appropriate types of singularities.

12.3.2.3. *Doublet*

A particularly interesting singularity is obtained by starting from a source and a sink of equal strength σ placed a distance l apart, and allowing them to approach each other. However, we assume that as l goes to zero, σ increases indefinitely in such a way that the product σl remains equal to a constant value μ. This particular solution of the Laplace equation, which is another type of singularity, is called a *doublet* or a *dipole* of strength μ. An axis of the doublet is also defined and is directed from the sink to the source. The streamlines corresponding to a doublet with the axis in the positive x direction are shown in Fig. 12.11.

If the application point is located at the origin of the reference system, it may be seen that the velocity potential of a doublet is given, in cylindrical coordinates, by the following expression

$$\phi(r,\theta) = -\frac{\mu}{2\pi}\frac{\cos\theta}{r}. \tag{12.36}$$

The velocity components in the same reference system are then

$$v_r = \frac{\partial\phi}{\partial r} = \frac{\mu\cos\theta}{2\pi r^2}, \tag{12.37}$$

$$v_\theta = \frac{1}{r}\frac{\partial\phi}{\partial\theta} = \frac{\mu\sin\theta}{2\pi r^2}, \tag{12.38}$$

and the stream function is

$$\psi(r,\theta) = \frac{\mu}{2\pi}\frac{\sin\theta}{r}. \tag{12.39}$$

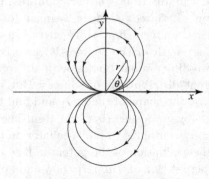

Fig. 12.11.   Flow due to a doublet.

It may be useful to write the expressions in Cartesian coordinates for the velocity potential and the velocity components of a doublet located at the generic point (x_0, y_0). We obtain

$$\phi(x,y) = -\frac{\mu}{2\pi} \frac{x - x_0}{(x - x_0)^2 + (y - y_0)^2}, \tag{12.40}$$

$$u(x,y) = \frac{\mu}{2\pi} \frac{(x - x_0)^2 - (y - y_0)^2}{[(x - x_0)^2 + (y - y_0)^2]^2}, \tag{12.41}$$

$$v(x,y) = \frac{\mu}{2\pi} \frac{2(x - x_0)(y - y_0)}{[(x - x_0)^2 + (y - y_0)^2]^2}. \tag{12.42}$$

We may now consider the superposition of a uniform flow Ue_x and a doublet located at the origin, with axis pointing in the negative x direction, i.e. with a strength $-\mu_0 < 0$. In cylindrical coordinates, the potential and the stream function of the global velocity field are then

$$\phi(r,\theta) = Ur\cos\theta + \frac{\mu_0}{2\pi}\frac{\cos\theta}{r}, \tag{12.43}$$

$$\psi(r,\theta) = Ur\sin\theta - \frac{\mu_0}{2\pi}\frac{\sin\theta}{r}, \tag{12.44}$$

where arbitrary constants have been omitted and we have considered that $x = r\cos\theta$ and $y = r\sin\theta$.

Consequently, the velocity components become

$$v_r = \frac{\partial\phi}{\partial r} = \frac{1}{r}\frac{\partial\psi}{\partial\theta} = \left(U - \frac{\mu_0}{2\pi r^2}\right)\cos\theta, \tag{12.45}$$

$$v_\theta = \frac{1}{r}\frac{\partial\phi}{\partial\theta} = -\frac{\partial\psi}{\partial r} = -\left(U + \frac{\mu_0}{2\pi r^2}\right)\sin\theta. \tag{12.46}$$

This flow may be seen as the limit, for mutually approaching source and sink, of the previously analysed case, which produced the flow around a Rankine oval. Therefore, the superposition of a uniform flow with a doublet gives rise to a flow with a closed streamline of circular shape as shown in Fig. 12.12. This may be checked by observing, from (12.45), that the radial velocity vanishes for all θ when

$$r = a = \sqrt{\frac{\mu_0}{2\pi U}}. \tag{12.47}$$

We thus find that the potential flow tangent to a circular cylinder of radius a placed in a uniform stream may be represented by introducing a

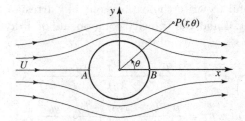

Fig. 12.12.   Flow due to a uniform flow and a doublet.

doublet whose axis opposes the stream and whose strength is such that

$$\mu_0 = 2\pi U a^2. \tag{12.48}$$

The velocity components are then derived by introducing this value of μ_0 into relations (12.45) and (12.46); we thus get

$$v_r = U\left(1 - \frac{a^2}{r^2}\right)\cos\theta, \tag{12.49}$$

$$v_\theta = -U\left(1 + \frac{a^2}{r^2}\right)\sin\theta. \tag{12.50}$$

To obtain the pressure distribution over the cylinder, we first evaluate the velocity components at $r = a$, to get

$$v_r = 0, \quad v_\theta = -2U\sin\theta, \tag{12.51}$$

and then use Bernoulli's theorem to derive the pressure. In particular, if we denote by p_∞ the value of the pressure in the undisturbed free stream, after simple manipulations we obtain the following distribution of the pressure coefficient around the cylinder:

$$C_p = \frac{p - p_\infty}{(1/2)\rho U^2} = \left(1 - \frac{v_\theta^2}{U^2}\right) = (1 - 4\sin^2\theta). \tag{12.52}$$

This pressure distribution is the same as that derived from the solution of the non-viscous fluid equations (see Fig. 5.2a). In effect, we have seen that the equations of the irrotational motion of an incompressible viscous fluid coincide with those that would apply if the fluid were considered to be non-viscous. However, we also know that this pressure distribution, whose integration over the cylinder surface gives rise to a zero resultant force, is not representative of the one that occurs in reality around a circular cylinder in cross-flow, because it is incompatible with the existence of an attached boundary layer over the body surface. Nonetheless, the above solution is particularly important because a mathematical technique exists, known as *conformal mapping*, through which the problem of the irrotational flow around a generic two-dimensional body may be reduced to that of the flow around a circular cylinder. The technique is based on the theory of functions of a complex variable, and its details may be found in many classical textbooks (see, e.g., Karamcheti, 1966). However, the use of conformal mapping has now become less popular due to the advent of numerical methods that permit the potential flow around two-dimensional and three-dimensional aerodynamic bodies of arbitrary shape to be easily obtained by means of other techniques.

12.3.2.4. *Point vortex*

The last singularity we consider is the point vortex, which was introduced in Chapter 11 and whose corresponding velocity field is shown in Fig. 11.3. The velocity components for a point vortex whose application point — also denoted as *centre* — is placed at the origin of a cylindrical coordinate system are then

$$v_r = \frac{\partial \phi}{\partial r} = \frac{1}{r}\frac{\partial \psi}{\partial \theta} = 0, \tag{12.53}$$

$$v_\theta = \frac{1}{r}\frac{\partial \phi}{\partial \theta} = -\frac{\partial \psi}{\partial r} = \frac{\Gamma}{2\pi r}, \tag{12.54}$$

where Γ is the strength of the vortex, which coincides with the value of the circulation around the circular streamlines and, in fact, around any closed circuit enclosing the vortex centre. Apart from a constant, the velocity potential and the stream function for this flow are then

$$\phi(r, \theta) = \frac{\Gamma}{2\pi}\theta, \tag{12.55}$$

$$\psi(r, \theta) = -\frac{\Gamma}{2\pi}\ln r. \tag{12.56}$$

As may be deduced from expression (12.55), the velocity potential is multivalued, i.e. it increases by Γ for each turn around the vortex application point. This is consistent with the fact that the circulation is non-zero, and that the flow field is doubly connected. In effect, the vortex centre is a singular point, and must then be excluded from the domain, for instance by means of a small circuit around it. As we have already mentioned, the solution of the Laplace equation with Neumann boundary conditions exists and is unique for doubly connected domains only provided the value of the circulation around irreducible circuits is given, or is specified by a condition on the velocity field (such as the Kutta condition). Note that, in a two-dimensional flow, all singularities produce multiply connected domains, because their application points must be outside the considered flow; however, one can readily check that the circulation around a source or a doublet is zero. Therefore, point vortices are the natural singularities to be used to model the first-order potential flow around lifting airfoils, around which we know a finite circulation to be present.

We now imagine adding a point vortex with centre at the origin to the irrotational flow around a circular cylinder of radius a, whose velocity components are given by relations (12.49) and (12.50). The result is the flow past a cylinder placed in a uniform cross-stream around which a circulation with the value of the vortex strength is present. Let us assume that the circulatory flow around the cylinder is clockwise, and is thus negative considering the adopted reference system (see Fig. 12.12). For convenience, we denote the vortex strength as $\Gamma = -\Gamma_0$. With this notation, the resulting flow is characterized by the same radial velocity component as for the case with no circulation, given by (12.49), and by the following tangential velocity component:

$$v_\theta = -U \left(1 + \frac{a^2}{r^2} \right) \sin\theta - \frac{\Gamma_0}{2\pi r}. \qquad (12.57)$$

It is easy to see that, as shown in Fig. 12.13, two stagnation points are present over the cylinder surface if $\Gamma_0 < 4\pi Ua$, and that when $\Gamma_0 = 4\pi Ua$ they join at point $r = a$, $\theta = 3\pi/2$. On the other hand, when $\Gamma_0 > 4\pi Ua$, there is no stagnation point over the cylinder surface, but there is one inside the flow field, and its coordinates are $\theta = 3\pi/2$ and $r > a$.

The above velocity field is still symmetrical with respect to the y axis, as was the case for a cylinder with no circulation around it. Therefore, the integration of the pressure distribution — obtained from the velocity tangent to the cylinder surface and from the application of Bernoulli's theorem — leads to a zero drag component again, in agreement with d'Alembert's paradox. However, the flow is no longer symmetrical with respect to the x axis, and the velocity magnitudes over the upper side of the cylinder are higher than those over the lower side. Consequently, the pressures are lower over the upper side than over the lower side, and their integration over the whole cylinder contour shows that now a lift force

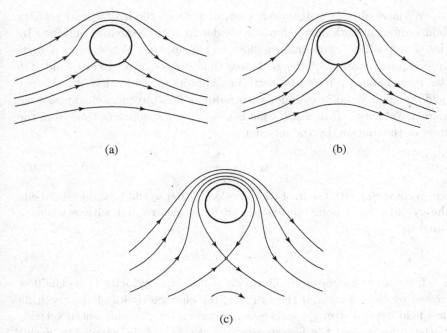

(a)   (b)

(c)

Fig. 12.13.   Flow due to a uniform free-stream, a doublet and a vortex. (a) $\Gamma_0 < 4\pi U a$;
(b) $\Gamma_0 = 4\pi U a$; (c) $\Gamma_0 > 4\pi U a$.

component is present and is equal to

$$F_y = L = \rho U \Gamma_0. \tag{12.58}$$

Considering that the free-stream velocity is directed rightwards and Γ_0 is the value of the *clockwise* circulation around the cylinder, formula (12.58) coincides with the Kutta–Joukowski theorem introduced through relation (12.6), and applies for the potential flow around bodies of generic shape. As we have done in Chapter 11, the formula may be put in a more general form by expressing a generic value of the circulation by means of a vector $\boldsymbol{\Gamma}$, with modulus equal to the absolute value of the circulation, directed inside the plane of the figure if the circulation is clockwise and outside if the circulation is counter-clockwise. Then, for any direction of the free-stream velocity vector $\boldsymbol{U}$, the global force acting on the cylinder is perpendicular to $\boldsymbol{U}$ and may be derived from the relation

$$\boldsymbol{F} = \rho \boldsymbol{U} \times \boldsymbol{\Gamma}. \tag{12.59}$$

We have already pointed out that, in general, the irrotational velocity field around a stationary circular cylinder in a cross-stream, obtained by the superposition of a uniform flow and a doublet, is not even a first-order approximation of the real flow that occurs in a viscous fluid. On the other hand, we have also seen, in Chapters 6 and 9, that the velocity field outside a circular cylinder with radius a which rotates clockwise with angular velocity $-\Omega$ in a *still* fluid is given, after a sufficient time from the start of the motion, by the relation

$$v_\theta = -\frac{\Omega a^2}{r}, \qquad (12.60)$$

which coincides with the irrotational velocity that would be induced outside the cylinder by a point vortex placed at its centre and with a strength equal to

$$\Gamma = -\Gamma_0 = -2\pi\Omega a^2. \qquad (12.61)$$

If we now superpose a uniform cross-flow with velocity U on the flow caused by the rotation of the cylinder, the characteristics of the resulting flow field depend strongly on the ratio between the circumferential velocity of the cylinder and the free-stream velocity, $\Omega a/U$. In particular, if this ratio is large enough, say $\Omega a/U > 4$, then the vorticity may remain confined in a region around the cylinder surface, and the streamlines may be not very different from those of Fig. 12.13(c). Conversely, when the surface circumferential velocity is comparable with or lower than the free-stream velocity, boundary layer separation will occur in the rear part of the cylinder, and a large wake will form. In that case, as happened for the cylinder with no rotation, the real flow will have no resemblance to the ideal irrotational flow shown in Fig. 12.13. Nonetheless, even when a separated wake is present, a rotation of the cylinder produces a flow that is not symmetrical with respect to the direction of the free-stream. In particular, over the upper side of the cylinder the velocity is generally higher — and thus the pressure is lower — than over the lower side; consequently, a cross-flow force arises. A circulation is then present around the cylinder, but it is no longer connected with the angular velocity through relation (12.61); in fact, even if its value cannot be easily predicted, it will generally be lower. However, considering the discussion in Section 11.4, it is interesting to note that, if this correct value of the circulation were available, then the cross-flow force acting on any two-dimensional body could be obtained from relation (11.75) which, incidentally, coincides with (12.59). The existence of a cross-flow force on a rotating and translating cylinder is usually known as the *Magnus effect*, after the German chemist and physicist Heinrich Gustav Magnus.

12.4. Outline of Thin Airfoil Theory

12.4.1. *Basic approximations and formulation*

Before analysing the problem of deriving the potential flow around a generic airfoil, it is convenient to describe briefly a simplified procedure that, in spite of some seemingly stringent assumptions as regards the airfoil geometry and type of motion, is capable of providing excellent first-order estimates of the lift forces. The method is widely known as *thin airfoil theory*, and is essentially due to the works of Munk (1922) and Glauert (1926).

We start by considering the flow past an airfoil of arbitrary shape placed in a flow field with free-stream velocity U and angle of attack α. The coordinate system is now chosen with the x axis along the chord and origin at the leading edge of the airfoil (see Fig. 12.14).

We adopt the formulation of the problem in terms of the perturbation potential ϕ, so that the velocity at a generic point in the flow is given by relation (12.15). If U is the modulus of U, the velocity potential of the global flow then becomes

$$\varphi(x,y) = (xU\cos\alpha + yU\sin\alpha) + \phi(x,y), \qquad (12.62)$$

and the mathematical problem takes the following form:

Field equation:

$$\frac{\partial^2\phi}{\partial x^2} + \frac{\partial^2\phi}{\partial y^2} = 0. \qquad (12.63)$$

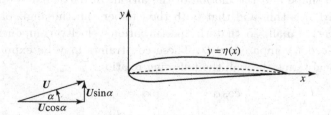

Fig. 12.14. Coordinates and symbols for the flow past an airfoil.

Boundary condition over the airfoil surface:

$$(\boldsymbol{U} + \text{grad}\phi) \cdot \boldsymbol{n} = 0. \tag{12.64}$$

Boundary condition at infinity:

$$\text{grad}\phi \to 0. \tag{12.65}$$

Kutta condition: the perturbation velocity must be finite and continuous at the trailing edge.

For convenience, we introduce the x and y components of the perturbation velocity, denoted as u and v respectively. The boundary condition to be applied at points over the airfoil surface then becomes

$$(U \cos \alpha + u)n_x + (U \sin \alpha + v)n_y = 0, \tag{12.66}$$

where n_x and n_y are the components of the unit vector normal to a generic point of the airfoil contour. Let us suppose now that the surface of the airfoil is described by the function $y = \eta(x)$, where $\eta = \eta_u(x)$ for the upper surface and $\eta = \eta_l(x)$ for the lower surface. Therefore, if we consider that over the airfoil surface we have

$$n_x dx + n_y dy = 0, \tag{12.67}$$

we may recast condition (12.66) in the following form:

$$v(x, \eta(x)) = \frac{d\eta}{dx}(U \cos \alpha + u) - U \sin \alpha, \cdot \tag{12.68}$$

which is valid for $0 \le x \le c$, where c is the airfoil chord.

It is then immediately seen that v is a function not only of the free-stream velocity, of the airfoil geometry and of its angle of attack, but also of the perturbation velocity component u. In this sense, this equation is often said to express the boundary condition in a non-linear form. Thus, in order to simplify the treatment, certain further assumptions are introduced as regards the shape and the motion of the airfoil. In particular, we assume that the airfoil is thin and that both the camber and the angle of attack are sufficiently small, so that the perturbation velocity components are of a lower order compared to U. These constraints may be expressed in mathematical terms through the following relations:

$$\cos \alpha = 1, \quad \sin \alpha = \alpha, \tag{12.69}$$

$$(U \cos \alpha + u)\frac{d\eta}{dx} = U\frac{d\eta}{dx}. \tag{12.70}$$

With these positions, Eq. (12.68) becomes

$$v(x, \eta(x)) = U\frac{d\eta}{dx} - U\alpha, \tag{12.71}$$

which is known as the *linearized* form of the boundary condition.

It may be observed that the assumption of small perturbations is not valid near any stagnation point, where the perturbation is obviously of the same order of magnitude as the free-stream velocity, and also near a rounded leading edge. Therefore, using the above simplified boundary condition we are assuming that this local inadequacy does not hinder the global validity of the model. Fortunately, this is indeed the case, and we shall see later that this simplified model may be extremely useful in providing a good first-order estimate of the lift on an airfoil.

In reality, thin airfoil theory is based on a further, and apparently even stronger, simplification, which consists in transferring the boundary conditions from the actual airfoil surface to the x axis. This is done by expanding $v(x, y)$ about points on the chord line, so that we write

$$v(x, \eta(x)) = v(x, 0) + \left(\frac{\partial v}{\partial y}\right)_{y=0} \eta(x) + O(\eta^2). \tag{12.72}$$

Consistently with our previous approximations, we then only retain the first term on the right-hand side of (12.72), and thus (12.71) becomes

$$v(x, 0) = U\frac{d\eta}{dx} - U\alpha. \tag{12.73}$$

The boundary condition may then be written separately for the upper and lower surfaces as follows:

$$v(x, 0^+) = U\frac{d\eta_u}{dx} - U\alpha, \tag{12.74a}$$

$$v(x, 0^-) = U\frac{d\eta_l}{dx} - U\alpha, \tag{12.74b}$$

where we denote by $(x, 0^+)$ and $(x, 0^-)$, respectively, the points lying on the upper and lower sides of the x axis strip $0 \le x \le c$.

Let us now describe the upper and lower surfaces of the airfoil by using a camber function $\eta_c(x)$ and a thickness function $\eta_t(x)$, defined as follows:

$$\eta_c = (\eta_u + \eta_l)/2, \tag{12.75}$$

$$\eta_t = (\eta_u - \eta_l)/2. \tag{12.76}$$

In other words, $y = \eta_c(x)$ represents the mean line of the airfoil and the function $y = \eta_t(x)$ is the thickness distribution to be added on both sides of the mean line to obtain the upper and lower surfaces. In effect, from the above relations, we obtain $\eta_u = \eta_c + \eta_t$ and $\eta_l = \eta_c - \eta_t$.

By introducing the thickness and camber functions, and returning to the expression of the velocity component v in terms of the perturbation potential, the simplified boundary conditions (12.74) may be recast as

$$\frac{\partial \phi}{\partial y}(x, 0^{\pm}) = U\frac{d\eta_c}{dx} \pm U\frac{d\eta_t}{dx} - U\alpha, \tag{12.77}$$

where $+$ is for the upper surface and $-$ is for the lower surface.

This expression of the boundary conditions suggests that the problem of the irrotational flow tangent to the original airfoil may be decomposed as the sum of three independent problems. The first concerns the flow past a symmetrical airfoil with upper and lower surfaces defined by the thickness distribution $y = \pm\eta_t(x)$, placed at zero angle of attack; we denote the corresponding velocity potential by ϕ_1. The second is the problem of the flow past a zero-thickness curved plate, defined by the camber function $y = \eta_c(x)$, also at zero angle of attack; let the relevant potential be ϕ_2. The last problem, with potential ϕ_3, refers to the flow past a flat plate at an angle of attack equal to α. The validity of this decomposition, illustrated schematically in Fig. 12.15, is totally dependent on the fact that we are using relations (12.74), namely the transfer of the linearized boundary conditions to the airfoil chord.

In effect, the boundary conditions for the three problems are

$$\frac{\partial \phi_1}{\partial y}(x, 0^{\pm}) = \pm U\frac{d\eta_t}{dx}, \tag{12.78}$$

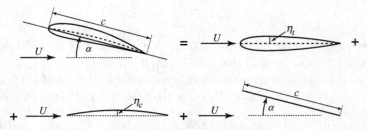

Fig. 12.15. Decomposition of the problem of an arbitrary thin airfoil into three problems.

$$\frac{\partial \phi_2}{\partial y}(x, 0^\pm) = U \frac{d\eta_c}{dx}, \tag{12.79}$$

$$\frac{\partial \phi_3}{\partial y}(x, 0^\pm) = -U\alpha. \tag{12.80}$$

Furthermore, the three functions ϕ_1, ϕ_2 and ϕ_3 represent the velocity potentials of three irrotational flow problems, and thus, besides satisfying the Kutta condition and the conditions at infinity, they must be solutions of the field equation (12.63). Therefore, due to the linearity of the Laplace equation, the function $\phi = \phi_1 + \phi_2 + \phi_3$ is also a solution of the same equation, and satisfies the sum of conditions (12.78)–(12.80) at points lying on the chord; in other words, it satisfies the boundary condition (12.77) for the flow past the original airfoil.

We may now enquire whether the same superposition procedure also applies to the pressure field, i.e. whether the pressures over the airfoil, and thus the forces acting on it, may also be obtained by adding the pressures derived from the three elementary problems. To verify this possibility, let us first denote by p_∞ the pressure in the free-stream, where the velocity is $\boldsymbol{U} \equiv (U \cos\alpha, U \sin\alpha)$, and as $\boldsymbol{V} = \boldsymbol{U} + \boldsymbol{q}$ the velocity in a generic point; therefore, $\boldsymbol{q} = \mathrm{grad}\phi \equiv (u, v)$ is the perturbation velocity. We then use Bernoulli's theorem to get the following expression for the pressure at a generic point in the flow:

$$p(x, y) = p_\infty + \frac{\rho}{2} U^2 \left(1 - \frac{V^2}{U^2}\right). \tag{12.81}$$

The square modulus of the velocity at a generic point may now be written in explicit form in terms of the perturbation

$$\begin{aligned}
V^2 &= (\boldsymbol{U} + \boldsymbol{q}) \cdot (\boldsymbol{U} + \boldsymbol{q}) = U^2 + 2\boldsymbol{U} \cdot \boldsymbol{q} + q^2 \\
&= U^2 + 2(uU \cos\alpha + vU \sin\alpha) + u^2 + v^2,
\end{aligned} \tag{12.82}$$

and we thus get

$$p = p_\infty - \frac{\rho}{2} U^2 \left(2\frac{u}{U} \cos\alpha + 2\frac{v}{U} \sin\alpha + \frac{u^2 + v^2}{U^2}\right). \tag{12.83}$$

We now introduce the thin airfoil theory approximations, namely that the thickness, curvature and angle of attack only produce small perturbations to the free-stream velocity. Consequently, we can use relations (12.69) and neglect all second-order terms in the ratio between perturbation velocities and U (including $v\alpha/U$, which is of order v^2/U^2). The pressure at a generic point in the field may then be approximated as

$$p = p_\infty - \rho U u, \tag{12.84}$$

and the corresponding expression of the pressure coefficient becomes

$$C_p = \frac{p - p_\infty}{\rho U^2/2} = -2\frac{u}{U} = -\frac{2}{U}\frac{\partial\phi}{\partial x}. \tag{12.85}$$

Therefore, in the thin airfoil theory, pressure is a linear function of the perturbation velocity u, and may thus be obtained by superposition of the pressures corresponding to the elementary problems. The same is then true for the forces and moments on the airfoil, considering that the pressure loads are the only ones that may be obtained from the first-order irrotational-flow model.

12.4.2. *Overview of the solution procedure and main results*

The basic procedure to obtain the loads on thin airfoils of generic cross-section is to take advantage of the linearity of the Laplace equation and to introduce elementary solutions so that the approximate boundary conditions (12.77) are satisfied. In particular, the perturbation velocity field produced by the airfoil may be represented as being caused by a distribution of singularities, like those that have been introduced in Section 12.3, choosing the most suitable ones according to the particular case. In the following, we give only a brief description of the solution procedure and concentrate on its main results. Further details may be found in classical texts on low-speed aerodynamics, such as Karamcheti (1966) or Katz and Plotkin (2001).

We start by observing that the main objective of the thin airfoil theory is to provide an estimate of the lift force generated by an airfoil and of its dependence on the airfoil angle of attack and geometrical features. However, considering the problem decomposition deriving from the approximate boundary conditions, it follows that the first of the problems depicted in Fig. 12.15 — namely the evaluation of the flow past a symmetrical airfoil at zero angle of attack having the same thickness distribution as the given airfoil — does not contribute to the generation of lift. Therefore, although the effect of a given thickness distribution $\eta_t(x)$ might be analysed by evaluating the corresponding strength of a continuous distribution of sources along the airfoil chord, thickness is generally neglected in the context of thin airfoil theory. On the other hand, the fact that, at first order, the thickness of an airfoil does not contribute to lift can actually be regarded as the first result of the approximations on which the theory is based. In reality, thickness does have an influence although, as will be seen later, a small and non-trivial one. In any case, neglecting it certainly does not prevent determination of the correct order of magnitude of lift.

Let us now consider the second and third problems of the decomposition, namely the mean line $y = \eta_c(x)$ at zero angle of attack, and a flat plate (representing the airfoil chord) placed at a given α. Both these problems contribute to lift, which may then be obtained by adding their separate results. However, we shall outline the procedure for the evaluation of the flow corresponding to their combined flows, and subsequently discuss the specific effects of α and of the camber function. In other words, we shall consider the problem of a zero-thickness airfoil (i.e. a curved plate) placed in a stream with a non-zero angle of attack.

The most natural solution procedure is to introduce a distribution of vortices along the curved plate. Indeed, vortices are elementary solutions of the irrotational flow problem and possess circulation, so that they are well suited to represent the perturbation field of a lifting body adequately. Furthermore, as happens for all singularities, the condition at infinity (12.65) is satisfied automatically by their induced velocity field. However, considering the assumptions that have been made to derive the approximate boundary conditions of thin airfoil theory, it is also customary to transfer the vortices from the mean line to the chord. In other words, the perturbation field of the curved zero-thickness airfoil is assumed to be represented adequately by the velocity induced by a suitable vortex sheet of variable strength $\gamma(x)$ placed along the chord.

The situation is thus as shown in Fig. 12.16, in which it is seen that, following the usual notation for airfoil analysis, the positive orientation of the vortices is now taken as clockwise.

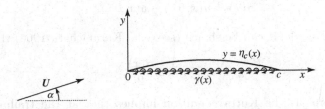

Fig. 12.16.   Vortex distribution representing the cambered airfoil.

The problem is then determining the function $\gamma(x)$ such that the following boundary condition is satisfied in the segment of the x axis defined by $0 \le x \le c$:

$$\frac{\partial \phi}{\partial y}(x, 0^{\pm}) = v(x, 0^{\pm}) = U \left(\frac{d\eta_c}{dx} - \alpha \right). \tag{12.86}$$

We now observe that, with the notation adopted in Fig. 12.16, the velocity induced at a generic point $(x, 0)$ by a vortex element with strength $\gamma(\xi)d\xi$ placed at the point $(\xi, 0)$ is equal to

$$dv(x) = -\frac{\gamma(\xi)d\xi}{2\pi(x - \xi)}. \tag{12.87}$$

Therefore, the boundary condition (12.86) becomes

$$-\frac{1}{2\pi} \int_0^c \frac{\gamma(\xi)d\xi}{(x - \xi)} = U \left(\frac{d\eta_c}{dx} - \alpha \right). \tag{12.88}$$

The integral in this equation is singular for $x = \xi$, but the singularity may be overcome by taking the Cauchy principal value of the integral.

Equation (12.88) is the basic integral equation from which the variable strength of the vortex sheet over the chord may be derived. However, the function $\gamma(x)$ must also be such that the Kutta condition is satisfied at the trailing edge. As we have seen, this condition requires the flow to leave the trailing edge smoothly and the velocity therein to be finite. In order to specify how the Kutta condition may be expressed in terms of the vorticity distribution, we recall from Chapter 11, and in particular from relation (11.58), that the local strength of a vortex sheet is equal to the jump of the tangential component of the velocity across the sheet. Considering that, in this case, the vorticity is directed perpendicularly to the plane of motion and that the velocity component tangent to the sheet is the x component u, we thus have

$$\gamma(x) = u(x, 0^+) - u(x, 0^-). \tag{12.89}$$

More precisely, it can be shown (see, e.g., Karamcheti, 1966) that

$$u(x, 0^{\pm}) = \pm \frac{\gamma(x)}{2}. \tag{12.90}$$

Therefore, as the Kutta condition implies that at the trailing edge the velocities tangent to the upper and lower surfaces must be equal, its

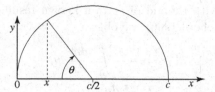

Fig. 12.17. Change of variables for the thin airfoil theory.

expression in terms of γ becomes simply

$$\gamma(c) = 0. \tag{12.91}$$

Considering relation (12.85), it is easy to see that (12.91) is also equivalent to imposing that there should be no pressure discontinuity at the trailing edge.

In order to solve the problem given by the integral equation (12.88), the classical approach (see, e.g., Glauert, 1926) is to expand $\gamma(x)$ as a function of suitable trigonometric terms. To this end, it is useful to introduce the following transformation (see Fig. 12.17):

$$x = \frac{c}{2}(1 - \cos\theta), \tag{12.92}$$

which implies that the leading and trailing edges are given by the coordinates $\theta = 0$ and $\theta = \pi$ respectively.

If we then introduce this new coordinate into Eq. (12.88), and write the integration variable as $\xi = (1 - \cos\vartheta)c/2$, we get

$$-\frac{1}{2\pi}\int_0^\pi \frac{\gamma(\vartheta)\sin\vartheta d\vartheta}{\cos\vartheta - \cos\theta} = U\left(\frac{d\eta_c}{dx} - \alpha\right). \tag{12.93}$$

The following expansion is now introduced for the vorticity distribution:

$$\gamma(\theta) = 2U\left[A_0\left(\frac{1 + \cos\theta}{\sin\theta}\right) + \sum_{n=1}^\infty A_n \sin(n\theta)\right]. \tag{12.94}$$

This expression implies that $\gamma(\pi) = 0$, so that the Kutta condition is satisfied identically. On the other hand, $\gamma(0) = \infty$, which means that for a zero-thickness airfoil an infinite velocity occurs at the leading edge, where the curvature of the tangent streamline is infinite unless the angle of attack is exactly the one that causes the front stagnation point to coincide with the leading edge. By introducing (12.94) into Eq. (12.93), after some manipulations and with the aid of certain known results for the singular integrals (see Karamcheti, 1966), we obtain

$$A_0 = \alpha - \frac{1}{\pi} \int_0^\pi \frac{d\eta_c}{dx} d\theta, \tag{12.95}$$

$$A_n = \frac{2}{\pi} \int_0^\pi \frac{d\eta_c}{dx} \cos n\theta \, d\theta. \tag{12.96}$$

The angle of attack and the camber function $\eta_c(x)$ are known for a given airfoil, and thus the coefficients A_i in (12.94) can be computed readily from relations (12.95) and (12.96). We are now able to evaluate, in terms of these coefficients, the loads acting on the airfoil. By using (12.84) and (12.89) we get first the pressure difference between points on the lower and upper surfaces of the airfoil:

$$\Delta p(x) = p_l(x) - p_u(x) = \rho U[u(x, 0^+) - u(x, 0^-)] = \rho U \gamma(x). \tag{12.97}$$

The resulting force normal to the chord is then

$$\int_0^c \Delta p(x) dx = \int_0^c \rho U \gamma(x) dx = \rho U \Gamma, \tag{12.98}$$

where we have introduced the global circulation around the airfoil:

$$\Gamma = \int_0^c \gamma(x) dx. \tag{12.99}$$

By comparing relation (12.98) with the Kutta–Joukowski theorem (12.6), and recalling the different sign convention for Γ, we see that the integration along the chord of the pressure difference between the lower and upper surfaces of the airfoil provides an estimate of the lift force. In fact, this is consistent with the approximations of the thin airfoil theory, in which the cosines of all the involved angles are put equal to unity. Therefore, the lift is evaluated from the relation

$$L = \rho U \int_0^c \gamma(x) dx = \rho U \frac{c}{2} \int_0^\pi \gamma(\theta) \sin \theta \, d\theta. \tag{12.100}$$

Now, by introducing the vorticity distribution (12.94) into (12.100), and observing that the integral of the function $\sin(n\theta) \sin \theta$ is equal to zero when $n \neq 1$, we obtain the following fundamental relation

$$L = \rho U^2 c\pi \left(A_0 + \frac{A_1}{2} \right). \tag{12.101}$$

Therefore, we find that just the first two terms of the vorticity expansion contribute to the lift. As for the lift coefficient, it becomes

$$C_L = \frac{L}{(1/2)\rho U^2 c} = 2\pi \left(A_0 + \frac{A_1}{2} \right). \tag{12.102}$$

With the chosen notation, the pitching moment relative to the leading edge of the airfoil is positive if clockwise, and is thus given by

$$M_0 = -\int_0^c \Delta p(x) \cdot x\,dx = -\rho U \frac{c^2}{4} \int_0^\pi \gamma(\theta)(1 - \cos\theta) \sin\theta\,d\theta. \tag{12.103}$$

By introducing (12.94) and using the properties of the trigonometric functions, we obtain

$$M_0 = -\rho U^2 \pi \frac{c^2}{4} \left(A_0 + A_1 - \frac{A_2}{2} \right). \tag{12.104}$$

This moment, plus the lift force applied at the leading edge, defines the load system on the airfoil. Obviously, we may change the reference point for the evaluation of the moment, and if we consider a generic point with coordinate x along the chord, the relevant moment is given by

$$M_x = M_0 + Lx. \tag{12.105}$$

The coordinate x_{cp} of the point about which the moment is zero may be seen as the point where the lift force acts, and is called the *centre of pressure* of the airfoil. From (12.105), we find

$$x_{cp} = -\frac{M_0}{L} = \frac{c}{4} \frac{(A_0 + A_1 - A_2/2)}{(A_0 + A_1/2)}. \tag{12.106}$$

Let us now consider the case in which $\eta_c = 0$, i.e. a symmetrical airfoil. With the approximations of thin airfoil theory, this type of airfoil reduces to its chord placed at a certain angle of attack. As follows from (12.95) and (12.96), in that case both A_1 and A_2 are zero and thus one readily obtains from (12.106) that the centre of pressure is positioned at a distance from the leading edge $x = c/4$. If we now return to the general case of an airfoil with curved mean line and evaluate the moment relative to that point, by using (12.105), (12.101) and (12.104) we get

$$M_{c/4} = M_0 + L\frac{c}{4} = -\rho U^2 c^2 \frac{\pi}{8} (A_1 - A_2). \tag{12.107}$$

The corresponding moment coefficient is then

$$C_{M_{c/4}} = \frac{M_{c/4}}{(1/2)\rho U^2 c^2} = -\frac{\pi}{4}(A_1 - A_2). \tag{12.108}$$

These relations show that the moment relative to the point $x = c/4$ does not depend on A_0 and thus, as follows from (12.95) and (12.96), is not affected by a variation of α. Therefore, within thin airfoil theory, the quarter-chord point is not only the centre of pressure of a symmetrical airfoil, but also the *aerodynamic centre* of any type of airfoil, which we have seen to be the point relative to which the moment does not vary when the angle of attack is changed. This is why it is customary to specify the load on an airfoil by giving the resultant lift force and the moment about the quarter-chord point.

By using the values of A_0 and A_1 obtained from (12.95) and (12.96), the expression for the lift coefficient (12.102) may be recast as follows:

$$C_L = 2\pi(\alpha - \alpha_0), \tag{12.109}$$

where α_0 is the *zero-lift angle*, i.e. the angle of attack for which the lift force on the airfoil is zero, and is readily seen to be equal to

$$\alpha_0 = \frac{1}{\pi} \int_0^\pi \frac{d\eta_c}{dx}(1 - \cos\theta)d\theta. \tag{12.110}$$

Relation (12.109) is perhaps the most important result of the thin airfoil theory, as it provides a rapid estimate of the lift of an airfoil once the shape of its mean line and its angle of attack are given. In particular, it provides a good prediction of the lift slope, which is defined as

$$C_{L\alpha} \equiv \frac{\partial C_L}{\partial \alpha} = 2\pi, \tag{12.111}$$

and gives the increase in lift coefficient for a hypothetical increase of one radian in α. Considering the small angles of attack for which the thin airfoil theory may be used and, actually, for which the potential flow model is a good first-order estimate of the real flow condition, it is useful to recall that $2\pi \, \text{rad}^{-1}$ corresponds to slightly less than 0.11 degree^{-1}.

12.5. Panel Methods

We shall see later on that, in spite of all its approximations, the thin airfoil theory provides very useful results as regards the prediction of the lift and moment acting on an airfoil and of their dependence on the shape of the mean line. However, it is not adequate as a first step in the iterative procedure for the evaluation of the interaction between irrotational flow and boundary layers that we described in Section 12.2. In order to use such a procedure, the method for evaluating the first-order potential flow must provide an adequate prediction of the front stagnation point position and of the pressure distribution over the whole airfoil contour. This is not the type of information that may be obtained from the thin airfoil theory, whose primary objective is only to give a first-order estimate of the pressure difference between the airfoil lower and upper surfaces and of its contribution to the global loads. In particular, the assumption of small perturbations with respect to the free-stream velocity precludes the possibility of obtaining good estimates of the flow near the front stagnation point.

We thus return to the problem of the prediction of the potential flow tangent to an arbitrary airfoil, without any simplifying assumption as regards its geometry or its motion. A straightforward approach to this problem consists in taking advantage of the linearity of the Laplace equation and superposing elementary solutions, which are then specified by imposing the boundary conditions. More precisely, different types of the singularities introduced in Section 12.3.2 may be distributed either inside the airfoil or over its contour, and their unknown strengths may be derived by requiring the flow to be tangent to the airfoil and the Kutta condition to be satisfied. In principle, one should apply the tangency boundary condition at the infinite number of points defining the airfoil contour, and thus, in the most general case, the strengths of an infinite number of singularities should be derived. In practice, the problem may be suitably discretized, so that a numerical solution is obtained by solving a linear system comprising a finite, albeit possibly high, number of equations. Subsequently, once the velocity field is available, the corresponding pressure distribution is derived from the application of Bernoulli's theorem.

The first step in the procedure is the geometrical representation of the body in a discrete form. The most common choice is to take a certain number of points along the airfoil contour and to connect them through segments, which are usually termed *panels*. The result is an inscribed polygon (see Fig. 12.18), which approaches the original airfoil geometry more and more as the number of panels is increased. However, given a certain number of panels, the geometrical representation is obviously improved if the panels are concentrated in the regions with higher curvature. A proper geometrical description of the airfoil geometry is important

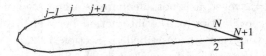

Fig. 12.18.   Discrete representation of an airfoil through flat panels.

because it is connected with the level of accuracy of the derived solution. In effect, the boundary conditions are also discretized, and the flow is required to be tangent to one point on each panel, which is called a control point or, more often, a *collocation point*.

The perturbation velocity field produced by the airfoil is then represented by means of a suitable distribution of singularities, so that both the Laplace equation (12.63) and the boundary condition at infinity (12.65) are satisfied automatically by the associated velocity potentials. The strengths of the introduced singularities are derived from the application not only of the tangency conditions at the collocation points, but also of a suitable explicit or implicit form of the Kutta condition. This means that the number of unknowns characterizing the singularity distribution must be finite and equal to the number of conditions that are imposed by means of apposite equations.

The next step is then choosing both the type of the singularities and their distribution in the field, and a wide variety of different choices are possible. In particular, one may use sources, doublets or vortices, or a combination of them, and the singularities may be distributed over the airfoil contour or inside it. A common solution is to place the singularities over the panels and to take their strength to be constant on each panel or to vary linearly along it. Furthermore, the position of the collocation points must be decided, the usual choice being the midpoint of the panels.

In order to satisfy, albeit in a discrete manner, the surface boundary condition of problem (12.17), it is necessary to evaluate, at each collocation point, the velocity component normal to each panel induced by all the singularities present in the field, and to equate it to the opposite of the normal component of the free-stream velocity. To this end, it is necessary to evaluate the basic computational elements of the method, which are the so-called *influence coefficients*, A_{ij}. The value of a generic influence coefficient A_{ij} is equal to the normal velocity component induced at the collocation point of the ith panel by a unitary value of the singularity strength associated with the jth panel or position.

As an example, let us assume that constant-strength distributions of sources (or sinks, depending on their sign) are placed over each of the N panels describing the airfoil contour. The velocity component normal to the collocation point placed on the ith panel due to a source distribution with strength σ_j placed on the jth panel, say $v_{n_{ij}}$, is given by

$$v_{n_{ij}} = A_{ij}\sigma_j. \tag{12.112}$$

However, for a lifting airfoil we cannot satisfy the Kutta condition by using sources only, because they do not produce any circulation around the airfoil. Therefore, one or more vortices, with their single or global strength representing the whole circulation around the airfoil, must be introduced. A possible choice might be to place one vortex of (unknown) strength γ inside the airfoil, perhaps on the mean line and near the maximum thickness position. By denoting $A_{i(N+1)}$ the normal velocity component induced at the ith-panel collocation point by such a vortex when its strength is unitary, the global normal velocity on that point will then be

$$v_{n_i} = \sum_{j=1}^{N} A_{ij}\sigma_j + A_{i(N+1)}\gamma. \tag{12.113}$$

Thus, if we have an airfoil immersed in a free-stream with velocity U, the tangency boundary condition on its ith panel, whose normal unit vector is denoted by n_i, becomes

$$\sum_{j=1}^{N} A_{ij}\sigma_j + A_{i(N+1)}\gamma = -U \cdot n_i. \tag{12.114}$$

The tangency boundary conditions are applied to all panels by letting i vary from 1 to N; we thus obtain N equations like (12.114) in which, however, $N+1$ unknowns are present. The last necessary equation is given by the Kutta condition, which may be expressed, for instance, by imposing that the *tangential* velocity components on the first and last panels (i.e. those adjacent to the trailing edge) be equal in magnitude and directed downstream. If we denote by B_{ij} the influence coefficients relating to the tangential velocity component and by t_i the unit vector tangent to the ith panel, directed from panel $(i-1)$ to panel $(i+1)$, we then get

$$\sum_{j=1}^{N} (B_{1j} + B_{Nj})\sigma_j + (B_{1(N+1)} + B_{N(N+1)})\gamma = -U \cdot t_1 - U \cdot t_N. \tag{12.115}$$

The obtained system of $N+1$ linear algebraic equations may be solved readily through conventional numerical methods. As a result, we get the strengths of the panel sources and the inner vortex that produce, together with the free-stream velocity, a potential flow in which the airfoil surface is a streamline — albeit in an approximate manner — and the Kutta condition is satisfied at the trailing edge.

Another common and effective panel method is based on the sole use of distributions of vortices over the panels, with a strength that varies linearly along each panel and is continuous at its edges (which are known as the *nodes* of the panel discretization). In this case the value of the vorticity strength at a generic point of the jth panel is a function of the values γ_j and γ_{j+1} of the vorticity strength at the nodes of that panel, as shown in Fig. 12.19(a). Therefore, if the number of panels is N, the number of unknown strengths is equal to the number of nodes, namely $N+1$. Consequently, the flow tangency boundary conditions at the panel collocation points produce the following N equations:

$$\sum_{j=1}^{N+1} A_{ij}\gamma_j = -\boldsymbol{U}\cdot\boldsymbol{n}_i \quad (i=1,N), \tag{12.116}$$

where the generic influence coefficient A_{ij} is equal to the normal velocity component induced at the collocation point of the ith panel by a vorticity distribution that has a unitary value at the jth node and decreases linearly to zero along the adjacent panels, as shown in Fig. 12.19(b).

The problem is closed through the Kutta condition, which, in this case, is expressed in a straightforward way. By considering that the velocity inside the airfoil is zero, it may be seen that the local value of γ at any point of the airfoil contour is equal to the tangent velocity component at that same

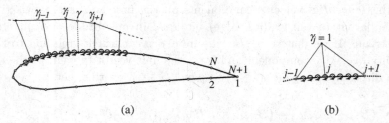

(a)   (b)

Fig. 12.19.   Vortex panel method. (a) Vorticity distribution; (b) inducing vorticity for the evaluation of the influence coefficients.

point. Consequently, if the trailing edge is taken to coincide with both the first and the last nodes, the Kutta condition becomes simply

$$\gamma_1 + \gamma_{N+1} = 0. \tag{12.117}$$

Furthermore, once the system of equations is solved and the values of γ_j at the nodal points are obtained, the vorticity is known at every point of the airfoil contour, and the corresponding value of the pressure coefficient is obtained, using Bernoulli's theorem, from the formula

$$C_p = \frac{p - p_\infty}{(1/2)\rho U^2} = \left(1 - \frac{\gamma^2}{U^2}\right). \tag{12.118}$$

To evaluate numerically the potential flow around airfoils it is possible to devise many other effective methods, whose relative merits are often more connected with practical aspects, such as computational efficiency or code robustness, rather than with accuracy, which is usually comparable for all methods if they are well implemented. For instance, in certain cases increasing the number of panels may not necessarily lead to higher accuracy due to possible ill-conditioning of the matrix of coefficients of the linear solving system. We shall not go through a detailed description of the various methods and of their merits and drawbacks, and the interested reader may refer to more specialized texts, such as Katz and Plotkin (2001), where a comprehensive account of the expressions of the influence coefficients for most of the commonly used distributions of singularities is also given. However, we emphasize again that panel methods are capable of providing, through straightforward numerical procedures, a very good prediction of the potential flow tangent to an airfoil, which we have seen to be the first-order estimate of the real flow when the airfoil is moving at moderate angles of attack. In particular, one obtains the position of the front stagnation point and the pressure distribution over the airfoil upper and lower surfaces, which may serve as an input for subsequent evaluations of the boundary layer. Panel methods may thus be used for all the steps requiring the calculation of the potential flow in the iterative procedure delineated in Section 12.2.2, and are then an essential tool for an accurate prediction of the lift and drag acting on an airfoil. As for the description of the various methods used to predict the evolution of laminar and turbulent boundary layers, which will not be considered in the present book, reference may be made to dedicated texts, such as the comprehensive book by Schlichting and Gersten (2000).

12.6. Typical Lift and Drag Characteristics of Airfoils

As we have seen in the previous sections, predictions of the aerodynamic
loads acting on airfoils may be obtained at different levels of accuracy
through theoretical and numerical procedures. In particular, for airfoils
of small thickness and curvature, the thin airfoil theory provides a first
estimate of the lift and moment and of their variation with angle of attack.
Furthermore, whatever the thickness and curvature of the airfoil, panel
methods may be used in conjunction with boundary layer calculations to
predict drag and obtain a second-order evaluation of lift. However, the basic
assumption of all these procedures is that the angle of attack be small
enough for the boundary layer to remain attached over the whole airfoil
surface. At most, and at the expense of using more refined procedures for
the prediction of the viscous–inviscid interaction, a separation confined to
limited regions might be accepted, provided it does not affect the global
character of the flow. Conversely, when the angles of attack are such that
massive separation is present, no reliable prediction procedures are available
yet for the high Reynolds numbers that are typical of practical applications.
Therefore, it is clear that experimental results will long be a source of
the necessary information for those flow conditions; in effect, only a very
substantial increase in computer performance might change the situation in
the future. In any case, experimental data will always allow an assessment to
be made of the accuracy and range of applicability of the different available
prediction techniques.

Considering the essential role of airfoils in the design of aeroplane
wings, their performance as lift-producing devices is clearly of primary
significance. Therefore, from the practical point of view, the most important
information derives from the curve giving the variation of the lift coefficient
with angle of attack, such as the typical one for a symmetrical airfoil shown
in Fig. 12.20. Several significant features are present in this figure, the
first being that, for small values of the angle of attack, the lift coefficient
is indeed a linear function of α, as predicted by the thin airfoil theory.
However, at a certain value of α, the slope of the curve starts decreasing
until a maximum of the lift coefficient, C_{Lmax}, is reached at a critical
value $\alpha = \alpha_s$.

This trend of the $C_L - \alpha$ curve is connected with the behaviour of
the boundary layer, which is subjected to increasingly adverse pressure
gradients with increasing angle of attack and lift coefficient, as may be
seen from Fig. 12.21. Consequently, at a certain value of α the boundary
layer starts separating from the rear part of the upper surface, and this is

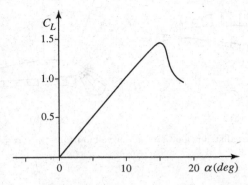

Fig. 12.20. Typical $C_L - \alpha$ curve of a symmetrical airfoil.

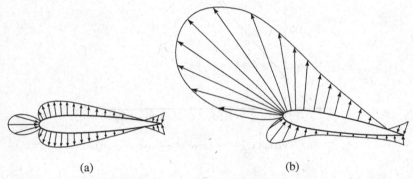

(a)   (b)

Fig. 12.21. Pressure distribution for a symmetrical airfoil. (a) $\alpha = 0°$, $C_L = 0$; (b) $\alpha = 10°$, $C_L = 1.2$.

detectable from the corresponding decrease of the slope in the $C_L - \alpha$ curve (and thus from the end of the linear range). Subsequently, the separation point moves upstream along the airfoil upper surface with increasing α and the slope continues decreasing until, at $\alpha = \alpha_s$, the separated region is so large that a further increase of the angle of attack gives rise to a reduction of the lift. The airfoil is then said to be *stalled*, and α_s is denoted as the *stall angle*.

In reality, the above stall phenomenology is typical of relatively thick airfoils, say with $t/c > 15\%$, and different scenarios occur for airfoils with lower values of the relative thickness, for which small or large separation bubbles may appear over the upper surface. In any case, it is clear that the

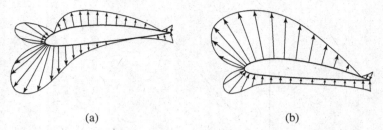

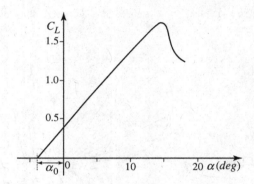

Fig. 12.22. Pressure distribution for a curved airfoil. (a) $\alpha = -4.2°$, $C_L = 0$; (b) $\alpha = 5.7°$, $C_L = 1.2$.

Fig. 12.23. Typical $C_L - \alpha$ curve of a cambered airfoil.

value of the maximum adverse pressure gradient for a given lift coefficient of the airfoil is a significant parameter as regards stall behaviour. Using camber, i.e. introducing a curvature of the mean line, may reduce the adverse pressure gradient for a given value of the lift coefficient. This may be clearly seen from Fig. 12.22, which shows, for an airfoil with 4% maximum camber, the pressure distributions at the same C_L values of the symmetrical airfoil of Fig. 12.21.

Therefore, a positive camber tends to delay stall, and thus to increase C_{Lmax}, as may be seen in Fig. 12.23. Note also that the expected leftward shift of the $C_L - \alpha$ curve, which is clear in that figure, is closely predicted by the zero-lift angle formula of the thin airfoil theory, Eq. (12.110).

There are other geometrical and fluid dynamical parameters that influence the maximum lift coefficient. In particular, the highest values of C_{Lmax} for classical airfoils occur at a percentage thickness of around 12%, as may be seen from the range of values shown in Fig. 12.24, which corresponds

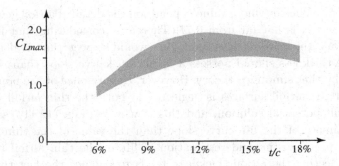

Fig. 12.24. Range of C_{Lmax} values for common airfoils.

to the data for different airfoils reported by Abbott and Von Doenhoff (1959). A point that also arises from Fig. 12.24 is that the values of C_{Lmax} that may be obtained vary from 1.0 to 2.0, depending on thickness and type of airfoil.

The Reynolds number based on the airfoil chord also has a significant effect on C_{Lmax}, due to its influence on the evolution of the boundary layers and on their consequent performance in overcoming adverse pressure gradients without separating. In fact, for $Re = Uc/\nu < 10^6$, the boundary layers may remain laminar, and thus undergo a premature separation. For higher Reynolds numbers, transition to turbulence occurs at positions that move progressively towards the leading edge, and a steady increase of C_{Lmax} with Re is observed. However, no further significant increase may be expected when the boundary layers are almost completely turbulent, as happens for $Re > 10^7$. It must also be considered that surface roughness, particularly in the leading edge region, may reduce the maximum lift by as much as 15 to 20 per cent.

As for the slope of the $C_L - \alpha$ curve, the first point to be noted is that the experimental value of $C_{L\alpha}$ for most types of airfoils is well approximated by the theoretical value of 2π rad^{-1} given by the thin airfoil theory, irrespective of thickness and curvature. This might seem surprising, considering that the value of $C_{L\alpha}$ predicted from the pressure distribution that corresponds to the potential flow tangent to the original airfoil — such as might be obtained through a panel method — is found to increase substantially with increasing thickness of the airfoil. This result is expressed, with good approximation, by the following formula:

$$C_{L\alpha} = 2\pi \left(1 + f\frac{t}{c} \right), \tag{12.119}$$

where f is a coefficient whose value depends on the detailed airfoil geometry but, in practice, is not far from 0.77. Therefore, considering that in thin airfoil theory thickness is neglected, one would expect that real airfoils with finite thickness should possess a higher lift–curve slope than the one predicted by that simplified theory. However, the presence of the boundary layers over the airfoil surfaces is neglected in both the thin airfoil theory and the full potential solution, and this is why formula (12.119) gives a poorer estimate of the lift–curve slope than the value of the thin airfoil theory. In practice, the underestimation of lift in the thin airfoil theory, due to neglecting the airfoil thickness, tends to balance the fact that the 'exact' potential-flow formula (12.119) overestimates the real-flow value because it does not consider the lift reduction caused by the presence of the boundary layers. This reduction is essentially due to the different evolution of the boundary layers over the upper and lower surfaces of a lifting airfoil. In effect, the upper surface boundary layer is subjected to a higher adverse pressure gradient, and thus its thickness in the trailing edge region can be expected to be larger than that of the lower boundary layer. The consequence is that the effective body 'seen' by the potential flow, which is obtained by adding the displacement thickness of the boundary layers over the airfoil surface, has a different effective mean line, evaluated as the line lying midway between the upper and lower surfaces of the modified body. In particular, as may be deduced from the sketch in Fig. 12.25 (in which the effect is purposely exaggerated), the effective angle of attack of the airfoil is slightly decreased and, perhaps more importantly, a fictitious upward curvature is produced near the trailing edge. The effect increases with increasing angle of attack, and the result is a lower value of $C_{L\alpha}$ with respect to the ideal case in which the presence of the boundary layers is neglected.

Airfoils are capable of producing high values of lift with low values of drag, and thus the variation of drag as a function of lift is certainly an important piece of information in airfoil design and optimization. A typical

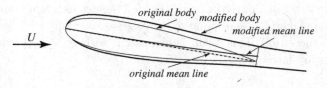

Fig. 12.25.   Influence of boundary layer on the geometry of the effective mean line.

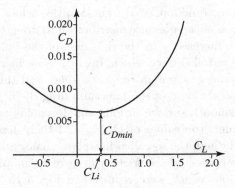

Fig. 12.26. Typical variation of drag as a function of lift for an airfoil.

curve of this type is shown in Fig. 12.26, where some important features may be observed. The first is that the curve has a minimum, which corresponds to zero lift for a symmetrical airfoil (i.e. to a zero angle of attack) and to a non-zero lift for a cambered airfoil. Apart from their good high-lift performance, this is actually one of the great advantages of cambered airfoils, which is particularly useful when the value of C_L for minimum drag, say C_{Li}, is equal or close to the value that corresponds to the cruise flight condition of an aircraft. In general, the minimum drag is obtained at the *ideal angle of attack*, which is the angle of attack for which the fore stagnation point is placed exactly at the leading edge of the airfoil. Obviously, for a symmetrical airfoil this happens when $\alpha = 0$ and the lift is zero, whereas for a cambered airfoil the ideal angle of attack (which may be still close to zero) corresponds to a non-zero C_L, whose value is of the order of $10h/c$. Thus, for a maximum camber of 4% of the chord, the value of C_{Li} is around 0.4. Furthermore, the curve may be quite flat in the region of minimum C_D, and thus similar values of drag may be obtained for a certain range of values of C_L, which might be those of maximum interest for a certain application. However, a drawback of excessive camber is the production of a high negative moment (i.e. nose-down) which, in aircraft design, implies that a balancing negative lift must be applied at the tail surface, thus introducing extra drag and a decrease of the global lift.

The minimum drag of an airfoil is mainly due to friction, i.e. to the action of the tangential viscous stresses; consequently, it is a function of the Reynolds number and of the relative extent of the laminar boundary layer region. Conversely, it is practically independent of camber and a moderately increasing function of thickness ratio, which influences the (smaller) contribution of the pressure distribution to drag. One may expect the pressure drag to increase with the thickness of the boundary layer on the rear part of the airfoil surface which, in turn, is an increasing function of the adverse pressure gradient in the same zone, and thus of the airfoil thickness ratio. Typical values of the minimum drag coefficient range from 0.004 to 0.006 for smooth surface airfoils and Reynolds numbers around 6×10^6. At first order, the values of C_{Dmin} and their dependence on Re may be estimated through the so-called flat-plate analogy, i.e. by using the curves giving the variation of the friction drag of a flat plate for various locations of transition, which were reported in Fig. 7.15. Obviously, the values of the flat-plate friction drag must be multiplied by 2 to consider the upper and lower surfaces of the airfoil, and may then be increased by a factor between 20% and 40% to take account of the airfoil thickness.

It must be pointed out that the range of C_L values corresponding to a low drag may be extended by shaping the airfoil properly, so that transition from laminar to turbulent boundary layers is delayed. This is usually obtained by moving the position of the maximum airfoil thickness backwards, so that the same happens to the minimum pressure and to the subsequent adverse pressure gradient. However, this type of airfoils may show a reduced high-lift performance; moreover, the advantages of the extended laminar region may be nullified by the presence of even moderate amounts of surface roughness.

When the lift departs appreciably from the values corresponding to the low-drag region, a significant increase in drag takes place. This is primarily due to the increase and upstream movement of the suction peak over the airfoil upper surface, which cause an enhanced and anticipated adverse pressure gradient. The consequences are an earlier transition and a thicker boundary layer in the aft region, which produce, respectively, higher friction stresses and a significant increase in pressure drag. In fact, most of the drag increase occurring at high lift coefficients is due to the pressure contribution. In other words, the relative percentage of pressure drag, which may be of the order of 10–15% in the low-drag region, increases significantly when the maximum values of lift are approached. In summary, the variation of the drag coefficient with lift coefficient may be expressed through the following formula (see Torenbeek, 1976):

$$C_D = C_{Dmin} + k \left(\frac{C_L - C_{Li}}{C_{Lmax} - C_{Li}} \right)^2, \tag{12.120}$$

where the coefficient k depends on Reynolds number and airfoil character-istics, such as C_{Lmax} and relative thickness t/c.

As may be easily supposed, one of the most important parameters defining the performance of an airfoil in aeronautical applications is its lift-to-drag ratio, i.e. its efficiency. Values of E above 100 are quite common for well-designed airfoils. However, it must be pointed out that equally important is the range of lift coefficients for which the efficiency remains near its maximum value. In practice, it is advantageous that this range be as wide as possible and correspond to the most suitable one for the considered application.

Finally, for almost all airfoils, experiments show that, provided the angles of attack are such that no separation of the boundary layers occurs, the aerodynamic centre is placed at a distance from the leading edge lying in the range $0.23 < x_{ac}/c < 0.27$. In fact, in many cases, the aerodynamic centre is very close indeed to the quarter chord point, thus confirming the soundness of thin airfoil theory, which, in spite of all its simplifying assumptions, is evidently capable of describing accurately the essential mechanisms of the generation of lift on an airfoil. Further details on airfoil characteristics and performance, as well as an extensive collection of experimental data, may be found in Abbott and Von Doenhoff (1959).

We end this chapter with a brief outline on high-lift devices, which are widely used during the take-off and landing of aircraft to reduce the necessary flight speeds. These devices increase the airfoil C_{Lmax} considerably, and this effect is obtained by using two mechanisms, namely an increase of the effective airfoil camber and an enhancement of the resistance to separation of the boundary layers. These goals are obtained, in the flight conditions in which lift enhancement is required, through two different types of devices, whose basic idea is to deflect downwards the trailing edge or the leading edge regions of the airfoil (see Fig. 12.27).

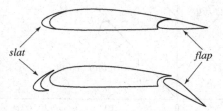

Fig. 12.27. Example of retracted and extended leading edge and trailing edge high-lift devices.

In order to understand the different consequence of the deflection of
the trailing edge or of the leading edge, it is instructive to refer to the
results of the thin airfoil theory again. In effect, relation (12.110), which
gives the zero-lift angle, shows clearly that the effect of an increased slope
of the mean line is higher near the trailing edge than near the leading
edge, due the fact that the function $(1 - \cos\theta)$ multiplying $d\eta_c/dx$ tends
to zero near the leading edge and to its maximum value near the trailing
edge. Therefore, the downward deflection of a rear *flap* enhances the global
curvature of the airfoil, and increases both the lift for a given α and the
value of C_{Lmax}, although to a lesser extent.

Conversely, a leading edge deflection, such as the one obtained with
a *slat*, does not have a great effect on the global curvature of the airfoil,
but produces a reduction of the adverse pressure gradients over the upper
surface and thus delays boundary layer separation and stall. The effects
of the separate and joint use of trailing edge and leading edge devices are
illustrated schematically in Fig. 12.28.

As can be seen, a trailing edge flap moves the $C_L - \alpha$ curve leftwards and
upwards, whereas a leading edge slat prolongs the curve and delays stall
up to higher angles of attack. The joint use of leading edge and trailing
edge devices permits a much higher C_{Lmax} to be obtained at roughly the
same value of α as the original airfoil. The performance of both leading
edge and trailing edge devices is improved significantly by introducing slots
between them and the main airfoil, which tend to energize the boundary

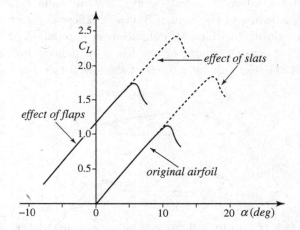

Fig. 12.28.   Effects of high-lift devices on $C_L - \alpha$ curve of a symmetrical airfoil.

layers and delay their separation; an example is shown in Fig. 12.27. Furthermore, a further enhancement of lift is obtained through multiple surface flaps with downstream extensions, which also produce a virtual increase of the airfoil chord and thus permit achieving values of C_{Lmax} as high as 4.

Chapter 13

FINITE WINGS IN INCOMPRESSIBLE FLOWS

13.1. Introduction

In the previous chapter, we analysed the aerodynamics of airfoils, i.e. of particular two-dimensional bodies that are capable of producing a significant lift force in conjunction with a much lower drag force. Also, in view of their typical application, we considered airfoils as the cross-sections of hypothetical infinite wings and, in Chapter 8, we described in detail the mechanisms causing the production of lift when an airfoil starts moving. However, we all know that infinite wings do not exist, and thus it is now essential to analyse which modifications, in terms of both flow features and fluid dynamic loads, are produced because wings are finite. As was the case for airfoils, we are concerned with incompressible flow conditions. However, we recall that the main qualitative features of the velocity and vorticity fields that will be described do not change significantly as long as the flow remains subsonic, i.e. with a Mach number that is lower than one everywhere.

In order to introduce the quantities that define the geometry of a wing, let us start by saying that the wing is assumed to be isolated, i.e. we do not consider the presence of a fuselage. In other words, the shape of the wing is obtained by continuing it inside any fuselage, up to a symmetry plane where the *wing root* cross-section lies (see Fig. 13.1). The origin of all usual reference systems is located on the symmetry plane, with the x axis in the direction either of the free-stream or of the root chord, the y axis perpendicular to the symmetry plane and the z axis in the upward vertical direction when the wing is moving horizontally.

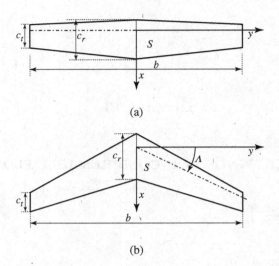

(a)

(b)

Fig. 13.1.   Airfoil geometry and notation (a) straight wing; (b) swept wing.

The extremities of the wing are called *wing tips*, and the distance from one wing tip to the other is the wing *span*, usually denoted as b. The ratio between the chords of the tip and root airfoils, $\lambda = c_t/c_r$, is called wing *taper ratio*. The projection of the wing on the x–y plane defines the wing *planform*, which is bounded by the leading and trailing edges and by the tips of the wing. The area of this projection, often simply denoted as the *wing area S*, is the usual reference surface for the definition of the wing force coefficients, and should not be confused with the total surface area of the wing, which is the area of the wetted surface, and thus includes both the upper and lower surfaces of the wing.

We shall see later that a fundamental parameter in the aerodynamics of wings is the *aspect ratio*, AR, which is defined as

$$AR = \frac{b^2}{S}, \tag{13.1}$$

and is thus equal to b/c for a rectangular wing with chord c. Furthermore, the mean geometrical chord of the wing is defined as $\bar{c} = S/b$.

For the sake of generality, a *swept wing* is shown in Fig. 13.1(b), although such geometry is not typical of incompressible flow conditions, being adopted to delay or modify the compressibility effects when flying at high Mach numbers. The *sweep angle* Λ is usually defined as the angle between the y axis and the *quarter-chord line*, i.e. the line joining the quarter-chord points along the wing planform. With such a definition, a straight wing is a wing with a straight quarter-chord line. Consequently, for a tapered straight wing, i.e. when $\lambda < 1$, the leading and trailing edges are inclined with respect to a line perpendicular to the free-stream direction, as may be seen in Fig. 13.1(a). Likewise, Fig. 13.1(b) shows that the inclinations of the leading and trailing edges of a tapered swept wing are different from Λ.

In general, the airfoils that constitute the wing sections may vary along the span, both in shape and angle of attack, and we shall see that this may be advantageous to optimize the wing performance. However, for the purpose of explaining the main differences between the flow fields of an airfoil and of a finite wing, in the following section we shall first refer to a simple rectangular wing, composed of airfoils that are all similar and orientated with the same angle of attack.

13.2. The Velocity and Vorticity Fields Around a Finite Wing

13.2.1. *General features*

Especially for high-aspect-ratio wings, it is easy to be convinced that the mechanisms causing the formation of circulation and lift in the two-dimensional case — namely the generation of vorticity by the no-slip viscous condition and the shedding of a starting vortex — still work for the single cross-sections of a finite wing. Incidentally, provided there is no lateral free-stream velocity component, the airfoil corresponding to the wing root section is immersed in a two-dimensional flow, and thus its aerodynamic features must be, at least qualitatively, analogous to those that characterize the flow around an infinite wing.

Let us assume that the wing starts moving in such a way that attached boundary layers are present over its surface and lift is produced. Consequently, the pressures acting on its upper surface are, relatively speaking, lower than those acting on the lower surface. If the wing span were infinite, the consequent flow would be two-dimensional, with no velocity components in the y direction. Conversely, for a finite wing, the pressure difference between lower and upper surfaces, and thus the lift, must inevitably go to zero at the tips. Note, however, that the lift — and the associated circulation — can never be uniform along the whole wing and go to zero abruptly at its tips, and a more gradual variation, such as the one schematically shown in Fig. 13.2, will then occur. Obviously, the exact distribution of lift along the span depends on the detailed shape of

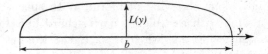

Fig. 13.2.   Spanwise variation of the lift along a finite wing.

Fig. 13.3.   Cross-flow due to wing finiteness.

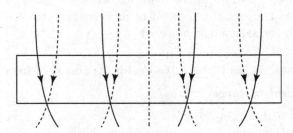

Fig. 13.4.   Schematic streamlines above and below a finite wing. Solid lines: above the wing; dashed lines: below the wing.

the wing but, in any case, the maximum spanwise gradient will occur at the wing tips, and the minimum one at the wing centreline. In any case, the spanwise pressure gradients arising along the two surfaces of the wing create a velocity field with components in the $y-z$ plane. In particular, the fluid particles from the lower surface will tend to round the wing tips, moving towards the upper surface, and producing the cross-flow schematically shown in Fig. 13.3. Actually, the spanwise velocity components may be expected to be larger where the pressure gradients are higher, i.e. near the wing tips, and to vanish at the midplane. The global velocity field may then be obtained by superposing this cross-flow on the (prevailing) longitudinal main flow. As a consequence, the streamlines above the wing will be curved towards the midplane and those below the wing will be curved towards the tips, as qualitatively shown in figure Fig. 13.4.

The above-described velocity field refers to the irrotational flow produced by the pressure gradients around the wing, just outside the boundary layers existing over its upper and lower surfaces. To complete the flow

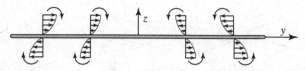

Fig. 13.5. Schematic velocity profiles within the wing boundary layers in the $y-z$ plane.

description, we must then analyse the effects of the wing finiteness on the velocity field inside the boundary layers.

To this end, let us consider the additional cross-flow due to the spanwise pressure gradients, which is depicted in Fig. 13.3. The necessity of satisfying the no-slip boundary condition implies that the consequent boundary layers present over the upper and lower surfaces of the wing also have velocity components in the spanwise direction, as schematically shown in Fig. 13.5, in a view from downstream of the wing. Obviously, this spanwise velocity field is additive to the prevailing one parallel to the $x-z$ plane, which is qualitatively similar to the two-dimensional boundary layer flow present around an airfoil. Consequently, the velocity profiles within the boundary layers over the surface of a finite wing are three-dimensional and, in general, skewed, i.e. such that the direction of the resultant velocity vector changes along a line normal to the surface.

The important point to be noted is that the velocity field in Fig. 13.5 corresponds to vorticity components that are parallel to the longitudinal $x-z$ plane, and whose orientation is indicated by the arrows in the figure. This additional vorticity is completely due to the wing finiteness, and is additive to the vorticity field in the y direction connected with the 'two-dimensional' velocity components parallel to the $x-z$ plane.

As can be appreciated from Fig. 13.5, the longitudinal vorticity field connected with the boundary layer cross-flow velocity is fundamentally different from the vorticity field in the y direction. In effect, at each spanwise position, the longitudinal vorticity has the same sign in the boundary layers present over the upper and lower surfaces. Furthermore, apart from particular flight conditions, the vorticity has the same sign for each half-wing, and is balanced by the opposite-sign vorticity present at the corresponding symmetric cross-section in the other half-wing.

It is then important to analyse the main features of the vorticity field left downstream, in the wake produced by the merging of the boundary layers originating from the upper and lower surfaces of the wing. From the above description of the velocity field, it is indeed evident that, besides

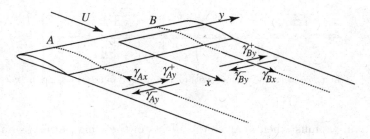

Fig. 13.6.   Schematic vorticity field downstream of a finite wing.

a spanwise vorticity that is similar to that of the infinite-wing case, longitudinal vorticity components are present in this wake, with opposite sign for each half-wing. The qualitative situation is shown in Fig. 13.6, which depicts a sketch of the wake of a straight wing. In this figure, the x axis is taken to be directed as the free-stream velocity and to lie on the wake midsurface; although we shall see that a real wake is neither plane nor parallel to the free-stream velocity, this assumption does not affect the present discussion. We also assume that the time elapsed from the start of the motion is such that the starting vortex is far enough downstream not to influence the flow around the wing.

We now consider two cross-sections of the wing, say A and B, which are symmetrical with respect to the wing centre-plane. We then analyse the vorticity field at two positions in the wake, defined by the two spanwise coordinates y_A and y_B that correspond to the cross-sections A and B, and by the same downstream coordinate. In each position, we have three vectors that represent the resultants of the vorticity of each sign and direction present in that location of the wake. Each vector is assumed to be positioned in the centre of gravity of the relevant vorticity component. More precisely, γ_{Ay}^{+} and γ_{Ay}^{-} are the resultants of the spanwise vorticities present at a lateral position $y = y_A$ and originating from the upper and lower surfaces of the wing respectively. Furthermore, these vectors are positioned at a distance from the wake mid-surface equal to the displacement thicknesses of the vorticity sheets generated by the boundary layers originating from the upper and lower surfaces. As already pointed out for the two-dimensional case, γ_{Ay}^{+} and γ_{Ay}^{-} are of opposite sign and equal magnitude and, due to the symmetry of the flow in the two half-wings, are equal to the corresponding vorticity vectors γ_{By}^{+} and γ_{By}^{-} present at the coordinate $y = y_B$. On the other hand, the resultants of the longitudinal vorticity, $\boldsymbol{\gamma}_{Ax}$ and $\boldsymbol{\gamma}_{Bx}$, originating from the cross-flow caused over the wing by its finiteness and lying on the two planes defined by the coordinates y_A and y_B, are equal and opposite.

Therefore, one can immediately see that, at any location in the wake, the equal and opposite vorticity vectors in the y direction are close to each other, being at a mutual distance of the order of the sum of the upper and lower displacement thicknesses at that location. Consequently, the velocity field induced by these vectors vanishes rapidly with increasing distance from the wake. In other words, at any point in the flow that is sufficiently far from the wake, the velocities induced by the spanwise vorticity components tend to be equal and opposite, and thus to cancel out. Conversely, the vorticity vectors in the longitudinal direction, γ_{Ax} and γ_{Bx}, which are also equal and opposite, are placed at a mutual distance that is generally large compared to the wake thickness, and may actually reach a maximum value of b. Therefore, these vorticity vectors induce significant velocities around the wing and the wake.

In effect, as sketched in Fig. 13.7, each pair of counter-rotating wake longitudinal vortices induces a velocity with a significant downward component, which is therefore usually denoted as *downwash*. This velocity acts over the wing, in front of it (albeit with decreasing magnitude moving in the upstream direction), and also on the wake itself. In reality, the wake is also subjected to the downward velocity induced by the transversal vorticity present over the wing boundary layers; in effect, that vorticity is the basis of the circulation around the wing cross-sections that produces the lift force over the wing.

As a consequence of this induced velocity field, the wake moves downwards and its midsurface does not remain plane. In fact, its shape changes moving downstream, in conjunction with a redistribution of the wake

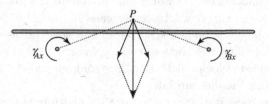

Fig. 13.7. Velocity induced by a couple of longitudinal wake vortices.

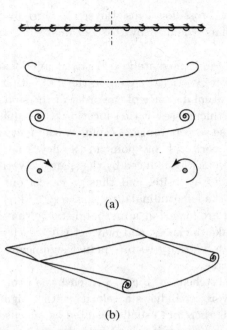

(a)

(b)

Fig. 13.8.   (a) Vorticity evolution in successive downstream cross-sections of the wake.
(b) Schematic view of the wake of a finite wing.

vorticity caused by its self-induced velocity field. As a result, the longitudinal vorticity tends rapidly to become concentrated in two tight counter-
rotating vortices, around which the wake vorticity sheet rolls up. Further
downstream, these vortices end up being positioned approximately at a
lateral coordinate coinciding with the centre of gravity of the longitudinal
vorticity shed by each half-wing. This progressive evolution is shown
schematically in Fig. 13.8(a), whereas a global qualitative view of the wake
is depicted in Fig. 13.8(b). The wake vortices may be visualized behind an
aircraft by the condensation of water vapour due to the low temperatures
existing inside their cores, where a decrease in pressure occurs with a
magnitude that depends on the strength of the vortices, as we have seen
in Chapter 9. In reality, the vortices trailing behind a large aeroplane may
produce an induced velocity field that is strong enough to pose a severe
hazard to following smaller aircraft.

However, the most important effect of the longitudinal vorticity present
in the wake is that the lift acting on a finite wing is lower than the

one that would act on an infinite wing with the same cross-sections and positioned at the same nominal angle of attack. This reduction is the consequence of the decreased effective angle of attack of the wing caused by the wake-induced downwash. Furthermore, the velocity field in planes perpendicular to the flight direction, associated with the wake longitudinal vorticity, corresponds to significant kinetic energy, which was not present in the flow field upstream of the wing. Therefore, this introduction of new kinetic energy in the flow must inevitably correspond to a work done on the fluid, and thus to a drag acting on the wing. This drag, which is usually called *induced drag*, is totally due to the finiteness of the wing, and is thus additive to the drag that would act on an infinite wing. Moreover, it may be expected to increase with the intensity of the wake vortices, and thus with the lift acting on the wing.

13.2.2. *Wake trailing vorticity*

From the above flow-field description, it should now be clear that, in order to develop methods for the prediction of the forces acting on finite wings, it is essential to connect the longitudinal vorticity shed in the wake from each wing cross-section with the circulation $\Gamma(y)$ present around the same cross-section. To this end, we apply Stokes' theorem to the surface S_{12} lying between two circuits C_1 and C_2 rounding the cross-sections of the wing at $y = y_1$ and $y = y_2$, as shown in Fig. 13.9. The two circuits C_1 and C_2 are assumed to be far enough from the wing surface to be outside the boundary layers developing over the wing, and to cross the wake close to the wing trailing edge. Consequently, on S_{12} vorticity is present only

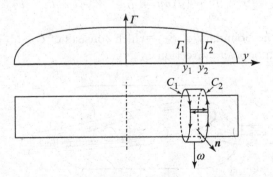

Fig. 13.9. Evaluation of the strength of the wake longitudinal vortices.

in that limited region where the surface intersects the wake, and only the longitudinal component of this vorticity contributes to a non-zero flux through that surface.

The circulation is assumed to be positive when it is clockwise looking in the positive direction of the y axis (i.e. with the free-stream velocity coming from the left). Furthermore, to obtain a reducible circuit, as required by Stokes' theorem, we introduce a cut from C_1 to C_2, which is followed twice in opposite directions when evaluating the circulation, and thus gives it a zero total contribution. We also note that, having assumed the normal unit vector to be directed outwards from the surface, as shown in Fig. 13.9, and imagining that we look in the positive y direction, circuit C_1 must be followed clockwise and circuit C_2 counter-clockwise.

Let us denote as $\gamma(y)$ the strength of the longitudinal vorticity leaving the wing trailing edge at a generic cross-section, per unit length along y. Therefore, $\gamma(y)$ is also the flux of vorticity through a circuit C around that cross-section as may be better seen from Fig. 13.10, which shows an example of such a circuit. In other words,

$$\gamma(y) = \int_C \boldsymbol{\omega} \cdot \boldsymbol{n}\, dl, \tag{13.2}$$

where dl is an infinitesimal element along the circuit and $\boldsymbol{\omega} \cdot \boldsymbol{n}$ is zero outside the small length in which C crosses the wake formed by the boundary layers emanating from the wing (see Fig. 13.10). Note that $\gamma(y)$ has the dimensions of vorticity times a length, i.e. the same as a velocity.

The application of Stokes' theorem to the considered surface S_{12}, i.e. to the strip between C_1 and C_2, may then be expressed as follows:

$$\int_{S_{12}} \boldsymbol{\omega} \cdot \boldsymbol{n}\, dS = \int_{y_1}^{y_2} \gamma(y)\, dy = \int_{\partial S_{12}} \boldsymbol{V} \cdot dl = \Gamma_1 - \Gamma_2, \tag{13.3}$$

where ∂S_{12} is the boundary of S_{12}, which consists of C_1, C_2 and the cut, counted twice in opposite directions.

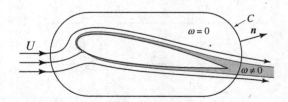

Fig. 13.10. Flux of vorticity through a circuit around a wing cross-section.

We now let the two cross-sections approach each other until they are separated by an infinitesimal distance dy, so that $y_2 = y_1 + dy$. We have

$$\Gamma_2 = \Gamma_1 + \frac{d\Gamma}{dy} dy, \tag{13.4}$$

and thus the strength of the global longitudinal vorticity being released into the wake across a strip of length dy placed around the wing is

$$\gamma(y)\, dy = -\frac{d\Gamma}{dy} dy. \tag{13.5}$$

Therefore, we finally get

$$\gamma(y) = -\frac{d\Gamma}{dy}. \tag{13.6}$$

Note that, as long as $\gamma(y)$ is defined by expression (13.2) and the relevant circuit C crosses the wake just downstream of the wing trailing edge, relation (13.6) is exact and does not rely on any simplifying assumption as regards the shape or the orientation of the wake.

It is now easy to verify that the complete vorticity field satisfies the conservation of the total vorticity. As a matter of fact, the wake vortex sheet is composed of pairs of opposite sign vortices, which start from the wing trailing edge and, in principle, end at the starting vortex, bending inside it in the cross-flow direction. Furthermore, the vorticity present in the boundary layers over the wing surface is equal and opposite to the vorticity left in the starting vortex. Therefore, the global integral of vorticity over the whole fluid domain is equal to zero, as it was before the start of the motion.

13.3. Evaluation of the Fluid Dynamic Loads

13.3.1. *Generalities*

The presence of longitudinal vorticity components in the wake is the most important effect of the finiteness of a wing, with significant consequences not only on the lift and drag forces acting on the wing, but also on the methods for their prediction.

Assuming that a sufficient time has elapsed from the outset of the motion, the starting vortex may be considered to be far enough downstream not to influence the velocity field around the wing, where the flow may also be assumed to be steady. Moreover, the Reynolds number is assumed to be high and the boundary layers to be thin and completely attached to the wing surface. Therefore, it is reasonable to presume that a satisfactory first-order model of the flow around a finite wing may be constructed similar to what has been done for airfoils. We can thus imagine squeezing the vorticity-containing regions, namely the boundary layers and the wake, to zero thickness. As a result, we end up with a flow that is tangent to the wing surface, and with a value of the circulation around each wing cross-section such that a Kutta condition along the trailing edge is satisfied; simultaneously, the thickness of the wake is also reduced to zero.

The above flow configuration is apparently analogous to that of the first-order model of the flow around two-dimensional airfoils; however, in reality a significant difference exists. Reducing the thickness of the wake to zero in the flow around an airfoil implied that the vorticity contained therein cancelled out, whereas for a finite wing this cancellation occurs for the transversal vorticity components only. Conversely, the longitudinal vorticity components of opposite sign are separated by a large cross-flow distance, which obviously does not decrease on squeezing the wake thickness. Therefore, we can only say that the wake becomes a vortex sheet, which is also a stream-surface, i.e. a surface to which the velocity vectors are locally tangent. Note, however, that even if the velocity vectors on the upper and lower sides of the wake have the same modulus — in agreement with the continuity of pressure — their direction is not equal. In effect, as shown by relation (11.57), the transversal velocity component jumps through the wake by a quantity equal to the local strength of the corresponding vortex sheet. More precisely, relations (11.56) show that the transversal velocity components on the two sides of the vortex sheet are opposite but with the same magnitude, equal to $\gamma/2$.

Consequently, even in a first-order potential-flow model, longitudinal ideal vortices of adequate strength — i.e. satisfying relation (13.6) — must be introduced to describe the basic features of the real flow around a finite wing and to predict, with a satisfactory approximation, the effects of the wake vorticity. Having assumed that the starting vortex is sufficiently far from the wing, the longitudinal trailing vortices may be assumed, in a first-order model, to extend downstream to infinity. As a result, the velocity field around the wing and the wake is a potential flow that may be described as the sum of the free-stream velocity and of a perturbation velocity, similar to what was expressed in relation (12.15) for the analysis of the flow around an airfoil.

The relevant mathematical problem is then apparently similar to the one expressed in (12.17). In effect, we have to find, again, a harmonic potential function ϕ such that the corresponding velocity field is tangent to certain surfaces, albeit in this case these surfaces comprise not only the contour of the considered body, namely the wing, but also the upper and lower sides of the vortex sheet representing the wake. Nevertheless, the analogy is only formal because now, in principle, we do not know in advance the shape and position of the surface representing the wake. We only know that it has the same orientation as the velocity field which, however, may only be obtained once the velocity potential is available. Therefore, even if the Laplace equation that must be solved to find the velocity potential is linear, and the strength of the vortex sheet may be expressed as a function of the wing vorticity through relation (13.6), the problem is actually non-linear because to solve it one should know the position and shape of the wake vorticity sheet, which influences the velocity field around the wing; however, this position may only be known after the problem has been solved.

The situation might seem discouraging, but, in practice, the above problem may be overcome without great difficulty in several ways. The first one is to devise an iterative procedure that starts with the assignment of a first tentative shape and position of the wake, so that the mathematical problem may be solved. Subsequently, the velocity component perpendicular to the given wake surface is evaluated at a sufficient number of judiciously chosen control points. The wake is then displaced at those points of a quantity proportional to the evaluated component, and a new shape is thus obtained. A second potential flow problem may then be solved, after which the check of the wake tangency is repeated and the procedure is iterated until the variation of the shape between two subsequent steps becomes lower than a given threshold. Through this method, the downstream shape variation of the wake mean surface, including its roll-up, may be satisfactorily predicted.

In order to reduce the number of iteration steps, the control should preferably be carried out on the quantity to be predicted through the procedure. For instance, very often interest is only in obtaining the lift acting on the wing and, in that case, the lift force would be the sensible quantity to be used for a comparison between the results of two successive iteration steps. Lift is obviously a function of the pressure field around the wing, which, in turn, depends on the velocity field in the same region. However, the flow around the wing depends on the longitudinal vorticity present in the wake through the downwash induced by each elementary vortex of intensity $\gamma(y)\,dy$ trailing downstream, as schematically shown in

Fig. 13.7. Now, as may be easily derived by applying formula (11.67) to the present flow configuration, with the notation of Fig. 11.4, the velocity induced by a segment of unitary length of this elementary vortex decreases rapidly as its downstream distance from the wing increases. Therefore, only the wake portion immediately adjacent to the wing trailing edge is really important for the evaluation of the velocity induced on the wing. Conversely, the exact shape and position of the more downstream part of the wake has a much lower influence on the global lift acting on the wing.

In reality, if the objective is the prediction of the lift on the wing, the limited importance of the exact shape and position of the downstream portion of the wake suggests an alternative and simpler way of overcoming the above-cited non-linearity of the mathematical problem. In practice, one may simply ignore the real evolution of the wake and simulate it as a vortex sheet with a given shape. Obviously, this shape should be chosen in a sensible way. If all the wing cross-sections are similar and at the same angle of attack (i.e. if the wing is not twisted), a simple and reasonable choice may be a plane surface continuing the direction of the mean wing surface at the trailing edge. If spanwise variations are instead present, a non-plane surface or a plane one with an average direction may be chosen. In reality, one may use any plausible shape, often obtaining satisfactory results even by choosing the wake surface to be plane and to lie in the direction of the free-stream velocity. Whatever the rationale behind the choice of the shape of a fixed wake vortex sheet, the result is a linear potential flow problem, in which the unknown is the vorticity distribution over the wing surface. However, the iterative procedure for an accurate prediction of the wake shape becomes necessary when the interest is in the evaluation of the flow field — and of the consequent loads — on a body placed downstream of the wing, like a tail surface. Indeed, the actual shape and position of the wake will become important, in that case, because the velocities induced by the wing wake vorticity may vary significantly with small variations of the distance between the trailing vortices and the downstream surface.

The evaluation of a second-order approximation to the flow around finite wings using the coupling between potential flow and boundary layer calculations is presently not very common. This is due essentially to the progressive spread of the use of codes for the solution of the RANS equations (see Chapter 7) and to the higher difficulty of developing numerical codes for an efficient prediction of the evolution of three-dimensional boundary layers for generic geometries. Nonetheless, further activities aimed at improving the performance of numerical methods for the evaluation of three-dimensional boundary layers may still be advantageous, particularly to obtain more reliable estimates of the friction drag of wings and, more generally, of three-dimensional aerodynamic bodies with complex configurations.

13.3.2. *Prediction of the potential flow*

The mathematical problem to be solved, either in each iteration step or in a single fixed-wake calculation, should now be clear: find the potential flow tangent to the wing surface in the presence of a wake vortex sheet having a given shape and a vorticity distribution that is a function of the circulation around the wing cross-sections, and thus of the cross-flow vorticity present over the wing surface. It is also necessary to impose the Kutta condition along the wing trailing edge in order to specify the circulation around each wing cross-section. In practice, the Kutta condition may be enforced either explicitly, by imposing that the flow is tangent to a given direction above and below the wake in points just behind the trailing edge, or implicitly, through relations that are equivalent to enforcing that the pressure difference between the upper and lower wing surfaces is zero at the trailing edge.

This problem may be solved in different ways and at different levels of accuracy, similar to what we saw for the potential flow around airfoils (see, e.g., Hess, 1990). In particular, to obtain a high-level evaluation of the potential flow around complex configurations, perhaps including not only wings but also fuselages, one may devise a variety of three-dimensional panel methods in which the various surfaces are discretized through polygonal panels and the superposition of elementary solutions is exploited. Therefore, three-dimensional singularities are used, which are the equivalents of those we introduced for the two-dimensional case, namely sources, doublets and vortex filaments, and the strengths of these singularities represent the unknowns of the discretized mathematical problem. Details on the potential flow fields associated with these singularities may be found, for instance, in Katz and Plotkin (2001), where various panel methods are also described. Conversely, in other methods (Morino and Kuo, 1974) the unknowns are directly the values of the perturbation potential on the panels, rather than the strengths of the elementary solutions.

In reality, the number of different panel methods that have been developed to predict the potential flow around wings and, more generally, around complex three-dimensional configurations, is very high, and their detailed description is beyond the scope of the present book. What we can say is that, with the currently available performance of electronic computers, most of these methods are capable of providing accurate results with comparable and definitely small computational costs. Therefore, they may also be advantageously introduced in optimization routines for the design of new aircraft configurations.

It must now be pointed out that panel methods provide, as an output, the pressure distribution over the wing surface due to the first-order potential flow model. They are then the basic first step for any procedure in which potential flow calculations are coupled with the solution of the boundary layer equations. However, similar to what has been done for the evaluation of the loads on airfoils, more simplified methods may be devised with the limited objective of obtaining only a satisfactory first-order prediction of the lift acting on the wing. In particular, the so-called *vortex-lattice method* may be considered as the equivalent of the thin airfoil theory. In this method also, the boundary conditions over the wing surface are linearized and transferred from the actual wing surface to the projection of the wing on a reference mean plane, which either contains or is close to the chords of the wing sections. The projected planform is then divided into quadrilateral panels and, once again, the vorticity distribution existing over the wing is represented in a discrete manner.

The basic ideal vorticity element used for this purpose is schematically illustrated in Fig. 13.11; due to its shape, it is generally known as a *horseshoe vortex*. As can be seen, this is a particular constant-strength line vortex composed of a vortex segment connecting two points and two semi-infinite line vortices starting from these points and extending downstream to infinity.

One horseshoe vortex of unknown strength Γ_k is then positioned on each panel, with the segment (often denoted as the *bound vortex*) in the spanwise direction and the two semi-infinite trailing vortices in the longitudinal direction. As can be seen from the sketch in Fig. 13.12 (in which

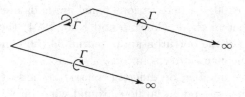

Fig. 13.11.   Horseshoe vortex.

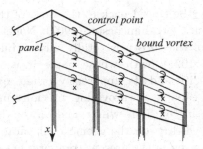

Fig. 13.12. Simplified layout of horseshoe vortices and control points over a wing.

the wing is divided into a small number of panels for the sake of clarity), the vortices trailing downstream from the various panels at the same spanwise position join behind the wing trailing edge and add up algebraically, forming a model of the vortex sheet representing the wake through a finite number of semi-infinite vortices, whose strength is easily verified to be a discrete version of relation (13.6). The vortex tails forming the wake are generally positioned on a fixed plane surface in the direction of the free-stream velocity. The velocity field induced by each horseshoe vortex may then be evaluated by applying relation (11.67), and the problem is closed by imposing the tangency boundary conditions at one control point for each panel.

The bound vortex is usually positioned at the quarter-chord line of each panel and the control point at the three-quarter-chord point. This choice has a theoretical justification; in effect, it may be shown that if a lifting flat plate is simulated through a single vortex placed at a distance from the leading edge $x = c/4$, then the result of thin airfoil theory may be recovered provided the tangency condition is enforced at the point positioned at $x = 3c/4$ (see, e.g., Katz and Plotkin, 2001). This result does not have an immediate connection with the role of the single panels in the vortex-lattice method but, in practice, the above positioning of vortices and control points is widely used and provides numerically robust procedures and satisfactory results. Finally, if an approximation of the true shape and position of the wake must be obtained, one may develop non-linear vortex-lattice methods in which the previously-described iteration procedure is applied until the wake vortex sheet becomes a stream-surface.

The vortex-lattice method is quite simple, and may be implemented without difficulty in effective numerical procedures. However, an even simpler method exists, namely the so-called *lifting-line theory*, whose fundamental ideas may be traced back to the work of the British engineer Frederick W. Lanchester (see Lanchester, 1907). Nonetheless, the mathematical formulation of the theory is totally due to Ludwig Prandtl (see Prandtl, 1918, 1919, 1921), with whose name it is rightly associated. In spite of its simplicity and its apparently crude approximations, the lifting-line theory is capable of providing precious information on the aerodynamic loads acting on finite wings, and on the dependence of both the lift and the induced drag on the geometrical parameters defining the configuration. In fact, even in the present computer era, its results may be used for first-order evaluations of the expected performance of finite wings. Therefore, considering its primary importance in aeronautical engineering applications, we dedicate the next section to a description of this method and its main results.

13.4. The Lifting-Line Theory

13.4.1. *Basic approximations and formulation*

The wings that are considered in the classical lifting-line theory developed by Prandtl are straight, i.e. their quarter-chord lines are straight segments. In fact, the theory was subsequently extended to swept wings (see Weissinger, 1947), but we shall limit ourselves to the case of a straight wing and, for added simplicity, we may even start by considering the wing to be rectangular. To highlight all the approximations implied in the lifting-line model, we shall follow a step-by-step procedure for its description.

As usual, we imagine that the vorticity-containing regions have already been squeezed to zero thickness. The wake is then assumed to be made up of semi-infinite line vortices with the same direction as the free-stream velocity U, and with a strength that is linked to the circulation existing around the wing cross-sections by relation (13.6). The global transversal vorticity present inside the boundary layers over the upper and lower surfaces of the wing, which is responsible for the circulation and lift around each wing cross-section, is imagined to be represented by a zero-thickness vortex sheet. This is further assumed to lie on a plane parallel to the free-stream direction, and to be modelled by continuing the wake vortices along the wing, to form a multiplicity of constant-intensity horseshoe vortices, as shown in Fig. 13.13.

We now introduce the basic assumption of the lifting-line theory: *each wing cross-section behaves as if it belonged to an infinite wing, except for the fact that it is immersed in a free-stream whose angle of attack is modified*

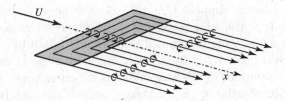

Fig. 13.13.   Horseshoe vortices representing the wing and the wake.

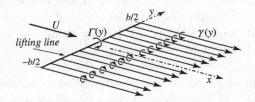

Fig. 13.14.   Lifting-line model of the flow past a wing.

by the downwash velocity field induced by the wake vortices produced by the finiteness of the wing.

This assumption implies that the only considered effect of the cross-flow existing over the wing is the production of the longitudinal vorticity present in the wake. Conversely, any modification that this flow may cause to the two-dimensional aerodynamic characteristics of the airfoils that coincide with the wing cross-sections is neglected. Note that this implies that we may utilize the results of any method for the evaluation of those characteristics, and the problem becomes the prediction of the variations in the angle of attack of each wing cross-section induced by the wake vortices.

In order to simplify the evaluation of the downwash effects, a further and even more drastic approximation is introduced, namely the chords of all the wing cross-sections are shrunk to zero. In other words, the cross-flow portions of the horseshoe vortices that represent the wing vorticity in Fig. 13.13 are condensed in a single vortex segment of variable strength $\Gamma(y)$. Hence, the surprisingly simple model on which this theory is based is the one shown in Fig. 13.14. As can be seen, the wing is modelled through a segment of line vortex — usually denoted as the 'lifting line' — whose strength represents the circulation existing around the wing cross-sections, and is thus variable in the spanwise direction. The downwash velocity induced by the wake vortices, and producing the variation in angle of attack, is then evaluated at points on the lifting line.

It is reasonable to suppose that the above strongly simplified model is applicable only if the wing geometry satisfies certain geometrical constraints. In effect, the wing is assumed to be characterized by a sufficiently large aspect ratio (say $AR > 6$), so that one may neglect, as is done in the model, the variations of the wake-induced velocities along the chord of each cross-section. Furthermore, the theory neglects completely the effects of the spanwise velocity components present over the wing. Now, these spanwise components are generally larger at the wing tips, where the pressure gradients are stronger. Therefore, when the wing aspect ratio is large, we may reasonably expect that any influence of the neglected spanwise flow will be restricted to a relatively small portion of the wing, thus producing a small global effect on the wing load.

In summary, in the lifting-line model, the effect of the finiteness of the wing is assumed to be completely described by a reduction of the angle of attack of each wing cross-section. Moreover, it is supposed that this reduction may be derived from the downwash velocity induced along the lifting line by the wake vorticity, which is represented through the simplified configuration of Fig. 13.14. The resulting situation is depicted in Fig. 13.15, which also shows the system of coordinates that will be used in the following. As can be seen, we take the z axis to be positive in the upwards direction, and the wake-induced velocity to be directed downwards, with a magnitude $w(y)$. Therefore, the resultant velocity at any point of the lifting line is equal to the vector sum of the free-stream velocity $\boldsymbol{U} = U\boldsymbol{e}_x$ and of the downwash $-w(y)\boldsymbol{e}_z$. We denote this resultant velocity by $\boldsymbol{U}_R(y)$ and note that, in general, it is variable along the lifting line.

Let us now denote by $\alpha_a(y)$ the *aerodynamic angle of attack* of each wing cross-section, i.e. the angle between the free-stream velocity and the zero-lift direction of the relevant airfoil, so that

$$\alpha_a(y) = \alpha(y) - \alpha_0(y). \tag{13.7}$$

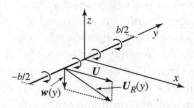

Fig. 13.15.   Coordinates system and velocities along the lifting line.

Note that, in general, α_a is a function of y, as neither α nor α_0 are assumed to be constant in the spanwise direction. In other words, the theory allows the wing to be *twisted*, i.e. with variable geometrical angle of attack, and to be constructed using different types of airfoils (and thus different values of α_0) along the span.

Each wing cross-section is then assumed to behave as if it were immersed in a two-dimensional flow with a free-stream velocity that is no longer U but the resultant velocity $U_R(y)$, i.e. in a flow whose direction is varied by the effect of the downwash velocity. Therefore, the effective angle of attack of $U_R(y)$ is $\alpha_R(y)$, which is different from the aerodynamic angle of attack $\alpha_a(y)$ of the nominal undisturbed free-stream velocity. As shown in Fig. 13.16, we have

$$\alpha_R(y) = \alpha_a(y) - \alpha_i(y), \tag{13.8}$$

where $\alpha_i(y)$ is the *induced angle of attack*, giving the variation in flow direction due to the effect of the wake vortices. For positive values of $w(y)$, i.e. when the induced velocity is indeed a downwash, this angle is positive and corresponds to a reduction in the angle of attack. Moreover, $\alpha_i(y)$ is assumed to be small, so that we have, with good approximation,

$$\alpha_i(y) \simeq \frac{w(y)}{U}. \tag{13.9}$$

We now observe that the function $\Gamma(y)$, giving the variation of the circulation along the lifting line, is the unknown function of the problem, and that (13.8) is the basic relation leading to its derivation. The induced velocity $w(y)$ at a generic point on the lifting line may be evaluated by adding up the velocities induced at that point by all the wake vortices. To do so we note that, as may be seen from Fig. 13.17, the downwash induced at point y by the elementary vortex with strength $\gamma(\eta)d\eta$ positioned at $y = \eta$, say $w_\eta(y)$, is equal to the velocity induced by a semi-infinite straight-line vortex positioned at a distance $(\eta - y)$ from the considered point.

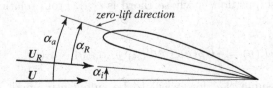

Fig. 13.16.   Angles of attack of the cross-section of a finite wing.

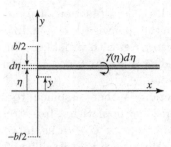

Fig. 13.17.   Evaluation of downwash induced on a point of the lifting line.

By recalling expression (11.68), the elementary contribution to the down-wash is then equal to

$$w_\eta(y) = \frac{\gamma(\eta)d\eta}{4\pi(\eta - y)}. \tag{13.10}$$

Consequently, the total downwash velocity $w(y)$ may be obtained by integrating expression (13.10) over the whole lifting line. If we use relation (13.5) for the strength of each vortex trailing downstream from an infinitesimal length of the lifting line, we get

$$w(y) = \frac{1}{4\pi} \int_{-b/2}^{+b/2} \frac{\gamma(\eta)}{(\eta - y)} d\eta = \frac{1}{4\pi} \int_{-b/2}^{+b/2} \frac{d\Gamma}{dy}(\eta)\frac{1}{(y - \eta)} d\eta, \tag{13.11}$$

in which the principal value of the integral is implied. With the present choice of the coordinate axes and of the positive direction of the circulation, we may now express the cross-flow force acting on a generic cross-section of the wing as follows:

$$F(y) = \frac{1}{2}\rho U_R^2(y)c(y)C_{L\alpha_0}(y)\alpha_R(y) = \rho U_R(y)\Gamma(y), \tag{13.12}$$

where $C_{L\alpha_0}(y)$ is the slope of the lift curve of the airfoil placed at the generic spanwise position y, whose chord is $c(y)$. From relation (13.8), we then obtain

$$\Gamma(y) = \frac{1}{2}U_R(y)c(y)C_{L\alpha_0}(y)[\alpha_a(y) - \alpha_i(y)]. \tag{13.13}$$

We now assume the downwash to be sufficiently small to allow the approximation $\cos \alpha_i \simeq 1$, so that we may put $U_R(y) \simeq U$. Consequently, by using (13.9) and (13.11), we get from relation (13.13)

$$\Gamma(y) = \frac{1}{2}c(y)C_{L\alpha_0}(y)\left[U\alpha_a(y) - \frac{1}{4\pi}\int_{-b/2}^{+b/2}\frac{d\Gamma}{dy}(\eta)\frac{1}{(y-\eta)}d\eta\right]. \quad (13.14)$$

This is an integro-differential equation for $\Gamma(y)$, and may be seen as the fundamental relation deriving from Prandtl's lifting-line theory. Note that in Eq. (13.14) all the geometrical parameters and the characteristics of the airfoils at the generic wing cross-sections, defined by the functions $c(y)$, $\alpha_a(y) = [\alpha(y) - \alpha_0(y)]$ and $C_{L\alpha_0}(y)$, are assumed to be known. In particular, a strong feature of the model is that the value of the slope of the airfoil lift curve to be introduced in the equation, $C_{L\alpha_0}(y)$, may derive from any prediction procedure. In other words, it may result not only from a potential-flow theory (such as the value 2π of thin airfoil theory), but also from a higher-order method that takes the boundary layers into account, or even from experimental data. Furthermore, the wing may be twisted (i.e. with variable α), tapered (variable c), and made up of different airfoils along the span (variable α_0 and $C_{L\alpha_0}$).

The description of the methods used to solve Eq. (13.14) is beyond the scope of the present book, and reference may be made to classical texts of aerodynamics, such as Karamcheti (1966). Suffice it here to say that one can introduce a suitable change of variables and express the circulation as a series of terms satisfying the boundary condition that Γ be zero at the wing tips, i.e. at the extremities of the lifting line. The various terms in the series may then be derived by imposing that Eq. (13.14) be satisfied at a sufficiently high number of spanwise cross-sections. We shall then assume in the following that $\Gamma(y)$ be available, and discuss in some detail the main results deriving from the theory.

13.4.2. *Main results and discussion*

Once the distribution of circulation along the wing span is available, the global forces acting on the wing may be easily derived. To this end, it is useful to refer to Fig. 13.18, which shows the forces acting on a wing cross-section (per unit spanwise length), neglecting the drag due to viscosity. From the Kutta–Joukowski theorem in the form (12.59), the resultant force is found to act in a direction that is perpendicular to the local modified free-stream velocity $U_R(y)$ seen by the wing cross-section. Therefore, considering that the lift and drag acting on the wing are, by definition,

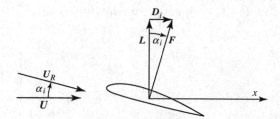

Fig. 13.18.   Lift and drag on a wing cross-section.

the force components that are respectively orthogonal and parallel to the real free-stream velocity U, the resultant force contributes not only to the lift but also to an added drag component, which is the induced drag that we have seen to be due to the finiteness of the wing.

Therefore, considering that the values of $\alpha_i(y)$ have been assumed to be small, we may express the local lift and induced drag as follows:

$$L(y) = F(y)\cos\alpha_i(y) = \rho U_R(y)\Gamma(y)\cos\alpha_i(y) = \rho U\Gamma(y), \qquad (13.15)$$

$$D_i(y) = F(y)\sin\alpha_i(y) = \rho U_R(y)\Gamma(y)\sin\alpha_i(y) = \rho w(y)\Gamma(y). \qquad (13.16)$$

Using relation (13.9), expression (13.16) may also be recast as

$$D_i(y) = L(y)\alpha_i(y). \qquad (13.17)$$

The global lift acting on the wing is thus

$$L = \rho U \int_{-b/2}^{b/2} \Gamma(y)\,dy, \qquad (13.18)$$

whereas the induced drag may be obtained from the relations

$$\begin{aligned} D_i &= \int_{-b/2}^{b/2} \alpha_i(y)L(y)\,dy \\ &= \rho U \int_{-b/2}^{b/2} \alpha_i(y)\Gamma(y)\,dy = \rho \int_{-b/2}^{b/2} w(y)\Gamma(y)\,dy. \end{aligned} \qquad (13.19)$$

The corresponding force coefficients may be immediately obtained by dividing the wing lift and drag by the quantity $(1/2\rho U^2 S)$.

One of the most important results of the lifting line theory is that the spanwise lift distribution giving the smallest induced drag, for given wing lift and span, is the elliptical distribution. From relation (13.15) it is seen

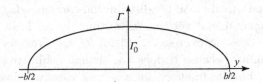

Fig. 13.19.   Elliptical spanwise distribution of the circulation.

that this corresponds to a distribution of circulation that is also elliptical, i.e. to a function $\Gamma(y)$ having the following expression:

$$\Gamma(y) = \Gamma_0 \sqrt{1 - (2y/b)^2}. \tag{13.20}$$

As shown in Fig. 13.19, the elliptical distribution is represented by the upper half of an ellipse, and Γ_0 is the maximum circulation at the wing midsection. The corresponding wing lift may thus be immediately derived from the area below the curve in the figure, and we obtain

$$L = \rho U \int_{-b/2}^{b/2} \Gamma(y)\, dy = \rho U \frac{1}{2}\pi \Gamma_0 \frac{b}{2} = \frac{\pi}{4}\rho U \Gamma_0 b. \tag{13.21}$$

Another characteristic and important feature of the elliptical lift distribution is that the relevant downwash velocity is constant along the span. In particular, one finds (see, e.g., Karamcheti, 1966)

$$w(y) = w_i = \frac{\Gamma_0}{2b}. \tag{13.22}$$

Consequently, the induced drag becomes

$$D_i = \rho w_i \int_{-b/2}^{b/2} \Gamma(y)\, dy = \frac{w_i}{U} L = \frac{\Gamma_0}{2b}\frac{L}{U} = \frac{\pi}{8}\rho \Gamma_0^2. \tag{13.23}$$

The corresponding lift and drag coefficients are then

$$C_L = \frac{L}{1/2\rho U^2 S} = \frac{\pi}{2}\frac{b}{S}\frac{\Gamma_0}{U}, \tag{13.24}$$

$$C_{Di} = \frac{D_i}{1/2\rho U^2 S} = \frac{\pi}{4S}\left(\frac{\Gamma_0}{U}\right)^2 = \frac{C_L^2 S}{\pi b^2} = \frac{C_L^2}{\pi AR}. \tag{13.25}$$

By means of relations (13.22) and (13.24), the constant induced angle of attack may also be written as

$$\alpha_i = \frac{w_i}{U} = \frac{C_L}{\pi AR}. \tag{13.26}$$

The result that an elliptical lift distribution corresponds to a minimum drag is a fundamental and very useful output of the lifting-line theory. Furthermore, the relevant expression (13.25) for the drag coefficient has a more general validity because it does not change qualitatively for a non-elliptical lift distribution. In effect, for a generic case, we have

$$C_{Di} = \frac{C_L^2}{\pi AR}(1 + \delta),$$ (13.27)

where δ is a small positive number (usually less than 0.1), which is zero for an elliptical lift distribution. The general conclusion we may draw is that the induced drag coefficient increases with the square of the lift coefficient and is inversely proportional to the aspect ratio, AR. The consequence is that to lower the induced drag of a wing it is convenient to have wings with high aspect ratio and to fly at low values of C_L.

However, it should also be pointed out that the wing lift is normally a prescribed design quantity because, in cruise conditions, it is the force necessary to counterbalance the weight of the aeroplane. Therefore, it may be convenient to use the formulas derived for a wing with elliptical lift distribution to express the direct dependence of the induced drag on the total lift. With the help of (13.23) and (13.21) one easily finds

$$D_i = \frac{2}{\pi} \frac{L^2}{\rho U^2 b^2},$$ (13.28)

from which we deduce that, for given values of the wing lift L and the kinetic pressure $1/2\rho U^2$, the induced drag is inversely proportional to the square of the span. However, one must recall that the span is limited by structural and practical considerations, and also that the same lift may be obtained with different values of C_L and S, and that both these quantities also have an influence on the viscous drag of the wing. This suggests that optimizing the design of an aircraft wing is a complex problem, in which many different factors play a significant role.

An elliptical lift distribution may be obtained with different spanwise distributions of chord, angle of attack and type of airfoil. This can be seen immediately by rewriting Eq. (13.14) for the particular case of an elliptical variation of the circulation, i.e. by introducing relations (13.20) and (13.22). We obtain the following relation:

$$\Gamma_0 \sqrt{1 - (2y/b)^2} = \frac{1}{2}c(y)C_{L\alpha_0}(y)\left[U\alpha_a(y) - \frac{\Gamma_0}{2b}\right],$$ (13.29)

which may be satisfied by several different combinations of $c(y)$, $C_{L\alpha_0}(y)$ and $\alpha_a(y) = [\alpha(y) - \alpha_0(y)]$. In particular, if we choose the wing to be composed of the same airfoils (which implies that the values of α_0 and $C_{L\alpha_0}$ are constant), and the geometrical angle of attack α to be the same along the whole span, relation (13.29) gives

$$c(y) = k\Gamma(y), \tag{13.30}$$

where k is a constant that depends on the values of Γ_0, U, α, b and $C_{L\alpha_0}$. Therefore, in order to obtain an elliptical lift distribution on an untwisted wing composed of a single type of airfoil, the spanwise distribution of the chord (and thus the wing planform) must be elliptical; such a type of wing is usually called an *elliptical wing*.

If we now assume the wing to be elliptical, a simple and very useful expression may be derived for the relation between the lift curve slope of the whole wing, say $C_{L\alpha}$, and the lift curve slope of the basic airfoil, $C_{L\alpha_0}$. In effect, considering the constancy of $C_{L\alpha_0}$, of α_a and of the induced angle of attack α_i, and using relation (13.26), we get

$$\begin{aligned} L(y) &= \frac{1}{2}\rho U^2 SC_L = \frac{1}{2}\rho U^2 C_{L\alpha_0}(\alpha_a - \alpha_i)\int_{-b/2}^{b/2} c(y)\,dy \\ &= \frac{1}{2}\rho U^2 SC_{L\alpha_0}(\alpha_a - \alpha_i) = \frac{1}{2}\rho U^2 SC_{L\alpha_0}\left(\alpha_a - \frac{C_L}{\pi AR}\right), \end{aligned} \tag{13.31}$$

from which we obtain

$$C_L\left(1 + \frac{C_{L\alpha_0}}{\pi AR}\right) = C_{L\alpha}\alpha_a\left(1 + \frac{C_{L\alpha_0}}{\pi AR}\right) = C_{L\alpha_0}\alpha_a, \tag{13.32}$$

and, finally,

$$C_{L\alpha} = \frac{C_{L\alpha_0}}{1 + C_{L\alpha_0}/(\pi AR)}. \tag{13.33}$$

This relation may be further simplified by introducing for $C_{L\alpha_0}$ the value given by the thin-airfoil theory, namely 2π. We thus obtain the following mnemonic formula:

$$\frac{C_{L\alpha}}{C_{L\alpha_0}} = \frac{AR}{AR + 2} = \frac{C_L}{C_{L_0}}, \tag{13.34}$$

which may be used to readily estimate the order of magnitude of the lift coefficient C_L of a finite wing having a certain aspect ratio and an aerodynamic angle of attack α_a, compared to the value C_{L_0} that would correspond to an infinite wing placed at the same α_a.

Instead of using an elliptical planform, which is a rather expensive solution, one may also obtain an almost elliptical load distribution by introducing taper, i.e. with a trapezoidal wing planform with a smaller chord at the wing tip than at the wing root (see Fig. 13.1). The good performance of tapered wings is demonstrated by the fact that the coefficient δ appearing in Eq. (13.27) may become as low as 0.01 for a taper ratio around 0.3, with a flat minimum for $0.2 \leq \lambda \leq 0.4$, which is indeed the range of values found in most commercial aircraft.

The predictions of the lifting-line theory are in very good agreement with experimental data, at least as long as the angles of attack are reasonably small, i.e. of the order of those corresponding to usual cruise conditions. Actually, if one compares the real vorticity field of a finite wing (see the sketch in Fig. 13.8b, for instance) and the simplified model of Fig. 13.14, the results of the lifting-line theory may be considered as beyond all possible expectations, and we may then wonder why this theory is able to give such satisfactory results. One explanation is that the lifting-line theory captures the essential physical effect of the wing finiteness, namely the virtual reduction in the angle of attack of the wing cross-sections. The fact that the downwash causing this reduction is evaluated with satisfactory approximation, using the model of Fig. 13.14, is a consequence of the excellent prediction of the strength of the wake vortices provided by relation (13.6) and of the prevailing influence of the portion of wake immediately downstream of the wing trailing edge.

In conclusion, the results of the lifting-line theory, together with those of the thin-airfoil theory, are the basic tools for rapid first-order estimates of the loads on wings. Obviously, a real optimized design would require other levels of accuracy, and thus the use of higher-order models and powerful computational resources. Nonetheless, these simplified models, all developed well before the present computer era, allow the prediction not only of the order of magnitude of the results, but also of their expected trends as a function of the involved parameters. This ability to provide such rapid predictions is fundamental in engineering practice, and is invaluable to maintain a critical attitude in the analysis of the outputs of the complex numerical procedures that may be currently included in more sophisticated design tools.

13.5. Wing Loads in Terms of Vorticity Dynamics

To conclude this chapter, we now show briefly how the forces acting on a finite wing may be expressed in terms of vorticity dynamics, using the relevant formulas derived in Chapter 11. To this end, we recall that we must evaluate the *time variation* of the moment of the vorticity introduced in the field by the motion of the body. Therefore, even for a wing in uniform translation and whatever the used reference system, the motion must be considered as unsteady because the relative dynamics of the wing and the starting vortex must be included in the analysis.

Let us then consider a finite wing in uniform translation with velocity $-U\,\boldsymbol{e}_x$ in a three-dimensional flow domain, and choose the y axis to be parallel to the wing span and the z axis in the vertical direction. As the wing velocity is constant in time, formula (11.74) becomes

$$\boldsymbol{F} = -\frac{1}{2}\rho\frac{d}{dt}\int_{R_\infty} \boldsymbol{x} \times \boldsymbol{\omega}\,dR. \tag{13.35}$$

We now try to recover the expressions of the lift and drag forces that we obtained using the lifting-line theory. We observe first that, as found from the above considerations, we cannot assume the wake to be of infinite length. Thus, even considering a simplified representation, similar to the one shown in Fig. 13.14, we must imagine the wake to be very long but finite, ending with a cross-flow vortex that represents the starting vortex and whose strength is equal and opposite to the circulation $\Gamma(y)$ on the lifting line. On the other hand, we reduce to zero the thickness of the vorticity-containing regions again, so that vorticity is confined to the lifting line, the wake surface S_w and to the starting vortex. However, for a reason connected with the evaluation of the induced drag, S_w cannot be represented as a flat surface parallel to $\boldsymbol{e}_x$, as was the case in the lifting-line model, and must be assumed to be a zero-thickness surface parallel to the local velocity. Nonetheless, the wake inclination angle is taken to be small everywhere, as a consequence of the limited value of the downwash. Furthermore, no rolling-up of the wake surface is considered, and thus each longitudinal vortex is still assumed to remain always parallel to the $x-z$ plane.

With these assumptions, the lift force component deriving from (13.35) may be written as

$$L = F_z = -\frac{1}{2}\rho\frac{d}{dt}\int_{S_w} (\omega_y x - \omega_x y)\,dS_w, \tag{13.36}$$

where ω_i indicates the global amount of the i-direction component of the vorticity contained in the wake and its boundaries, which is now squeezed over S_w.

The two terms appearing in the integral in (13.36) may be seen as the contributions to the time variation of the z component of the vorticity moment deriving, respectively, from the progressive increase of the distance between the lifting line and the starting vortex and from the continuous lengthening of the longitudinal vortices trailing from the wing. In order to evaluate the first contribution, we consider the wake to be sufficiently long that we can assume the starting vortex to move downstream, relative to the wing, with the velocity U, which thus also coincides with the time variation of its x coordinate. Furthermore, the vorticity in the starting vortex is a function of the y coordinate and is equal and opposite to the vorticity in the lifting line; in other words, we have there $\omega_y(y) = -\Gamma(y)$. Consequently, we obtain

$$-\frac{1}{2}\rho \frac{d}{dt} \int_{S_w} \omega_y x \, dS_w = \frac{1}{2}\rho U \int_{-b/2}^{b/2} \Gamma(y) \, dy. \qquad (13.37)$$

On the other hand, the time variation of the vorticity moment due to the longitudinal trailing vortices is caused exclusively by the lengthening of the wake. Now, if we analyse the flow from the wing (or the lifting-line), the length of each longitudinal wake vortex with strength $\gamma(y)$ may be assumed to increase, in a time interval dt, by $U dt$, which is the value of the downstream displacement of the starting vortex. In other words, it is as if a portion of length $U dt$ had been added to the wake in the region adjacent to the starting vortex. Furthermore, considering the small inclination of the wake relative to e_x, we may put $\omega_x(y) = \gamma(y)$. Therefore, the contribution of each longitudinal wake vortex of strength $\gamma(y)$ to the opposite of the variation in unit time of the z component of the vorticity moment is given by $U\gamma(y)y$.

By using expression (13.6) for the strength of the wake vortices, integrating by parts and recalling that $\Gamma(\pm b/2) = 0$, we then get

$$\frac{1}{2}\rho \frac{d}{dt} \int_{S_w} \omega_x y \, dS_w = -\frac{1}{2}\rho U \int_{-b/2}^{b/2} \frac{d\Gamma(y)}{dy} y \, dy$$

$$= \frac{1}{2}\rho U \int_{-b/2}^{b/2} \Gamma(y) \, dy, \qquad (13.38)$$

and finally, by introducing (13.37) and (13.38) into (13.36) we get the expression for the lift acting on the wing:

$$L = \rho U \int_{-b/2}^{b/2} \Gamma(y) \, dy, \qquad (13.39)$$

which coincides clearly with relation (13.18).

The induced drag force is given by the x component of expression (13.35), namely

$$D_i = F_x = -\frac{1}{2}\rho\frac{d}{dt}\int_{S_w} (\omega_z y - \omega_y z)\, dS_w. \tag{13.40}$$

The first term in the integral in (13.40) may be interpreted as the contribution of the increasing length of the wake to the variation of the x direction vorticity moment. As already seen, the increase in length of each longitudinal wake vortex in a time interval dt is due to the downstream movement of the starting vortex, which is equal to Udt. Considering the small inclination of the wake, which coincides with the local induced angle of attack, $\alpha_i(y) = w(y)/U$ (positive when the induced velocity is directed downwards), the contribution of the increased portion of longitudinal vortex to the variation of the z component of vorticity in the field is then $-Udt\,\gamma(y)\,\alpha_i(y) = -\gamma(y)w(y)\,dt$. We also observe that, with our model, the value of $w(y)$ in a region adjacent to the starting vortex is the same as that along the lifting line because, in both positions, the induced downwash is that due to all the semi-infinite vortices in the wake. If we then assume $\Gamma(y)$ to vanish at the wing tips in such a way that $\Gamma(\pm b/2)w(\pm b/2) = 0$, we get

$$
\begin{aligned}
-\frac{1}{2}\rho\frac{d}{dt}\int_{S_w}\omega_z y\, dS_w &= -\frac{1}{2}\rho\int_{-b/2}^{b/2}\frac{d\Gamma(y)}{dy}w(y)y\,dy \\
&= -\frac{1}{2}\rho\int_{-b/2}^{b/2}\frac{d[y\Gamma(y)]}{dy}w(y)\,dy \\
&\quad +\frac{1}{2}\rho\int_{-b/2}^{b/2}\Gamma(y)w(y)\,dy \\
&= \frac{1}{2}\rho\int_{-b/2}^{b/2}y\Gamma(y)\frac{dw(y)}{dy}\,dy + \frac{1}{2}\rho\int_{-b/2}^{b/2}\Gamma(y)w(y)\,dy.
\end{aligned}
$$
$$\tag{13.41}$$

As for the second term in the integral (13.40), again it is due to the movement of the starting vortex relative to the lifting line, this time in the vertical direction. This displacement is caused by the downwash acting on the starting vortex, which we have seen to be the same as that along the lifting line. Considering that, on the starting vortex, we have $\omega_y(y) = -\Gamma(y)$

and $dz(y)/dt = -w(y)$, we obtain

$$\frac{1}{2}\rho\frac{d}{dt}\int_{S_w} \omega_y z\, dS_w = \frac{1}{2}\rho\int_{-b/2}^{b/2} \Gamma(y)w(y)dy. \tag{13.42}$$

Finally, by introducing (13.41) and (13.42) into (13.40) we obtain, for the induced drag, the following expression derived from the formula giving the forces in terms of variation of the vorticity moment:

$$D_i = \rho\int_{-b/2}^{b/2} w(y)\Gamma(y)\, dy + \frac{1}{2}\rho\int_{-b/2}^{b/2} y\Gamma(y)\frac{dw(y)}{dy}dy. \tag{13.43}$$

The first integral in (13.43) coincides with the result of the lifting-line theory, given by formula (13.19). The second integral is thus an additive term, and we now try to explain both its origin and why it is not present in the lifting line theory. The first point we note is that it vanishes when the downwash velocity is constant and thus $dw/dy = 0$. We then observe that, having assumed that the wake surface is tangent to the local velocity, $-dw(y)/dy$ gives the variation in unit time of the inclination of the wake in a $x = constant$ plane. Therefore, considering that the vorticity of the starting vortex is $-\Gamma(y)$, we may interpret the term $dy[\Gamma(y)dw(y)/dy]$ as a new vorticity component in the $-z$ direction, introduced in unit time by the different movement of the extremities of a portion of the starting vortex with length dy. The consequent contribution to the opposite of the variation of the x component of the moment of vorticity is $ydy[\Gamma(y)dw(y)/dy]$, and its spanwise integration gives the second term in (13.43), which is thus totally due to a three-dimensional effect connected with the variation of the wake curvature in a $x = constant$ plane. Therefore, it cannot be derived from the lifting-line model, in which the wake is assumed to be plane. Furthermore, it vanishes when the downwash is constant because in that case no lateral curvature arises.

It is now interesting to note that the value of the circulation on the symmetry plane of the wing, say $\Gamma_0 = \Gamma(0)$, also gives the absolute value of the resultant vorticity shed from each half-wing. In effect, we have

$$\int_{-b/2}^{0} \gamma(y)\, dy = \int_{-b/2}^{0} -\frac{d\Gamma(y)}{dy}dy = -\Gamma_0, \tag{13.44}$$

$$\int_{0}^{b/2} \gamma(y)\, dy = \int_{0}^{b/2} -\frac{d\Gamma(y)}{dy}dy = \Gamma_0. \tag{13.45}$$

Let us then further analyse the case in which $w(y) = constant = w_i$. From (13.41), we see that, in this case, we have

$$-\frac{1}{2}\rho\frac{d}{dt}\int_{S_w}\omega_z y \, dS_w = \frac{1}{2}\rho w_i \int_{-b/2}^{b/2}\Gamma(y)\, dy$$

$$= -\frac{1}{2}\rho w_i \int_{-b/2}^{b/2}\frac{d\Gamma(y)}{dy}y\,dy = \frac{1}{2}\rho w_i \int_{-b/2}^{b/2}\gamma(y)y\,dy$$

$$= \frac{1}{2}\rho w_i \int_{-b/2}^{0}\gamma(y)y\,dy + \frac{1}{2}\rho w_i \int_{0}^{b/2}\gamma(y)y\,dy.$$

$$(13.46)$$

Consequently, the last two integrals in (13.46) are equivalent to the values of the resultant vorticities shed from each half-wing multiplied by the coordinates of their centres of gravity. Considering that $\gamma(y)$ is negative for $y < 0$ and positive for $y > 0$, we also deduce that the value of the two integrals is the same.

We consider now the case of an elliptical load distribution. By using the following change of variables

$$y = -\frac{b}{2}\cos\theta, \qquad (13.47)$$

relation (13.20) becomes

$$\Gamma(y) = \Gamma_0 \sin\theta, \qquad (13.48)$$

and we find, after straightforward calculations, that the coordinates of the centres of gravity of the vorticities shed from the two half-wings are

$$y_{cg}^{\pm} = \pm\frac{\pi}{8}b, \qquad (13.49)$$

where the signs + and − refer to the right and left half-wings respectively.

Therefore, the distance between the vector resultants of the vorticities with opposite sign shed from the two half wings is equal to

$$b_0 = \frac{\pi}{4}b. \qquad (13.50)$$

Thus, the contribution of the increasing length of the wake to the rate of variation of the x component of the moment of vorticity may be seen to be given by the lengthening of these two resultant vortices. Their increase in length is approximately $U dt$, and we must consider that they are inclined at an angle $-\alpha_i = -w_i/U$. If we now use for w_i the value given by the lifting-line theory for elliptical lift, Eq. (13.22), we get

$$-\frac{1}{2}\rho\frac{d}{dt}\int_{S_w}\omega_z y\,dS_w = \frac{1}{2}\rho U \Gamma_0 \frac{w_i}{U}b_0$$

$$= \frac{1}{2}\rho w_i \Gamma_0 b_0 = \frac{1}{2}\rho\frac{\Gamma_0}{2b}\Gamma_0\frac{\pi}{4}b = \frac{\pi}{16}\rho\Gamma_0^2. \qquad (13.51)$$

As for the contribution of the downward motion of the starting vortex, when the induced velocity is constant relation (13.42) becomes

$$\frac{1}{2}\rho\frac{d}{dt}\int_{S_w}\omega_y z\,dS_w = \frac{1}{2}\rho w_i \int_{-b/2}^{b/2}\Gamma(y)\,dy. \qquad (13.52)$$

If we now consider that, for an elliptical lift distribution, we have

$$\int_{-b/2}^{b/2}\Gamma(y)dy = \Gamma_0\frac{\pi b}{4}, \qquad (13.53)$$

we obtain

$$\frac{1}{2}\rho\frac{d}{dt}\int_{S_w}\omega_y z\,dS_w = \frac{1}{2}\rho w_i \Gamma_0\frac{\pi b}{4} = \frac{1}{2}\rho w_i \Gamma_0 b_0 = \frac{\pi}{16}\rho\Gamma_0^2. \qquad (13.54)$$

As could be expected, by adding up the final terms in (13.51) and (13.54) we recover the value of the induced drag given in (13.23).

If we now carefully analyse the contribution to the induced drag given by the various expressions appearing in (13.54), we see that it may be interpreted as deriving from the rate of variation of the x component of the moment of vorticity that would be given by the downward motion, with velocity w_i, of a vortex of length b_0 and constant strength Γ_0. Therefore, we may conclude that the value of the induced drag provided by the lifting-line theory for an elliptically loaded wing may also be obtained through formula (13.35), i.e. in terms of vorticity dynamics, using a simplified wake model in which all the vortices trailing behind the wing gather far downstream in a horseshoe vortex having strength Γ_0 (see Fig. 13.20). The longitudinal portions of this vortex are placed at lateral coordinates corresponding to the centres of gravity of the vorticities shed from each half-wing, and the transversal one contains all the vorticity present in the starting vortex. Furthermore, this transversal vortex segment moves in the x direction with the free-stream velocity U and in the z direction with the velocity w_i given by relation (13.22).

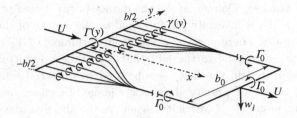

Fig. 13.20. Model of far wake.

Compared to the wake shape assumed in the lifting-line model, the representation of the far-wake given in Fig. 13.20 might seem closer to the situation occurring behind a real wing, in which the vorticities shed from the half wings do gather in two vortices located around their centres of gravity. In fact, this similarity is only superficial because the vortices trailing downstream of an aircraft tend to diffuse, as does the starting vortex. However, the objective of the procedures we have been considering is not to describe the actual flow field, especially as regards the far wake, but rather to provide satisfactory predictions of the loads acting on the wing or, more precisely, of the influence of the wing finiteness on those loads. And thus the fundamental issue is that the essential physical elements on which the loads depend be taken into account.

In particular, we have found that the induced drag is basically dependent on the magnitude Γ_0 of the resultant of the longitudinal vorticities of opposite sign released into the wake from the wing, which coincides with the circulation around the cross-section in the wing symmetry plane and with the strength of the two vortices in which those vorticities concentrate more downstream. This agrees with the view that the induced drag derives from the work that is necessary to generate the kinetic energy associated with the trailing vortices. Indeed, we have seen in Chapter 11 that the kinetic energy of two opposite-sign Rankine vortices is proportional to the square of their strengths, and this same dependence on Γ_0 is given for the induced drag by relation (13.23). Incidentally, it is interesting to recall that, for a given value of the lift, the induced drag is inversely proportional to the square of the span, as shown by relation (13.28). We now find that this is reasonable indeed because, as shown by relation (13.21), the lift is proportional to the product between the circulation in the symmetry plane and the wing span. Therefore, if the lift is held constant, Γ_0 is inversely proportional to the wing span, and thus the same is true for the strength of the trailing vortices. Consequently, the induced drag, being proportional to the square of Γ_0, is inversely proportional to the square of the span.

More generally, an important point from formula (13.35) is that no forces arise on a wing due to the fact that a certain amount of vorticity, previously introduced in the flow, is subjected to a spatial redistribution such that its centre of gravity does not change. In effect, no variation occurs, in the moment of vorticity, when a certain quantity of space-distributed vorticity, after concentrating around its centre of gravity, diffuses subsequently in the flow field. The irrelevance, as regards the drag experienced by a body, of the diffusion of vorticity occurring downstream in its wake may be interpreted in energetic terms by recalling that vorticity diffusion is associated with dissipation of kinetic energy. Therefore, when vorticity already introduced in the flow by the motion of a body starts diffusing, the consequence is simply that part of the previously produced kinetic energy is converted to internal energy. But, as we have seen, what has to be 'paid' by the work of the opposite of the drag force is the sum of the kinetic and internal energies introduced in the flow, and this global value does not change due to the process of diffusion.

The consequence of the above is that to predict the forces acting on a wing it is not essential that the physical mechanisms leading to a possible redistribution of vorticity, with constant centre of gravity, be accurately described. On the other hand, we have seen that the forces are connected with the variation of the vorticity moment, caused either by the introduction of new vorticity or by the relative motion of the centres of gravity of vorticities of opposite sign. Therefore, even an extremely simplified wake model may be adequate, but only provided it is used in a procedure that permits correct prediction of the amount of new vorticity of each sign introduced in the flow and of the subsequent evolution of its centre of gravity.

Chapter 14

AN OUTLINE OF BLUFF-BODY AERODYNAMICS

14.1. The Role of Geometrical and Flow Parameters

The distinctive feature of a bluff body, which distinguishes it from an aerodynamic body, is that the boundary layers over its surface do not remain completely attached over the whole extent of the body, but undergo a more or less premature separation. Obviously, this definition is wide enough to include in this class a great variety of bodies, which may significantly differ as regards their shape, the flow field they produce and also the type and magnitude of the consequent fluid dynamic loads. For instance, the region with separated flow may extend for a large or a small fraction of the body surface, and when the body is immersed in a time-constant free-stream the flow and the resulting loads may be either steady or highly unsteady. Also, the width of the wake may be significantly larger or appreciably smaller than a reference cross-flow dimension of the body. All these geometrical and flow features might be taken into account in order to define a parameter specifying a sort of 'degree of bluffness' of a body, especially considering that, as may be deduced from the phenomenological and theoretical analyses of the previous chapters, quite different fluid dynamic loads, and in particular drag forces, act on bodies that are substantially different as regards the above-cited characteristics.

However, as we have repeatedly pointed out, the distinction between aerodynamic and bluff bodies is primarily connected with the very different procedures that may be used to predict the fluid dynamic loads. In fact, the occurrence of boundary layer separation, irrespective of its extension, implies that, in principle, we can no longer apply the various procedures that were described in the last two chapters for the evaluation, at different levels of accuracy, of the fluid dynamic loads acting on aerodynamic bodies. Consequently, we must resort either to experiments or to the numerical solution of the full Navier–Stokes equations, and thus the difficulties related to the prediction of the flow around bluff bodies are definitely higher than those arising in the study of aerodynamic bodies. This is certainly not a satisfactory situation, considering the wide range of application of bluff-body aerodynamics. Indeed, bluff bodies immersed in a flow stream or in motion in a fluid are ubiquitous in engineering practice. We may cite, for instance, the problem of the safe and effective design of buildings, bridges and other types of civil structures subjected to wind, or the improvement of performance and reduction of fuel consumption for practically all types of ground vehicles. A brief outline of the main features of the loads acting on bluff bodies and of their origin was given in Chapter 8; in this chapter we shall describe and discuss, in deeper detail, certain more specific aspects of bluff-body aerodynamics.

We start our analysis by observing that a fundamental classification of bluff bodies may be made on the basis of the type of separation of the boundary layers from their surfaces. More precisely, according to its geometry, a bluff body may be such that separation is fixed or free. In effect, we have seen in Chapter 7 that sharp convex corners with small included angles always cause the separation of a boundary layer, irrespective of its flow condition, i.e. of its being laminar or turbulent. Consequently, when sharp convex corners are present over its surface, a bluff body is said to be a body with fixed separation. Conversely, when the body surface is smoothly curved everywhere, so that separation occurs as a consequence of the local conditions of the boundary layers, we speak of bodies with free separation. Therefore, for such bodies, the position of separation may change according to the flow conditions. In particular, separation will generally be delayed when transition from laminar to turbulent boundary layers occurs.

We now recall that, as already discussed in previous chapters, the width of the wake has a great influence on the drag of a body. A wider wake generally implies a higher level of energetic perturbation in the fluid, and also a greater time variation of the component of the vorticity moment in the free-stream direction, which we have seen to be linked to the drag acting on a body. Incidentally, we may anticipate that an accurate description of the evolution of the vorticity released from the separation points of a bluff body is a key factor for understanding the origin of the fluid dynamic loads and for devising means for their control.

The importance of the wake width for the drag of a body suggests clearly that the above classification of bluff bodies according to a geometrical characteristic, namely the presence or not of sharp corners over their surface, should correspond to different curves giving the drag coefficient as a function of the Reynolds number. In effect, the Reynolds number is the fundamental parameter influencing the transition from laminar to turbulent flow, and may thus be expected to influence the wake width — and so the drag — of bluff bodies with free separation. Conversely, when the body has fixed separation points, no influence of the Reynolds number on the wake width and on the drag should occur.

This different behaviour is depicted in Fig. 14.1, where two C_D–Re curves, derived from the data of Delany and Sorensen (1953), are shown. In the figure both parameters are evaluated using the free-stream velocity and the cross-flow dimension of the body. As may be seen, the drag coefficient of a body with fixed separation, such as a square cylinder, is practically independent of the Reynolds number; conversely, a non-constant trend characterizes the C_D–Re curve of an elliptical cylinder. In this case, as happens for all bluff bodies without sharp corners, a drop in the drag coefficient takes place at a certain value of the Reynolds number, as a consequence of the change from a laminar to a turbulent boundary layer. In effect, for low values of Re, transition occurs in the shear layers *after* separation, in a zone that moves progressively upstream with increasing Reynolds number. Therefore, a critical value of Re exists at which the transition reaches the separation point of the laminar boundary layer, which then becomes turbulent and is capable of remaining attached for a larger portion of surface. As a result, the separation point moves downstream, a narrower wake forms, and a higher recompression takes

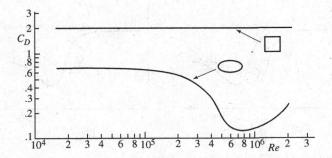

Fig. 14.1.   Typical C_D–Re curves for bluff bodies with fixed and free separation.

place before separation, with a consequent significant decrease of the drag coefficient. The above-described transition phenomenology is typical of free-separation bodies, although sometimes more complex situations may occur; for instance, we shall see that for a circular cylinder a separation bubble may be present over a certain range of Reynolds numbers.

It is then clear that the C_D–Re curve of a free-separation bluff body depends on all the parameters that may influence boundary layer transition. In particular, an increase of either the incoming turbulence level or the surface roughness of the body has the effect of anticipating the transition of a boundary layer, and thus may be expected to anticipate the consequent drop in the drag coefficient.

This is indeed what is found from experimental tests; however, as shown in the sketches of Fig. 14.2, the effect of increasing the free-stream turbulence level is not exactly the same as that of increasing the surface roughness of the body. As a matter of fact, a higher turbulence level has the sole effect of shifting the C_D–Re curve leftwards, i.e. the critical range of Reynolds numbers in which the variation of C_D takes place is anticipated without any variation of the values of the drag coefficient for higher Reynolds numbers. Conversely, increasing the surface roughness anticipates the critical region, but also has the consequence of increasing the value of the drag coefficient in the subsequent flow regime. This is actually reasonable, because the surface roughness causes a decrease of the momentum of the boundary layer, whose separation is consequently anticipated compared to what would happen for a turbulent boundary layer developing over a smooth surface. Both the minimum value and the high-Re value of the drag coefficient are then found to be increasing functions of the surface roughness.

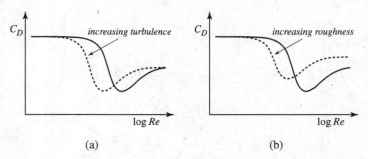

Fig. 14.2.   Effect of free-stream turbulence (a) and surface roughness (b) on the C_D–Re curve of a free-separation bluff body.

It is evident from Fig. 14.2(b) that, given a certain free-separation body, a range of Reynolds numbers exists for which the drag is lower when the body surface is rough rather than smooth. Incidentally, this is why the surface of golf balls is covered with dimples. The remarkable influence of the boundary layer condition on the flow around spheres is clearly illustrated by the visualizations of Fig. 14.3. Transition may also be induced by a local disturbance placed before the location of laminar separation; this produces a lower disturbance than distributed roughness, but can be used only if the direction of the flow is constant and known.

As previously mentioned in Chapter 8, when the shape of the body and/or the free-stream direction is such that the flow is not symmetrical, a bluff body may also be subjected to cross-flow forces, sometimes termed lift forces. We have also seen that, due to the comparatively high values of the drag, the lift-to-drag ratio is always much lower than for an aerodynamic body. However, a very peculiar feature is that certain fixed-separation bluff bodies with an upwind face bounded by sharp corners, such as a rectangular cylinder, may be characterized by a negative slope of the C_L–α curve. In other words, when the free-stream is inclined upwards, a downward force may act on the body. As may be seen from the sketch in Fig. 14.4(a), this is due to the higher convex curvature of the lower boundary of the wake, with consequent higher velocities and lower pressures in that region compared to the upper wake boundary. Even if the pressures on the lateral faces of the body are not equal to those on the wake boundaries, as would happen for an attached boundary layer, they may be expected to be connected to them, and thus a downward force may arise due to the pressure difference

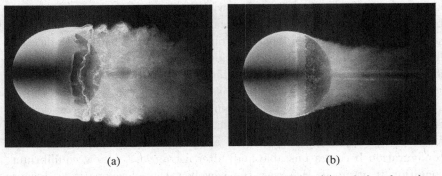

(a)   (b)

Fig. 14.3.   Flow over a sphere. (a) Laminar boundary layer; (b) turbulent boundary layer. (Copyright H. Werlé/ONERA: The French Aerospace Lab, reprinted with permission.)

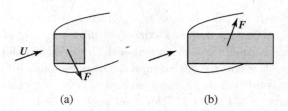

(a)   (b)

Fig. 14.4. Flow field around a rectangular cylinder.

between the two lateral faces. Obviously, if the ratio between the sides of the rectangular cross-section is such that the length of the body is large enough, the lower shear layer may reattach, as shown in Fig. 14.4(b), causing a higher pressure on the lower face and an upward global force.

When the cross-section of a structural member is such that, for a certain range of free-stream directions, the slope of its lift curve is sufficiently negative, it may become subjected, above a certain critical velocity of the incoming wind, to self-induced cross-flow oscillations, known as *galloping* oscillations (see, e.g., Simiu and Scanlan, 1986).

14.2. Vortex Shedding

14.2.1. *Two-dimensional bodies*

Let us consider a two-dimensional bluff body having a plane of symmetry in the flow direction, such as a circular cylinder or a rectangular cylinder with one face normal to the flow. When the Reynolds number — defined as $Re = UW/\nu$, where W is the cross-flow dimension of the body and ν the kinematic viscosity — is sufficiently low, a steady recirculation region forms behind the body, with two symmetrical counter-rotating vortices. In this situation, the shear layers separate from the body and become the closed boundary of the recirculation region, joining downstream to form a steady low-velocity wake of limited lateral extent. The length of the closed vortical region is found to increase with increasing Re.

However, when the Reynolds number is increased above a certain critical value that depends on the shape of the body (it is around 47 for a circular cylinder, but 120 for a blunt trailing edge), the steady configuration becomes unstable and, after a transient, a new equilibrium condition is attained. However, the new flow that develops is no longer steady; conversely, it is characterized by a regular alternate shedding of vortices from the two sides of the body (see Figs. 8.11(b) and 14.5) with

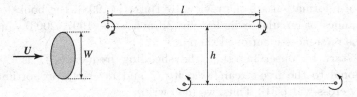

Fig. 14.5. Alternate vortex shedding from a bluff body and relevant notation.

a well-defined frequency f_v that depends on the shape of the body. In practice, the flow becomes periodic, with period $T = 1/f_v$.

The alternate shedding of vortices behind bluff bodies, and particularly from two-dimensional circular cylinders, is perhaps one of the most studied phenomena in fluid mechanics. Hundreds of related papers have been published in the scientific literature in the past century, and may now overwhelm any researcher consulting an electronic database with the keywords 'vortex shedding'. Although this phenomenon had attracted the attention of acute observers of physical reality — such as Leonardo da Vinci and earlier medieval painters (see Mizota *et al.*, 2000) — and had been studied by some researchers (e.g. Mallock, 1907, and Bénard, 1908), the person who first stimulated this widespread interest was certainly Theodore von Kármán and, in his honour, the double row of vortices of opposite sign shed from a cylinder soon became known as the *Kármán vortex street*.

In effect, von Kármán (1911, 1912) analysed the stability of a vortex within two parallel rows of ideal vortices of opposite sign, finding that the configuration was always unstable, except for a particular anti-symmetrical configuration having a critical spacing ratio $h/l \simeq 0.2806$, where h is the lateral separation between the two rows and l is the distance between consecutive vortices in a row (see Fig. 14.5). Later on, other researchers analysed the stability of rows of vortices with finite cores (Hooker 1936, Kida 1982, Saffman and Schatzman 1982), finding that a finite range of stable spacing ratios exists, at least for infinitesimal disturbances. However, if the centres of vorticity are considered, the spacing ratios of the vortices in the wake observed experimentally are not far from the value predicted by von Kármán.

A detailed discussion on the physical mechanisms at the basis of the alternate shedding of vortices from a bluff body is beyond the scope of the present book, and the reader is referred to the abundant specialized literature on the subject. Here, we shall only briefly describe some of the most important features of this phenomenon; unless otherwise specified, the sources of information are the review by Buresti (1998), the books on the aerodynamics of circular cylinders by Zdravkovich (1997, 2003), and the numerous references mentioned therein.

We start by observing that the shedding frequency f_v is directly proportional to the free-stream velocity U and inversely proportional to the body cross-width W. Thus, we may write

$$f_v = S\frac{U}{W}, \tag{14.1}$$

where the coefficient of proportionality S is the *Strouhal number*, which may be seen as a non-dimensional frequency and is a function of body shape, free-stream direction and, in general, of the flow conditions. In practice, the shedding frequency — and thus the value of S — is inversely proportional to the width of the wake rather than to the cross-flow dimension of the body. As a consequence, for bodies with free separation, the Strouhal number is found to be a function of the Reynolds number and of all the parameters influencing boundary layer transition, with a trend that is practically opposite to that of the drag coefficient, i.e. it increases when the wake width decreases. Conversely, for bodies with fixed separation, the Strouhal number does not depend on *Re*.

The alternate shedding of vortices produces time-varying asymmetric pressure distributions over the body surface, and thus fluctuating lift and drag forces to be added to their mean values. Therefore, the global forces acting on a two-dimensional bluff body are usually written as

$$F_L = \frac{1}{2}\rho U^2 W[C_{Lm} + C_{Lf}\sin(2\pi f_v t)], \tag{14.2}$$

$$F_D = \frac{1}{2}\rho U^2 W[C_{Dm} + C_{Df}\sin(4\pi f_v t)], \tag{14.3}$$

where C_{Lm} and C_{Dm} are the mean lift and drag coefficients, respectively. Note that C_{Lm} is non-zero only when the mean flow configuration is not symmetrical. When this is the case, generally vortex shedding still occurs but its features may change and differences may occur between the two shed vortices, as regards both the formation pattern and the interaction with the vorticity over the body.

As for the fluctuating forces, the lift component fluctuates at the frequency f_v because the relevant flow configuration changes in an anti-symmetric manner when a vortex is shed from the upper or from the lower surface. Conversely, as regards the force component in the free-stream

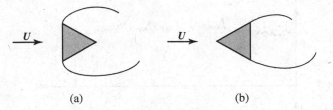

Fig. 14.6.   Dependence of fluctuating lift on geometry: (a) large C_{Lf}; (b) small C_{Lf}.

direction, it is irrelevant whether a vortex is separating from the upper or the lower side, and thus the drag fluctuates at a frequency that is twice that of vortex shedding.

The magnitude of the fluctuating lift force, which is expressed in relation (14.2) by the coefficient C_{Lf}, depends strongly on the shape of the body and, in particular, it is an increasing function of the longitudinal extent of the afterbody (see the example in Fig. 14.6). For this reason, the fluctuating lift force acting on bodies with free separation depends on the Reynolds number and on all the parameters influencing boundary layer transition. Indeed, delaying separation also implies a reduction of the longitudinal extent of the afterbody and thus, in such cases, C_{Lf} shows a trend similar to that of the mean drag coefficient C_{Dm}. This happens, for instance, for a circular cylinder, which is also characterized by fluctuating lift forces due to vortex shedding that are one order of magnitude larger than the analogous fluctuating drag forces.

As an example, the time signals of the sectional lift and drag forces acting on a circular cylinder with laminar separation, measured during the tests described by Buresti and Lanciotti (1992), are shown in Fig. 14.7. Several interesting features arise from this figure, in which the same scale is used for the drag and the lift. The first is that the drag fluctuations at twice the frequency of vortex shedding are so small that they are almost completely masked by the random ones due to the upstream turbulence, which was slightly less than 1% of the free-stream velocity. Conversely, the lift clearly shows large fluctuations at the shedding frequency, with peaks of the same order as the mean drag force. However, the other remarkable feature is that the amplitude of the lift fluctuations is far from constant: considerable and irregular modulations are indeed present. These are due to the fact that, although the flow is nominally two-dimensional, vortex shedding is neither perfectly equal nor exactly simultaneous along the cylinder, and thus significant spanwise variations are present in the wake.

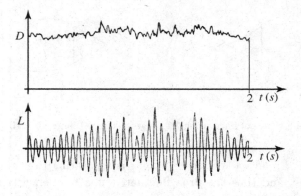

Fig. 14.7.   Experimental time variation of drag and lift over a circular cylinder.

The first effect of the three-dimensional character of the wake is that the coefficients C_{Lf} and C_{Df}, which appear in (14.2) and (14.3), are not really time-constant. Therefore, when those formulas are used, it is normally implied that the rms values of the coefficients should be introduced. The other important feature is that the correlation between the sectional fluctuating forces acting on two cross-sections of the cylinder decreases as the mutual spanwise distance between the cross-sections increases. Consequently, this lack of correlation along the cylinder gives rise to an appreciable decrease of the fluctuating forces when they are integrated over cylinder portions of increasing length.

One of the most important consequences of vortex shedding is that when the frequency of the related fluctuating forces approaches a natural frequency of a cylindrical structural member, significant oscillations may arise. The peculiar feature of the dynamic phenomena induced by vortex shedding is that they are non-linear, i.e. the oscillations influence the fluctuating forces from which they originate. For instance, considering a circular cylinder, as soon as it starts oscillating appreciably in the cross-flow direction due to vortex shedding, the spanwise correlation of the lift force rapidly increases and, at variance with the situation of Fig. 14.7, both the sectional and the integrated lift forces vary in time as a constant-amplitude sinusoid with a larger amplitude. Furthermore, the oscillation may occur even if the vortex shedding frequency given by relation (14.1) is not exactly equal to the natural frequency of the structure, and a range of free-stream velocities exists in which the frequency of vortex shedding becomes equal to the natural frequency of the structural member, giving rise to a phenomenon called *synchronization* or *lock-in*. Fortunately, the lift does not continue to increase indefinitely with the amplitude of oscillation. In fact, it becomes negative, i.e. it opposes the motion, when the amplitude exceeds a certain value, which is a function of the sectional shape but may be estimated to be of the order of the cross-sectional dimension of the body. Therefore, the oscillations induced by vortex-shedding are self-limiting, i.e. they have a limit cycle. However, these oscillations may be unacceptable, in many cases, either from the structural or from the functional points of view. The description of the complex features of vortex-induced oscillations, of their dependence on the geometrical and flow parameters, and of the methods for their limitation or avoidance are beyond the scope of the present book, and reference should be made to the specialized literature (e.g., Sarpkaya, 2004).

As might be expected, alternate vortex shedding is strictly connected with a high value of the mean drag acting on the body producing the wake. This is indeed understandable, since the appearance in the wake of two new concentrated vortices of opposite sign in each shedding cycle corresponds to a significant increase in the kinetic energy of the fluid, and thus to a related work being done by the opposite of the drag force acting on the body. Consequently, whenever the vortex shedding process is hindered, either naturally or artificially, and the vorticity is distributed more irregularly within the wake width, a decrease of the drag occurs due to the reduced kinetic energy introduced in the flow.

In order to identify the flow conditions that may cause irregular vortex shedding or even its disappearance, it is important to observe that, for vortex shedding to take place regularly along the span of a cylinder, it is necessary that the separation line be rectilinear (Naumann *et al.*, 1966). Therefore, when this does not occur, vortex shedding either disappears or becomes less regular. For instance, this explains why the C_{Dm}–Re curve for a circular cylinder is characterized not only by a drop caused by the change from laminar to turbulent boundary layer separation, but also by a minimum, as shown in Fig. 14.8.

Three characteristic regimes are indicated in the figure, and are denoted as subcritical, transitional and postcritical. Unfortunately, this nomenclature is not univocal and different terms have been used in the literature, even with the same term indicating different regimes (see Zdravkovich, 1997, for a discussion and a more detailed subdivision of the transitional regime). The values of the Reynolds numbers appearing in the figure correspond to the extremes of the transitional regime and to

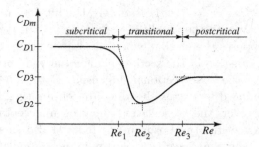

Fig. 14.8.   Characteristic values of C_{Dm} and Re for a circular cylinder.

the minimum of the drag coefficient. Analogously, the three values of the mean drag coefficient correspond to the subcritical and postcritical regimes and to the minimum in the transitional regime. Obviously, all these values depend on the surface roughness and on the turbulence intensity of the incoming flow. However, particularly interesting is the case of a smooth cylinder, for which the values of C_{D1}, C_{D2} and C_{D3} are respectively of the order of 1.2, 0.26 and 0.56. In that case, for subcritical Reynolds numbers the boundary layer is laminar at separation, which occurs at around 80°–82° from the front stagnation point. The same is true in the transitional regime, but the transition to turbulence in the free shear layer is so close to the separation point that reattachment occurs and a recirculation bubble is formed. The turbulent boundary layer then separates much further downstream (around 140°), and a sharp reduction of the wake width and of the drag coefficient occurs. With increasing Re, the bubble shrinks and finally it disappears in the postcritical regime, in which the boundary layer on the cylinder surface becomes turbulent before the separation point.

Regular vortex shedding takes place not only for subcritical but also for postcritical conditions whereas, in the transitional regime, it either disappears or becomes very irregular. In effect, when a bubble is present on the surface, the flow conditions are extremely unstable, and even small spanwise differences in the surface roughness or the presence of turbulence in the incoming stream may lead to a bubble with a highly oscillating reattachment line, which causes a similar oscillation of the subsequent separation line. This unstable character of the bubble may be easily observed through surface flow visualizations, and leads to a very irregular vortex shedding, which can thus be detected only when the free-stream turbulence level is very low. Therefore, the lack of regular vortex shedding decreases the kinetic energy introduced in the wake and is most probably the cause of the very large drop in drag coefficient, which cannot be explained just with the reduction in wake width. This physical explanation is in perfect agreement both with the energetic interpretation of the origin of drag and with the experimental findings related to the decreases in drag that may be obtained by introducing three-dimensional perturbations along the separation line of two-dimensional bluff bodies.

Finally, we note that for rectangular cylinders interactions between the vortices shed from the front and rear sharp corners may arise, so that the wake structure depends significantly on the ratio between the sides of the cross-section (see, e.g., Hourigan *et al.*, 2001, Shi *et al.*, 2010).

14.2.2. *Three-dimensional effects*

Up to this point, we have been concerned with vortex shedding from nominally two-dimensional bodies and we have seen that, even in these conditions, significant three-dimensional features do appear in the flow and influence the forces acting on the bodies. In the present section we shall describe briefly the influence on vortex shedding of explicit three-dimensional features, either in the geometry or in the incoming flow.

The first three-dimensional condition we analyse is the one produced when a geometrically two-dimensional bluff body — as could be considered a cylinder delimited by end walls — is immersed in a shear flow with a velocity U that varies along its axis, as shown in the sketch of Fig. 14.9(a).

A typical feature of this condition is that the vortex shedding frequency does not vary continuously along the cylinder, as would be required by the Strouhal law (14.1), but takes place in spanwise cells, with a constant frequency in each cell. The cells are separated by three-dimensional structures, called 'vortex dislocations', that allow the vortices to adapt their

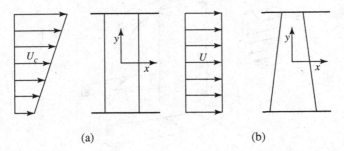

(a)   (b)

Fig. 14.9.   (a) Cylinder in shear flow. (b) Tapered cylinder in uniform flow.

frequency to that of the adjacent cells. The length and stability of the cells
are a function of the shear parameter

$$\beta = \frac{W}{U_c}\left(\frac{dU}{dy}\right), \tag{14.4}$$

where U_c is the upstream velocity at the mid-span of the cylinder. An
interesting feature is that, due to the varying incoming velocity, opposite
spanwise pressure gradients are present in the front and rear parts of the
cylinder. Indeed, along the front stagnation line the pressure increases
in the direction of increasing velocity, while the opposite happens at the
base, where higher suctions are connected with higher upstream velocities.
Consequently, secondary spanwise flows are present, whose intensity may
become an appreciable fraction of the free-stream velocity for high values
of β. This generates further components of vorticity in the flow, adding
complexity to the three-dimensional flow features.

An analogous and interesting condition is the flow around tapered
cylinders, shown in Fig. 14.9(b), in which the incoming flow is uniform, but
the cross-flow dimension of the cylinder varies in the spanwise direction.
As might be expected, also in this case vortex shedding in the wake may
occur in constant-frequency cells whose spanwise length decreases as the
taper ratio increases.

From the practical point of view, a particularly important three-
dimensional flow field arises when a finite-length cylinder is placed vertically
on a plane and is immersed in a cross-flow, as shown schematically
in Fig. 14.10. The first significant feature of this configuration is that

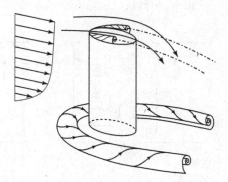

Fig. 14.10.   Schematic flow around a finite cylinder.

the incoming boundary layer present over the plane undergoes a three-dimensional separation due to the adverse pressure gradient caused by the presence of the cylinder. The consequence is the formation of a vortex, which rounds the base of the cylinder and is usually called a *horseshoe vortex*. This vorticity structure may be important in certain engineering applications; for instance, it may cause dangerous scouring around the base of a submerged bridge pier. Moreover, it generates two counter-rotating vortices downstream of the body, which interact with the lower region of the wake.

Nonetheless, the most remarkable aspects of the flow depicted in Fig. 14.10 are connected with the finiteness of the body. The presence of a free-end gives rise to an intense local flow, which passes over the body tip and is deflected within the separated wake by the low pressures that are present therein. This flow penetrates inside the wake, widens it, and delays the rolling up of the vortices detaching from the lateral sides. At variance with what happens in the two-dimensional case, the widening of the wake and the consequent decrease in the vortex shedding frequency occur, in this case, with a reduction of drag due to the increase of base pressure caused by the flow entering the wake, and to the lower value of the perturbation energy induced in the wake by the lower-intensity shed vortices. Depending on the shape of the cylinder cross-section, the upward flow near the tip of the body may also produce two intense counter-rotating vortices directed downstream, which interact further with the vortex shedding process. This happens, for instance, for circular cylinders or for prisms with isosceles triangular cross-section and the apex edge directed upstream.

As a general rule, if H is the height of the body, W its cross-flow width and δ the thickness of the incoming boundary layer over the plane, the flow field is a function of the parameters H/δ and H/W. The relative importance of the horseshoe vortex is obviously a function of H/δ; however, the influence of the incoming boundary layer on the flow near the free-end and on the characteristics of the vortex shedding process may be neglected when $H/\delta > 3.0$. As for the effect of body height, we shall refer prevailingly to finite circular cylinders, for which more data are available. However, many qualitative features should also apply for bodies with different cross-sections.

For cylinders of high aspect ratio, say $H/d > 10$ (where d is the cylinder diameter), pressure and velocity measurements have shown the presence of clear vortex shedding from most of the cylinder span, with Strouhal numbers of the same order as those of the two-dimensional case. However, a decrease of the frequency was found in a zone approaching the upper end of the cylinder, in apparent agreement with the widening of the wake in that region and with the increase in the formation length of the shed vortices. The relative extent of this region, which is probably a lower-frequency cell, increases with decreasing aspect ratio, and for $H/d \lesssim 5$ the vortex shedding from the whole cylinder takes place at a lower frequency than in the corresponding two-dimensional case. However, in the very upper region of the body, i.e. within the last diameter from the free-end, for all aspect ratios a clear peak at the vortex shedding frequency can no longer be detected in the pressure and velocity fluctuations. Conversely, fluctuations at lower frequencies appear, which were also found for prisms of triangular cross-section and shown to be connected with the dynamics of different vortical structures present in the wake (see Buresti and Iungo, 2010).

It is interesting to note that fluctuations with a clear dominating frequency and with opposite phase on the two sides of the wake of finite cylinders with different cross-sections have been found even for very low aspect ratios, i.e. down to values of H/W as low as one. These fluctuations were associated with alternate vortex shedding, although there is also evidence of the shedding of symmetrical arch-type vortices, specially at low Reynolds numbers. In any case, it must be emphasized, again, that the regularity of vortex shedding decreases with decreasing aspect ratio, and the same is true for the consequent fluctuating forces. In practice, for low values of H/W, the vortices shed from the sides of the body interact in a complicated manner with the vorticity structures originating from the free-end and with the horseshoe vortices caused by the separation of the incoming boundary layers. Furthermore, all these features are strongly dependent on the shape of the body cross-section and on the characteristics of the incoming free-stream, in terms not only of orientation and vertical profile of the mean velocity, but also of intensity and distribution of the turbulence. In particular, strongly concentrated vorticity structures may originate at the top or at the base of the body, and produce high local suctions over its surface. As a consequence of this flow complexity, present knowledge is not yet sufficient to permit the prediction of all the possible different conditions that may arise in the flow around finite cylinders. This is unfortunate, because these configurations are of interest in many engineering applications, for instance in the design of buildings subjected to wind. Therefore, further systematic experimental and numerical investigations are needed to increase our current predictive capabilities.

14.3. Axisymmetric and Elongated Bodies

Axisymmetric bluff bodies are quite common in applications, and it is thus interesting to analyse briefly the main peculiar features of the flow field and of the consequent loads for this type of bodies. From a geometrical point of view, the simplest axisymmetric configuration involving a bluff body is a sphere immersed in a uniform free-stream. However, from the fluid dynamic point of view, a sphere is certainly far from simple. We already know that it is a free-separation body, and thus the relevant flow field depends on the Reynolds number and on all the parameters influencing the conditions of the boundary layer, similar to what happens for the flow around a circular cylinder. Nonetheless, the different configurations that characterize the wake of a sphere, when the Reynolds number is varied, are more numerous and definitely more complex than those occurring for a circular cylinder. Thus, as reported in the review by Kiya *et al.* (2001), the wake is steady and axisymmetric at first, with a toroidal stationary vortex. However, it loses its axisymmetry above, say, $Re = 210$, and two longitudinal counter-rotating vortices trail downstream lying in a preferred plane that is neither fixed nor necessarily constant in time. For Reynolds numbers above 280–300 the flow becomes unsteady and, for increasing values of Re, vortices are shed with different shapes (such as rings, helices, hairpins and loops) and with various levels of regularity. For instance, axisymmetric rings with azimuthal perturbations are seen to be shed from the sphere in Fig. 14.3(a). The Strouhal numbers of the dominating frequencies of the wake fluctuations vary significantly with Reynolds number, and above $Re = 3.7 \times 10^5$ the wake becomes almost steady again. Even if more investigations are needed to fully understand the involved physical mechanisms, it is known that the wake fluctuations may induce oscillations of a sphere when it is elastically constrained or tethered in a flow (see, e.g., Govardhan and Williamson, 2005).

Another geometrically simple axisymmetric body is a circular disc normal to a free-stream, which, compared to the sphere, has the advantage of fixed separation and thus of a lower dependence on the Reynolds number, at least for values of Re corresponding to unsteady conditions. In this case, the wake fluctuations are dominated by the shedding of helical vorticity structures from the disc contour. On the other hand, elliptical and rectangular plates normal to the main flow are characterized by the shedding of opposite-sign hairpin vortices in the plane containing the minor axis and, further downstream, by a phenomenon of axis switching. More details on these complex flow fields my be found in Kiya *et al.* (2001).

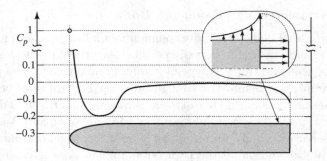

Fig. 14.11. Pressure distribution over an axisymmetric body.

Let us now analyse the case of bluff bodies that are elongated in the free-stream direction and have smaller cross-flow dimensions. Such bodies can be found in several applications and may be regarded as simplified archetypes of road vehicles. We consider the simplest body belonging to this category, namely an axisymmetric body with a rounded forebody, a constant-section main body and a sharp-edged flat base perpendicular to the free-stream direction (see Fig. 14.11). The geometry of the forebody is assumed to be such that the boundary layer remains attached up to the base contour, where it separates irrespective of its being laminar or turbulent. If the forebody is sufficiently elongated, its pressure drag may be very small because the overpressures acting around the front stagnation point are balanced by the suctions in the subsequent rounded part. This may happen, for instance, for an ogival shape. Considering that the cylindrical main body only contributes to the friction drag, the pressure drag of the body is almost completely dominated by the suctions acting on the base. We must then inquire into the factors influencing the value of the pressure acting over the base.

From the pressure distribution shown in Fig. 14.11, we observe that the pressure coefficient decreases from the maximum value of 1 at the stagnation point to a minimum value at the end of the forebody, which depends on the shape of this portion of the body. Subsequently, pressure recovery occurs, and the maximum value reached over the lateral surface depends on its length. In particular, if the main body is sufficiently long, the velocity outside the boundary layer over the lateral surface tends to the free-stream value, and thus the pressure coefficient tends to zero. However, the important point is that whatever the maximum value over the lateral surface, a decrease of the pressure occurs as the end of the body is approached, and another minimum is thus present at the base contour, where separation occurs.

Now, as already pointed out in Chapter 8, the pressure coefficient over the base is almost constant, and nearly equal to the value at the separation point. As a result, the base drag, which we have seen to be an essential portion of the total drag of the body, is strictly connected with the pressure decrease taking place over the lateral surface of the body, just before the base contour. In turn, from Bernoulli's theorem and relation (8.14), the pressure coefficient at separation depends on the velocity outside the boundary layer at the same point. In other words, the value of the base pressure is directly related to the acceleration of the flow taking place outside the boundary layer over the last portion of the lateral surface. This acceleration, whose origin is not immediately obvious (considering that the body cross-section is constant), is caused by the shape of the contour of the wake just after separation. In effect, as can be appreciated from the flow visualization of Fig. 14.12, the mean *near wake* is characterized by a decreasing cross-section. Consequently, the streamlines just outside the separating boundary layer and the near wake bend towards the axis, and this produces a convex curvature and an acceleration of the flow. More specifically, higher curvature of the streamlines bounding the near wake region implies lower base pressure and higher drag. This explains the origin of the base suction and suggests that any modification of the curvature of the streamlines at separation may have a noticeable effect on base pressure and drag.

Indeed, a change of this flow configuration, as well as of the resultant pressure distribution and drag experienced by the body, may be obtained through a simple geometrical modification known as *boat-tailing*, which

Fig. 14.12. Visualization of the mean flow in the near wake of an axisymmetric body. (Copyright H. Werlé/ONERA: The French Aerospace Lab, reprinted with permission.)

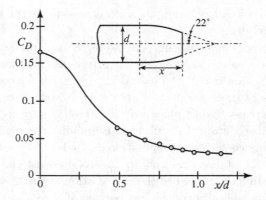

Fig. 14.13.   Effect of boat-tailing on drag (from the results of Mair, 1969).

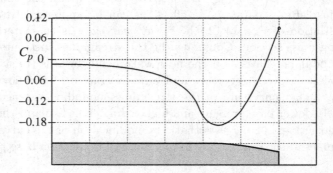

Fig. 14.14.   Typical pressure distribution over a boat-tail.

consists in a gradual reduction of the body cross-section before the base. In Fig. 14.13 an example is shown of both the geometry of a boat-tail and the pressure drag reduction that may be obtained by increasing its length. Obviously, for good performance of a boat-tail in reducing drag, its length and the ratio between the final and initial diameters — and thus the angle between the lateral and base surfaces — must be such that the boundary layer remains attached up to the sharp-edged base contour.

The mechanism that causes this effect is perhaps better understood by analysing the typical pressure distribution along the boat-tail surface reported in Fig. 14.14, which is the result of a numerical simulation. As can be seen, starting from slightly before the beginning of the boat-tail, the pressure decreases significantly and reaches a minimum value. This trend corresponds to an acceleration of the flow just outside the boundary layer, caused by the convex curvature of the streamlines. Considering that the outward normal along the boat-tail lateral surface has a positive component in the free-stream direction, this pressure decrease would produce an increase of the drag acting on the body (as well as of the cross-flow force along each generatrix of the body contour). However, when the boundary layer separates at the end of the body and becomes the contour of the wake, it must bend in the opposite direction, and this causes a change in the curvature of its outer streamlines. In reality, the change to concave curvature of the streamlines occurs even before the end of the boat-tail, due to the increase of the boundary layer thickness, and this produces a rapid and significant decrease of the velocity and increase of the pressure. The positive effect on drag of this pressure increase is greater than the negative one caused by the previous acceleration; in effect, the resulting pressure on the base — which, as already pointed out, is practically equal to the pressure at separation — is found to be much higher than the one that would be present without a boat-tail. In fact, even positive pressure coefficients may be obtained, as is indeed the case in Fig. 14.14. Finally, we point out that boat-tailing is an effective drag reduction method also for two-dimensional bodies (see Maull and Hoole, 1967).

One might wonder whether rounding the sharp-edged contour of the base of an axisymmetric body might produce a drag reduction, as happens using boat-tailing. However, this is not necessarily the case, as was found by Buresti *et al.* (1997), who measured the afterbody pressure drag of bodies similar to the one shown in Fig. 14.11, but with rounded base contours. The bodies had an ellipsoidal forebody, and tests were carried out for several values of the ratio r/d between the radius of curvature of the base contour and the diameter of the body cross-section, and for two different values of the Reynolds number $Re_L = UL/\nu$ (where L is the total length of the body). Furthermore, the transition of the boundary layer was either left to occur naturally or forced just downstream of the forebody, thus obtaining four different conditions of the boundary layer over the lateral surface of the body.

The results are reported in Fig. 14.15, and show that for all curves a maximum of the drag coefficient is present at a certain critical value of r/D. Even if the discrete values of the radius of curvature do not permit a precise estimation of this critical value, it seems to be a function of both Re_L and the type of boundary layer transition. A possible minimum for very small values of the curvature also seems to be present, at least for the

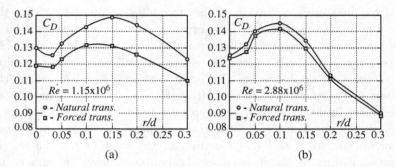

Fig. 14.15.   Variation of afterbody pressure drag coefficient with radius of base contour. (a) $Re_L = 1.15 \times 10^6$, (b) $Re_L = 2.88 \times 10^6$ (data from Buresti *et al.*, 1997).

lower Reynolds number. The increase in C_D at the critical r/d ranges from 11% to 16% for the various cases. As expected, the drag decreases for high values of the radius of curvature, albeit more rapidly for the higher value of Re_L.

This non-obvious behaviour of the drag coefficient was shown to derive from the insufficient pressure recovery before separation that occurs for small values of r/d. In effect, the rounding causes, once again, a decrease of the pressure — and a contribution to drag — due to the associated acceleration of the flow but, at variance with what happens for boat-tails, this drag increase is not compensated by the subsequent pressure recovery before separation. Furthermore, both the suction peaks and the values of the pressure recovery depend on the conditions of the boundary layer approaching the base, and this explains the different values of r/d corresponding to the maximum pressure drag.

Another very interesting result that may be seen in Fig. 14.15 is that for a sharp-edged base, i.e. for $r/D = 0$, different values of C_D were found for the various cases. In particular, the drag seems to decrease with increasing thickness of the boundary layer. Apparently, this result confirms the high sensitivity of the flow and the pressure distribution to even small variations in the curvature of the streamlines at separation, which might be induced by different evolutions of the boundary layer. In effect, it is reasonable to infer that a thinner boundary layer would correspond to a higher curvature. Unfortunately, a detailed characterization of the boundary layer for the flow conditions of the tests of Fig. 14.15 is not available, and further investigations are necessary to gain a deeper understanding of all the involved physical features.

Considering now, in more detail, the connection between various wake structures and the drag of elongated bodies, the first point that must be made is that, although we have seen that fluctuations are present in the wakes of such bodies due to the shedding and dynamics of different vorticity structures, such fluctuations are generally significantly smaller than those that characterize the wakes of two-dimensional bluff bodies. In particular, the vorticity is shed with a lower regularity and remains more diffused, i.e. less concentrated in tight structures. As a result, a lower introduction of kinetic energy in the fluid occurs and, thus, a lower drag acts on the body.

Nonetheless, there are conditions in which, even for these bodies, a higher level of organization and concentration of vorticity may occur as a consequence of even small geometrical or flow variations. A striking example is an axisymmetric cylinder aligned with the flow, with a slanted sharp-edged base making different angles ϕ to the free-stream. This case was studied by Morel (1978) and Bearman (1980), and the drag coefficients they obtained are shown in Fig. 14.16. As can be seen, by increasing ϕ the drag coefficient increases first almost linearly up to very high values, and then shows an abrupt decrease, remaining almost constant for higher slant angles. As is apparent from the two different curves in Fig. 14.16, which were obtained in two different wind tunnels, the critical value of ϕ corresponding to the discontinuity is also seen to be sensitive to small variations in flow conditions.

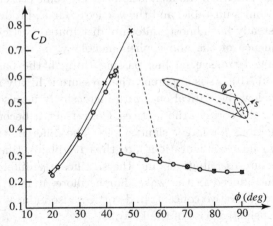

Fig. 14.16. Drag coefficient of an axisymmetric body with slanted base (data from Bearman, 1980).

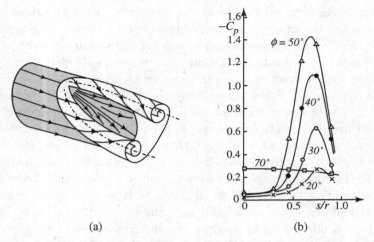

(a)   (b)

Fig. 14.17.   (a) Near wake flow field for axisymmetric body with slanted base and small values of ϕ. (b) Pressure distribution along a transversal line through slanted base centre (data from Bearman, 1980).

This largely unpredictable trend is due to a sudden switch between two completely different flow regimes. For values of ϕ below the critical one, the vorticity introduced in the wake concentrates in two narrow cores, as shown in the sketch in Fig. 14.17(a), similar to what happens for a delta wing at high angles of attack. The intensity of these vortices increases with increasing ϕ, but when the inclination reaches some critical value, this configuration becomes unstable, and the vortices 'burst', producing a wake with a more unsteady but almost uniformly distributed vorticity.

The consequence of the above phenomenology, in terms of pressure distribution along a transversal line passing through the base centre, is shown in Fig. 14.17(b). As is apparent, the pressure is highly non-uniform for values of ϕ below the critical one, with increasingly intense suctions in the regions below the cores of the vortices. Conversely, it becomes uniform and with low suctions for larger slant angles. The different trends of the drags of the two flow configurations are thus readily justified, in terms of both the pressure distributions and the significantly different levels of kinetic energy introduced in the wake. Furthermore, it may be observed that the presence of the two concentrated vortices also induces a significant lift force over the slanted base of the body.

The first interesting outcome of this example is the recognition that bluff-body flows may undergo large and sudden discontinuities when one parameter controlling the configuration is varied. Consequently, the extrapolation of experimental data may be very uncertain and even potentially dangerous, as well exemplified by Fig. 14.16.

In addition, an even more important lesson is that, once the physical mechanisms causing a certain behaviour of the forces are understood, one may devise means to modify the flow configuration in a favourable way. Considering the observed limited stability of the above-described vortical structures when they are near the critical condition, it is quite natural to infer that it may be possible to force the establishment of the diffused-vorticity configuration by interfering with the formation of the concentrated vortices. As can be seen in Fig. 14.18, this is indeed what is obtained by introducing in the base region small lateral or transversal plates, whose seemingly limited disturb is actually sufficient to avoid the concentration of the shed vorticity into tight cores, and thus to reduce the drag significantly.

All the above phenomena regarding boat-tailing, rounding of the base contour and variation of the base slant angle are not restricted to axisymmetric bodies. For instance, qualitatively similar results are also found for bodies with rectangular cross-sections. Thus, the importance for automotive applications of the above-described results should be quite evident (see, e.g., Hucho and Sovran, 1993), and the reader with an interest in that field may easily recognize that the relevant knowledge has indeed been widely used to decrease the drag of modern road vehicles.

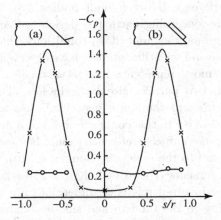

Fig. 14.18. Effect of a spoiler (a) or a strake (b) on the pressure distribution along a transversal line through slanted base. x, plain base; o, with device added (data from Bearman, 1980).

14.4. Interference Effects

The complexity of the possible flow configurations that may characterize isolated bluff bodies suggests that even more complicated conditions may be found when several bodies are placed close enough for interference effects to become significant. In many practical applications, such as the design of civil structures subjected to wind, situations with closely spaced bodies are definitely common.

One of the main consequences of such situations is that, in many cases, the asymmetries that originate in the flow due to interference may produce mean cross-flow forces of considerable magnitude even for bodies which, thanks to their symmetrical shapes, do not experience such loads when they are isolated. Significant increases or decreases of the drag forces and of the moments may also occur due to interference. Furthermore, the fluctuating forces may be considerably altered by interference, and new instabilities may derive from proximity effects. The opposite may also be true, i.e. certain flow features leading to fluctuating loads and high drag, such as vortex shedding, may be changed and even suppressed by interference effects.

Many different configurations involving interfering bluff bodies may be encountered in engineering practice, and correspond to innumerable complex flow fields and consequent loads. Therefore, we can only briefly outline the main features of some of the most common situations and point out that present knowledge is still far from being sufficient to cover all possible practical cases. Many — if not most — of the research activities carried out on interference between bluff bodies may be considered as *ad hoc* investigations concerning particular design problems. Nonetheless, there are certain geometries and flow conditions that, due to their ubiquitous occurrence and scientific interest, have deserved more systematic investigations. The most representative of this class is certainly the interference between two parallel circular cylinders. The complexity of the relevant flow fields and the consequent loads has led researchers to dedicate significant activity to this configuration. For a detailed account of present knowledge on the subject, reference should be made to the book by Zdravkovich (2003) and to the review by Sumner (2010).

The basic arrangements of two cylinders in cross-flow are shown in Fig. 14.19, where the involved reference dimensions are also defined. The available data relating to the tandem and side-by-side configurations are more numerous but recently significant research activity has also been devoted to the case of staggered cylinders, which in practice is also the most common one.

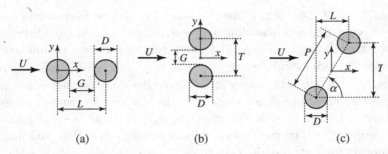

Fig. 14.19. Geometrical arrangements of two circular cylinders in cross-flow: (a) tandem, (b) side-by-side, (c) staggered.

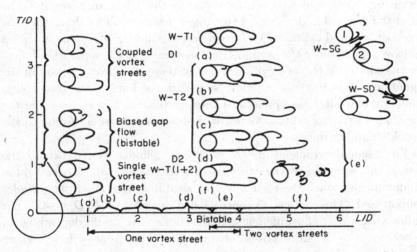

Fig. 14.20. Flow regimes for two parallel circular cylinders. Tandem regimes: (a) single slender body; (b) alternate reattachment; (c) quasi-steady reattachment; (d) intermittent shedding; (e) bistable jump between (d) and (f); (f) coupled vortex shedding. (From Zdravkovich, 1987, reprinted with permission.)

A first classification of the different flow regimes that may arise according to the mutual position of the cylinders was proposed by Zdravkovich (1987), and is schematically reported in Fig. 14.20.

In the case of side-by-side cylinders, three regimes may be recognized as a function of the distance between the axes of the cylinders, T:

- When the spacing between the cylinders is small ($1 \leq T/D < 1.2$), the two cylinders behave as a single bluff body with base bleed, and a single alternate vortex street is formed downstream.

- When $1.2 < T/D < 2.2$, a bistable condition occurs: narrow and wide wakes, with different shedding frequencies, are formed, divided by a biased flow through the gap. The wakes can interchange intermittently between the two cylinders.

- When $T/D > 2.2$, both vortex streets have the same frequency, but are coupled in an out-of-phase mode, i.e. the vortices are simultaneously formed and shed first on the gap side and then on the outer sides. The coupling gradually decreases with increasing separation, and finally disappears for $T/D > 4-5$.

In the bistable regime, the two cylinders experience different mean and fluctuating forces, which switch from one to the other. In particular, the sum of the low and high values of the drag is always smaller than twice the drag of an isolated cylinder. The mean across-wind (or lift) force is always repulsive, as a result of the asymmetrical movement of the separation points over the surface of the two cylinders and of the consequent rotation of the stagnation points towards the gap. Note that the level of the free-stream turbulence may alter the boundaries between the different regimes and the magnitude of the fluctuating forces, with different effects according to the Reynolds number range.

The tandem arrangement, i.e. with one cylinder placed behind the other, is characterized by the existence of a critical distance L between the cylinders below which the regular vortex shedding from the front cylinder is suppressed. This critical distance is of the order of $L/D = 3.8$. For smaller values of L/D, different configurations may occur depending on the behaviour of the shear layers separating from the front cylinder, which may or may not reattach, steadily or intermittently, to the sides of the downstream cylinder. As for the forces, below the critical distance the downstream cylinder is subjected to a negative mean drag, and even on the upstream cylinder the mean drag is lower than that acting on an isolated cylinder. In this condition, the fluctuating forces are negligible for the upstream cylinder, and relatively high for the downstream one. For distances beyond the critical one, both cylinders shed alternate vortices in their wakes and the downstream one experiences a very large fluctuating drag, due to the strongly fluctuating flow in which it is immersed.

For the staggered arrangements, the possible flow conditions are obviously even more numerous and complex, and most of the relevant data, as regards both the forces and the characteristics of the flow field, have been derived only recently. Their detailed analysis is beyond the scope of the present book, and the interested reader may refer to the review by Sumner (2010).

The above-described interference phenomena are also present, at least qualitatively, for cylinders with different cross-sections. As regards square cylinders in the tandem arrangement, for instance, the critical distance below which there is no vortex shedding from the upstream cylinder, but only from the downstream one, marks the onset of sudden and dramatic changes in all the force components, both mean and fluctuating (Sakamoto et al., 1987). On the other hand, the presence of sharp corners forcing separation is the origin of certain differences between the results for square cylinders and those for circular cylinders. A clear example is the side-by-side configuration (see Alam et al., 2011), for which the lift forces were found to be attractive, rather than repulsive, due to the lack of movement of the separation points and to the jet-like flow existing in the gap, which produces high velocities and low pressures.

Further information regarding clusters of multiple cylinders in cross-flow may be found in Zdravkovich (2003), where data on interference effects between a plane surface and a circular cylinder are also reported. This latter case is interesting for applications, and it is found that vortex shedding may be suppressed when the ratio between the distance of the cylinder from the plane and the cylinder diameter is lower than 0.3–0.4. This leads to significant variations of the mean and fluctuating forces acting on the cylinder which, depending on the thickness of the incoming boundary layer over the plane, may also show non-monotonic trends with increasing gap value (see also Buresti and Lanciotti, 1992).

Finally, we observe that interference may also have favourable consequences. In particular, it is possible to reduce the drag of a bluff body if another one is suitably placed in its proximity.

A striking example of this desirable interference effect is the drag reduction obtained by Prasad and Williamson (1997) by placing a small flat plate in front of a circular cylinder. By changing the distance g of the plate from the cylinder and the ratio between its cross-flow dimension and the cylinder diameter, P/D, they found an optimal configuration whose total drag — i.e. the sum of the drags acting on the plate and on the cylinder — was estimated to be 38% of the drag of a single isolated cylinder. The geometrical configuration corresponding to this maximum drag reduction is defined by the values $P/D = 0.34$ and $g/D = 1.5$, and is shown in Fig. 14.21. In the same figure, a curve is also reported which gives, for each

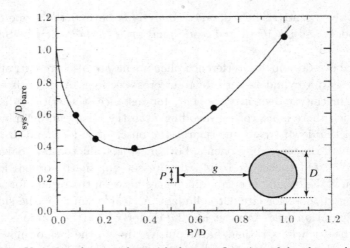

Fig. 14.21.   Variation of system drag with the cross-flow sizes of the plate and cylinder, and optimum geometrical configuration. (From Prasad and Williamson, 1997, reprinted with permission.)

value of P/D, the minimum value of the ratio between the total system drag and the drag of an isolated cylinder.

The physical reason for this drastic drag reduction is the interference between the wakes of the plate and the cylinder. In particular, regular alternate vortex shedding from the cylinder is either suppressed or strongly hampered, with a consequent strong reduction of the perturbation energy introduced in the wake. Similar results, both as regards the modifications of the wake flow field and the reduction of drag and fluctuating lift, were obtained by introducing a small rod in front of a square cylinder (Igarashi, 1997), or in the wake of a circular cylinder (Dalton *et al.*, 2001).

14.5. A Short Discussion on the Control of Bluff-Body Loads

The objective of the control of the fluid dynamic loads acting on bluff bodies is often the reduction of drag. However, at least as important is the suppression or reduction of the fluctuating forces which, besides producing possibly dangerous structural fatigue phenomena, may cause serious self-excited oscillations or unacceptable noise levels. A review on the various methods developed for the control of the flow over bluff bodies is given by Choi *et al.* (2008), whose reference list may serve as a guide to anyone more interested in this topic and — unless otherwise specified — is the source of information for what is described hereafter.

An important point to be highlighted is that any research on load control, i.e. aimed at devising methods to modify the loads favourably, should be based on a preliminary effort to achieve a deep comprehension of the physical mechanisms causing those loads. In other words, although it is definitely useful to know that a certain action has a positive effect for a desired design objective, this information is certainly less valuable than knowing *why* this happens.

For instance, the existence of the phenomenon of regular alternate vortex shedding from a two-dimensional bluff body provides an immediate physical justification for the consequent fluctuating forces. However, we also know that the drag of a body is connected with the level of perturbation energy introduced in the flow, and that this level increases when strong and distant opposite-sign vortices are shed into the wake. Therefore, this explains why modifications in the geometry of a body, or in the flow around it, that alter or suppress vortex shedding are also found to reduce drag significantly. Furthermore, once we are aware of this connection between drag and vortex shedding, we may concentrate on the best method to inhibit vortex shedding with prescribed constraints on the allowed modifications.

One of the best known methods to avoid vortex shedding is to use a so-called *splitter plate*, i.e. a plate placed in the near wake parallel to the free-stream. If the length of the plate is sufficient (for instance, five diameters), a reattachment of the shear layers on the plate occurs, with the formation of two almost steady recirculation regions and a total suppression of vortex shedding (see Fig. 14.22).

The consequent drag reduction may be as high as 50%. In addition, as described in more detail by Buresti (1998), even very short splitter plates, if properly positioned in the near wake, may reduce drag significantly. This happens because the plates disturb the 'cross-talk' between the shear layers that separate from the two sides of the body, and thus hamper the regular vortex shedding.

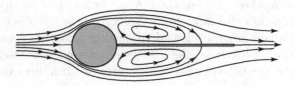

Fig. 14.22.   Flow field around cylinder with splitter plate.

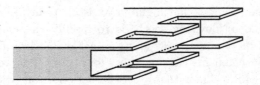

Fig. 14.23.   Ventilated base cavity.

Significant drag reductions may also be obtained by disorganizing vortex shedding, thus causing a diffusion of the wake vorticity and reducing the perturbation energy. The basic method of achieving this result is to generate various degrees of spanwise three-dimensionality by preventing the boundary layer separation from occurring along a straight line. This may be obtained, for instance, by introducing waviness along the trailing edge or on the whole cylinder, by distributing bumps over the body surface or by adding small tabs perpendicular to the flow along the trailing edge. These geometrical modifications introduce into the wake localized streamwise vortices and, more generally, significant three-dimensional vorticity structures. Another geometrical configuration that may provide a drag reduction as high as that produced by a splitter plate is the so-called 'ventilated cavity', first introduced by Nash (1967) and sketched in Fig. 14.23. Obviously, as happens for other solutions, the shape of the body must be such that the flow remains attached up to its blunt base.

All these geometrical modifications, as well as the favourable interference effects between bodies described at the end of the previous section, belong to the category of passive controls. Incidentally, we may note that, without adequate knowledge of the physical aspects of bluff-body aerodynamics, the effectiveness of all these methods would probably not seem obvious.

The other fundamental class of control methods is that of active controls, in which some power is introduced in order to obtain the required effect on the loads. In most cases, the flow is modified by means of blowing or suction through some portions of the body surface. One of the best known of these methods is *base bleed*, which consists in the introduction of a flux of fluid into the wake from the body base. Another possibility is distributing a non-uniform blowing or suction along the span of a two-dimensional body. In all these cases, the favourable effect is essentially due to interference with the evolution of the vorticity structures in the wake, and the active control is open-loop, i.e. the actuation is set *a priori* and not driven by any information deriving from the response of the system. Conversely, when the control actuation is continuously modified according to the response of the flow, then we have a closed-loop control. In that case, one of the problems is to devise a suitable sensor giving feedback control to the actuator. Anyhow, it must be recalled that all active controls must be not only effective, i.e. capable of producing the desired load control, but also efficient, i.e. the spent power must be sufficiently low, so that we may have a global gain, when the objective is drag reduction or, more generally, an acceptable operating cost.

Much research work is presently being carried out as regards the general topic of flow control, both passive and active. In particular, the management of the flow around bluff bodies and of the consequent loads is one of the most interesting fields of application of this activity. Furthermore, a peculiar feature of flow control is that it exemplifies perfectly a research area in which significant progress is being obtained thanks to a synergic use of experimental, numerical and theoretical approaches. As an introduction to this field, the interested reader may refer, for instance, to Gad-el-Hak (2000), Choi *et al.* (2008) and Sipp *et al.* (2010).

Finally, it is worth pointing out, once again, that the multiplicity of complicated flow features deriving from boundary layer separation, and thus characterizing bluff-body aerodynamics, challenges the present prediction capabilities of researchers and designers, and may thus be expected to attract the efforts of dedicated investigators for a long time to come. In spite of the conciseness of the outline given in this chapter, it should be evident that this is simultaneously a fascinating and a very complex research area. Therefore, previous experience and the ability to carry out attentive scrutiny of the essential physical mechanisms playing a role in each different flow configuration are the basic tools that may lead to a deeper understanding and to successful actions.

For instance, the high Reynolds numbers that characterize most of the applications involving bluff bodies imply that numerical simulations encounter significant difficulties in the prediction of fluid dynamic loads. However, even low Reynolds number simulations, perhaps together with flow visualizations, may prove invaluable to identify the possible existence of dominating structures. This information may then be used not only to plan dedicated experimental tests and to interpret their results, but also to develop new ingenious simplified models that may provide satisfactory predictions for higher Reynolds numbers.

In any case, systematic experimental and numerical analyses are necessary to gather the information that is indispensable to enhance our present level of understanding of the aerodynamics of bluff bodies and, thus, to develop acceptable design procedures, without having to resort to *ad hoc* tests for each different configuration encountered in practice.

Chapter 15

ONE-DIMENSIONAL COMPRESSIBLE FLOWS

15.1. Introduction

In this chapter we analyse some of the effects of the compressibility of fluids, which we have neglected in most of the previous chapters. However, we do so by using an apparently drastic simplification for the description of motion, namely that the flow be one-dimensional. Strictly speaking, this assumption implies that all quantities must be a function of only one independent space variable, which is normally taken to be the coordinate in the direction of motion, hereafter denoted by x.

It is easy to be convinced that a perfectly one-dimensional flow is impossible, the only exception being the uninteresting case of a completely space-uniform flow. When the flow occurs inside a duct, for instance, the no-slip condition acts at the duct walls and thus, even when very thin boundary layers are present, the velocity cannot be exactly uniform over the whole cross-section. Furthermore, even if one neglects viscosity, a variation of the duct cross-section area would imply an inclination of the duct walls, where the direction of the tangent velocity vector would be different from that on the axis. Thus, the flow would be one-dimensional only for a duct with constant cross-section but, as we shall see, the assumption of non-viscosity would lead, in that case, to a trivial result with little relevance to realistic conditions.

Nonetheless, if certain conditions are satisfied, a one-dimensional treatment is capable of highlighting the main effects of compressibility in many cases of practical importance, providing also extremely useful results from both the qualitative and quantitative points of view.

The most important situation in which the assumption of one-dimensional flow may be satisfactorily used is in the study of the motion in ducts. In that case, the values of the various quantities at some coordinate x — such as velocity, pressure, density and temperature — may be considered as suitable mean values evaluated over the cross-section corresponding to that coordinate. Thus, for instance, the value of the velocity obtained along a duct through this treatment, say V, is usually different from the one that might be measured at a generic point of the relevant cross-section. Nevertheless, we shall see that valuable predictions may be obtained for the mass flow rate and the variations of Mach number and pressure along the duct. Moreover, when the conditions of motion are such that the thickness of the boundary layer over the duct inner wall is much smaller than the duct diameter, then V would even be a good first-order approximation of the velocity actually existing over most of the cross-section. Finally, the one-dimensional treatment may also be extended to the analysis of the motion in small streamtubes with variable cross-sections inside three-dimensional flows. In the limit of an extremely small cross-section, the analysis will thus give the variation of the quantities along a streamline.

However, as already mentioned, the assumption of one-dimensional flow may provide a satisfactory approximation of the real flow only provided certain conditions are satisfied, namely:

- the variation of the duct cross-section area must be sufficiently gradual, i.e. sudden contractions or enlargements are not allowed;
- the radius of curvature of the duct axis must be much greater than the duct diameter.

In effect, the above conditions ensure that the velocity and the pressure do not vary significantly over the tube cross-section. For instance, the configuration shown in Fig. 15.1 would be acceptable, and the relevant flow

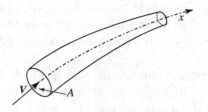

Fig. 15.1.   One-dimensional flow inside a streamtube.

could be analysed adequately with a one-dimensional treatment. It must also be pointed out that some relations connecting the conditions in two different sections, in which the flow may be assumed as one-dimensional, are valid even when the flow *between* these two sections is such that it cannot be approximated as being one-dimensional.

In the following sections we shall derive first the equations of motion of the compressible one-dimensional flow, with particular reference to the motion of a perfect gas inside ducts with varying cross-section. Subsequently, some further assumptions restricting the validity of our analysis will be introduced. In particular, we shall consider steady flows and analyse the case in which viscous effects are neglected, thus highlighting the specific significant effects of compressibility when the Mach number becomes sufficiently high. Furthermore, we shall see that surfaces can exist through which discontinuous changes of the flow quantities occur, and we shall demonstrate that their appearance is connected with the necessity of satisfying the boundary conditions. The particular influence of friction will be analysed by considering the motion in a duct of constant cross-section, which is a case that has remarkable practical importance and would correspond, if the fluid were assumed to be non-viscous, to a flow with no variation along the duct.

15.2. The Equations of One-Dimensional Flow

15.2.1. *Mass balance*

The equations of one-dimensional motion may be derived by starting from the integral form of the general equations of motion for a compressible fluid, which were derived in Chapter 4. We refer to the case of motion inside a duct having a cross-section area $A(x)$, and we neglect the contribution of the body forces. Unless otherwise specified, the considered fluid is a perfect gas.

The control volume, shown in Fig. 15.2, is delimited by the inner duct walls and by two cross-flow surfaces at the coordinates x_1 and x_2, where the areas are A_1 and A_2 respectively. We may then use the mass balance equation in its integral form given by Eq. (4.4c), noting that the elementary volume may be written as $dv = A dx$ and recalling that the normal vectors are directed outwards from the volume. We get

$$\int_{x_1}^{x_2} \frac{\partial \rho}{\partial t} A dx + (\rho_2 V_2 A_2) - (\rho_1 V_1 A_1) = 0. \qquad (15.1)$$

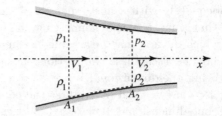

Fig. 15.2.   Control volume for one-dimensional flow balances.

If the flow is steady, we obtain simply

$$\rho_2 V_2 A_2 = \rho_1 V_1 A_1. \tag{15.2}$$

Note that this equation is also valid for a streamtube, and it only requires that the flow be uniform in the two considered cross-sections, and not necessarily between them. Conversely, if the flow is assumed to be one-dimensional at every section, and we denote the mass flow rate with the symbol w, we have

$$\rho V A = w = constant. \tag{15.3}$$

This equation may be differentiated to obtain the differential form of the steady continuity equation:

$$\frac{d(\rho V A)}{dx} = 0. \tag{15.4}$$

It is easy to see that this equation may also be expressed in the form

$$\frac{d\rho}{\rho} + \frac{dV}{V} + \frac{dA}{A} = 0, \tag{15.5}$$

which can be obtained from Eq. (15.4) by carrying out the derivation, dividing the result by w, and noting that, for a steady flow, the various quantities are a function of the same single independent variable. Alternatively, Eq. (15.5) may be directly obtained from Eq. (15.3) by applying logarithmic differentiation, which consists in taking first the logarithm of w and then differentiating the result.

We observe that Eq. (15.5) provides the relation that must be satisfied between the variations of area, density and velocity along the duct. Obviously, it is also valid for an incompressible flow, in which case we have $d\rho/\rho = 0$ and we find the well-known result that the average velocity over the cross-section is inversely proportional to A.

15.2.2. *Energy balance and adiabatic flow*

We know from Chapter 5 that in an incompressible flow the energy equation is decoupled from the mass and momentum equations and so may be disregarded when the objective of the analysis is the prediction of the velocity and pressure fields. Conversely, in the description of the motion of a compressible fluid, density is one of the unknown functions and strong coupling of all the balance equations occurs. In particular, in the one-dimensional treatment of the motion in ducts, the energy balance equation plays a crucial role, and it is then advantageous to introduce it before the momentum balance equation.

We then consider the integral energy balance equation in the form (4.56), and rewrite it for the present case, neglecting the body forces and noting that the velocity flux is non-zero only at A_1 and A_2; we thus get

$$\int_{x_1}^{x_2} \frac{\partial \rho(e + V^2/2)}{\partial t} A\,dx + \rho_2 V_2 A_2 \left(h_2 + \frac{V_2^2}{2} \right) - \rho_1 V_1 A_1 \left(h_1 + \frac{V_1^2}{2} \right)$$

$$= \int_{S} \boldsymbol{\tau}_n \cdot \boldsymbol{V}\,dS - \int_{S} \boldsymbol{q} \cdot \boldsymbol{n}\,dS = W_\tau + Q, \qquad (15.6)$$

where the surface integrals are extended over the two cross-sections and over the lateral surface of the duct. For a steady flow, we obtain

$$\rho_2 V_2 A_2 \left(h_2 + \frac{V_2^2}{2} \right) - \rho_1 V_1 A_1 \left(h_1 + \frac{V_1^2}{2} \right) = W_\tau + Q. \qquad (15.7)$$

Therefore, considering the constancy of the mass flow rate, we may divide Eq. (15.7) by w and obtain

$$\left(h_2 + \frac{V_2^2}{2} \right) - \left(h_1 + \frac{V_1^2}{2} \right) = w_{\tau 12} + q_{12}, \qquad (15.8)$$

where we have denoted by $w_{\tau 12}$ and q_{12} the work of the viscous forces and the heat flux respectively, both per unit mass flow rate, over the surfaces bounding the control volume.

We now observe that, in a real duct, the velocity of the fluid is zero at the duct inner surface due to the no-slip condition, and thus the work of the viscous forces is zero therein. Furthermore, the work of the *normal* viscous stresses over the two cross-sections is definitely negligible compared to the work of the pressure forces (which is included in the enthalpy variations). This is true in general, but it is particularly so when the motion is such that it may be analysed approximately through a one-dimensional treatment, because the assumed gradual variations of the cross-section also imply small velocity gradients, and thus small normal viscous stresses. Therefore, putting $w_{\tau 12} = 0$ in Eq. (15.8), when applying it to the analysis of the motion in a duct, does not imply neglecting the action of viscosity over the duct surface, but only the work of the viscous stresses over the duct cross-sections, which is quite an acceptable assumption. Note that the same would not be necessarily true for the motion inside a generic streamtube in a three-dimensional flow because, in that case, the velocity would not be zero over the lateral surface. Therefore, the cross-flow velocity variations in that region must be sufficiently small for the viscous stresses to be negligible and, thus, for the assumption $w_{\tau 12} = 0$ to be representative of the real situation.

In summary, Eq. (15.8) shows that the variation of the total enthalpy in a duct flow is essentially due to the introduction of heat through the surface of the considered control volume. Therefore, the energy equation for an *adiabatic* one-dimensional flow in a duct becomes simply

$$h_0 = h + \frac{V^2}{2} = constant. \tag{15.9}$$

This is a general equation but, as we have done at the beginning of Chapter 5, we apply it to the motion of a thermally and calorically perfect gas, i.e. a gas in which pressure, density and temperature are linked by Eq. (2.12) and such that its specific heats may be taken as constant. We thus get Eq. (5.23), which we now rewrite as

$$T_0 = T + \frac{V^2}{2C_p}. \tag{15.10}$$

As already pointed out in Section 5.1, this relation defines the total temperature T_0 as the temperature that would be reached if the velocity were brought to zero through an adiabatic process. Obviously, it is not necessary that the velocity be really brought to zero to define the total temperature in a section of the duct, as we may just *imagine* carrying out an adiabatic deceleration, and derive the value of T_0 directly from (15.10), once the temperature and velocity in that section are known. In practice, the total temperature quantifies the available global energy in the motion, and may be considered as an index of the adiabaticity of the flow. In other words, if T_0 is constant along the duct, then the motion is adiabatic. Conversely, if heat is introduced between two sections, then T_0 increases and the opposite occurs if heat is subtracted from the duct.

We might also imagine T_0 to be the temperature in a reservoir where the velocity is zero and which is connected to a region with lower pressure through the adiabatic-wall duct we are analysing. Then, relation (15.10) shows that, whatever the difference between the pressures inside and outside the reservoir, there exists a maximum velocity that may theoretically be reached by the flow in the duct, and this is equal to

$$V_{\max} = \sqrt{2C_p T_0}. \tag{15.11}$$

Obviously, this velocity is not attainable, as it would correspond to zero absolute temperature, and the fluid would cease behaving as a perfect gas well before approaching that value. We now recall that

$$C_p = \frac{\gamma R}{\gamma - 1},$$

and thus Eq. (15.10) may be written as

$$\frac{\gamma R T_0}{\gamma - 1} = \frac{\gamma R T}{\gamma - 1} + \frac{V^2}{2}. \tag{15.12}$$

Using the expression of the speed of sound in a perfect gas, we may define the speed of sound in total conditions as

$$a_0 = \sqrt{\gamma R T_0}, \tag{15.13}$$

so that, after straightforward manipulations, Eq. (15.12) may be recast as

$$\frac{a^2}{a_0^2} + \frac{\gamma - 1}{2} \frac{V^2}{a_0^2} = 1. \tag{15.14}$$

This relation may be readily interpreted as the equation of an ellipse in the $V - a$ plane (see Fig. 15.3), whose semi-axes are a_0 and

$$a_0 \sqrt{\frac{2}{\gamma - 1}} = \sqrt{\frac{2\gamma R T_0}{\gamma - 1}} = \sqrt{2C_p T_0} = V_{\max}. \tag{15.15}$$

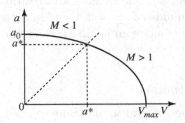

Fig. 15.3. Ellipse of adiabatic flow conditions.

Incidentally, we note that for $V = V_{\max}$ we have $M = \infty$. It may also be seen that the intersection of the bisector of the first quadrant with the ellipse divides it into two parts, corresponding, respectively, to $M < 1$ and to $M > 1$. Furthermore, it defines the so-called *critical flow conditions*, in which $V = a = a^*$ and $M = 1$. From (15.14) it is easy to find that

$$\frac{a^{*2}}{a_0^2} = \frac{T^*}{T_0} = \frac{2}{\gamma + 1}. \tag{15.16}$$

For a diatomic gas, such as air, we have $\gamma = 1.4$, and thus the ratio in (15.16) is equal to 0.8333. Both a_0 and a^* (or T_0 and T^*) identify the particular ellipse describing all the conditions of motion that may be reached adiabatically with a certain global energetic content.

From (15.12) we may also derive relation (5.26), linking the total temperature T_0 to the static temperature and to the Mach number in an adiabatic flow; for convenience, we rewrite and renumber it here

$$\frac{a_0^2}{a^2} = \frac{T_0}{T} = 1 + \frac{\gamma - 1}{2} M^2. \tag{15.17}$$

As discussed in Section 5.1, and as we shall further appreciate in the remainder of this chapter, this is an extremely important relation. In particular, we see that the variation of the temperature for an adiabatic flow, in which T_0 is constant, may be derived from knowledge of the variation of the Mach number along the duct. In effect, if we know the temperature T_1 and the Mach number M_1 at a certain section of the duct, and the Mach number M_2 at another generic one, then the temperature T_2 can be found from the relation

$$\frac{T_2}{T_1} = \frac{T_2}{T_0} \frac{T_0}{T_1} = \frac{1 + \frac{\gamma - 1}{2} M_1^2}{1 + \frac{\gamma - 1}{2} M_2^2}. \tag{15.18}$$

In Appendix A.2 the values of the inverse of relation (15.17) are given as a function of the Mach number for $\gamma = 1.4$, and may thus be used to predict the temperature variation along an adiabatic duct once the variation of M is available.

15.2.3. *Momentum balance*

Let us now consider the integral momentum balance equation applied to the control volume of Fig. 15.2. By neglecting the body forces, the balance equation may be expressed in the following form:

$$\int_{x_1}^{x_2} \frac{\partial(\rho V)}{\partial t} A dx + \int_S (\rho V) \boldsymbol{V} \cdot \boldsymbol{n} \, dS = -\int_S p n_x dS + \int_S \tau_x dS, \quad (15.19)$$

where $\boldsymbol{n}$ is the outward-directed unit vector normal to the surfaces bounding the control volume, τ_x is the x component of the viscous stress vector and the velocity $\boldsymbol{V}$ is always in the x direction. Again, the velocity flux is non-zero only at the two duct cross-sections where, however, the normal viscous stresses may be neglected.

For a steady flow, and denoting by S_L the lateral surface of the control volume, we thus obtain

$$\rho_2 V_2^2 A_2 - \rho_1 V_1^2 A_1 = p_1 A_1 - p_2 A_2 - \int_{S_L} p n_x dS + \int_{S_L} \tau_x dS. \quad (15.20)$$

It is easy to see that, over the lateral surface, one has

$$-n_x dS = dA, \quad (15.21)$$

so that, also considering the mass balance (15.2), we get

$$\rho V A (V_2 - V_1) = p_1 A_1 - p_2 A_2 + \int_{x_1}^{x_2} p dA + \int_{S_L} \tau_x \, dS. \quad (15.22)$$

The first integral on the right-hand side may be also expressed in terms of a mean pressure p_m, so that we obtain

$$\rho V A (V_2 - V_1) = p_1 A_1 - p_2 A_2 + p_m (A_2 - A_1) + \int_{S_L} \tau_x dS. \quad (15.23)$$

We thus see that, at variance with what occurred for the energy equation, the viscous stresses acting on the inner lateral surfaces of the duct do give a contribution to the momentum equation. Viscous effects may thus be taken into account in a one-dimensional treatment, and they will indeed be considered in Section 15.6, albeit for the particular (but important) case of the flow in a constant-area duct. Nonetheless, we now assume that wall friction forces can be neglected, so that the last term on the right-hand side of Eq. (15.23) may be dropped. This assumption is an acceptable approximation for short ducts with very thin boundary layers and with varying cross-sections, such as the nozzles that may be found in wind tunnels or in rocket engines. In such cases, the variations in flow quantities due to the variations of the cross-section are usually much larger and more rapid than those connected with the effects of wall friction. Therefore, neglecting viscosity does not affect the accuracy of the predictions significantly, and the results of the analysis may be useful in practical applications, at least as long as the flow conditions are such that the one-dimensional approximation remains an acceptable one. Furthermore, this assumption permits highlighting the specific effects due to the fluid compressibility, and we shall soon realize that these effects are remarkable indeed and, to a certain extent, even unpredictable.

If we now assume the two cross-sections to be separated by an infinitesimal distance dx, so that the corresponding variations in velocity and area are dV and dA respectively, we may easily derive the differential form of the momentum equation for one-dimensional motion. Neglecting the viscous term, Eq. (15.23) becomes

$$\rho VAdV = -d(pA) + pdA = -Adp, \tag{15.24}$$

and we thus obtain

$$VdV = -\frac{1}{\rho}dp. \tag{15.25}$$

This is known as Euler's equation, and it is readily seen that, with the assumptions of steady flow and negligible body forces and viscous stresses, it may be obtained easily from the general momentum equation we derived in Chapter 4, for instance in the form (4.40), by applying it to one-dimensional flow.

It is now useful to derive the differential form of the energy equation for adiabatic flow, which may be obtained by differentiating Eq. (15.9); the result is

$$dh + VdV = 0. \tag{15.26}$$

Then, we write the differential form of the entropy equation, starting from the fundamental equation (4.78) and introducing the enthalpy. We have

$$TdS = de + pd\left(\frac{1}{\rho}\right) = dh - \frac{1}{\rho}dp. \tag{15.27}$$

Therefore, by introducing (15.25) in this equation and using (15.26), we find $dS = 0$. In other words, the one-dimensional adiabatic flow in a duct in which the effects of viscosity are negligible is also isentropic. This finding should not come as a surprise because the absence of viscosity renders the flow inside the duct a reversible process. Nonetheless, it is an extremely important result because it permits using all the relations linking pressure, temperature and density that apply for an isentropic process. As we shall see in next section, this allows the variation of all quantities in a duct to be obtained from knowledge of the geometry of the duct and suitable boundary conditions.

15.3. Isentropic Flow in Ducts

In order to analyse the isentropic motion of a perfect gas in a duct, it is useful to recall the definitions of total pressure, p_0, and total density, ρ_0, which were first introduced in Section 5.1. The total pressure (density) is the pressure (density) that would be attained in a one-dimensional flow if the velocity were brought to zero through an *adiabatic and isentropic* process. By using the relations linking pressure and density to temperature in an isentropic process of a perfect gas, and relation (15.17), we obtain the following expressions in terms of Mach number

$$\frac{p_0}{p} = \left(1 + \frac{\gamma - 1}{2}M^2\right)^{\frac{\gamma}{\gamma - 1}}, \tag{15.28}$$

$$\frac{\rho_0}{\rho} = \left(1 + \frac{\gamma - 1}{2}M^2\right)^{\frac{1}{\gamma - 1}}. \tag{15.29}$$

Similar to what we observed for the total temperature, it is not necessary that the velocity be actually brought to zero to define, say, the total pressure in a section of the duct. Given a section of the duct in which the pressure and the Mach number are known, again we may *imagine* carrying out an isentropic deceleration to zero velocity. The pressure that would be obtained with that process is the value of p_0 given by (15.28). In practice, the total pressure may be considered as an index of the isentropicity of the flow. In effect, if the value of p_0, evaluated through (15.28) from the values of p and M along the duct, is constant, then the motion is isentropic. Conversely, if an irreversible process occurs between two sections, then the entropy increases and p_0 decreases. This may be seen more clearly by deriving the relation between the variations of entropy and total pressure in an adiabatic flow. To this end, we firstly elaborate on relation (15.27):

$$dS = \frac{1}{T}\left(dh - \frac{1}{\rho}dp\right) = C_p\frac{dT}{T} - R\frac{dp}{p} = \frac{\gamma R}{\gamma - 1}\frac{dT}{T} - R\frac{dp}{p}. \tag{15.30}$$

The requested connection between S and p_0 may then be obtained through further manipulations, using both relation (15.28) and the constancy of T_0,

$$\frac{dS}{R} = d\left(\ln\frac{T^{\frac{\gamma}{\gamma-1}}}{p}\right) = d\left[\ln\frac{\left(\frac{T}{T_0}\right)^{\frac{\gamma}{\gamma-1}} T_0^{\frac{\gamma}{\gamma-1}}}{p}\right]$$

$$= d\left(\ln\frac{T_0^{\frac{\gamma}{\gamma-1}}}{p_0}\right) = -d(\ln p_0).$$

(15.31)

Therefore, if we integrate between two duct sections, we get

$$\frac{S_2 - S_1}{R} = -\ln\frac{p_{02}}{p_{01}} = \ln\frac{p_{01}}{p_{02}}.$$

(15.32)

We thus see that, as anticipated, any irreversible process causing an increase of entropy in an adiabatic flow, such as viscous friction, also produces a decrease of the total pressure.

As we have done for the temperature in an adiabatic flow, we may define the critical values of pressure and density, p^* and ρ^*, as those that would be reached if the velocity were brought to coincide with the local speed of sound through an isentropic process. Their relations to the total values are easily found by putting $M = 1$ in (15.28) and (15.29):

$$\frac{p^*}{p_0} = \left(\frac{2}{\gamma+1}\right)^{\frac{\gamma}{\gamma-1}} \quad (= 0.5283 \text{ for } \gamma = 1.4),$$

(15.33)

$$\frac{\rho^*}{\rho_0} = \left(\frac{2}{\gamma+1}\right)^{\frac{1}{\gamma-1}} \quad (= 0.6339 \text{ for } \gamma = 1.4).$$

(15.34)

Once again, both p_0 and p^* (or ρ_0 and ρ^*) identify all the conditions of motion that may be reached isentropically with a certain global energetic content, which is specified by the value of T_0. The variations of pressure and density in an isentropic flow, in which p_0 and ρ_0 are constant, may then be derived from knowledge of the variation of the Mach number along the duct. Indeed, we may write

$$\frac{p_2}{p_1} = \frac{p_2}{p_0}\frac{p_0}{p_1},$$

(15.35)

$$\frac{\rho_2}{\rho_1} = \frac{\rho_2}{\rho_0}\frac{\rho_0}{\rho_1},$$

(15.36)

and use relations (15.28) and (15.29), whose values for $\gamma = 1.4$ are also reported, as a function of Mach number, in Appendix A.2.

A fundamental step in our analysis is the derivation of the expression giving the connection between the variations in velocity and in cross-section area in an isentropic flow. Firstly, we note that Euler's equation (15.25) implies that the pressure always decreases in an accelerating flow and increases in a decelerating flow, irrespective of the value of the Mach number. Let us now rewrite Eq. (15.25) by introducing some manipulations and the definition of speed of sound (2.16):

$$\frac{dV}{V} = -\frac{1}{V^2}\frac{d\rho}{\rho}\frac{dp}{d\rho} = -\frac{1}{V^2}\frac{d\rho}{\rho}\left(\frac{\partial p}{\partial \rho}\right)_S = -\frac{a^2}{V^2}\frac{d\rho}{\rho} = -\frac{1}{M^2}\frac{d\rho}{\rho}. \qquad (15.37)$$

If we now derive $d\rho/\rho$ from (15.37) and introduce it into Eq. (15.5), we easily obtain the so-called *velocity–area relation*:

$$\frac{dA}{A} = (M^2 - 1)\frac{dV}{V}. \qquad (15.38)$$

This equation is a fundamental one because it clearly highlights the essential role of the Mach number in flow evolution. In particular, it shows that an area change produces completely different effects for subsonic and supersonic conditions. As a matter of fact, we find that for $M < 1$ an increase (decrease) in the area corresponds to a decrease (increase) in velocity, similar to what happens for an incompressible flow. However, the opposite occurs for a supersonic flow ($M > 1$), and this is certainly a non-obvious result. The consequence is that, to accelerate a flow in a duct from subsonic to supersonic conditions, it is necessary that the duct be first convergent and then divergent. Furthermore, the sonic conditions ($M = 1$) may be attained only at the location of the minimum cross-section (the *throat* of the duct).

Hence, supersonic conditions cannot be reached from subsonic ones in a continuously converging duct. In other words, if a duct with such geometry connects a reservoir — where a still gas with pressure p_0 is present — to an external environment with pressure p_e, the maximum possible value of M at its exit section is 1, and the minimum pressure therein is equal to p^*, even when the external pressure is below that value. In fact, if $p_e < p^*$, the expansion will continue outside the duct, but in such a way that it can no longer be analysed with the assumption of one-dimensional flow. Thus, in spite of all its simplifying assumptions, our analysis has been able to show that compressibility can indeed play a crucial role in flow development. In particular, it should be clear from the above example that a *choking* of the mass flow may occur, i.e. whatever the value of the pressure difference driving the flow, the mass flow per unit area cannot increase above a certain limit.

In order to identify the parameters on which the above mentioned phenomena depend, it is now important to derive the relation between the area distribution along the duct and the Mach number. To this end, it is useful to refer to the critical conditions again, defining a reference area A^* as the one at which the sonic conditions would be reached through an isentropic flow, irrespective of whether this actually happens or not in the considered duct. Then, the mass balance equation implies that

$$\rho V A = \rho^* a^* A^*, \tag{15.39}$$

which we may recast as

$$\frac{A}{A^*} = \frac{\rho^* a^*}{\rho V} = \frac{\rho^*}{\rho_0} \frac{\rho_0}{\rho} \frac{a^*}{a_0} \frac{a_0}{a} \frac{a}{V}. \tag{15.40}$$

By introducing relations (15.34), (15.29), (15.16) and (15.17), together with the definition of Mach number, we easily obtain the following fundamental relation:

$$\frac{A}{A^*} = \frac{1}{M} \left[\left(\frac{2}{\gamma+1} \right) \left(1 + \frac{\gamma-1}{2} M^2 \right) \right]^{\frac{\gamma+1}{2(\gamma-1)}}. \tag{15.41}$$

The values of A/A^* as a function of M, corresponding to the particular case $\gamma = 1.4$, are reported in both Appendix A.2 and Fig. 15.4. Note that,

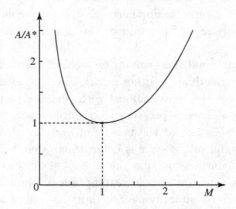

Fig. 15.4.   Area ratio A/A^* as a function of Mach number in an isentropic flow.

as is the case for the critical pressure and density, A^* is also an index of isentropicity and identifies the conditions of motion that may be reached in a certain isentropic flow. In other words, if a non-isentropic process occurs along the duct, then A^* changes, and we shall show that it actually increases.

As can be seen, A/A^* is always ≥ 1 and the limit value corresponds to $M = 1$. Again, this shows that the critical conditions may be reached only at the throat of a duct. Furthermore, for each value of $A/A^* > 1$, two values of the Mach number are possible, one for subsonic flow and the other for supersonic flow. Finally, from the area relation (15.41), one can derive the variation of the Mach number along a whole duct of given geometry from the knowledge of the value $M = M_1$ at a certain section where the area is $A = A_1$, provided a completely isentropic flow may indeed be possible in that duct. In effect, considering another generic duct section identified by the suffix 2, one may write

$$\frac{A_2}{A_1} = \frac{A_2}{A^*} \frac{A^*}{A_1}, \tag{15.42}$$

and thus, considering that A_2/A_1 is given and A_1/A^* is obtained by inserting the value M_1 into (15.41), one may derive A_2/A^*, and thus M_2. However, the above relation gives a possible value only if $A_2/A^* > 1$, which corresponds to $A_2/A_1 > A^*/A_1$. Furthermore, if $A_2/A^* > 1$, then we see from Fig. 15.4 that two values are possible for M_2, and the one that effectively occurs depends on the geometry of the duct. For example, if $M_1 < 1$ and the duct is converging, then M_2 must also be subsonic. On the other hand, if $M_1 < 1$ and the duct is converging–diverging, the flow at section 2 may be either subsonic or supersonic, depending on the ratio between the pressures at the inlet and outlet of the duct. This will be discussed in more detail in Section 15.5, where it will be shown that, given a converging–diverging duct, there are values of that pressure ratio such that no one-dimensional isentropic flow may exist in the duct. On the other hand, we shall see that, in that case, in the diverging part of the duct the flow may pass through a surface of discontinuity which involves an entropy increase and joins two duct portions with isentropic flow. These surfaces of discontinuity are the shock waves that will be analysed in next section. It will be seen that, in all cases, once the variation of the Mach number along the duct is obtained, the variation of all the remaining quantities may be derived from formulas expressing their link to M.

An important point to be noted is that the x coordinate is no longer present in the formulas that refer to an isentropic flow. In effect, the Mach number — and thus all other quantities — depends on the variation of the cross-section area but not on the longitudinal distribution of the areas along the duct. In other words, the duct length does not enter into the analysis and thus, for instance, the value of M at a certain section where the area is equal to A would not change if the duct were stretched. This result should not be surprising because it is perfectly consistent with the fact that friction over the inner duct wall has been neglected. We shall see in Section 15.6 that the coordinate along the duct axis does indeed reappear when friction is taken into account.

Finally, for isentropic flow of a perfect gas, an important relation may be obtained connecting the specific mass flow rate w/A (i.e. the mass flow per unit time and area) with the Mach number, the total pressure and the total temperature. We may write

$$\frac{w}{A} = \frac{\rho V A}{A} = \frac{\rho^* a^* A^*}{A} = \frac{\rho^*}{\rho_0} \rho_0 \frac{a^*}{a_0} a_0 \frac{A^*}{A}, \tag{15.43}$$

and, from the equation of state of a perfect gas and the definition of the speed of sound, we have

$$\rho_0 a_0 = \frac{p_0}{RT_0} \sqrt{\gamma R T_0} = p_0 \sqrt{\frac{\gamma}{RT_0}}. \tag{15.44}$$

Therefore, by introducing this relation into (15.43), together with (15.34), (15.16) and (15.41), we get

$$\frac{w}{A} = p_0 \sqrt{\frac{\gamma}{RT_0}} M \left(1 + \frac{\gamma - 1}{2} M^2\right)^{-\frac{(\gamma+1)}{2(\gamma-1)}}. \tag{15.45}$$

It is easy to see that the specific mass flow reaches its maximum value for $M = 1$, i.e. in a possible cross-section with minimum area A^*, and that this value is

$$\left(\frac{w}{A}\right)_{max} = \frac{w}{A^*} = p_0 \sqrt{\frac{\gamma}{RT_0} \left(\frac{2}{\gamma + 1}\right)^{\frac{\gamma+1}{\gamma-1}}}. \tag{15.46}$$

These relations have an important practical application in the analysis of the flow of a gas emanating from a reservoir connected to an outer environment by a short duct (i.e. a nozzle) in which the flow may be considered as adiabatic and with no friction, and thus isentropic. In that case, the temperature and pressure in the reservoir may be assumed to correspond to the total values T_0 and p_0 in the flow through the nozzle, and the value of the mass flow may be obtained directly from knowledge of the Mach number and the area A in any cross-section of the duct. Furthermore, if the ratio between the external pressure and the reservoir pressure, p_e/p_0, is lower than the value corresponding to sonic conditions in the minimum nozzle cross-section — which is the outlet cross-section if the nozzle area decreases continuously, but an intermediate one if the nozzle first converges and then diverges — then the mass flow is *choked* to the maximum value given by (15.46). This limit value is a linear function of p_0 if the reservoir temperature is kept constant, and this implies that control of the mass flow can be obtained by controlling the reservoir pressure.

Finally, if a localized non-isentropic process occurs in a duct, the maximum specific mass flow rate decreases due to the decrease of the total pressure. Consequently, assuming adiabatic flow — and thus constant T_0 — the fact that the mass flow rate w cannot change along the duct implies that the critical area A^* must increase in such a way that the quantity $p_0 A^*$ remains constant.

15.4. Normal Shock Waves

In Section 15.2.3 we derived the differential form of the momentum equation (15.25), which is valid when the friction effects on the lateral walls of the duct may be neglected. In doing so, we implicitly assumed, as always done when deriving a differential equation from an integral one, that no discontinuities of the various quantities were present in the considered control volume. In that case we found that, if adiabatic conditions can be assumed, then the flow is also isentropic. Now, we shall show that, in reality, we cannot exclude the presence of surfaces of discontinuity in a flow in which the effects of the viscous stresses may be neglected. Furthermore, we shall specify the conditions for the occurrence of these discontinuities and derive the relations giving the jumps in the values of all the flow quantities across them.

To this end, let us consider a one-dimensional flow in a duct and a control volume that may be derived from that in Fig. 15.2 by imagining the two cross-sections at x_1 and x_2 to approach each other in such a way that both the viscous stresses on the lateral walls and the variation in area may be neglected. Our objective is to derive the possible relations between the quantities on the two cross-sections. In practice, this will correspond to writing the integral balance equations for the flow in a constant-area duct in which we may neglect both heat transfer and viscous effects. Therefore, the equations expressing the balances of mass, momentum and energy, derived from Eqs. (15.2), (15.22) and (15.8) with the introduction of the above assumptions, are

$$\rho_1 V_1 = \rho_2 V_2 = G, \tag{15.47}$$

$$p_1 + \rho_1 V_1^2 = p_2 + \rho_2 V_2^2 = I, \tag{15.48}$$

$$h_1 + \frac{V_1^2}{2} = h_2 + \frac{V_2^2}{2} = h_0, \tag{15.49}$$

where we have introduced the specific mass flow rate G and the so-called *impulse function* I, which are both constant in this motion.

All the above equations are trivially satisfied when ρ, V, p and h do not vary between sections 1 and 2, and we thus enquire whether other solutions are possible. We refer, for simplicity, to the case in which the fluid is a perfect gas, so that, with the help of Eqs. (15.14) and (15.16), the energy equation may be written as follows:

$$a_1^2 + \frac{\gamma - 1}{2} V_1^2 = a_2^2 + \frac{\gamma - 1}{2} V_2^2 = \frac{\gamma + 1}{2} a^{*2}. \tag{15.50}$$

If we now divide the two sides of the momentum equation (15.48) respectively by $\rho_1 V_1$ and $\rho_2 V_2$, which are equal from (15.47), introduce the equation of state and the definition of speed of sound, we get

$$V_1 - V_2 = \frac{a_2^2}{\gamma V_2} - \frac{a_1^2}{\gamma V_1}. \tag{15.51}$$

By introducing in this equation the values of a_1 and a_2 that may be derived, in terms of a^*, from Eq. (15.50), we get, after straightforward manipulations,

$$V_1 - V_2 = (V_1 - V_2) \frac{a^{*2}}{V_1 V_2}. \tag{15.52}$$

This relation is satisfied not only by an unperturbed flow $V_2 = V_1$, but also by a flow in which there is a sudden jump to a $V_2 \neq V_1$ such that

$$V_1 V_2 = a^{*2}. \tag{15.53}$$

This is usually known as the *Prandtl–Meyer relation*, and gives the link between the velocities on the two sides of a discontinuity surface that, in theory, may exist in a flow in which the previously described assumptions are satisfied. In an actual flow, one would not expect that real discontinuity surfaces could exist because of the diffusion effects that tend to smooth out all excessive variations in the flow quantities. Nonetheless, as we have already mentioned and as may be actually seen from flow visualizations at high Mach numbers, flow regions leading to extremely sudden variations of the flow quantities do occur in real flows, and are known as *shock waves*. The thickness of such flow structures is extremely small, of the order of the mean free path of the molecules, and thus for all practical purposes they may indeed be considered as discontinuity surfaces. We are generally not interested in the variations occurring in the interior of a shock wave (whose evaluation through a continuous fluid model might even be questionable), we are rather concerned with the changes in flow properties *across* the wave. As we are considering only one-dimensional flows, we are dealing with *normal* shock waves, but we point out that oblique shock waves do occur in two-dimensional and three-dimensional flows, and they are flow structures allowing sudden variations of the flow direction to take place. For their analysis and for a description of flow problems in which they play a role, reference should be made to textbooks devoted explicitly to the treatment of compressible flows, such as Shapiro (1953) or Liepmann and Roshko (1957), where more details on the physical mechanisms generating shock waves may also be found.

Relation (15.53) is often expressed in an alternative way by defining a modified Mach number, which is usually denoted by $M^* = V/a^*$. In spite of this possibly confusing notation, M^* should not be interpreted as the Mach number in critical conditions, which is obviously always 1. It is just another convenient parameter, which is linked to the usual Mach number but does not tend to infinity as the velocity approaches the maximum value given by (15.15). In fact, one may derive the function connecting M^* and M by using their definitions and relations (15.16) and (15.17), so that one easily obtains

$$M^{*2} = \frac{V^2}{a^{*2}} = \frac{V^2}{a^2}\frac{a^2}{a_0^2}\frac{a_0^2}{a^{*2}} = \frac{(\gamma+1)M^2}{(\gamma-1)M^2+2}. \tag{15.54}$$

It is then readily seen that the following equivalences hold:

$$M < 1 \Rightarrow M^* < 1,$$

$$M = 1 \Rightarrow M^* = 1,$$

$$M > 1 \Rightarrow M^* > 1,$$

$$M \to \infty \Rightarrow M^* \to \sqrt{\frac{\gamma+1}{\gamma-1}}.$$

Therefore, we may now express the Prandtl–Meyer relation as

$$M_1^* M_2^* = 1, \tag{15.55}$$

which shows immediately that across a normal shock wave the flow passes from supersonic to subsonic conditions, or vice versa; however, we shall see that only one of the two processes is possible.

Relation (15.55) may be written in terms of the usual Mach number through relation (15.54); after simple manipulations, we find the following relation between the downstream and upstream values of M:

$$M_2^2 = \frac{2 + (\gamma - 1)M_1^2}{2\gamma M_1^2 - (\gamma - 1)}. \tag{15.56}$$

It is now easy to derive the variations of the remaining flow quantities across a normal shock wave. In effect, using (15.53), we get

$$\frac{V_2}{V_1} = \frac{V_1 V_2}{V_1^2} = \frac{a^{*2}}{V_1^2} = \frac{1}{M_1^{*2}}. \tag{15.57}$$

From the mass balance and from (15.54), we thus obtain

$$\frac{\rho_2}{\rho_1} = \frac{V_1}{V_2} = M_1^{*2} = \frac{(\gamma + 1)M_1^2}{(\gamma - 1)M_1^2 + 2}. \tag{15.58}$$

The pressure jump may be derived from the momentum equation which, taking the mass balance into account, may be written

$$\frac{p_2 - p_1}{p_1} = \frac{\rho_1 V_1^2}{p_1}\left(1 - \frac{V_2}{V_1}\right) = \gamma M_1^2\left(1 - \frac{V_2}{V_1}\right). \tag{15.59}$$

By introducing the expression for the velocity jump, we get

$$\frac{p_2 - p_1}{p_1} = \frac{2\gamma}{\gamma + 1}(M_1^2 - 1), \tag{15.60}$$

and thus, finally,

$$\frac{p_2}{p_1} = 1 + \frac{2\gamma}{\gamma + 1}(M_1^2 - 1). \tag{15.61}$$

As for the temperature variation across a normal shock wave, it may be derived by noting that the total temperature is constant, and thus relation (15.18) is still valid. Therefore, using relation (15.56), we obtain, after some rearrangement,

$$\frac{T_2}{T_1} = 1 + \frac{2(\gamma - 1)}{(\gamma + 1)^2} \frac{(1 + \gamma M_1^2)}{M_1^2} (M_1^2 - 1). \tag{15.62}$$

Finally, it is important to check whether an entropy variation occurs across a normal shock wave. Considering that the total temperature is constant, we may use relation (15.32) and derive the entropy jump from that of the total pressure. The latter may be obtained by writing

$$\frac{p_{02}}{p_{01}} = \frac{p_{02}}{p_2} \frac{p_2}{p_1} \frac{p_1}{p_{01}}, \tag{15.63}$$

so that, with the help of (15.28), (15.56) and (15.61) we get, after some manipulations,

$$\frac{p_{02}}{p_{01}} = \left[\frac{(\gamma + 1)M_1^2}{2 + (\gamma - 1)M_1^2}\right]^{\frac{\gamma}{\gamma - 1}} \left[1 + \frac{2\gamma}{\gamma + 1}(M_1^2 - 1)\right]^{-\frac{1}{\gamma - 1}}. \tag{15.64}$$

A careful scrutiny of Eqs. (15.64) and (15.32) shows that the entropy variation $S_2 - S_1$ across a normal shock wake is positive when $M_1 > 1$ and negative when $M_1 < 1$. Therefore, considering that in an adiabatic flow the entropy cannot decrease, this means that only shock waves through which the flow jumps from supersonic to subsonic conditions may exist, and this is indeed consistent with the experimental evidence. We thus conclude that the supersonic flow in a duct may suddenly become subsonic through a normal shock wave, with a consequent decrease of the total pressure and a corresponding increase of the critical area A^*, so that the quantity $p_0 A^*$ remains constant. We shall see in the next section that the shock may be considered as an additional mechanism to satisfy the boundary conditions at the exit of the duct.

All the relations providing the variations of flow quantities across a normal shock wake, namely (15.56), (15.58), (15.61), (15.62) and (15.64) are reported in Appendix A.3 in the form of a table giving the corresponding values as a function of the upstream Mach number, M_1. An attentive analysis of this table is extremely instructive. Indeed, one can note that the decrease in total pressure is quite limited for low supersonic values of M_1, whereas it becomes significant when M_1 is large. For instance, the loss in total pressure for $M_1 = 1.2$ is still below 1%, and thus the compression through a shock wave is not very different from an isentropic one. Conversely, almost 28% of the total pressure is lost for a normal shock wave with $M_1 = 2.0$.

15.5. Flow in Converging–Diverging Ducts

In this section we analyse the one-dimensional flow in a duct connecting a reservoir, where the pressure is fixed and equal to p_0, to an external region where the pressure, p_e, is variable. We assume that the geometry is assigned and that the flow may be considered adiabatic; thus, along the whole duct the value of the total temperature remains equal to the temperature T_0 existing in the reservoir. Furthermore, we assume the duct to be short enough for wall friction effects to be negligible, so that we may predict the variations of the flow quantities by using the formulas for isentropic flow and normal shock waves that we derived in the previous sections.

In all cases, the fluid starts flowing through the duct towards the external region only when $p_e/p_0 < 1$. If the duct is a converging one, the flow rate through the duct will increase as p_e decreases, and the pressure in the exit section will coincide with the external pressure. However, as we have already seen in Section 15.3, when $p_e/p_0 = p^*/p_0$, the Mach number in the exit section becomes equal to one and the flow rate reaches the maximum value given by (15.46). For lower external pressure, no further variations of the flow occur in the duct and, in the exit section, the Mach number remains equal to one and the pressure to p^*.

Let us now consider the case of a converging–diverging duct, a configuration that is quite common in several applications, such as the measurement of flow rate in subsonic flows, and the design of nozzles for rockets or supersonic wind tunnels. We denote by A_e and A_t the areas of the exit and throat cross-sections of the duct respectively. The variation of pressure and Mach number along the duct for decreasing values of the ratio between the external and the reservoir pressures are qualitatively shown in Fig. 15.5. For values of p_e/p_0 slightly less than unity, a completely subsonic flow occurs in the duct, with an acceleration and a pressure decrease in the converging part of the duct, and a subsequent deceleration and recompression in the diverging part; the pressure in the exit section is equal to p_e (case a in Fig. 15.5). Thus, the minimum pressure occurs at the throat, where the velocity and the Mach number reach a maximum value which, in any case, is always subsonic. A duct working in this condition is often called a *Venturi tube*.

With further decrease of the external pressure, the velocities inside the duct increase and so does the flow rate, until a first critical pressure is reached, say $p_e = p_{cr1}$, such that in the throat section we have $M_t = 1$ and $p_t = p^*(= 0.528\,p_0$ for $\gamma = 1.4)$. Even in this limit condition (case b

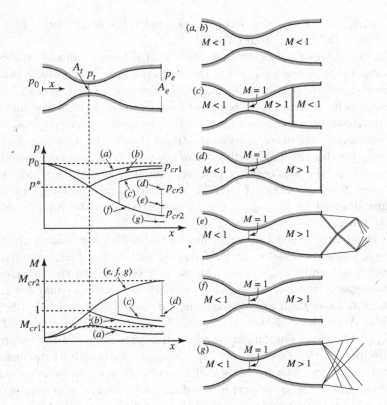

Fig. 15.5. Flow through a converging–diverging duct.

in Fig. 15.5), in the diverging part of the duct we have subsonic flow and recompression up to $p = p_e = p_{cr1}$ at the exit section. The pressure p_{cr1} may be predicted from the given duct geometry, and in particular from the ratio A_e/A_t. In effect, when the critical conditions are reached at the throat, we have $A_e/A_t = A_e/A^*$ and thus, from relation (15.41) or Fig. 15.4, we can obtain the value of the *subsonic* Mach number corresponding to an isentropic flow in the diverging part of the duct, say M_{cr1}. The critical pressure p_{cr1}/p_0 is then found by introducing this value of M into Eq. (15.28). We also note that for any external pressure such that $p_e/p_0 \geq p_{cr1}/p_0$, knowledge of the pressure at the duct exit allows the values of M and p in any generic cross-section with area A to be obtained. In effect, the Mach number in the exit section is known from p_e/p_0 and (15.28), and thus, from (15.41), one also derives the value of A_e/A^*. Therefore, the value of A^*/A may be obtained as the only unknown in the expression

$A_e/A = (A_e/A^*)(A^*/A)$ and this permits, finally, the derivation of first M from (15.41) and then p from (15.28).

For any value of the external pressure that is lower than p_{cr1}, the flow conditions in the converging part of the duct do not vary, and at the throat we always have $M_t = 1$. In other words, choking occurs and the mass flow rate remains fixed to the value obtained from expression (15.46) by putting $A^* = A_t$. However, we also note that, for the given value of $A_e/A_t = A_e/A^*$, there exists only one other possible isentropic flow in the diverging part of the duct, besides the one leading to $p_e = p_{cr1}$. This second possibility corresponds to the *supersonic* exit Mach number, say M_{cr2}, which may be derived from (15.41) or Fig. 15.4; again, the corresponding external pressure, denoted by $p_e = p_{cr2}$, can be predicted from the value of M_{cr2} and Eq. (15.28).

In conclusion, when the conditions at the throat are sonic, only two isentropic flows are possible in the diverging part of the duct, namely the subsonic flow leading to $p_e = p_{cr1}$ at the exit section and the supersonic flow corresponding to $p_e = p_{cr2}$. It is easy to see that p_{cr2} may even be considerably lower than p_{cr1}. For instance, if the value of A_e/A_t is 1.3, we find from Appendix A.2 that $M_{cr1} \simeq 0.52$, $M_{cr2} \simeq 1.66$, $p_{cr1} \simeq 0.831\,p_0$ and $p_{cr2} \simeq 0.215\,p_0$. One might wonder what happens when the external pressure lies in the range $p_{cr2} < p_e < p_{cr1}$, for which there are no isentropic solutions. The answer is that a non-isentropic process takes place somewhere in the diverging part of the duct. In particular, a normal shock wave may occur in a specific section of the diverging part of the duct, thus connecting an isentropic supersonic flow with a subsonic isentropic flow where, subsequently, compression takes place so that the pressure at the exit section coincides with the external pressure (case c in Fig. 15.5). In fact, the solution corresponding to this flow regime may be found without great difficulty by noting that the position of the shock wave is determined once the Mach number in front of it, say M_1, is specified. In effect, from (15.41), we see that only one value of the cross-section area exists, say A_1, such that an isentropic flow leads from the reservoir to that supersonic Mach number; the other unknown is the exit Mach number M_e. The assigned quantities are the ratios A_e/A_t and p_e/p_0, which may be written in the following form, in which we indicate with subscript 2 the conditions immediately downstream of the shock wave,

$$\frac{A_e}{A_t} = \frac{A_e}{A_2}\frac{A_2}{A_1}\frac{A_1}{A_t} = \frac{A_e}{A_2^*}\frac{A_2^*}{A_2} \cdot 1 \cdot \frac{A_1}{A_1^*}, \tag{15.65}$$

$$\frac{p_e}{p_0} = \frac{p_e}{p_2} \frac{p_2}{p_1} \frac{p_1}{p_0} = \frac{p_e}{p_{02}} \frac{p_{02}}{p_2} \frac{p_2}{p_1} \frac{p_1}{p_0}. \tag{15.66}$$

Note that, although not strictly necessary, we have distinguished between the values of A^* and p_0 upstream and downstream of the shock wave. A careful analysis of these two expressions shows that, after the introduction of relations (15.28), (15.41), (15.56) and (15.61), they provide two equations in the unknowns M_1 and M_e, from which the shock wave position can be found, together with the flow conditions along the whole duct.

Therefore, as the external pressure drops slightly below the value p_{cr1}, a shock wave forms just downstream of the duct throat. This wave is very weak, as the upstream Mach number is only slightly larger than 1. With the further progressive decrease of the external pressure, the shock wave moves gradually downstream and its strength increases, until a new limit condition is reached (case d in Fig. 15.5), in which the shock wave is positioned exactly at the exit cross-section of the duct. In this situation, the flow inside the duct is the completely isentropic flow that leads from the reservoir to the Mach number M_{cr2} and to the pressure p_{cr2} at the downstream extreme cross-section of the duct, just in front of the shock wave. The corresponding external pressure, which we denote as p_{cr3}, is easily found, as it coincides with the pressure behind a shock wave that has a flow upstream with the Mach number M_{cr2}.

For external pressures that are lower than p_{cr3}, no further variations occur inside the duct. In the range $p_{cr2} < p_e < p_{cr3}$ (case e in Fig. 15.5), the duct is said to be *overexpanded* because the pressure at its exit is lower than the external pressure, and the compression to p_e occurs outside the duct, through a system of oblique shock waves that cannot be described through a one-dimensional analysis. When $p_e = p_{cr2}$, the duct works in its *design* condition, as an expansion nozzle leading to supersonic flow conditions (case f in Fig. 15.5) and, finally, when $p_e < p_{cr2}$, the duct is said to be *underexpanded*, and the further expansion to the external pressure occurs outside it through oblique expansion waves, which, similarly, cannot be treated in one-dimensional terms (case g in Fig. 15.5).

Although we have not taken the effects of friction into account, the above phenomenology closely corresponds to what is seen in experiments, apart from possible lower recompressions found in reality when a shock wave is present, due to its interaction with the boundary layers. In particular, the choking of the flow quantities within the duct is indeed observed, and this may be considered as a significant result provided by the one-dimensional treatment, in spite of all its simplifications.

15.6. Flow in Constant-Area Ducts with Friction

15.6.1. *Generalities on the variation of the flow quantities*

The analysis of the motion of gases in constant-area ducts is of interest in many engineering fields, such as the transport of fluids in chemical plants or of natural gas in long pipelines. In these cases, viscous friction at the duct walls and the introduction of heat must generally be taken into account because they may have significant effects on the flow evolution.

Therefore, let us write the one-dimensional equations of motion for the control volume shown in Fig. 15.6. The balance of mass is still given by Eq. (15.47), in which the specific mass flow rate $G = w/A$ is constant along the duct.

Starting from Eq. (15.20), and considering that now the duct area is constant, we may write the balance of momentum in the following form:

$$(p_2 + \rho_2 V_2^2) - (p_1 + \rho_1 V_1^2) = \Delta I = -f_{r_{12}}, \qquad (15.67)$$

where we have introduced the impulse function I, defined in (15.48), and the function f_{r12}, representing the magnitude of the total friction force, per unit cross-section area, acting on the duct walls between the cross-sections 1 and 2. Having assumed that the friction force opposes the motion, we thus see that the action of friction is to reduce the impulse of the motion which, conversely, remains constant if the effects of friction are neglected.

The balance of energy is expressed by putting $w_{\tau12} = 0$ in Eq. (15.8), which we have seen to be a perfectly acceptable assumption. Considering the expressions of enthalpy and speed of sound for a perfect gas, introducing

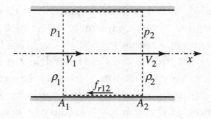

Fig. 15.6.   Control volume for flow in a constant-area duct.

Eq. (15.12) and using relations (15.13) and (15.16), we get

$$\left(\frac{a_2^2}{\gamma-1} + \frac{V_2^2}{2}\right) - \left(\frac{a_1^2}{\gamma-1} + \frac{V_1^2}{2}\right) = \frac{\gamma+1}{2(\gamma-1)}[(a_2^*)^2 - (a_1^*)^2] = q_{12}.$$

(15.68)

Here, we have expressed the total enthalpy at a generic duct cross-section in terms of the critical speed of sound a^*, but we might have used the speed of sound in total conditions a_0 or the total temperature T_0. Thus, a^* is constant in an adiabatic flow, but it increases or decreases if heat is introduced into or extracted from the flow.

Before deriving and utilizing the consequent differential equations of motion, it is convenient to elaborate on the significance of the parameters we have introduced, namely G, I and a^*. In particular, we may wonder whether these quantities may be assigned arbitrarily or, conversely, some constraints between their values must exist for the flow to be possible.

To this end, let us write again the fundamental relations defining the three parameters:

$$\rho V = G,$$

(15.69)

$$p + \rho V^2 = I,$$

(15.70)

$$\frac{2a^2}{\gamma-1} + V^2 = \frac{\gamma+1}{\gamma-1}a^{*2}.$$

(15.71)

By multiplying Eq. (15.70) by γ/ρ and recalling that $a^2 = \gamma p/\rho$, we get

$$a^2 = \frac{\gamma I}{\rho} - \gamma V^2,$$

(15.72)

which may be introduced in Eq. (15.71) together with the value $V = G/\rho$ derived from Eq. (15.69). After some rearrangements, we obtain

$$\frac{1}{\rho^2} - \frac{2\gamma}{\gamma+1}\frac{I}{G^2}\frac{1}{\rho} + \frac{a^{*2}}{G^2} = 0,$$

(15.73)

which is a quadratic equation in the unknown $1/\rho$, whose solutions are

$$\frac{1}{\rho} = \frac{\gamma}{\gamma+1}\frac{I}{G^2}\left[1 \pm \sqrt{1 - \left(\frac{\gamma+1}{\gamma}\right)^2\left(\frac{a^*G}{I}\right)^2}\right].$$

(15.74)

We thus find that two solutions are possible, and it is easy to verify that they correspond to the conditions existing before and after a normal shock wave. This is reasonable because assuming the constancy of G, I and a^*

corresponds to seeking the solution to the problem given by Eqs. (15.47)–(15.49), which was analysed in Section 15.4. However, the most important result deriving from (15.74) is that a constraint exists on the possible values of G, I and a^* in a flow, namely we must have

$$\frac{I}{a^*G} \geq \frac{\gamma+1}{\gamma}. \tag{15.75}$$

The limit condition in (15.75) is found by expressing G and I through (15.69) and (15.70), and using (15.54). We thus get

$$\left(M + \frac{1}{M\gamma}\right)\sqrt{\frac{\gamma+1}{2+(\gamma-1)M^2}} - \frac{\gamma+1}{\gamma} = 0, \tag{15.76}$$

which is readily seen to be satisfied by $M = 1$, i.e. by sonic conditions. Let us consider, then, a possible flow, either subsonic or supersonic, in a certain section of a duct. We necessarily have therein

$$\frac{I}{a^*G} > \frac{\gamma+1}{\gamma}. \tag{15.77}$$

Now, we have seen that, in a general motion, I and a^* vary along the duct. In particular, friction effects cause I to decrease in the downstream direction, whereas an increase of a^* occurs when heat is introduced through the duct walls. As a result, both these processes tend to reduce the value of the parameter $I/(a^*G)$, and thus the flow quantities along the duct vary in such a way that the limit condition is approached. In other words, irrespective of the flow being subsonic or supersonic, in a constant-area duct with friction and/or heat addition, the flow tends to sonic conditions. Thus, it is clear that, starting from given flow values in a certain section, a maximum duct length exists that is compatible with the existence of the flow conditions assigned in that section. Furthermore, at the exit section of the duct with maximum length, we have $M = 1$. One may then wonder what happens if the length of the duct is larger than this maximum value or, more interestingly, what happens if another portion of duct is added at its downstream extremity. The answer is that a rearrangement of the flow conditions occurs, involving either a decrease of the specific mass flow rate, so that condition (15.76) is satisfied, or, for supersonic flow, the appearance of a shock wave. Conversely, we note that subtraction of heat, involving a decrease of a^*, increases the distance from the critical conditions.

Let us now fix all the flow quantities in a duct section, say section 1, and construct the curve corresponding to all conditions that may be reached by keeping I constant, but allowing a^* to vary by introduction or removal of heat. If we choose a particular value for V_2, we may derive ρ_2 from (15.47), and p_2 by putting $\Delta I = 0$ in (15.67). Then, the temperature is obtained from the equation of state, and the enthalpy and entropy in the new flow condition may be derived consequently. Thus, we may draw a curve in the h–S plane (known as the *Rayleigh line*), giving all the possible conditions corresponding to frictionless one-dimensional motion in a constant-area duct with heat exchange.

Alternatively, one may be interested in the conditions in a duct in which friction cannot be neglected, and thus the impulse I decreases along the duct, but the flow is adiabatic, so that a^* is constant (as is any parameter defining the total energy content, such as h_0 or T_0). In this case, starting from the conditions in one section, and choosing the velocity V_2 in another section, we get ρ_2 from (15.47), and a_2 by putting $q_{12} = 0$ in (15.68). Then, using the equation of state, the values of pressure, enthalpy and entropy may be easily obtained. Again, one can construct a curve in the h–S plane (known as the *Fanno line*), giving all the possible conditions corresponding to adiabatic one-dimensional motion in a constant-area duct with friction.

The Rayleigh and Fanno lines are qualitatively shown Fig. 15.7. The curves have two points in common, at which G, I and a^* have the same values. Therefore, these intersection points correspond to the flow conditions on the two sides of a normal shock wave and, considering the relevant entropy values, we may deduce that, in both curves, the lower branch corresponds to supersonic flow and the upper one to subsonic flow. In fact, it may be easily shown that the extreme points of the two curves, with maximum entropy, correspond to a flow with $M = 1$. The entropy

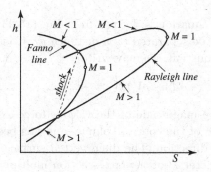

Fig. 15.7.   The Fanno and Rayleigh lines.

increases when the flows tend to the critical conditions, and this is the only direction of variation of the flow quantities for the Fanno line, and the one corresponding to heat introduction in the Rayleigh line. Finally, note that if the mass flow rate G decreases, the curves move rightwards, thus allowing higher values of entropy to be attained.

15.6.2. *The Fanno problem*

Due to its practical importance, we now describe in more detail the so-called *Fanno problem*, which corresponds to the adiabatic steady motion of a gas in a constant-area duct with friction. The description of the *Rayleigh problem*, concerning the flow in a frictionless duct with heat transfer, may be found, for instance, in Shapiro (1953), where flows with combined friction, heat exchange and area variation are also considered, together with more general situations.

Let us now derive the equations for the Fanno flow. The mass balance equation (15.5) becomes

$$\frac{d\rho}{\rho} + \frac{dV}{V} = 0. \tag{15.78}$$

The energy equation is obtained by differentiating Eq. (15.9), and by considering that we are analysing a perfect gas. We get

$$dh + VdV = C_p dT + VdV = \frac{a^2}{(\gamma - 1)} \frac{dT}{T} + V^2 \frac{dV}{V} = 0, \tag{15.79}$$

and, therefore,

$$\frac{dT}{T} + (\gamma - 1)M^2 \frac{dV}{V} = 0. \tag{15.80}$$

The momentum equation is derived from Eq. (15.23) by assuming the two cross-sections to be separated by a distance dx. Considering that both $G = \rho V$ and A are constant, we may write

$$A dp + \rho V A dV = -\tau_w dA_w, \tag{15.81}$$

where $-\tau_w dA_w$ is the magnitude of the viscous force exerted on the fluid by the lateral surface of the control volume, and has been assumed to act opposite to the flow direction. The duct wetted surface, denoted by dA_w, is equal to the perimeter P of the cross-section multiplied by dx. We now introduce the hydraulic diameter of the duct, defined as four times the ratio between the cross-sectional area and its perimeter,

$$D_i = \frac{4A}{P} = \frac{4A}{dA_w/dx} = \frac{4A}{dA_w}dx. \tag{15.82}$$

We also define the friction coefficient as

$$f = \frac{\tau_w}{\rho V^2/2}, \tag{15.83}$$

so that Eq. (15.81) may be finally recast as

$$dp + \rho V^2 \frac{dV}{V} = -\frac{1}{2}\rho V^2 \frac{4f}{D_i}dx. \tag{15.84}$$

By observing that $\rho V^2 = \gamma p M^2$ and dividing Eq. (15.84) by p, we get

$$\frac{dp}{p} + \gamma M^2 \frac{dV}{V} + \frac{\gamma M^2}{2}\frac{4f}{D_i}dx = 0. \tag{15.85}$$

Furthermore, the perfect gas equation of state gives

$$\frac{dp}{p} = \frac{d\rho}{\rho} + \frac{dT}{T}, \tag{15.86}$$

and from the definition of the Mach number, with $a^2 = \gamma RT$, we obtain

$$\frac{dM}{M} = \frac{dV}{V} - \frac{1}{2}\frac{dT}{T}. \tag{15.87}$$

Therefore, we have a set of five equations, namely Eqs. (15.78), (15.80), (15.85), (15.86) and (15.87), which relate six differential variables: $d\rho/\rho$, dV/V, dT/T, dp/p, dM/M and $4fdx/D_i$. Considering that we are analysing the effect of friction, we initially choose $4fdx/D_i$ as the independent variable. The remaining variables may be expressed solely in terms of $4fdx/D_i$ and of the Mach number by a process of successive elimination. After lengthy but straightforward manipulations, the following equations are thus obtained

$$\frac{dp}{p} = -\frac{\gamma M^2[1 + (\gamma - 1)M^2]}{2(1 - M^2)}\frac{4f}{D_i}dx, \tag{15.88}$$

$$\frac{dV}{V} = -\frac{d\rho}{\rho} = \frac{\gamma M^2}{2(1 - M^2)}\frac{4f}{D_i}dx, \tag{15.89}$$

$$\frac{dT}{T} = -\frac{\gamma(\gamma - 1)M^4}{2(1 - M^2)}\frac{4f}{D_i}dx, \tag{15.90}$$

$$\frac{dM}{M} = \frac{\gamma M^2[2 + (\gamma - 1)M^2]}{4(1 - M^2)}\frac{4f}{D_i}dx. \tag{15.91}$$

It is useful to derive analogous equations also for the impulse and for the total pressure. To this end, we apply the logarithmic differentiation to relations (15.28) and (15.70), and then use Eqs. (15.88) and (15.91), so that the following equations are obtained:

$$\frac{dp_0}{p_0} = -\frac{\gamma M^2}{2}\frac{4f}{D_i}dx,$$ (15.92)

$$\frac{dI}{I} = -\frac{\gamma M^2}{2(1 + \gamma M^2)}\frac{4f}{D_i}dx.$$ (15.93)

Finally, the variation of entropy may be derived from that of the total pressure from relation (15.31). Thus, using (15.92) and (5.25), we get

$$\frac{dS}{C_p} = \frac{(\gamma - 1)M^2}{2}\frac{4f}{D_i}dx.$$ (15.94)

By convention, dx is positive in the direction of the flow and thus, considering that $\gamma > 1$ and that in an adiabatic process entropy may not decrease, we deduce from Eq. (15.94) that the friction coefficient f must always be positive. Consequently, we conclude that the friction force acts opposite to the flow direction, as we assumed in (15.81). The trends of the variations of the flow quantities along the duct are then derived from the above equations, and are shown using arrows in Table 15.1. Apart from the total pressure and the impulse, which always decrease, the other flow quantities are seen to have different trends for subsonic and for supersonic conditions, due to the presence of the term $(1 - M^2)$ in the denominator of the relevant equations. In particular, the Mach number always tends to unity, which is the limit flow condition. It may be seen from Table 15.1 that the signs of the variations of p, V, T, ρ and M are the same as occur in an isentropic flow for a converging duct.

Starting from the flow conditions in a certain cross-section, one could now integrate Eqs. (15.88)–(15.94) along the duct to obtain the various quantities. However, we have seen that, for given conditions at an initial section, a maximum length of the duct exists that is compatible with

Table 15.1. Variation of flow quantities in Fanno flow.

	p	V	T	ρ	M	p_0	I
Subsonic	⇓	⇑	⇓	⇓	⇑	⇓	⇓
Supersonic	⇑	⇓	⇑	⇑	⇓	⇓	⇓

those initial conditions, and for which the exit Mach number is exactly unity. We may then use this limit case as a reference condition, and derive the maximum allowable length corresponding to the initial conditions. To this end, it is convenient to use the Mach number as the independent variable, rather than the streamwise coordinate x, and this is done by using Eq. (15.91) to express dx in terms of dM. Thus, the maximum length of duct that may exist downstream of a section where the Mach number has the generic value M is obtained from the relation

$$\int_0^{L_{\max}} \frac{4f}{D_i} dx = \int_M^1 \frac{4(1-M^2)}{\gamma M^3 [2 + (\gamma - 1)M^2]} dM. \tag{15.95}$$

We may carry out the integration by introducing the mean value of the friction coefficient along the considered duct length, f_m, to get

$$\frac{4f_m L_{\max}}{D_i} = \frac{1-M^2}{\gamma M^2} + \frac{\gamma + 1}{2\gamma} \ln \frac{(\gamma + 1)M^2}{2 + (\gamma - 1)M^2}. \tag{15.96}$$

Note that, since $4f_m L_{\max}/D_i$ is a function of M only, the duct length L required for the flow to pass from M_1 to M_2 is found from the expression

$$\frac{4f_m L}{D_i} = \left(\frac{4f_m L_{\max}}{D_i}\right)_{M_1} - \left(\frac{4f_m L_{\max}}{D_i}\right)_{M_2}. \tag{15.97}$$

The conditions at the exit of the maximum length duct may now be used as reference conditions. Considering that $M = 1$ in that section, these conditions are denoted as *critical* and, again, indicated with starred symbols, to which we add the subscript F to specify that they refer to the Fanno problem. The ratio between the various quantities in a generic section, where the Mach number is M, and the critical values at the exit of the maximum-length duct are found by using Eq. (15.91) to express all the variations as a function of the Mach number, and then integrating from M to 1. We thus obtain the following expressions:

$$\frac{p}{p_F^*} = \frac{1}{M} \sqrt{\frac{\gamma + 1}{2 + (\gamma - 1)M^2}}, \tag{15.98}$$

$$\frac{V}{V_F^*} = \frac{\rho_F^*}{\rho} = M \sqrt{\frac{\gamma + 1}{2 + (\gamma - 1)M^2}}, \tag{15.99}$$

$$\frac{T}{T_F^*} = \frac{\gamma + 1}{2 + (\gamma - 1)M^2}, \tag{15.100}$$

$$\frac{p_0}{p_{0F}^*} = \frac{1}{M}\left[\frac{2+(\gamma-1)M^2}{\gamma+1}\right]^{\frac{\gamma+1}{2(\gamma-1)}}, \tag{15.101}$$

$$\frac{I}{I_F^*} = \frac{1}{M}\frac{1+\gamma M^2}{\sqrt{(\gamma+1)[2+(\gamma-1)M^2]}}. \tag{15.102}$$

Finally, one must recall that the total temperature is constant along the duct because the Fanno flow is adiabatic. By using the constancy of the critical conditions along the duct, from the above relations we can easily obtain the change in any stream property between two generic sections where the Mach numbers are M_1 and M_2 respectively. For instance, if we are interested in the pressure variation, we may write

$$\frac{p_2}{p_1} = \frac{(p/p_F^*)_{M_2}}{(p/p_F^*)_{M_1}}, \tag{15.103}$$

and similar relations may be used for the remaining quantities.

As an example, the maximum duct lengths for several initial values of M and $f_m = 0.0025$, which is a plausible value, are given in Table 15.2. It is interesting to note that $L_{\max}/D_i$ has a finite limit value for $M = \infty$.

Let us now consider the adiabatic flow in a long constant-area duct with friction, fed by a converging nozzle that is short enough for the flow to be considered as isentropic, with total pressure equal to p_0 (Fig. 15.8).

Table 15.2. Maximum duct lengths in Fanno flow for $f_m = 0.0025$.

M	0	0.25	0.5	0.75	1	1.5	2	3	∞
$L_{\max}/D_i$	∞	850	110	12	0	14	31	52	82

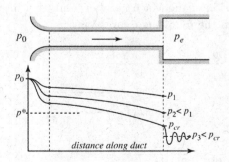

Fig. 15.8.   Long constant-area duct fed by converging nozzle.

With decreasing external pressure p_e, the flow in the nozzle and in the duct increases, always remaining subsonic and such that the pressure at the duct exit coincides with p_e. However, by further decreasing p_e, a limit value p_{cr} for which we have $M = 1$ at the duct exit is reached, and the flow becomes choked. This critical external pressure is derived easily by considering that it coincides with the Fanno critical pressure and that, for every given length of the duct, the Mach number M_i at the duct inlet corresponding to $M = 1$ at the exit can be derived immediately from relation (15.96). We thus have

$$\frac{p_{cr}}{p_0} = \frac{p_F^*}{p_0} = \frac{p_F^*}{p_i} \frac{p_i}{p_0}, \qquad (15.104)$$

where p_i is the pressure at the duct inlet, and the ratios on the right-hand side are obtained, as a function of M_i, from (15.98) and (15.28). It is easy to check that, for any finite length of the constant-area duct with friction, the critical external pressure is lower than that given by (15.33), which corresponds to the completely isentropic flow that would occur if the constant-area duct were either absent or frictionless. Furthermore, the mass flow rate is also lower because it is the one that can be obtained from (15.45) for $M = M_i$ whereas, if the flow were isentropic, it would be the one given by (15.46). Thus, even if from Table 15.1 one might have the impression that, in subsonic conditions, friction causes an acceleration of the flow, in reality this would not be a correct way of analysing the effect of friction. In fact, what one should do is to compare the results of the Fanno flow with those of an isentropic flow, and we have just seen that in the latter we would have constant conditions along the duct, with a higher Mach number and a higher velocity everywhere apart, perhaps, from the last section. Thus, friction causes the maximum flow rate to be lower and to be attained with a higher pressure difference. More generally, for the same pressure difference between the extremities of the duct, the mass flow rate is lower with friction than without friction, and this is indeed what one might expect.

On the other hand, when a constant-area duct with friction is fed by a frictionless converging–diverging nozzle, the situation is much more complex because several different flow conditions may arise depending on the value of the downstream external pressure; for a detailed analysis the reader may refer, for instance, to the book by Shapiro (1953). Here, we only consider the case in which the Mach number at the inlet of the constant-area duct, M_i, is equal to the supersonic value that corresponds to sonic conditions at the throat of the feeding nozzle and isentropic flow in its diverging part.

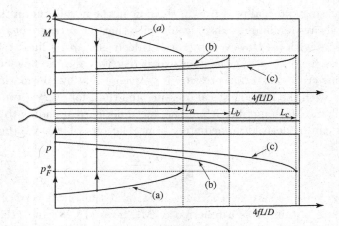

Fig. 15.9. Long constant-area duct fed by converging–diverging nozzle.

Furthermore, the external pressure is assumed to be as low as necessary for critical conditions to be attained at the end of the duct.

Therefore, let us start by assuming that the length of the duct is exactly the maximum one that allows the condition $M = M_i$ to be present at its inlet; for instance, $M_i = 2$ in Fig. 15.9. Consequently, the Mach number will decrease progressively along the duct and become unity at the exit section, as shown in Fig. 15.9, curve a. If we now increase the length of the constant-area duct, we cannot have a decrease of the flow rate, as occurs in the subsonic case, because sonic conditions are present at the throat of the feeding nozzle, so that variations occurring downstream are not felt in the converging part of the nozzle and the flow rate is choked. Therefore, what happens is that a shock wave appears inside the duct, in a position that is perfectly defined by the new length of the duct. In effect, as shown in Fig. 15.9, curve b, the value of the Mach number downstream of the shock must coincide exactly with the subsonic inlet value of a maximum-length duct equal to the remaining portion of the duct. If the length of the duct continues to be increased, the shock wave moves progressively upstream until it reaches the duct inlet (curve c in Fig. 15.9). A further progressive lengthening of the duct causes the shock to move into the diverging part of the nozzle, so that the Mach number at the inlet of the duct becomes subsonic, and decreases as the increase in duct length causes the shock to move further into the nozzle. Finally, the shock reaches the throat and vanishes, the flow becomes subsonic in the whole feeding nozzle, and further increases in length produce reductions in flow rate.

Appendix A

TABLES

A.1. Physical Properties of Air and Water

A.1.1. *Dry air*

Properties at the pressure of one atmosphere $= 1.01325 \times 10^5 \, \text{N/m}^2$.

Temperature T [°C]	Density ρ [kg/m^3]	Dynamic viscosity μ [kg/(ms)]	Kinematic viscosity ν [m^2/s]	Thermal conductivity k [J/(msK)]	Prandtl number Pr
−100	2.04	1.16×10^{-5}	0.57×10^{-5}	0.0158	0.74
−50	1.582	1.46	0.92	0.0201	0.73
0	1.293	1.72	1.33	0.0241	0.72
10	1.247	1.76	1.41	0.0249	0.71
15	1.225	1.78	1.45	0.0253	0.71
20	1.205	1.81	1.50	0.0257	0.71
25	1.185	1.84	1.55	0.0260	0.71
30	1.165	1.86	1.60	0.0264	0.71
40	1.127	1.91	1.69	0.0272	0.71
50	1.092	1.95	1.79	0.0279	0.70
60	1.060	2.00	1.89	0.0287	0.70
70	1.030	2.05	1.99	0.0294	0.70
80	1.000	2.09	2.09	0.0302	0.70
90	0.973	2.13	2.19	0.0309	0.70
100	0.946	2.17	2.30	0.0316	0.69
200	0.746	2.57	3.45	0.0385	0.68
300	0.616	2.93	4.75	0.0449	0.68
500	0.456	3.58	7.85	0.0563	0.69

A.1.2. *Pure water*

Properties at the pressure of one atmosphere $= 1.01325 \times 10^5$ N/m^2.

Temperature T [°C]	Density ρ [kg/m^3]	Dynamic viscosity μ [kg/(ms)]	Kinematic viscosity ν [m^2/s]	Thermal conductivity k [J/(msK)]	Prandtl number Pr
0	999.8	1.792×10^{-3}	1.792×10^{-6}	0.563	13.46
5	1000.0	1.520	1.520	0.573	11.21
10	999.8	1.308	1.308	0.582	9.46
15	999.2	1.139	1.140	0.591	8.08
20	998.3	1.003	1.005	0.600	7.11
25	997.1	0.891	0.894	0.609	6.12
30	995.7	0.798	0.801	0.616	5.47
40	992.3	0.653	0.658	0.630	4.33
50	988.0	0.547	0.554	0.642	3.57
60	983.1	0.467	0.475	0.652	3.01
70	977.6	0.404	0.413	0.661	2.56
80	971.6	0.355	0.365	0.668	2.22
90	965.1	0.315	0.326	0.672	1.98
100	958.1	0.282	0.294	0.676	1.77

A.2. Isentropic Flow Relations

Perfect gas with $\gamma = 1.4$.

M	T/T_0	p/p_0	ρ/ρ_0	A/A^*	M^*
0.00	1	1	1	∞	0
0.01	0.999980	0.999930	0.999950	57.873844	0.010954
0.02	0.999920	0.999720	0.999800	28.942131	0.021908
0.03	0.999820	0.999370	0.999550	19.300543	0.032860
0.04	0.999680	0.998881	0.999200	14.481486	0.043811
0.05	0.999500	0.998252	0.998751	11.591444	0.054759
0.06	0.999281	0.997484	0.998202	9.665911	0.065703
0.07	0.999021	0.996578	0.997554	8.291525	0.076644
0.08	0.998722	0.995533	0.996807	7.261610	0.087580
0.09	0.998383	0.994351	0.995961	6.461341	0.098510
0.10	0.998004	0.993031	0.995017	5.821829	0.109435
0.11	0.997586	0.991576	0.993976	5.299230	0.120353
0.12	0.997128	0.989985	0.992836	4.864318	0.131265
0.13	0.996631	0.988259	0.991600	4.496859	0.142168
0.14	0.996095	0.986400	0.990267	4.182400	0.153063
0.15	0.995520	0.984408	0.988838	3.910343	0.163948

(Continued)

(*Continued*)

M	T/T_0	p/p_0	ρ/ρ_0	A/A^*	M^*
0.16	0.994906	0.982285	0.987314	3.672739	0.174824
0.17	0.994253	0.980030	0.985695	3.463509	0.185690
0.18	0.993562	0.977647	0.983982	3.277926	0.196544
0.19	0.992832	0.975135	0.982176	3.112259	0.207387
0.20	0.992063	0.972497	0.980277	2.963520	0.218218
0.21	0.991257	0.969733	0.978286	2.829294	0.229036
0.22	0.990413	0.966845	0.976204	2.707602	0.239840
0.23	0.989531	0.963835	0.974032	2.596812	0.250630
0.24	0.988611	0.960703	0.971771	2.495563	0.261405
0.25	0.987654	0.957453	0.969421	2.402710	0.272166
0.26	0.986660	0.954085	0.966984	2.317287	0.282910
0.27	0.985629	0.950600	0.964460	2.238471	0.293637
0.28	0.984562	0.947002	0.961851	2.165554	0.304348
0.29	0.983458	0.943291	0.959157	2.097928	0.315041
0.30	0.982318	0.939470	0.956380	2.035065	0.325715
0.31	0.981142	0.935540	0.953521	1.976507	0.336371
0.32	0.979931	0.931503	0.950580	1.921851	0.347007
0.33	0.978684	0.927362	0.947560	1.870745	0.357623
0.34	0.977402	0.923118	0.944460	1.822876	0.368219
0.35	0.976086	0.918773	0.941283	1.777969	0.378794
0.36	0.974735	0.914330	0.938029	1.735778	0.389347
0.37	0.973350	0.909790	0.934700	1.696086	0.399877
0.38	0.971931	0.905156	0.931297	1.658696	0.410385
0.39	0.970478	0.900430	0.927821	1.623433	0.420870
0.40	0.968992	0.895614	0.924274	1.590140	0.431331
0.41	0.967474	0.890711	0.920657	1.558673	0.441768
0.42	0.965922	0.885722	0.916971	1.528905	0.452180
0.43	0.964339	0.880651	0.913217	1.500718	0.462566
0.44	0.962723	0.875498	0.909398	1.474005	0.472927
0.45	0.961076	0.870267	0.905513	1.448672	0.483261
0.46	0.959398	0.864960	0.901566	1.424628	0.493569
0.47	0.957689	0.859580	0.897556	1.401795	0.503849
0.48	0.955950	0.854128	0.893486	1.380097	0.514102
0.49	0.954180	0.848607	0.889357	1.359468	0.524327
0.50	0.952381	0.843019	0.885170	1.339844	0.534522
0.51	0.950552	0.837367	0.880927	1.321168	0.544689
0.52	0.948695	0.831654	0.876629	1.303388	0.554826
0.53	0.946808	0.825881	0.872278	1.286454	0.564934
0.54	0.944894	0.820050	0.867876	1.270321	0.575011
0.55	0.942951	0.814165	0.863422	1.254948	0.585057
0.56	0.940982	0.808228	0.858920	1.240294	0.595072
0.57	0.938985	0.802241	0.854371	1.226326	0.605055
0.58	0.936961	0.796206	0.849775	1.213007	0.615006
0.59	0.934911	0.790127	0.845135	1.200308	0.624925
0.60	0.932836	0.784004	0.840452	1.188200	0.634811

(*Continued*)

(*Continued*)

M	T/T_0	p/p_0	ρ/ρ_0	A/A^*	M^*
0.61	0.930735	0.777841	0.835728	1.176654	0.644664
0.62	0.928609	0.771639	0.830963	1.165645	0.654483
0.63	0.926458	0.765402	0.826160	1.155151	0.664269
0.64	0.924283	0.759131	0.821320	1.145148	0.674020
0.65	0.922084	0.752829	0.816443	1.135616	0.683737
0.66	0.919862	0.746498	0.811533	1.126535	0.693419
0.67	0.917616	0.740140	0.806590	1.117887	0.703066
0.68	0.915349	0.733758	0.801616	1.109655	0.712677
0.69	0.913059	0.727353	0.796612	1.101822	0.722252
0.70	0.910747	0.720928	0.791579	1.094373	0.731792
0.71	0.908414	0.714485	0.786519	1.087294	0.741295
0.72	0.906060	0.708025	0.781434	1.080571	0.750761
0.73	0.903685	0.701552	0.776324	1.074192	0.760190
0.74	0.901291	0.695068	0.771191	1.068144	0.769582
0.75	0.898876	0.688573	0.766037	1.062417	0.778936
0.76	0.896443	0.682070	0.760863	1.056999	0.788253
0.77	0.893991	0.675562	0.755670	1.051881	0.797531
0.78	0.891520	0.669050	0.750460	1.047053	0.806772
0.79	0.889031	0.662536	0.745234	1.042505	0.815974
0.80	0.886525	0.656022	0.739992	1.038230	0.825137
0.81	0.884001	0.649509	0.734738	1.034219	0.834261
0.82	0.881461	0.643000	0.729471	1.030464	0.843346
0.83	0.878905	0.636496	0.724193	1.026959	0.852392
0.84	0.876332	0.630000	0.718905	1.023696	0.861399
0.85	0.873744	0.623512	0.713609	1.020669	0.870366
0.86	0.871141	0.617034	0.708306	1.017871	0.879292
0.87	0.868523	0.610569	0.702997	1.015297	0.888179
0.88	0.865891	0.604117	0.697683	1.012941	0.897026
0.89	0.863245	0.597680	0.692365	1.010798	0.905832
0.90	0.860585	0.591260	0.687044	1.008863	0.914598
0.91	0.857913	0.584858	0.681722	1.007131	0.923323
0.92	0.855227	0.578476	0.676400	1.005597	0.932007
0.93	0.852529	0.572114	0.671079	1.004258	0.940650
0.94	0.849820	0.565775	0.665759	1.003108	0.949253
0.95	0.847099	0.559460	0.660443	1.002145	0.957814
0.96	0.844366	0.553170	0.655130	1.001365	0.966334
0.97	0.841623	0.546905	0.649822	1.000763	0.974813
0.98	0.838870	0.540669	0.644520	1.000337	0.983250
0.99	0.836106	0.534460	0.639225	1.000084	0.991646
1.00	0.833333	0.528282	0.633938	1	1
1.01	0.830551	0.522134	0.628660	1.000083	1.008312
1.02	0.827760	0.516018	0.623391	1.000330	1.016583
1.03	0.824960	0.509935	0.618133	1.000738	1.024812
1.04	0.822152	0.503886	0.612887	1.001305	1.032999
1.05	0.819336	0.497872	0.607653	1.002029	1.041144

(*Continued*)

(*Continued*)

M	T/T_0	p/p_0	ρ/ρ_0	A/A^*	M^*
1.06	0.816513	0.491894	0.602432	1.002907	1.049248
1.07	0.813683	0.485952	0.597225	1.003938	1.057309
1.08	0.810846	0.480047	0.592033	1.005119	1.065328
1.09	0.808002	0.474181	0.586856	1.006449	1.073306
1.10	0.805153	0.468354	0.581696	1.007925	1.081241
1.11	0.802298	0.462567	0.576553	1.009547	1.089134
1.12	0.799437	0.456820	0.571427	1.011312	1.096985
1.13	0.796572	0.451115	0.566320	1.013219	1.104794
1.14	0.793701	0.445451	0.561232	1.015267	1.112561
1.15	0.790826	0.439829	0.556164	1.017454	1.120286
1.16	0.787948	0.434251	0.551116	1.019780	1.127969
1.17	0.785065	0.428716	0.546090	1.022242	1.135610
1.18	0.782179	0.423225	0.541085	1.024840	1.143209
1.19	0.779290	0.417778	0.536102	1.027573	1.150766
1.20	0.776398	0.412377	0.531142	1.030440	1.158281
1.21	0.773503	0.407021	0.526205	1.033440	1.165754
1.22	0.770606	0.401711	0.521292	1.036572	1.173185
1.23	0.767707	0.396446	0.516403	1.039835	1.180575
1.24	0.764807	0.391229	0.511539	1.043229	1.187923
1.25	0.761905	0.386058	0.506701	1.046753	1.195229
1.26	0.759002	0.380934	0.501888	1.050406	1.202493
1.27	0.756098	0.375857	0.497102	1.054189	1.209716
1.28	0.753194	0.370828	0.492342	1.058100	1.216897
1.29	0.750289	0.365847	0.487608	1.062138	1.224037
1.30	0.747384	0.360914	0.482903	1.066304	1.231136
1.31	0.744480	0.356029	0.478225	1.070598	1.238193
1.32	0.741576	0.351192	0.473575	1.075018	1.245209
1.33	0.738672	0.346403	0.468953	1.079565	1.252184
1.34	0.735770	0.341663	0.464361	1.084239	1.259118
1.35	0.732869	0.336971	0.459797	1.089038	1.266011
1.36	0.729970	0.332328	0.455263	1.093964	1.272864
1.37	0.727072	0.327733	0.450758	1.099015	1.279675
1.38	0.724176	0.323187	0.446283	1.104193	1.286447
1.39	0.721282	0.318690	0.441838	1.109496	1.293177
1.40	0.718391	0.314241	0.437423	1.114926	1.299867
1.41	0.715502	0.309840	0.433039	1.120481	1.306517
1.42	0.712616	0.305488	0.428686	1.126162	1.313127
1.43	0.709733	0.301185	0.424363	1.131969	1.319697
1.44	0.706854	0.296929	0.420072	1.137903	1.326227
1.45	0.703977	0.292722	0.415812	1.143963	1.332717
1.46	0.701105	0.288563	0.411583	1.150150	1.339168
1.47	0.698236	0.284452	0.407386	1.156463	1.345579
1.48	0.695372	0.280388	0.403220	1.162904	1.351951
1.49	0.692511	0.276372	0.399086	1.169471	1.358283
1.50	0.689655	0.272403	0.394984	1.176167	1.364576

(*Continued*)

(*Continued*)

M	T/T_0	p/p_0	ρ/ρ_0	A/A^*	M^*
1.51	0.686804	0.268481	0.390914	1.182991	1.370831
1.52	0.683957	0.264607	0.386876	1.189943	1.377047
1.53	0.681115	0.260779	0.382870	1.197024	1.383224
1.54	0.678279	0.256997	0.378897	1.204234	1.389362
1.55	0.675448	0.253262	0.374955	1.211574	1.395462
1.56	0.672622	0.249573	0.371045	1.219043	1.401524
1.57	0.669801	0.245930	0.367168	1.226644	1.407548
1.58	0.666987	0.242332	0.363323	1.234376	1.413534
1.59	0.664178	0.238779	0.359510	1.242239	1.419483
1.60	0.661376	0.235271	0.355730	1.250235	1.425393
1.61	0.658579	0.231808	0.351982	1.258364	1.431267
1.62	0.655789	0.228389	0.348266	1.266626	1.437103
1.63	0.653006	0.225014	0.344582	1.275022	1.442902
1.64	0.650229	0.221683	0.340930	1.283553	1.448664
1.65	0.647459	0.218395	0.337311	1.292219	1.454389
1.66	0.644695	0.215150	0.333723	1.301021	1.460078
1.67	0.641939	0.211948	0.330168	1.309960	1.465730
1.68	0.639190	0.208788	0.326644	1.319037	1.471346
1.69	0.636448	0.205670	0.323152	1.328252	1.476926
1.70	0.633714	0.202593	0.319693	1.337606	1.482471
1.71	0.630986	0.199558	0.316264	1.347100	1.487979
1.72	0.628267	0.196564	0.312868	1.356735	1.493452
1.73	0.625555	0.193611	0.309502	1.366511	1.498889
1.74	0.622851	0.190697	0.306169	1.376430	1.504292
1.75	0.620155	0.187824	0.302866	1.386492	1.509659
1.76	0.617467	0.184990	0.299595	1.396698	1.514991
1.77	0.614787	0.182195	0.296354	1.407049	1.520289
1.78	0.612115	0.179438	0.293145	1.417546	1.525552
1.79	0.609451	0.176720	0.289966	1.428190	1.530781
1.80	0.606796	0.174040	0.286818	1.438982	1.535976
1.81	0.604149	0.171398	0.283701	1.449923	1.541137
1.82	0.601511	0.168792	0.280614	1.461013	1.546265
1.83	0.598881	0.166224	0.277557	1.472255	1.551358
1.84	0.596260	0.163691	0.274530	1.483648	1.556419
1.85	0.593648	0.161195	0.271533	1.495194	1.561446
1.86	0.591044	0.158734	0.268566	1.506894	1.566440
1.87	0.588450	0.156309	0.265628	1.518750	1.571401
1.88	0.585864	0.153918	0.262720	1.530761	1.576329
1.89	0.583288	0.151562	0.259841	1.542930	1.581226
1.90	0.580720	0.149240	0.256991	1.555257	1.586089
1.91	0.578162	0.146951	0.254169	1.567743	1.590921
1.92	0.575612	0.144696	0.251377	1.580391	1.595721
1.93	0.573072	0.142473	0.248613	1.593200	1.600489
1.94	0.570542	0.140283	0.245877	1.606172	1.605226
1.95	0.568020	0.138125	0.243170	1.619309	1.609931

(*Continued*)

(*Continued*)

M	T/T_0	p/p_0	ρ/ρ_0	A/A^*	M^*
1.96	0.565508	0.135999	0.240490	1.632611	1.614605
1.97	0.563006	0.133905	0.237839	1.646080	1.619248
1.98	0.560513	0.131841	0.235215	1.659717	1.623860
1.99	0.558029	0.129808	0.232618	1.673523	1.628442
2.00	0.555556	0.127805	0.230048	1.687500	1.632993
2.01	0.553091	0.125831	0.227506	1.701649	1.637514
2.02	0.550637	0.123888	0.224990	1.715971	1.642005
2.03	0.548192	0.121973	0.222500	1.730467	1.646466
2.04	0.545756	0.120087	0.220038	1.745139	1.650898
2.05	0.543331	0.118229	0.217601	1.759989	1.655299
2.06	0.540915	0.116399	0.215190	1.775017	1.659672
2.07	0.538509	0.114597	0.212805	1.790225	1.664015
2.08	0.536113	0.112823	0.210446	1.805614	1.668330
2.09	0.533726	0.111075	0.208112	1.821187	1.672616
2.10	0.531350	0.109353	0.205803	1.836944	1.676873
2.11	0.528983	0.107658	0.203519	1.852886	1.681101
2.12	0.526626	0.105989	0.201259	1.869016	1.685302
2.13	0.524279	0.104345	0.199025	1.885334	1.689474
2.14	0.521942	0.102726	0.196814	1.901843	1.693619
2.15	0.519615	0.101132	0.194628	1.918543	1.697736
2.16	0.517298	0.099562	0.192466	1.935437	1.701825
2.17	0.514991	0.098017	0.190327	1.952525	1.705887
2.18	0.512694	0.096495	0.188212	1.969810	1.709922
2.19	0.510407	0.094997	0.186120	1.987292	1.713930
2.20	0.508130	0.093522	0.184051	2.004975	1.717911
2.21	0.505863	0.092069	0.182005	2.022858	1.721866
2.22	0.503606	0.090640	0.179981	2.040944	1.725794
2.23	0.501359	0.089232	0.177980	2.059234	1.729696
2.24	0.499122	0.087846	0.176001	2.077731	1.733572
2.25	0.496894	0.086482	0.174044	2.096435	1.737422
2.26	0.494677	0.085139	0.172110	2.115349	1.741246
2.27	0.492470	0.083817	0.170196	2.134473	1.745044
2.28	0.490273	0.082515	0.168304	2.153811	1.748817
2.29	0.488086	0.081234	0.166433	2.173362	1.752565
2.30	0.485909	0.079973	0.164584	2.193131	1.756288
2.31	0.483741	0.078731	0.162755	2.213117	1.759986
2.32	0.481584	0.077509	0.160946	2.233323	1.763659
2.33	0.479437	0.076306	0.159158	2.253750	1.767308
2.34	0.477300	0.075122	0.157390	2.274401	1.770933
2.35	0.475172	0.073957	0.155642	2.295277	1.774533
2.36	0.473055	0.072810	0.153914	2.316380	1.778109
2.37	0.470947	0.071681	0.152206	2.337712	1.781661
2.38	0.468850	0.070570	0.150516	2.359275	1.785190
2.39	0.466762	0.069476	0.148846	2.381070	1.788695
2.40	0.464684	0.068399	0.147195	2.403100	1.792176

(*Continued*)

(*Continued*)

M	T/T_0	p/p_0	ρ/ρ_0	A/A^*	M^*
2.41	0.462616	0.067340	0.145563	2.425366	1.795635
2.42	0.460558	0.066297	0.143950	2.447870	1.799070
2.43	0.458509	0.065271	0.142354	2.470615	1.802482
2.44	0.456471	0.064261	0.140777	2.493602	1.805872
2.45	0.454442	0.063267	0.139218	2.516833	1.809239
2.46	0.452423	0.062288	0.137677	2.540309	1.812584
2.47	0.450414	0.061326	0.136154	2.564035	1.815907
2.48	0.448414	0.060378	0.134648	2.588010	1.819207
2.49	0.446425	0.059445	0.133159	2.612237	1.822485
2.50	0.444444	0.058528	0.131687	2.636719	1.825742
2.51	0.442474	0.057624	0.130232	2.661457	1.828977
2.52	0.440513	0.056736	0.128794	2.686453	1.832190
2.53	0.438562	0.055861	0.127373	2.711709	1.835382
2.54	0.436620	0.055000	0.125968	2.737228	1.838553
2.55	0.434688	0.054153	0.124579	2.763012	1.841703
2.56	0.432766	0.053319	0.123206	2.789063	1.844832
2.57	0.430853	0.052499	0.121849	2.815382	1.847941
2.58	0.428949	0.051692	0.120507	2.841972	1.851028
2.59	0.427055	0.050897	0.119182	2.868836	1.854096
2.60	0.425170	0.050115	0.117871	2.895975	1.857143
2.61	0.423295	0.049346	0.116575	2.923392	1.860170
2.62	0.421429	0.048589	0.115295	2.951089	1.863177
2.63	0.419572	0.047844	0.114029	2.979068	1.866164
2.64	0.417725	0.047110	0.112778	3.007331	1.869131
2.65	0.415887	0.046389	0.111542	3.035881	1.872079
2.66	0.414058	0.045679	0.110320	3.064720	1.875007
2.67	0.412239	0.044980	0.109112	3.093850	1.877916
2.68	0.410428	0.044292	0.107918	3.123274	1.880806
2.69	0.408627	0.043616	0.106738	3.152993	1.883677
2.70	0.406835	0.042950	0.105571	3.183011	1.886529
2.71	0.405052	0.042295	0.104418	3.213330	1.889362
2.72	0.403278	0.041650	0.103279	3.243951	1.892177
2.73	0.401513	0.041016	0.102152	3.274878	1.894973
2.74	0.399757	0.040391	0.101039	3.306113	1.897751
2.75	0.398010	0.039777	0.099939	3.337658	1.900511
2.76	0.396272	0.039172	0.098851	3.369515	1.903252
2.77	0.394543	0.038577	0.097777	3.401688	1.905976
2.78	0.392822	0.037992	0.096714	3.434179	1.908682
2.79	0.391111	0.037415	0.095664	3.466989	1.911370
2.80	0.389408	0.036848	0.094626	3.500123	1.914041
2.81	0.387714	0.036290	0.093601	3.533581	1.916694
2.82	0.386029	0.035741	0.092587	3.567368	1.919330
2.83	0.384352	0.035201	0.091585	3.601484	1.921949

(*Continued*)

(Continued)

M	T/T_0	p/p_0	ρ/ρ_0	A/A^*	M^*
2.84	0.382684	0.034669	0.090594	3.635934	1.924550
2.85	0.381025	0.034146	0.089616	3.670719	1.927135
2.86	0.379374	0.033631	0.088648	3.705842	1.929703
2.87	0.377732	0.033124	0.087692	3.741305	1.932255
2.88	0.376098	0.032625	0.086747	3.777113	1.934790
2.89	0.374473	0.032134	0.085813	3.813267	1.937308
2.90	0.372856	0.031651	0.084889	3.849769	1.939810
2.91	0.371248	0.031176	0.083977	3.886622	1.942296
2.92	0.369648	0.030708	0.083075	3.923829	1.944766
2.93	0.368056	0.030248	0.082183	3.961394	1.947220
2.94	0.366472	0.029795	0.081302	3.999317	1.949658
2.95	0.364897	0.029349	0.080431	4.037604	1.952081
2.96	0.363330	0.028910	0.079571	4.076255	1.954487
2.97	0.361771	0.028479	0.078720	4.115274	1.956879
2.98	0.360220	0.028054	0.077879	4.154665	1.959255
2.99	0.358677	0.027635	0.077048	4.194428	1.961615
3.00	0.357143	0.027224	0.076226	4.234568	1.963961
3.10	0.342231	0.023449	0.068517	4.657310	1.986608
3.20	0.328084	0.020228	0.061654	5.120958	2.007859
3.30	0.314663	0.017477	0.055541	5.628646	2.027812
3.40	0.301932	0.015125	0.050093	6.183700	2.046560
3.50	0.289855	0.013111	0.045233	6.789620	2.064187
3.60	0.278396	0.011385	0.040894	7.450110	2.080774
3.70	0.267523	0.009903	0.037017	8.169066	2.096393
3.80	0.257202	0.008629	0.033549	8.950584	2.111111
3.90	0.247402	0.007532	0.030445	9.798974	2.124991
4.00	0.238095	0.006586	0.027662	10.718750	2.138090
4.10	0.229253	0.005769	0.025164	11.714651	2.150461
4.20	0.220848	0.005062	0.022921	12.791637	2.162154
4.30	0.212857	0.004449	0.020903	13.954906	2.173214
4.40	0.205255	0.003918	0.019087	15.209868	2.183683
4.50	0.198020	0.003455	0.017449	16.562195	2.193600
4.60	0.191132	0.003053	0.015971	18.017794	2.203001
4.70	0.184570	0.002701	0.014635	19.582825	2.211918
4.80	0.178317	0.002394	0.013427	21.263714	2.220383
4.90	0.172354	0.002126	0.012333	23.067122	2.228424
5.00	0.166667	0.001890	0.011340	25.000000	2.236068
6.00	0.121951	0.000633	0.005194	53.179783	2.295276
7.00	0.092593	0.000242	0.002609	104.142860	2.333333
8.00	0.072464	0.000102	0.001414	190.109375	2.359071
9.00	0.058140	0.000047	0.000815	327.189301	2.377218
10.00	0.047619	0.000024	0.000495	535.937500	2.390457
∞	0	0	0	∞	2.449490

A.3. Normal Shock Relations

Perfect gas with $\gamma = 1.4$.

M_1	M_2	p_2/p_1	ρ_2/ρ_1 and V_1/V_2	T_2/T_1	p_{02}/p_{01}
1.00	1	1	1	1	1
1.01	0.990132	1.023450	1.016694	1.006645	0.999999
1.02	0.980520	1.047133	1.033441	1.013249	0.999990
1.03	0.971154	1.071050	1.050240	1.019814	0.999967
1.04	0.962026	1.095200	1.067088	1.026345	0.999923
1.05	0.953125	1.119583	1.083982	1.032843	0.999853
1.06	0.944445	1.144200	1.100921	1.039312	0.999751
1.07	0.935977	1.169050	1.117903	1.045753	0.999611
1.08	0.927713	1.194133	1.134925	1.052170	0.999431
1.09	0.919647	1.219450	1.151985	1.058564	0.999204
1.10	0.911770	1.245000	1.169082	1.064938	0.998928
1.11	0.904078	1.270783	1.186213	1.071294	0.998599
1.12	0.896563	1.296800	1.203377	1.077634	0.998213
1.13	0.889219	1.323050	1.220571	1.083960	0.997768
1.14	0.882042	1.349533	1.237793	1.090274	0.997261
1.15	0.875024	1.376250	1.255041	1.096577	0.996690
1.16	0.868162	1.403200	1.272315	1.102872	0.996052
1.17	0.861451	1.430383	1.289610	1.109159	0.995345
1.18	0.854884	1.457800	1.306927	1.115441	0.994569
1.19	0.848459	1.485450	1.324262	1.121719	0.993720
1.20	0.842170	1.513333	1.341615	1.127994	0.992798
1.21	0.836014	1.541450	1.358983	1.134268	0.991802
1.22	0.829986	1.569800	1.376364	1.140541	0.990731
1.23	0.824083	1.598383	1.393757	1.146816	0.989583
1.24	0.818301	1.627200	1.411160	1.153094	0.988359
1.25	0.812636	1.656250	1.428571	1.159375	0.987057
1.26	0.807085	1.685533	1.445989	1.165661	0.985677
1.27	0.801645	1.715050	1.463412	1.171953	0.984219
1.28	0.796312	1.744800	1.480839	1.178251	0.982682
1.29	0.791084	1.774783	1.498267	1.184558	0.981067
1.30	0.785957	1.805000	1.515695	1.190873	0.979374
1.31	0.780929	1.835450	1.533122	1.197198	0.977602
1.32	0.775997	1.866133	1.550546	1.203533	0.975752
1.33	0.771159	1.897050	1.567965	1.209880	0.973824
1.34	0.766412	1.928200	1.585379	1.216239	0.971819
1.35	0.761753	1.959583	1.602785	1.222612	0.969737
1.36	0.757181	1.991200	1.620182	1.228998	0.967579
1.37	0.752692	2.023050	1.637569	1.235398	0.965344
1.38	0.748286	2.055133	1.654945	1.241814	0.963035
1.39	0.743959	2.087450	1.672307	1.248246	0.960652
1.40	0.739709	2.120000	1.689655	1.254694	0.958194
1.41	0.735536	2.152783	1.706988	1.261159	0.955665
1.42	0.731436	2.185800	1.724303	1.267643	0.953063

(*Continued*)

(Continued)

M_1	M_2	p_2/p_1	ρ_2/ρ_1 and V_1/V_2	T_2/T_1	p_{02}/p_{01}
1.43	0.727408	2.219050	1.741600	1.274144	0.950390
1.44	0.723451	2.252533	1.758878	1.280665	0.947648
1.45	0.719562	2.286250	1.776135	1.287205	0.944837
1.46	0.715740	2.320200	1.793370	1.293765	0.941958
1.47	0.711983	2.354383	1.810583	1.300346	0.939012
1.48	0.708290	2.388800	1.827770	1.306948	0.936001
1.49	0.704659	2.423450	1.844933	1.313571	0.932925
1.50	0.701089	2.458333	1.862069	1.320216	0.929787
1.51	0.697578	2.493450	1.879177	1.326884	0.926586
1.52	0.694125	2.528800	1.896257	1.333574	0.923324
1.53	0.690729	2.564383	1.913308	1.340288	0.920003
1.54	0.687388	2.600200	1.930327	1.347026	0.916624
1.55	0.684101	2.636250	1.947315	1.353787	0.913188
1.56	0.680867	2.672533	1.964270	1.360573	0.909697
1.57	0.677685	2.709050	1.981192	1.367384	0.906151
1.58	0.674553	2.745800	1.998079	1.374220	0.902552
1.59	0.671471	2.782784	2.014931	1.381081	0.898901
1.60	0.668437	2.820000	2.031746	1.387969	0.895200
1.61	0.665451	2.857450	2.048524	1.394882	0.891450
1.62	0.662511	2.895133	2.065264	1.401822	0.887653
1.63	0.659616	2.933050	2.081965	1.408789	0.883809
1.64	0.656765	2.971200	2.098627	1.415783	0.879921
1.65	0.653958	3.009583	2.115248	1.422804	0.875988
1.66	0.651194	3.048200	2.131827	1.429853	0.872014
1.67	0.648471	3.087050	2.148365	1.436930	0.867999
1.68	0.645789	3.126133	2.164860	1.444035	0.863944
1.69	0.643147	3.165450	2.181311	1.451168	0.859851
1.70	0.640544	3.205000	2.197719	1.458331	0.855721
1.71	0.637979	3.244783	2.214081	1.465521	0.851556
1.72	0.635452	3.284800	2.230398	1.472742	0.847356
1.73	0.632962	3.325050	2.246669	1.479991	0.843124
1.74	0.630508	3.365533	2.262893	1.487270	0.838860
1.75	0.628089	3.406250	2.279070	1.494579	0.834565
1.76	0.625705	3.447200	2.295199	1.501918	0.830242
1.77	0.623354	3.488383	2.311279	1.509287	0.825891
1.78	0.621037	3.529800	2.327310	1.516687	0.821513
1.79	0.618753	3.571450	2.343292	1.524117	0.817111
1.80	0.616501	3.613333	2.359223	1.531577	0.812684
1.81	0.614281	3.655450	2.375104	1.539069	0.808234
1.82	0.612091	3.697800	2.390934	1.546592	0.803763
1.83	0.609931	3.740384	2.406712	1.554147	0.799271
1.84	0.607802	3.783200	2.422439	1.561732	0.794761
1.85	0.605701	3.826250	2.438112	1.569349	0.790232
1.86	0.603629	3.869533	2.453733	1.576998	0.785686
1.87	0.601585	3.913050	2.469301	1.584679	0.781125

(Continued)

(*Continued*)

M_1	M_2	p_2/p_1	ρ_2/ρ_1 and V_1/V_2	T_2/T_1	p_{02}/p_{01}
1.88	0.599569	3.956800	2.484814	1.592393	0.776549
1.89	0.597579	4.000783	2.500274	1.600138	0.771959
1.90	0.595616	4.045000	2.515679	1.607916	0.767357
1.91	0.593680	4.089450	2.531030	1.615726	0.762743
1.92	0.591769	4.134133	2.546325	1.623568	0.758119
1.93	0.589883	4.179050	2.561565	1.631444	0.753486
1.94	0.588022	4.224200	2.576749	1.639352	0.748844
1.95	0.586185	4.269584	2.591877	1.647294	0.744195
1.96	0.584372	4.315200	2.606949	1.655268	0.739540
1.97	0.582582	4.361050	2.621964	1.663276	0.734879
1.98	0.580816	4.407134	2.636922	1.671317	0.730214
1.99	0.579072	4.453450	2.651823	1.679392	0.725545
2.00	0.577350	4.500000	2.666667	1.687500	0.720874
2.01	0.575650	4.546783	2.681453	1.695642	0.716201
2.02	0.573972	4.593800	2.696181	1.703817	0.711527
2.03	0.572315	4.641050	2.710851	1.712027	0.706853
2.04	0.570679	4.688533	2.725463	1.720271	0.702180
2.05	0.569063	4.736250	2.740016	1.728548	0.697508
2.06	0.567467	4.784200	2.754511	1.736860	0.692839
2.07	0.565890	4.832383	2.768947	1.745206	0.688174
2.08	0.564334	4.880800	2.783325	1.753586	0.683512
2.09	0.562796	4.929450	2.797643	1.762001	0.678855
2.10	0.561277	4.978333	2.811902	1.770450	0.674203
2.11	0.559776	5.027450	2.826102	1.778934	0.669558
2.12	0.558294	5.076799	2.840243	1.787453	0.664919
2.13	0.556830	5.126384	2.854324	1.796006	0.660288
2.14	0.555383	5.176200	2.868345	1.804595	0.655665
2.15	0.553953	5.226251	2.882307	1.813218	0.651052
2.16	0.552541	5.276534	2.896209	1.821876	0.646447
2.17	0.551145	5.327050	2.910052	1.830569	0.641853
2.18	0.549766	5.377800	2.923834	1.839297	0.637269
2.19	0.548403	5.428783	2.937557	1.848061	0.632697
2.20	0.547056	5.480000	2.951220	1.856860	0.628136
2.21	0.545725	5.531450	2.964822	1.865694	0.623588
2.22	0.544409	5.583134	2.978365	1.874563	0.619053
2.23	0.543108	5.635050	2.991848	1.883468	0.614531
2.24	0.541822	5.687200	3.005271	1.892409	0.610023
2.25	0.540552	5.739583	3.018634	1.901385	0.605530
2.26	0.539295	5.792200	3.031936	1.910396	0.601051
2.27	0.538053	5.845050	3.045179	1.919444	0.596588
2.28	0.536825	5.898133	3.058362	1.928527	0.592140
2.29	0.535612	5.951450	3.071485	1.937646	0.587709
2.30	0.534411	6.005000	3.084548	1.946800	0.583295
2.31	0.533224	6.058783	3.097551	1.955991	0.578897
2.32	0.532051	6.112800	3.110495	1.965218	0.574517
2.33	0.530890	6.167049	3.123378	1.974481	0.570154

(*Continued*)

(*Continued*)

M_1	M_2	p_2/p_1	ρ_2/ρ_1 and V_1/V_2	T_2/T_1	p_{02}/p_{01}
2.34	0.529743	6.221533	3.136202	1.983779	0.565810
2.35	0.528608	6.276249	3.148966	1.993114	0.561484
2.36	0.527486	6.331200	3.161671	2.002485	0.557177
2.37	0.526376	6.386383	3.174316	2.011892	0.552889
2.38	0.525278	6.441801	3.186902	2.021336	0.548621
2.39	0.524192	6.497450	3.199429	2.030816	0.544372
2.40	0.523118	6.553334	3.211896	2.040332	0.540144
2.41	0.522055	6.609450	3.224304	2.049884	0.535936
2.42	0.521004	6.665801	3.236653	2.059473	0.531748
2.43	0.519964	6.722383	3.248943	2.069098	0.527581
2.44	0.518936	6.779201	3.261174	2.078760	0.523435
2.45	0.517918	6.836250	3.273347	2.088459	0.519311
2.46	0.516911	6.893534	3.285461	2.098194	0.515208
2.47	0.515915	6.951050	3.297516	2.107965	0.511126
2.48	0.514929	7.008800	3.309514	2.117774	0.507067
2.49	0.513954	7.066783	3.321452	2.127618	0.503030
2.50	0.512989	7.125000	3.333333	2.137500	0.499015
2.51	0.512034	7.183450	3.345156	2.147418	0.495022
2.52	0.511089	7.242133	3.356921	2.157373	0.491052
2.53	0.510154	7.301050	3.368629	2.167366	0.487105
2.54	0.509228	7.360200	3.380279	2.177394	0.483181
2.55	0.508312	7.419583	3.391871	2.187460	0.479280
2.56	0.507406	7.479200	3.403407	2.197563	0.475402
2.57	0.506509	7.539050	3.414885	2.207702	0.471547
2.58	0.505620	7.599133	3.426306	2.217879	0.467715
2.59	0.504741	7.659450	3.437671	2.228092	0.463907
2.60	0.503871	7.719999	3.448979	2.238343	0.460123
2.61	0.503010	7.780783	3.460231	2.248631	0.456362
2.62	0.502157	7.841799	3.471427	2.258955	0.452625
2.63	0.501313	7.903051	3.482567	2.269318	0.448912
2.64	0.500477	7.964534	3.493651	2.279717	0.445223
2.65	0.499649	8.026251	3.504679	2.290153	0.441557
2.66	0.498830	8.088201	3.515651	2.300626	0.437916
2.67	0.498019	8.150384	3.526569	2.311137	0.434298
2.68	0.497216	8.212800	3.537431	2.321685	0.430705
2.69	0.496421	8.275451	3.548239	2.332270	0.427136
2.70	0.495634	8.338334	3.558991	2.342892	0.423590
2.71	0.494854	8.401450	3.569689	2.353552	0.420069
2.72	0.494082	8.464800	3.580333	2.364249	0.416572
2.73	0.493317	8.528383	3.590923	2.374984	0.413099
2.74	0.492560	8.592200	3.601458	2.385756	0.409650
2.75	0.491810	8.656250	3.611940	2.396565	0.406226
2.76	0.491068	8.720533	3.622369	2.407412	0.402825
2.77	0.490332	8.785049	3.632744	2.418296	0.399449
2.78	0.489604	8.849800	3.643066	2.429218	0.396096
2.79	0.488882	8.914783	3.653335	2.440177	0.392768

(*Continued*)

(*Continued*)

M_1	M_2	p_2/p_1	ρ_2/ρ_1 and V_1/V_2	T_2/T_1	p_{02}/p_{01}
2.80	0.488167	8.980000	3.663551	2.451173	0.389464
2.81	0.487459	9.045449	3.673715	2.462208	0.386184
2.82	0.486758	9.111133	3.683827	2.473279	0.382927
2.83	0.486064	9.177050	3.693887	2.484389	0.379695
2.84	0.485376	9.243199	3.703894	2.495536	0.376486
2.85	0.484694	9.309583	3.713850	2.506720	0.373302
2.86	0.484019	9.376200	3.723755	2.517942	0.370141
2.87	0.483350	9.443049	3.733608	2.529202	0.367003
2.88	0.482687	9.510134	3.743411	2.540500	0.363890
2.89	0.482030	9.577451	3.753163	2.551835	0.360800
2.90	0.481380	9.645000	3.762864	2.563208	0.357733
2.91	0.480735	9.712784	3.772514	2.574618	0.354690
2.92	0.480096	9.780801	3.782115	2.586066	0.351670
2.93	0.479463	9.849051	3.791666	2.597552	0.348674
2.94	0.478836	9.917534	3.801167	2.609076	0.345701
2.95	0.478215	9.986250	3.810619	2.620638	0.342750
2.96	0.477599	10.055201	3.820021	2.632237	0.339823
2.97	0.476989	10.124384	3.829374	2.643874	0.336919
2.98	0.476384	10.193800	3.838679	2.655549	0.334038
2.99	0.475785	10.263450	3.847935	2.667262	0.331180
3.00	0.475191	10.333333	3.857143	2.679012	0.328344
3.10	0.469534	11.044999	3.946612	2.798603	0.301211
3.20	0.464349	11.780001	4.031496	2.921992	0.276229
3.30	0.459586	12.538333	4.112020	3.049191	0.253276
3.40	0.455200	13.320001	4.188406	3.180208	0.232226
3.50	0.451154	14.125000	4.260870	3.315051	0.212948
3.60	0.447413	14.953333	4.329621	3.453727	0.195312
3.70	0.443948	15.805000	4.394864	3.596244	0.179194
3.80	0.440732	16.680000	4.456790	3.742604	0.164470
3.90	0.437742	17.578335	4.515586	3.892813	0.151027
4.00	0.434959	18.500000	4.571429	4.046875	0.138756
4.10	0.432363	19.445000	4.624484	4.204793	0.127556
4.20	0.429938	20.413332	4.674911	4.366570	0.117334
4.30	0.427669	21.405003	4.722861	4.532211	0.108002
4.40	0.425545	22.420000	4.768473	4.701715	0.099481
4.50	0.423552	23.458334	4.811881	4.875086	0.091698
4.60	0.421680	24.519999	4.853211	5.052325	0.084586
4.70	0.419920	25.604998	4.892580	5.233435	0.078086
4.80	0.418263	26.713335	4.930100	5.418417	0.072140
4.90	0.416701	27.845001	4.965874	5.607271	0.066699
5.00	0.415227	29.000000	5.000000	5.800000	0.061716
6.00	0.404162	41.833332	5.268293	7.940587	0.029651
7.00	0.397360	57.000000	5.444445	10.469388	0.015351
8.00	0.392890	74.500000	5.565217	13.386719	0.008488
9.00	0.389799	94.333336	5.651163	16.692730	0.004964
10.00	0.387575	116.500000	5.714286	20.387501	0.003045
∞	0.377964	∞	6.000000	∞	0

Appendix B

OPERATORS AND EQUATIONS IN DIFFERENT COORDINATE SYSTEMS

B.1. Cylindrical Coordinates

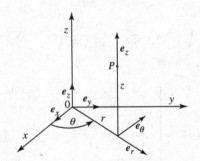

Fig. B.1. Cylindrical coordinate system.

Relations with Cartesian coordinates

$$x = r \cos \theta; \quad y = r \sin \theta; \quad z = z.$$

$$a = a_x e_x + a_y e_y + a_z e_z = a_r e_r + a_\theta e_\theta + a_z e_z.$$

$$\begin{cases} a_r = a_x \cos \theta + a_y \sin \theta, \\ a_\theta = -a_x \sin \theta + a_y \cos \theta, \\ a_z = a_z. \end{cases}$$

Divergence of a vector a

$$\mathrm{div}\, a = \nabla \cdot a = \frac{1}{r} \frac{\partial}{\partial r}(r a_r) + \frac{1}{r} \frac{\partial a_\theta}{\partial \theta} + \frac{\partial a_z}{\partial z}.$$

Gradient of a scalar φ

$$\text{grad}\phi = \nabla\phi = \frac{\partial\phi}{\partial r}\boldsymbol{e}_r + \frac{1}{r}\frac{\partial\phi}{\partial\theta}\boldsymbol{e}_\theta + \frac{\partial\phi}{\partial z}\boldsymbol{e}_z.$$

Curl of a vector $\boldsymbol{a}$

$$\text{curl}\boldsymbol{a} = \nabla\times\boldsymbol{a} = \left(\frac{1}{r}\frac{\partial a_z}{\partial\theta} - \frac{\partial a_\theta}{\partial z}\right)\boldsymbol{e}_r + \left(\frac{\partial a_r}{\partial z} - \frac{\partial a_z}{\partial r}\right)\boldsymbol{e}_\theta$$

$$+ \frac{1}{r}\left[\frac{\partial(ra_\theta)}{\partial r} - \frac{\partial a_r}{\partial\theta}\right]\boldsymbol{e}_z.$$

Laplacian of a scalar φ

$$\nabla^2\phi = \frac{1}{r}\frac{\partial\phi}{\partial r} + \frac{\partial^2\phi}{\partial r^2} + \frac{1}{r^2}\frac{\partial^2\phi}{\partial\theta^2} + \frac{\partial^2\phi}{\partial z^2}.$$

Laplacian of a vector $\boldsymbol{a}$

$$\nabla^2\boldsymbol{a} = \left(\nabla^2 a_r - \frac{a_r}{r^2} - \frac{2}{r^2}\frac{\partial a_\theta}{\partial\theta}\right)\boldsymbol{e}_r$$

$$+ \left(\nabla^2 a_\theta - \frac{a_\theta}{r^2} + \frac{2}{r^2}\frac{\partial a_r}{\partial\theta}\right)\boldsymbol{e}_\theta + (\nabla^2 a_z)\boldsymbol{e}_z.$$

Gradient of a vector $\boldsymbol{a}$

$$[(\text{grad}\boldsymbol{a})_{ij}] = [(\nabla\boldsymbol{a})_{ij}] = \begin{bmatrix} \dfrac{\partial a_r}{\partial r} & \dfrac{\partial a_\theta}{\partial r} & \dfrac{\partial a_z}{\partial r} \\ \left(\dfrac{1}{r}\dfrac{\partial a_r}{\partial\theta} - \dfrac{a_\theta}{r}\right) & \left(\dfrac{1}{r}\dfrac{\partial a_\theta}{\partial\theta} + \dfrac{a_r}{r}\right) & \dfrac{1}{r}\dfrac{\partial a_z}{\partial\theta} \\ \dfrac{\partial a_r}{\partial z} & \dfrac{\partial a_\theta}{\partial z} & \dfrac{\partial a_z}{\partial z} \end{bmatrix}.$$

If $\boldsymbol{a}$ and $\boldsymbol{b}$ are two vectors,

$$\boldsymbol{a}\cdot\text{grad}\boldsymbol{b} = \boldsymbol{a}\cdot\nabla\boldsymbol{b} = \left(\boldsymbol{a}\cdot\nabla b_r - \frac{a_\theta b_\theta}{r}\right)\boldsymbol{e}_r$$

$$+ \left(\boldsymbol{a}\cdot\nabla b_\theta + \frac{a_\theta b_r}{r}\right)\boldsymbol{e}_\theta + (\boldsymbol{a}\cdot\nabla b_z)\boldsymbol{e}_z.$$

Convective acceleration with velocity vector $\boldsymbol{V} = v_r\boldsymbol{e}_r + v_\theta\boldsymbol{e}_\theta + v_z\boldsymbol{e}_z$

$$\boldsymbol{V}\cdot\text{grad}\boldsymbol{V} = \boldsymbol{V}\cdot\nabla\boldsymbol{V} = \left(\boldsymbol{V}\cdot\nabla v_r - \frac{v_\theta^2}{r}\right)\boldsymbol{e}_r$$

$$+ \left(\boldsymbol{V}\cdot\nabla v_\theta + \frac{v_\theta v_r}{r}\right)\boldsymbol{e}_\theta + (\boldsymbol{V}\cdot\nabla v_z)\boldsymbol{e}_z.$$

Components of the rate of strain tensor

$$E_{rr} = \frac{\partial v_r}{\partial z}; \quad E_{r\theta} = \frac{1}{2}\left[r\frac{\partial}{\partial r}\left(\frac{v_\theta}{r}\right) + \frac{1}{r}\frac{\partial v_r}{\partial \theta}\right]; \quad E_{rz} = \frac{1}{2}\left(\frac{\partial v_z}{\partial r} + \frac{\partial v_r}{\partial z}\right);$$

$$E_{\theta\theta} = \frac{1}{r}\frac{\partial v_\theta}{\partial \theta} + \frac{v_r}{r}; \quad E_{\theta z} = \frac{1}{2}\left(\frac{1}{r}\frac{\partial v_z}{\partial \theta} + \frac{\partial v_\theta}{\partial z}\right); \quad E_{zz} = \frac{\partial v_z}{\partial z}.$$

Equations of momentum balance for an incompressible flow
   (Body force: $\boldsymbol{f} = f_r\boldsymbol{e}_r + f_\theta\boldsymbol{e}_\theta + f_z\boldsymbol{e}_z$)

$$\frac{\partial v_r}{\partial t} + \boldsymbol{V}\cdot\nabla v_r - \frac{v_\theta^2}{r} = f_r - \frac{1}{\rho}\frac{\partial p}{\partial r} + \nu\left(\nabla^2 v_r - \frac{v_r}{r^2} - \frac{2}{r^2}\frac{\partial v_\theta}{\partial \theta}\right),$$

$$\frac{\partial v_\theta}{\partial t} + \boldsymbol{V}\cdot\nabla v_\theta + \frac{v_\theta v_r}{r} = f_\theta - \frac{1}{\rho r}\frac{\partial p}{\partial \theta} + \nu\left(\nabla^2 v_\theta - \frac{v_\theta}{r^2} + \frac{2}{r^2}\frac{\partial v_r}{\partial \theta}\right),$$

$$\frac{\partial \dot{v}_z}{\partial t} + \boldsymbol{V}\cdot\nabla v_z = f_z - \frac{1}{\rho}\frac{\partial p}{\partial z} + \nu\nabla^2 v_z.$$

B.2. Spherical Coordinates

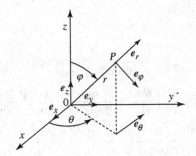

Fig. B.2.   Spherical coordinate system.

Relations with Cartesian coordinates

$$x = r\sin\varphi\cos\theta; \quad y = r\sin\varphi\sin\theta; \quad z = r\cos\varphi.$$

$$\boldsymbol{a} = a_x\boldsymbol{e}_x + a_y\boldsymbol{e}_y + a_z\boldsymbol{e}_z = a_r\boldsymbol{e}_r + a_\varphi\boldsymbol{e}_\varphi + a_\theta\boldsymbol{e}_\theta.$$

$$\begin{cases} a_r = a_x\sin\varphi\cos\theta + a_y\sin\varphi\sin\theta + a_z\cos\varphi, \\ a_\varphi = a_x\cos\varphi\cos\theta + a_y\cos\varphi\sin\theta - a_z\sin\varphi, \\ a_\theta = -a_x\sin\theta + a_y\cos\theta. \end{cases}$$

*Divergence of a vector **a***

$$\operatorname{div}\boldsymbol{a} = \nabla \cdot \boldsymbol{a} = \frac{1}{r^2}\frac{\partial}{\partial r}(r^2 a_r) + \frac{1}{r\sin\varphi}\frac{\partial}{\partial\varphi}(a_\varphi\sin\varphi) + \frac{1}{r\sin\varphi}\frac{\partial a_\theta}{\partial\theta}.$$

Gradient of a scalar ϕ

$$\operatorname{grad}\phi = \nabla\phi = \frac{\partial\phi}{\partial r}\boldsymbol{e}_r + \frac{1}{r}\frac{\partial\phi}{\partial\varphi}\boldsymbol{e}_\varphi + \frac{1}{r\sin\varphi}\frac{\partial\phi}{\partial\theta}\boldsymbol{e}_\theta.$$

*Curl of a vector **a***

$$\operatorname{curl}\boldsymbol{a} = \nabla\times\boldsymbol{a} = \frac{1}{r\sin\varphi}\left[\frac{\partial}{\partial\varphi}(a_\theta\sin\varphi) - \frac{\partial a_\varphi}{\partial\theta}\right]\boldsymbol{e}_r$$
$$+ \frac{1}{r}\left[\frac{1}{\sin\varphi}\frac{\partial a_r}{\partial\theta} - \frac{\partial(ra_\theta)}{\partial r}\right]\boldsymbol{e}_\varphi + \frac{1}{r}\left[\frac{\partial(ra_\varphi)}{\partial r} - \frac{\partial a_r}{\partial\varphi}\right]\boldsymbol{e}_\theta.$$

Laplacian of a scalar ϕ

$$\nabla^2\phi = \frac{1}{r^2}\frac{\partial}{\partial r}\left(r^2\frac{\partial\phi}{\partial r}\right) + \frac{1}{r^2\sin\varphi}\frac{\partial}{\partial\varphi}\left(\sin\varphi\frac{\partial\phi}{\partial\varphi}\right) + \frac{1}{r^2\sin^2\varphi}\frac{\partial^2\phi}{\partial\theta^2}.$$

*Laplacian of a vector **a***

$$\nabla^2\boldsymbol{a} = \left\{\nabla^2 a_r - \frac{2}{r^2}\left[a_r + \frac{1}{\sin\varphi}\frac{\partial}{\partial\varphi}(a_\varphi\sin\varphi) + \frac{1}{\sin\varphi}\frac{\partial a_\theta}{\partial\theta}\right]\right\}\boldsymbol{e}_r$$
$$+ \left[\nabla^2 a_\varphi + \frac{2}{r^2}\left(\frac{\partial a_r}{\partial\varphi} - \frac{a_\varphi}{2\sin^2\varphi} - \frac{\cos\varphi}{\sin^2\varphi}\frac{\partial a_\theta}{\partial\theta}\right)\right]\boldsymbol{e}_\varphi$$
$$+ \left[\nabla^2 a_\theta + \frac{2}{r^2\sin\varphi}\left(\frac{\partial a_r}{\partial\theta} + \cot\varphi\frac{\partial a_\varphi}{\partial\theta} - \frac{a_\theta}{2\sin\varphi}\right)\right]\boldsymbol{e}_\theta.$$

*Gradient of a vector **a***

$[(\operatorname{grad}\boldsymbol{a})_{ij}]$
$= [(\nabla\boldsymbol{a})_{ij}]$

$$= \begin{bmatrix} \dfrac{\partial a_r}{\partial r} & \dfrac{\partial a_\varphi}{\partial r} & \dfrac{\partial a_\theta}{\partial r} \\[2mm] \dfrac{1}{r}\dfrac{\partial a_r}{\partial\varphi} - \dfrac{a_\varphi}{r} & \dfrac{1}{r}\dfrac{\partial a_\varphi}{\partial\varphi} + \dfrac{a_r}{r} & \dfrac{1}{r}\dfrac{\partial a_\theta}{\partial\varphi} \\[2mm] \dfrac{1}{r\sin\varphi}\dfrac{\partial a_r}{\partial\theta} - \dfrac{a_\theta}{r} & \dfrac{1}{r\sin\varphi}\dfrac{\partial a_\varphi}{\partial\theta} - \dfrac{a_\theta}{r}\cot\varphi & \dfrac{1}{r\sin\varphi}\dfrac{\partial a_\theta}{\partial\theta} + \dfrac{a_r}{r} + \dfrac{a_\varphi}{r}\cot\varphi \end{bmatrix}$$

If $\boldsymbol{a}$ and $\boldsymbol{b}$ are two vectors,

$$\boldsymbol{a} \cdot \mathrm{grad}\boldsymbol{b} = \boldsymbol{a} \cdot \nabla \boldsymbol{b} = \left(\boldsymbol{a} \cdot \nabla b_r - \frac{a_\varphi b_\varphi}{r} - \frac{a_\theta b_\theta}{r} \right) \boldsymbol{e}_r$$

$$+ \left(\boldsymbol{a} \cdot \nabla b_\varphi + \frac{a_\varphi b_r}{r} - \frac{a_\theta b_\theta}{r} \cot \varphi \right) \boldsymbol{e}_\varphi$$

$$+ \left(\boldsymbol{a} \cdot \nabla b_\theta + \frac{a_\theta b_r}{r} + \frac{a_\theta b_\varphi}{r} \cot \varphi \right) \boldsymbol{e}_\theta.$$

Convective acceleration with velocity vector $\boldsymbol{V} = v_r \boldsymbol{e}_r + v_\varphi \boldsymbol{e}_\varphi + v_\theta \boldsymbol{e}_\theta$

$$\boldsymbol{V} \cdot \mathrm{grad}\boldsymbol{V} = \boldsymbol{V} \cdot \nabla \boldsymbol{V} = \left(\boldsymbol{V} \cdot \nabla v_r - \frac{v_\varphi^2}{r} - \frac{v_\theta^2}{r} \right) \boldsymbol{e}_r$$

$$+ \left(\boldsymbol{V} \cdot \nabla v_\varphi + \frac{v_\varphi v_r}{r} - \frac{v_\theta^2}{r} \cot \varphi \right) \boldsymbol{e}_\varphi$$

$$+ \left(\boldsymbol{V} \cdot \nabla v_\theta + \frac{v_\theta v_r}{r} + \frac{v_\theta v_\varphi}{r} \cot \varphi \right) \boldsymbol{e}_\theta.$$

Components of the rate of strain tensor

$$E_{rr} = \frac{\partial v_r}{\partial z}; \quad E_{\varphi\varphi} = \frac{1}{r} \frac{\partial v_\varphi}{\partial \varphi} + \frac{v_r}{r}; \quad E_{\theta\theta} = \frac{1}{r \sin \varphi} \frac{\partial v_\theta}{\partial \theta} + \frac{v_r}{r} + \frac{v_\varphi \cot \varphi}{r};$$

$$E_{r\varphi} = \frac{1}{2} \left[r \frac{\partial}{\partial r} \left(\frac{v_\varphi}{r} \right) + \frac{1}{r} \frac{\partial v_r}{\partial \varphi} \right]; \quad E_{r\theta} = \frac{1}{2} \left[r \frac{\partial}{\partial r} \left(\frac{v_\theta}{r} \right) + \frac{1}{r \sin \varphi} \frac{\partial v_r}{\partial \theta} \right];$$

$$E_{\varphi\theta} = \frac{1}{2} \left[\frac{\sin \varphi}{r} \frac{\partial}{\partial \varphi} \left(\frac{v_\theta}{\sin \varphi} \right) + \frac{1}{r \sin \varphi} \frac{\partial v_\varphi}{\partial \theta} \right].$$

Equations of momentum balance for an incompressible flow
    (Body force: $\boldsymbol{f} = f_r \boldsymbol{e}_r + f_\varphi \boldsymbol{e}_\varphi + f_\theta \boldsymbol{e}_\theta$)

$$\frac{\partial v_r}{\partial t} + \boldsymbol{V} \cdot \nabla v_r - \frac{v_\varphi^2}{r} - \frac{v_\theta^2}{r}$$

$$= f_r - \frac{1}{\rho} \frac{\partial p}{\partial r} + \nu \left\{ \nabla^2 v_r - \frac{2}{r^2} \left[v_r + \frac{1}{\sin \varphi} \frac{\partial}{\partial \varphi} (v_\varphi \sin \varphi) + \frac{1}{\sin \varphi} \frac{\partial v_\theta}{\partial \theta} \right] \right\},$$

$$\frac{\partial v_\varphi}{\partial t} + \boldsymbol{V} \cdot \nabla v_\varphi + \frac{v_\varphi v_r}{r} - \frac{v_\theta^2}{r} \cot \varphi$$

$$= f_\varphi - \frac{1}{\rho r} \frac{\partial p}{\partial \varphi} + \nu \left[\nabla^2 v_\varphi + \frac{2}{r^2} \left(\frac{\partial v_r}{\partial \varphi} - \frac{v_\varphi}{2 \sin^2 \varphi} - \frac{\cos \varphi}{\sin^2 \varphi} \frac{\partial v_\theta}{\partial \theta} \right) \right],$$

$$\frac{\partial v_\theta}{\partial t} + \boldsymbol{V} \cdot \nabla v_\theta + \frac{v_\theta v_r}{r} + \frac{v_\theta v_\varphi}{r} \cot \varphi$$

$$= f_\theta - \frac{1}{\rho r \sin \varphi} \frac{\partial p}{\partial \theta}$$

$$+ \nu \left[\nabla^2 v_\theta + \frac{2}{r^2 \sin \varphi} \left(\frac{\partial v_r}{\partial \theta} + \cot \varphi \frac{\partial v_\varphi}{\partial \theta} - \frac{v_\theta}{2 \sin \varphi} \right) \right].$$

REFERENCES

Abbott, I. A. and Von Doenhoff, A. E. (1959). *Theory of wing sections*, Dover Publications, New York.

Alam, M. M., Zhou, Y. and Wang, X. W. (2011). The wake of two side-by-side square cylinders, *J. Fluid Mech.*, 669, pp. 432–471.

Ballabh, R. (1940). Superposable fluid motions, *Proc. Benares Math. Soc.*, 2, pp. 69–79.

Batchelor, G. K. (1967). *An introduction to fluid dynamics*, Cambridge University Press, Cambridge.

Bearman, P. W. (1980). Bluff body flows applicable to vehicle aerodynamics, *J. Fluids Eng. Trans. ASME*, 102, pp. 265–274.

Bénard, H. (1908). Formation des centres de giration à l'arrière d'un obstacle en mouvement, *C. R. Acad. Sci. Paris*, 147, pp. 839–842.

Berker, R. (1963). Intégration des équations du mouvement d'un fluide visqueux incompressible, in *Handbuch der physik*, Flügge, S. (Ed.), VIII(2), Springer-Verlag, Berlin, pp. 1–384.

Bhatnagar, P. L. and Verma, P. D. (1957). On superposable flows, *Proc. Math. Sci.*, 45, pp. 281–292.

Blasius, H. (1908). Grenzschichten in Flüssigkeiten mit kleiner Reibung, *Z. Math. Phys.*, 56, pp. 1–37; 60, pp. 397–398.

Bobyleff, D. (1873). Einige Betrachtungen über die Gleichungen der Hydrodynamik, *Math. Ann.*, 6, pp. 72–84.

Buresti, G. (1998). Vortex shedding from bluff bodies, in *Wind effects on buildings and structures*, Riera, J. D. and Davenport, A. G. (Eds.), Balkema, Rotterdam, pp. 61–95.

Buresti, G. (2008). Notes on incompressible flows, *Atti del Dipartimento di Ingegneria Aerospaziale di Pisa*, ADIA 2008-2, ETS Editrice, Pisa.

Buresti, G. and Iungo, G. V. (2010). Experimental investigation on the connection between flow fluctuations and vorticity dynamics in the near wake of a triangular prism placed vertically on a plane, *J. Wind Eng. Ind. Aerodyn.*, 98, pp. 253–262.

Buresti, G. and Lanciotti, A. (1992). Mean and fluctuating forces on a circular cylinder in cross-flow near a plane surface, *J. Wind Eng. Ind. Aerodyn.*, 41–44, pp. 639–650.

Buresti, G., Fedeli, R. and Ferraresi, A. (1997). Influence of afterbody rounding on the pressure drag of an axisymmetrical bluff body, *J. Wind Eng. Ind. Aerodyn.*, 69–71, pp. 179–188.

Choi, H., Jeon, W.-P. and Kim, J. (2008). Control of flow over a bluff body, *Annu. Rev. Fluid Mech.*, 40, pp. 113–139.

Cisotti, U. (1924). Rotazioni viscose, *Rend. Accad. Naz. Lincei, Cl. Sci. Fis. Mat. Nat.*, 33, pp. 161–167.

Coenen, E. G. M. (2001). Viscous-inviscid interaction with the quasi-simultaneous method for 2D and 3D aerodynamic flow, Ph.D. Thesis, Rijksuniversiteit Groningen. Available at: http://irs.ub.rug.nl/ppn/227769082.

Cousteix, J. (1988). *Couche limite laminaire*, Cepadues Editions, Toulouse.

Craig, P. P. (1959). Observations of perfect potential flow and critical velocities in superfluid helium II, Ph.D. Thesis, California Institute of Technology. Available at http://thesis.library.caltech.edu/455/.

Dalton, C., Xu, Y. and Owen, J. C. (2001). The suppression of lift on a circular cylinder due to vortex shedding at moderate Reynolds numbers, *J. Fluids Struct.*, 15, pp. 617–628.

Delany, N. K. and Sorensen, N. E. (1953). *Low-speed drag of cylinders of various shapes*, NACA TN 3038.

Drazin, P. G. and Reid, W. H. (2004). *Hydrodynamic stability*, 2nd edition, Cambridge University Press, Cambridge.

Drazin, P. G. and Riley, N. (2006). *The Navier-Stokes equations: a classification of flows and exact solutions*, Cambridge University Press, Cambridge.

Dukhin, A. S. and Goetz, P. J. (2009). Bulk viscosity and compressibility measurements using acoustic spectroscopy, *J. Chem. Phys.*, 130, 124519.

Emanuel, G. (1992). Effect of bulk viscosity on a hypersonic boundary layer, *Phys. Fluids A*, 4, pp. 491–495.

Emanuel, G. and Argrow, B. M. (1994). Linear dependence of the bulk viscosity on shock wave thickness, *Phys. Fluids*, 6, pp. 3203–3205.

Erdogan, M. E. (2002). On the unsteady unidirectional flows generated by impulsive motion of a boundary or sudden application of a pressure gradient, *Int. J. Non-Linear Mech.*, 37, pp. 1091–1106.

Erdogan, M. E. (2003). On the flows produced by sudden application of a constant pressure gradient or by impulsive motion of a boundary, *Int. J. Non-Linear Mech.*, 38, pp. 781–797.

Ergun, A. N. (1949). Some cases of superposable fluid motions, *Commun. Fac. Sci. Univ. Ankara*, 2, pp. 48–88.

Forsyth, A. R. (1880). On the motion of a viscous incompressible fluid, *Messenger of Mathematics*, 9, pp. 134–139.

Fransson, J. H. M., Talamelli, A., Brandt, L. and Cossu, C. (2006). Delaying transition to turbulence by a passive mechanism, *Phys. Rev. Lett.*, 96, pp. 064501-1–064501-4.

Gad-el-Hak, M. (1995). Questions in fluid mechanics: Stokes' hypothesis for a Newtonian, isotropic fluid, *J. Fluids Eng.*, 117, pp. 3–5.

Gad-el-Hak, M. (1999). The fluid mechanics of microdevices — The Freeman Scholar Lecture, *J. Fluids Eng.*, 121, pp. 5–33.

Gad-el-Hak, M. (2000). *Flow control: passive, active and reactive flow management*, Cambridge University Press, Cambridge.

Glauert, H. (1926). *The elements of aerofoil and airscrew theory*, Cambridge University Press, Cambridge.

Goldstein, S. (1957). *Modern developments in fluid dynamics, Vol. II*, Clarendon Press, Oxford.

Govardhan, R. N. and Williamson, C. H. K. (2005). Vortex-induced vibrations of a sphere, *J. Fluid Mech.*, 531, pp. 11–47.

Graziani, G. and Bassanini, P. (2002). Unsteady viscous flows about bodies: vorticity release and forces, *Meccanica*, 37, pp. 283–303.

Gresho, P. M. (1991). Incompressible fluid dynamics: some fundamental formulation issues, *Annu. Rev. Fluid Mech.*, 23, pp. 413–453.

Hess, J. L. (1990). Panel methods in computational fluid dynamics, *Annu. Rev. Fluid Mech.*, 22, pp. 255–274.

Hinze, J. O. (1975). *Turbulence*, 2nd edition, McGraw-Hill, New York.

Hooker, S. G. (1936). On the action of viscosity in increasing the spacing ratio of a vortex street, *Proc. R. Soc. London A*, 154, pp. 67–89.

Hourigan, K., Thompson, M. C. and Tan, B. T. (2001). Self-sustained oscillations in flows around long blunt plates, *J. Fluids Struct.*, 15, pp. 387–398.

Hucho, W.-H. and Sovran, G. (1993). Aerodynamics of road vehicles, *Annu. Rev. Fluid Mech.*, 25, pp. 485–537.

Igarashi, T. (1997). Drag reduction of a square prism by flow control using a small rod, *J. Wind Eng. Ind. Aerodyn.*, 69–71, pp. 141–153.

Karamcheti, K. (1966). *Principles of ideal-fluid aerodynamics*, Krieger Publishing Company, Malabar, FL.

Kármán, Th. von (1911). Über den Mechanismus den Widerstands, den ein bewegter Korper in einer Flussigkeit erfahrt, 1, *Göttingen Nachr. Math. Phys. Kl.*, pp. 509–517.

Kármán, Th. von (1912). Über den Mechanismus den Widerstands, den ein bewegter Korper in einer Flussigkeit erfahrt, 2, *Göttingen Nachr. Math. Phys. Kl.*, pp. 547–556.

Katz, J. and Plotkin, A. (2001). *Low-speed aerodynamics*, 2nd edition, Cambridge University Press, Cambridge.

Kida, S. (1982). Stabilizing effects of finite core on Karman vortex street, *J. Fluid Mech.*, 122, pp. 487–504.

Kiya, M., Ishikawa, H. and Sakamoto, H. (2001). Near-wake instabilities and vortex structures of three-dimensional bluff bodies: a review, *J. Wind Eng. Ind. Aerodyn.*, 89, pp. 1219–1232.

Lamb, H. (1932). *Hydrodynamics*, 6th edition, Cambridge University Press, Cambridge.

Lanchester, F. W. (1907). *Aerodynamics*, A. Constable & Co., London.

Lesieur, M. (2008). *Turbulence in fluids*, 4th edition, Springer, Berlin.

Liepmann, H. W. and Roshko, A. (1957). *Elements of gasdynamics*, John Wiley & Sons, New York.

Lighthill, M. J. (1956). Introductory remarks to the meeting of the Physical Society on "The physics of gas flow at very high speeds", *Nature*, 178, p. 343.

Lighthill, M. J. (1958). On displacement thickness, *J. Fluid Mech.*, 4, pp. 383–392.

Lighthill, M. J. (1963). Introduction. Boundary layer theory, in *Laminar boundary layers, Part II*, Rosenhead, L. (Ed.), Oxford University Press, Oxford, pp. 46–113.

Lock, R. C. and Williams, B. R. (1987). Viscous-inviscid interactions in external aerodynamics, *Prog. Aerosp. Sci.*, 24, pp. 51–171.

Mair, W. A. (1969). Reduction of base drag by boat-tailed afterbodies in low-speed flow, *Aeronaut. Q.*, XX, pp. 307–320.

Mallick, D. D. (1957). Nonuniform rotation of an infinite circular cylinder in an infinite viscous liquid, *Z. Angew. Math. Mech.*, 37, pp. 385–392.

Mallock, A. (1907). On the resistance of air, *Proc. R. Soc. London A*, 79, pp. 262–273.

Maull, D. J. and Hoole, B. J. (1967). The effect of boat-tailing on the flow around a two-dimensional blunt-based aerofoil at zero incidence, *J. R. Aero. Soc.*, 71, pp. 854–858.

Mizota, T., Zdravkovich, M., Graw, K.-U. and Leder, A. (2000). St. Christopher and the vortex, *Nature*, 404, p. 226.

Morel, T. (1978). The effect of base slant on the flow pattern and drag of three-dimensional bodies with blunt ends, in *Aerodynamic drag mechanisms of bluff bodies and road vehicles*, Sovran, G., Morel T. and Mason W. T. (Eds.), Plenum Press, New York, pp. 191–217.

Morino, L. and Kuo, C.-C. (1974). Subsonic potential aerodynamics for complex configurations: a general theory, *AIAA J.*, 12, pp. 191–197.

Munk, M. M. (1922). *General theory of thin wing sections*, NACA Report 142.

Nash, J. F. (1967). *A discussion of two-dimensional turbulent base flow*, Aeronautical Research Council R. & M. 3468.

Naumann, A., Morsbach, M. and Kramer, C. (1966). The conditions of separation and vortex formation past cylinders, in *Separated flows*, AGARD CP-4, pp. 547–574.

Navier, C.-L. (1827). Mémoire sur les lois du mouvement des fluides, *Mem. Acad. Sci. Inst. France*, 6, pp. 389–440.

Newton, I. (1687). *Philosophiae naturalis principia mathematica*, Royal Society, London.

Noca, F., Shiels, D. and Jeon, D. (1999). A comparison of methods for evaluating time-dependent fluid dynamic forces on bodies, using only velocity fields and their derivatives, *J. Fluids Struct.*, 13, pp. 551–578.

Panton, R. L. (1984). *Incompressible flow*, John Wiley & Sons, New York.

Petrila, T. and Trif, D. (2005). *Basics of fluid mechanics and introduction to computational fluid dynamics*, Springer, Berlin.

Poisson, S.-D. (1831). Mémoire sur les équations générales de l'équilibre et du mouvement des corps solides élastique et des fluides, *J. Ec. Polytech.*, 13(20), pp. 1–174.

Pope, S. B. (2000). *Turbulent flows*, Cambridge University Press, Cambridge.

Prandtl, L. (1904). Über Flüssigkeitsbewegung bei sehr kleiner Reibung, *Verh. III Int. Math. Kong., Heidelberg*, Teubner, Leipzig, pp. 484–491. Translation available as: Motion of fluids with very little viscosity, NACA-TM-452 (March 1928).

Prandtl, L. (1918). Tragflügeltheorie. I Mitteilung, *Nachr. Ges. Wiss. Göttingen. Math.-Phys. Kl.*, pp. 451–477.

Prandtl, L. (1919). Tragflügeltheorie. II Mitteilung, *Nachr. Ges. Wiss. Göttingen. Math.-Phys. Kl.*, pp. 107–137.

Prandtl, L. (1921). *Applications of modern hydrodynamics to aeronautics. Part 1: Fundamental concepts and the most important theorems. Part 2: Applications*, NACA-TR-116.

Prasad, A. and Williamson, C. H. K. (1997). A method for the reduction of bluff body drag, *J. Wind Eng. Ind. Aerodyn.*, 69–71, pp. 155–167.

Remorini, G. (1983a). Sulla sovrapponibilità dei moti idrodinamici, *Rend. Mat.*, 3, Ser. VII, pp. 239–248.

Remorini, G. (1983b). Ancora sulla sovrapponibilità dei moti idrodinamici, *Boll. Un. Mat. Ital., Fis. Mat.*, 2, pp. 173–185.

Remorini, G. (1989). Sull'integrale di Bernoulli, *Note Mat.*, IX, pp. 65–75.

Reynolds, O. (1883). An experimental investigation of the circumstances which determine whether the motion of water shall be direct and sinuous, and of the law of resistance in parallel channels, *Philos. Trans. R. Soc. London A*, 174, pp. 935–982.

Reynolds, O. (1895). On the dynamical theory of incompressible viscous fluids and the determination of the criterion, *Philos. Trans. R. Soc. London A*, 186, pp. 123–164.

Rosenhead, L. (1954). A discussion on the first and second viscosities of fluids, *Proc. R. Soc. London A*, 226, pp. 1–69.

Saffman, P. G. (1992). *Vortex dynamics*, Cambridge University Press, Cambridge.

Saffman, P. G. and Schatzman, J. C. (1982). Stability of a vortex street of finite vortices, *J. Fluid Mech.*, 117, pp. 171–185.

Saint-Venant, A.-J.-C. B. de (1843). Mémoire sur la dynamique des fluides, *C. R. Acad. Sci. Paris*, 17, pp. 1240–1242.

Sakamoto, H., Haniu, H. and Obata, Y. (1987). Fluctuating forces acting on two square prisms in a tandem arrangement, *J. Wind Eng. Ind. Aerodyn.*, 26, pp. 85–103.

Sarpkaya, T. (2004). A critical review of the intrinsic nature of vortex-induced vibrations, *J. Fluids Struct.*, 19, pp. 389–447.

Sarpkaya, T. (2010). *Wave forces on offshore structures*, Cambridge University Press, Cambridge.

Sarpkaya, T. and Isaacson, M. (1981). *Mechanics of wave forces on offshore structures*, Van Nostrand Reinhold, New York.

Schlichting, H. and Gersten, K. (2000). *Boundary layer theory*, 8th edition, Springer-Verlag, Berlin.

Schmid, P. J. and Henningson, D. S. (2002). *Stability and transition in shear flows*, Springer, Berlin.

Serrin, J. (1959). Mathematical principles of classical fluid mechanics, in *Handbuch der physik*, Flügge, S. (Ed.), VIII(1), Springer-Verlag, Berlin, pp. 125–263.

Shapiro, A. H. (1953). *The dynamics and thermodynamics of compressible fluid flow, Vol. 1*, John Wiley & Sons, New York.

Shi, L. L., Liu, Y. Z. and Yu, J. (2010). PIV measurements of separated flow over a blunt plate with different chord-to-thickness rations, *J. Fluids Struct.*, 26, pp. 644–657.

Simiu, E. and Scanlan, R. (1986). *Wind effects on structures: an introduction to wind engineering*, John Wiley & Sons, New York.

Sipp, D., Marquet, O., Meliga, P. and Barbagallo, A. (2010). Dynamics and control of global instabilities in open-flows: a linearized approach, *Appl. Mech. Rev.*, 63, 030801-1–030801-26.

Stokes, G. G. (1845). On the theories of the internal friction of fluids in motion, and of the equilibrium and motion of elastic solids, *Trans. Cambridge Philos. Soc.*, 8, pp. 287–319.

Stokes, G. G. (1846). Report on recent researches in hydrodynamics, *Rep. British Assoc.*, pp. 1–20.

Strang, J. A. (1948). Superposable fluid motions, *Commun. Fac. Sci. Univ. Ankara*, 1, pp. 1–32.

Sumner, D. (2010). Two circular cylinders in cross-flow: a review, *J. Fluids Struct.*, 26, pp. 849–899.

Tani, I. (1977). History of boundary layer theory, *Annu. Rev. Fluid Mech.*, 9, pp. 87–111.

Tennekes, H. and Lumley, J. L. (1972). *A first course in turbulence*, MIT Press, Cambridge, MA.

Torenbeek, E. (1976). *Synthesis of subsonic airplane design*, Delft University Press, Delft.

Trevena, D. H. (1975). *The liquid phase*, Wykeham Publications, London.

Truesdell, C. (1950). Bernoulli's theorem for viscous compressible fluids, *Phys. Rev.*, 77, pp. 535–536.

Truesdell, C. (1953). Notes on the history of the general equations of hydrodynamics, *Am. Math. Monthly*, 60, pp. 445–458.

Truesdell, C. (1954). *The kinematics of vorticity*, Indiana University Press, Bloomington, IN.

Vincenti, W. G. and Kruger, C .H. (1965). *Introduction to physical gas dynamics*, John Wiley & Sons, New York.

Wang, C. Y. (1989). Exact solutions of the unsteady Navier-Stokes equations, *Appl. Mech. Rev.*, 42, pp. S269–S282.

Wang, C. Y. (1991). Exact solutions of the steady-state Navier-Stokes equations, *Annu. Rev. Fluid Mech.*, 23, pp. 159–177.

Weissinger, J. (1947). *The lift distribution of swept-back wings*, NACA–TM–1120.

Wu, J. C. (1981). Theory for aerodynamic force and moment in viscous flows, *AIAA J.*, 19, pp. 432–441.

Wu, J. C. (2005). *Elements of vorticity aerodynamics*, Tsinghua University Press, Beijing.

Wu, J. Z. and Wu, J. M. (1993). Interactions between a solid surface and a viscous compressible flow field, *J. Fluid Mech.*, 254, pp. 183–211.

Wu, J. Z. and Wu, J. M. (1996). Vorticity dynamics on boundaries, *Adv. Appl. Mech.*, 32, pp. 119–275.

Wu, J. Z., Lu, X. Y. and Zhuang, L. X. (2007). Integral force acting on a body due to local flow structures, *J. Fluid Mech.*, 576, pp. 265–286.

Wu, J. Z., Ma, H. Y. and Zhou, M. D. (2006). *Vorticity and vortex dynamics*, Springer, Berlin.

Wu, J. Z., Pan, Z. L. and Lu, X. Y. (2005). Unsteady fluid-dynamic force solely in terms of control-surface integral, *Phys. Fluids*, 17, 098102.

Zdravkovich, M. M. (1987). The effects of interference between circular cylinders in cross flow, *J. Fluids Struct.*, 1, pp. 239–261.

Zdravkovich, M. M. (1997). *Flow around circular cylinders. Vol. 1: Fundamentals*, Oxford University Press, Oxford.

Zdravkovich, M. M. (2003). *Flow around circular cylinders. Vol. 2: Applications*, Oxford University Press, Oxford.

Zeytounian, R. Kh. (2001). A historical survey of some mathematical aspects of Newtonian fluid flows, *Appl. Mech. Rev.*, 54, pp. 525–562.

Zhu, G., Bearman, P. W. and Graham, J. M. R. (2002). Prediction of drag and lift using velocity and vorticity fields, *Aero. J.*, 106, pp. 547–554.

INDEX